Lecture Notes in Computer Sci

T0238173

Commenced Publication in 1973
Founding and Former Series Editors:
Gerhard Goos, Juris Hartmanis, and Jan van Leeuwen

José Mira   José R. Álvarez (Eds.)

Mechanisms, Symbols, and Models Underlying Cognition

First International Work-Conference on the Interplay
Between Natural and Artificial Computation, IWINAC 2005
Las Palmas, Canary Islands, Spain, June 15-18, 2005
Proceedings, Part I

 Springer

Volume Editors

José Mira
José R. Álvarez
Universidad Nacional de Educación a Distancia
E.T.S. de Ingeniería Informática
Departamento de Inteligencia Artificial
Juan del Rosal, 16, 28040 Madrid, Spain
E-mail: {jmira, jras}@dia.uned.es

Library of Congress Control Number: 2005927486

CR Subject Classification (1998): F.1, F.2, I.2, G.2, I.4, I.5, J.3, J.4, J.1

ISSN   0302-9743
ISBN-10   3-540-26298-9 Springer Berlin Heidelberg New York
ISBN-13   978-3-540-26298-5 Springer Berlin Heidelberg New York

Springer is a part of Springer Science+Business Media

springeronline.com

© Springer-Verlag Berlin Heidelberg 2005
Printed in Germany

Typesetting: Camera-ready by author, data conversion by Scientific Publishing Services, Chennai, India
Printed on acid-free paper   SPIN: 11499220   06/3142   5 4 3 2 1 0

Preface

The computational paradigm is considered here as a conceptual, theoretical and formal framework situated above machines and living creatures (two instantiations), sufficiently solid, and still non-exclusive, to allow us:

1. To help Neuroscientists to formulate intentions, questions, experiments, methods and explanation mechanisms assuming that neural circuits are the psychological support of calculus.
2. To help Scientists and Engineers from the fields of Artificial Intelligence (AI) and Knowledge Engineering (KE) to model, formalize and program the computable part of human knowledge.
3. To establish an interaction framework between Natural System Computation (NSC) and Artificial System Computation (ASC) in both directions, from ASC to NSC (in Computational Neuroscience), and from NSC to ASC (in Bioinspired Computation).

Under these global purpose, we have organized IWINAC2005, the first "International Work-conference on the Interplay between Natural and Artificial Computation" that took place in Las Palmas de Gran Canaria, Canary Islands (Spain), during June 15-18, 2005, trying to contribute to both directions of the interplay:

I: From Artificial to Natural Computation. What can Computation, Artificial Intelligence (AI) and Knowledge Engineering (KE) contribute to the understanding of Nervous System, Cognitive Processes and Social Behavior?. This is the scope of Computational Neuroscience and Cognition, which uses the computational paradigm to model and improve our understanding of natural science.

II: From Sciences of Natural to Computation, AI&KE. How can computation, AI and KE find inspiration in the behavior and internal functioning of physical, biological and social systems to conceive, develop and build-up new concepts, materials, mechanisms and algorithms of potential value in real world applications? This is the scope of the new Bionics, known as Bioinspired Engineering and Computation, as well as of Natural Computing.

To address the two questions exposed in the scope of IWINAC-2005, we have made use of the "building of and for knowledge" concepts that distinguish three levels of description in each calculus: The Physical Level (PL), where the hardware lives, the Symbol Level (SL) where the programs live and a third level, introduced by Newell and Marr, situated above the symbol level and named by Newell "the Knowledge Level" (KL) and by Marr the level of "the theory of calculus". We seek the interplay between the natural and the artificial at each one of these three levels (PL, SL, KL).

1. In the interplay at the **Physical Level** we consider:
   - **Computational Neuroscience**. *Tools*: Conceptual, formal, and computational tools and methods in the modelling of neuronal processes and neural nets: individual and collective dynamics. *Mechanisms*: Computational modelling of neural mechanisms at the architectural level: oscillatory/regulatory feedback loops, lateral inhibition, reflex arches, connectivity and signal routing networks, distributed central-patterns generators. Contributions to library of neural circuitry. *Plasticity*: Models of memory, adaptation, learning and other plasticity phenomena. Mechanisms of reinforcement, self-organization, anatomo-physiological coordination and structural coupling.
   - **Bio-inspired Circuits and Mechanisms**. *Electronics*: Bio-inspired electronics and computer architectures. Advanced models for ANN. Evolvable hardware (CPLDs, FPGAs, ...). Adaptive cellular automata. Redundancy, parallelism and fault-tolerant computation. Retinotopic organizations. *Non-conventional (Natural) Computation:* Biomaterials for computational systems. DNA, cellular and membrane computing. *Sensory and motor prostheses:* Signal processing, artificial cochlea, audio-tactile vision substitution. Artificial sensory and motor systems for handicapped people. Intersensory transfer and sensory plasticity.

2. In the Interplay at the **Symbol Level** we consider:
   - **Neuro-informatics**. *Symbols:* From neurons to neurophysiological symbols (regularities, synchronization, resonance, dynamics binding and other potential mechanisms underlying neural coding). Neural data structures and neural "algorithms". *Brain databases:* Neural data analysis, integration and sharing. Standardization, construction and use of databases in neuroscience and cognition. *Neurosimulators:* Development and use of biologically oriented Neurosimulators. Contributions to the understanding of the relationships between structure and function in biology.
   - **Bio-inspired Programming Strategies**. *Behavior based computational methods:* Reactive mechanisms. Self-organizative optimization. Collective emergent behavior (ant colonies). Ethology and Artificial Life. *Evolutionary computation*: Genetic algorithms, evolutionary strategies, evolutionary programming and genetic programming. Macroevolution and the interplay between evolution and learning. *Hybrid approaches:* Neuro-symbolic integration. Knowledge-based ANN and connectionist KBS. Neuro-fuzzy systems. Hybrid adaptation and learning at the symbol level.

3. In the Interplay at the **Knowledge Level** we consider:
   - **Computational approach to Cognition**. *AI&KE:* Use of AI&KE concepts, tools, and methods in the modelling of mental processes and behavior. Contribution to the AI debate on paradigms for knowledge representation and use: symbolic (representational), connectionist, situated, and hybrid. *Controversies:* Open questions and controversies in

AI&Cognition (semantics versus syntax, knowledge as mechanisms that knows, cognition without computation,...). Minsky, Simon, Newell, Marr, Searly, Maturana, Clancey, Brooks, Pylyshyn, Fodor, and more. *Knowledge Modelling:* Reusability of components in knowledge modelling (libraries of tasks, methods, inferences and roles). Ontologies (generic, domain specific, object oriented, methods, and tasks). Knowledge representation Methodologies and Knowledge edition tools.

– **Cognitive Inspiration in AI&KE**. *Synthetic cognition:* Bio-inspired modelling of cognitive tasks. Perception, decision-making, planning and control. Biologically plausible (user sensitive) man-machine interfaces. Natural language programming attempts. Social organizations, distributed AI, and multi-agent systems. *Bio-inspired solutions to engineering, computational and social problems in different application domains:* Medicine, Image understanding, KBS's and ANN's for diagnoses, therapy planning, and patient follow-up. Telemedicine. Robotic paradigms. Dynamic vision. Path planning, map building, and behavior based navigation methods. Anthropomorphic robots. Health biotechnology. Bio-inspired solutions for sustainable growth and development.

IWINAC2005 was organized by the Universidad Nacional de Educación a Distancia (UNED) in cooperation with the Instituto Universitario de Ciencias y Tecnologías Cibernéticas de la Universidad de Las Palmas de Gran Canaria and Las Palmas UNED Associated Center.

Sponsorship was obtained from the Spanish Ministerio de Ciencia y Tecnología and the organizing universities (UNED and Las Palmas de Gran Canaria).

The chapters of these two books of proceedings correspond to the talks delivered at the IWINAC2005 Conference. After the referee process, 118 papers were accepted for oral or poster presentation, according to the authors preferences. We have organized these papers into two volumes basically following the topics list previously mentioned. The first volume, entitled *"In Search of Mechanisms, Symbols, and Models underlying Cognition"*, includes all the contributions mainly related with the methodological, conceptual, formal, and experimental developments in the fields of Neurophysiology and cognitive science.

In the second volume, *"Artificial Intelligence and Knowledge Engineering Applications: A Bioinspired Approach"*, we have collected the papers related with bioinspired programming strategies and all the contributions related with the computational solutions to engineering problems in different application domains.

And now, is the time for acknowledgements. A task like this of organizing a work-conference with a well defined scope cannot be achieved without the active engagement of a broad set of colleagues that share with us the conference principles, foundations and objectives. First, let me express my sincere gratitude to all the Scientific and Organizing Committees, in particular, the members of these committees that have acted as effective and efficient referees and as promoters and managers of pre-organized sessions on autonomous and relevant topics under the IWINAC global scope. Thanks also to the invited speakers, Joost N. Kok, Dana Ballard, and Juan Vicente Sánchez Andrés for their capacity of synthesis in preparing the plenary lectures. Finally, thanks to all the authors for their interest in our call and the effort in preparing the papers, condition *sine qua non* for these proceedings.

My debt of gratitude with José Ramón Alvarez and Félix de la Paz goes further the frontiers of a preface. Without their collaboration IWINAC2005 would not be possible, in strict sense. And the same is true concerning Springer-Verlag and Dr. Alfred Hofmann, for the continuous receptivity and collaboration in all our editorial join ventures on the interplay between Neuroscience and Computation from the first IWANN in Granada (1991, LNCS 540), to the successive meetings in Sitges (1993, LNCS 686), Torremolinos (1995, LNCS 930), Lanzarote (1997, LNCS 1240), Alicante (1999, LNCS 1606 and 1607), again in Granada (2001, LNCS 2084 and 2085) then in Maó (2003, LNCS 2686 and 2687) and now, the first IWINAC in Las Palmas.

June 2005   José Mira

Organization

General Chairman

José Mira, UNED (Spain)

Organizing Committee

José Ramón Álvarez Sánchez, UNED (Spain)
Félix de la Paz López, UNED (Spain)

Local Organizing Committee

Roberto Moreno-Díaz, jr., Univ. Las Palmas de Gran Canaria (Spain)
Alexis Quesada, Univ. Las Palmas de Gran Canaria (Spain)
José Carlos Rodriguez, Univ. Las Palmas de Gran Canaria (Spain)
Cristobal García Blairsy, UNED (Spain)
José Antonio Muñoz, Univ. Las Palmas de Gran Canaria (Spain)

Invited Speakers

Joost N. Kok, Leiden University (The Netherlands)
Dana Ballard, University of Rochester (USA)
Juan Vicente Sánchez Andrés, University of La Laguna (Spain)

Field Editors

Eris Chinellato, Universitat Jaume-I (Spain)
Carlos Cotta, University of Málaga (Spain)
Angel P. del Pobil, Universitat Jaume-I (Spain)
Antonio Fernández-Caballero, Universidad de Castilla-La Mancha (Spain)
Oscar Herreras, Hospital Ramón y Cajal (Spain)
Heinz Hügli, University of Neuchâtel (Switzerland)
Roque Marín, Universidad de Murcia (Spain)
Carlos Martin-Vide, Universitat Rovira i Virgili (Spain)
Victor Mitrana, Rovira i Virgili University of Tarragona (Spain)
José T. Palma Méndez, University of Murcia (Spain)
Miguel Angel Patricio Guisado, Universidad de Alcalá (Spain)
Eduardo Sánchez Vila, Universidad de Santiago de Compostela (Spain)
Ramiro Varela Arias, Universidad de Oviedo (Spain)

Scientific Committee (Referees)

Ajith Abraham, Chung Ang University (Korea (South))
Igor Aleksander, Imperial College of Sci., Tech. and Med. (United Kingdom)

José Ramón Álvarez Sánchez, UNED (Spain)
Margarita Bachiller Mayoral, UNED (Spain)
Antonio Bahamonde, Universidad de Oviedo (Spain)
Emilia I. Barakova, RIKEN (Japan)
Alvaro Barreiro, Univ. A Coruña (Spain)
Senen Barro Ameneiro, University of Santiago de Compostela (Spain)
Luc Berthouze, AIST (Japan)
Joanna J. Bryson, University of Bath (United Kingdom)
Lola Cañamero, University of Hertfordshire (United Kingdom)
Joaquín Cerdá Boluda, Univ. Politécnica de Valencia (Spain)
Enric Cervera Mateu, Universitat Jaume I (Spain)
Eris Chinellato, Universitat Jaume-I (Spain)
Carlos Cotta, University of Málaga (Spain)
Paul Cull, Oregon State University (United States)
Kerstin Dautenhahn, Univ. Hertfordshire (United Kingdom)
Félix de la Paz López, UNED (Spain)
Ana E. Delgado García, UNED (Spain)
Javier de Lope, Universidad Politécnica de Madrid (Spain)
Angel P. del Pobil, Universitat Jaume-I (Spain)
Jose Dorronsoro, Universidad Autónoma de Madrid (Spain)
Richard Duro, Universidade da Coruña (Spain)
Juan Pedro Febles Rodriguez, Centro Nacional de Bioinformática (Cuba)
Antonio Fernández-Caballero, Universidad de Castilla-La Mancha (Spain)
Jose Manuel Ferrández, Univ. Politécnica de Cartagena (Spain)
Nicolas Franceschini, Université de la Méditerranée (France)
Marian Gheorghe, University of Sheffield (United Kingdom)
Karl Goser, Univ. Dortmund (Germany)
Carlos G. Puntonet, Universidad de Granada (Spain)
Manuel Graña Romay, Universidad Pais Vasco (Spain)
John Hallam, University of Southern Denmark (Denmark)
Denise Y. P. Henriques, York University (Canada)
Oscar Herreras, Hospital Ramón y Cajal (Spain)
Juan Carlos Herrero, (Spain)
Heinz Hügli, University of Neuchâtel (Switzerland)
Shahla Keyvan, University of Missouri - Columbia (United States)
Kostadin Koroutchev, Universidad Autónoma de Madrid (Spain)
Elka Korutcheva, UNED (Spain)
Max Lungarella, University of Tokyo (Japan)
Francisco Maciá Pérez, Universidad de Alicante (Spain)
george Maistros, The University of Edinburgh (United Kingdom)
Dario Maravall, Universidad Politécnica de Madrid (Spain)
Roque Marín, Universidad de Murcia (Spain)
Rafael Martínez Tomás, UNED (Spain)
Jose del R. Millan, IDIAP (Switzerland)
José Mira, UNED (Spain)

Victor Mitrana, Rovira i Virgili University of Tarragona (Spain)
Roberto Moreno-Díaz, Univ de Las Palmas de G.C. (Spain)
Lucas Paletta, Joanneum Research (Austria)
José T. Palma Méndez, University of Murcia (Spain)
Miguel Angel Patricio Guisado, Universidad de Alcalá (Spain)
Mario J. Pérez Jiménez, Universidad de Sevilla (Spain)
Franz Pichler, Johannes Kepler University (Austria)
Luigi M. Ricciardi, Università di Napoli Federico II (Italy)
Mariano Rincón Zamorano, UNED (Spain)
Camino Rodríguez Vela, Universidad de Oviedo (Spain)
Ulrich Rückert, Univ. Paderborn (Germany)
Daniel Ruiz Fernández, Univ. de Alicante (Spain)
Ramón Ruiz Merino, Universidad Politécnica de Cartagena (Spain)
Eduardo Sánchez Vila, Universidad de Santiago de Compostela (Spain)
José Santos Reyes, Universidade da Coruña (Spain)
Juan A. Sigüenza, Universidad Autónoma de Madrid (Spain)
Wolf Singer, Max Planck Institute for Brain Research (Germany)
Mikhail M. Svinin, RIKEN (Japan)
Mª.Jesus Taboada, Univ. Santiago de Compostela (Spain)
Ramiro Varela Arias, Universidad de Oviedo (Spain)
Marley Vellasco, Pontifical Catholic University of Rio de Janeiro (Brazil)
Barbara Webb, University of Edinburgh (United Kingdom)
Stefan Wermter, University of Sunderland (United Kingdom)
Tom Ziemke, University of Skövde (Sweden)

Table of Contents - Part I

Computational Neuroscience

Bioinspired Computation

Tasks Modelling at the Knowledge Level

On the Use of the Computational Paradigm in Neurophysiology and Cognitive Science

José Mira Mira

Dpto. de Inteligencia Artificial, ETS Ing. Informática, UNED, Madrid, Spain
jmira@dia.uned.es

Abstract. Virtually from its origins, with Turing and McCulloch's formulations, the use of the computational paradigm as a conceptual and theoretical framework to explain neurophysiology and cognition has aroused controversy. Some of the objections raised, relating to its constitutive and formal limitations, still prevail. We believe that others stem from the assumption that its objectives are different from those of a methodological approximation to the problem of knowing.

In this work we start from the hypothesis that it is useful to look at the neuronal circuits assuming that they are the neurophysiological support of a calculus, whose full description requires considering three nested levels of organization, one of circuits, other of neurophysiological symbols and another of knowledge, and two description domains, the intrinsic to each level and that of the external observer.

1   Introduction

The general aim of this work is to present a wider and more comprehensive view of the Computational Paradigm (CP) than is usual and propose a way of using it that makes it possible: (1) To help Neuroscience and Cognitive Science, including the possibility of explaining the latter as a result of the former. (2) To establish an interaction framework between Natural System Computation (NSC) and Artificial System Computation (ASC), thereby posing a series of appropriate questions in both directions of the interaction, from ASC to NSC (in Computational Neuroscience), and from NSC to ASC (in Bioinspired Computation).

The principle of the cross-fertilization potential between ASC and NSC was initially proposed by Wiener [30] and McCulloch's [13, 12] Neurocybernetics, when they identified that both living creatures and machines, can be understood using the same experimental methodology, the same analysis procedures, the same formal tools and the same organizational and structural principles: Those of Physics, Logic, Mathematics and, Engineering and, embracing all of them those of the Computational Paradigm, interpreted the paradigm concept according to Thomas Kuhn [7]. In both instances, living creatures and machines, we are talking of systems determined structurally by their constituent elements and by the stable causal relations between these elements inside the system and between the system and its environment, as they have evolved jointly over time.

J. Mira and J.R. Álvarez (Eds.): IWINAC 2005, LNCS 3561, pp. 1–15, 2005.

We start from the hypothesis that it is useful to look at the nervous system circuits assuming that they are the physiological support of a calculus and that the CP helps us interpret the results of this calculus at the physical, symbol and knowledge levels. The rejection of the CP in some sectors of Neuroscience and Cognitive Psychology is possibly a result of a simplistic interpretation of what computation really means by just associating it with the physical machine supporting it. This limited interpretation of the knowledge involved in a calculus, when this calculus is described at only one level and without reference to the external observer, provides a reductionist view of computation and consequent apparent impoverished analogies of the type "Brain-Hardware", "Mind-Software" and the conclusion that "Thinking is equivalent to Computing". Nevertheless, by considering the three levels of description of a calculus (the physical machine, symbols and programming language, and conceptual and formal models) and then distinguishing the descriptions in each level "own" domain from these other descriptions in the same level but now in the domain of an external observer, the CP substantially expands its capacity to explain Neurophysiological processes from which Cognition emerges.

This wider view of the CP, where the meaning of a calculus at three abstraction levels and in two causally distinguishable phenomenological domains is described and interpreted, with the corresponding emergence and reduction processes between levels, allows us to accommodate different theoretical positions, like emergentism or physicalist monism. The CP thus becomes a methodological approximation (set of intentions, questions, experiments, methods, and explanation mechanisms) to the problem of knowing. In other words, it becomes a tool and test bed for evaluating hypotheses, theories and experimental results in Neuroscience and Computation.

In accordance with Maturana and Varela [11, 10, 27], a key point in the right use of the CP in ASC and NSC is that you need to "know how to do the accounting well" and not mix entities, semantics and causality in the description of a calculus in a specific domain and level with others substantially different that correspond to the description of the same calculus in another domain or level. Only an external observer who knows and distinguishes the constituent entities, relations, language, causality and semantic tables of all the levels in all the domains, and also the reduction and emergence processes linking them, could have a global view of the calculus.

The rest of the paper is organized as follows. Section 2 is a summary of the building of the knowledge involved in the calculus carried out by a neural network. Sections 3, 4, and 5 describe the ascending path common to reverse neurophysiology. This ascending path allows us to consider the different knowledge components associated with a neuronal network, the need to model abstraction processes between the different levels of description and the additional knowledge that the external observer needs to obtain a total understanding of the meaning of the calculus carried out by a neuronal circuit. Finally, in section 6 we end by considering right and wrong objectives in the use of the CP.

The theses that we defend in this work are: (1) That knowledge (cognition) that each level (physical, symbol or cognitive) can accommodate is associated with the network of sensory, motor and association mechanisms that characterize this level as a class, as an autonomous organization, with its own language [9]. (2) That, consequently, just knowledge of the physical level (PL) of the NSC is not enough to understand cognition. It is necessary to inject the knowledge associated with the neurophysiological symbol level and associated with the last emergence of the calculus at the knowledge level (the cognitive model) [14, 16]. (3) Additionally, it is crucial to understand that all this injected knowledge (abstractions, symbols and emergences) only exists in the external observer's language.

2   The Building of the Knowledge Involved in the Calculus Carried Out by a Neural Network

This building has three levels and two domains in each level (figure 1). The three levels ("storeys") of this building are the physical level where the anatomical circuits live, the symbol level where the neural programs live (the neurophysiological symbols) and the third level, located over the symbol level and which Newell called "the knowledge level" [20] and Marr [8] "the theory of calculus" level. This third level is the house of natural language and cognition.

The basic idea behind this architectural metaphor is that the first floor, where the physical circuit lives, is not sufficient enough to complete the description of a calculus. We also need to have the description of the neurophysiological symbols ("the program") and the emergent conceptual model.

In order to complete our understanding of the meaning of a calculus at the three levels described above we use a new distinction inside each level between two description domains: the level's *own domain* (OD) "apartments to the right", and the *external observer domain* (EOD), "apartments to the left". The introduction of the figure of the external observer and the differentiation between a mechanism and its description comes from physics and has been reintroduced and elaborated in biology by Maturana [10] and Varela [27] and in computation by Mira and Delgado [15, 18, 17].

When we acknowledge the existence of an observer external to the computation in natural systems we are introducing the idea of different reference systems in which entities, relations, and their meanings are represented. Everything that happens in the descriptions in the level's own domain is causal. What "has to happen", happens because there is a coincidence of structure and function. In the physical level's OD, processes cannot be separated from the processors (neuronal mechanism) realizing them. Inverters invert, and adders add, and in every mechanism the semantics is intrinsic.

A frequent source of misunderstanding, both in ASC and NSC, is the mixing of OD entities and relations with others in the EOD with which there is no causal relation. Varela [27] clearly distinguishes between the explanations in the EOD and the operational and causal explanations in the OD. In both instances,

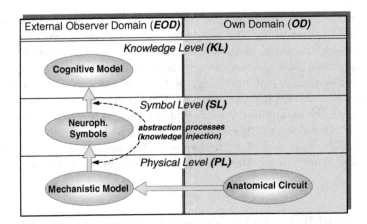

Fig. 1. The function of each neural network can be described at three levels (PL, SL, KL) and in two domains (EOD, OD). The reverse neurophysiology pathway starts at the PL in the OD (1st right "apartment") and ends in the KL in the EOD (3rd left "apartment")

the level entities and processes are described in two languages, which are independently consistent and subject to crossreferencing. The difference lies in the fact that in the operational descriptions (in the OD), the terms used to refer to processes cannot leave the domain in which they operate either in syntax or in semantics. On the other hand, the terms used in descriptions in the EOD belong to the natural language of an external observer and refer to the domain knowledge, although their referents are not obliged to follow the EOD laws. The EOD links do not operate in the OD unless we limit ourselves to the models that the observer possesses at the physical or symbol level. That is to say, unless we limit ourselves to electrical signals, differential equations, electronics, logic or automata theory. The rest of the meanings, including the concepts of neural code, neurophysiological symbols and knowledge itself, are always in the EOD, in the observer's language [9].

3   The CP at PL: Mechanistic Models in a Language of Signals

When we use the CP at the PL to represent the functions of the nervous system we are associating the calculus with the neurophysiological mechanisms supporting it. In other words, we model the neuronal function in terms of the local function done by the simple *constituent units* in each sublevel and by the interconnection schemes specifying the architecture of the composed units in this sublevel which, in turn, are simple constituents of the upper sublevel. The simple constituent units of a mechanistic model can be individual units (ionic channels, synaptic contacts, neurons) or collective entities (neural ensembles). Similarly, the composed units can be circuits of synaptic contacts or neurons, result of the interconnection of individual units, or aggregates of neural ensembles [26, 3].

From the point of view of the EOD, the process of modeling by mechanisms begins with an initial distinction between environment and system to be able to talk of "inputs", "outputs" and system dynamics. Or, expressed differently, to be able to talk of mutual disturbances and structural coupling [10, 27].

In all the cases mentioned above, the formal underlying model to modeling at the physical level is a three-dimensional hierarchical, recursive and recurrent graph where each *node* can unfold into a new graph with nodes of less granularity and, in turn, can form part of the network specifying a node of greater granularity (Fig. 2). Each node accepts an *external* (functional) description and another *internal* (structural) with the explicit and detailed specification of the mechanism (composed unit) from which the observed functionality emerges. Similarly, each arc of the graph accepts an external description of its function of modulated transmission of the information, and another internal description specifying the learning mechanism from which the plasticity observed in this connection emerges.

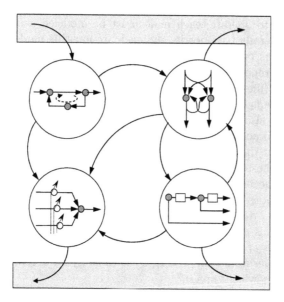

Fig. 2. Hierarchical, recurrent and recursive three-dimensional graph underlying all our mechanistic models of neurons and neural nets at the physical level

The equivalence between graphs and finite state automata (FSA) means that it is possible to model recursivity in terms of deterministic or stochastic measurements. Each internal state of a given level of description (macrostate) can be interpreted as the result of a measurement over the set of microstates, which specify this FSA at a lower level of granularity.

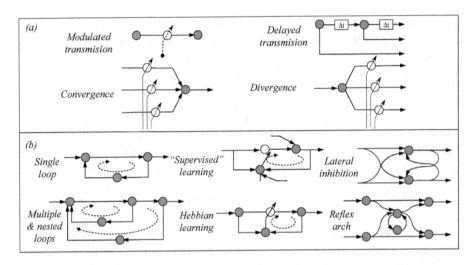

Fig. 3. (a) Basic structural elements, (b) components for a library of neural mechanisms

For the graph to operate, each node must be modeled using a function of local calculus, generally parametric, and each arc must be modeled using a new function of calculus of the temporal evolution of the value of this connection ("learning"). The repertory of analogical and logical functions of local calculus is well known (linear and non-linear differential equations, convolution-like functionals, and sequential logic). The repertory of supervised and correlational learning models is also well known [16].

The capacity to accommodate knowledge at the physical level is a result of the high quality of the nervous tissue (of its constituent elements) and of the different levels of nested organizations that evolution has superimposed [9, 19]. In other words, the neuronal calculus at the PL emerges from the successive layers of superimposed architectural mechanisms which, always preserving the organizations supporting them, have complicated the language of biochemical and electrical signals and converted them into a language of *mechanisms*. From each new stable set of these mechanisms (from each superimposed architecture, the external observer would say) emerges the capacity for accommodating again a new "common knowledge space."

From the topological point of view, all the neuronal circuits can be modeled using a very limited set of basic structural elements, as can be seen in figure 3.a (modulated transmission, delays, signal routing convergent and divergent circuits). From the composition of these structural elements many other autonomous mechanisms can be synthesized of a higher level of integration, as can be seen in figure 3.b.

A mechanism (single and multiple, positive and negative feedback loops, "hebbian" and "supervised" learning circuits, recurrent lateral inhibition or associative reflex arches) becomes stable and autonomous when it provides functionality (control, rhythm generation, plasticity, spatial and/or temporal contrast detection or accumulation of coincidences) relevant for survival. We should however remember that all these concepts (control, plasticity, ...) only exist in the EOD, in the observer's language.

4   In Search of a Library of Neurophysiological Symbols (NSs)

At the end of the analysis and modeling stage of neuronal circuits at the PL we have the following knowledge:

1. A set of patterns of concurrent temporal sequences of biochemical and electrical signals (slow potentials and all-or-nothing action potentials).
2. A set of local non-linear operators that fits reasonably well the continuous transformations of these sequences of signals.
3. A sketch of a language of neurophysiological signals (neural "code") to describe membrane phenomena, excitatory and inhibitory processes, adaptive thresholding, oscillatory and regulatory processes and some plasticity phenomena.
4. A library of the most frequently found neural mechanisms and connectivity patterns.

Let us assume for one moment that we have a complete theory of the physical level. In other words, that we know everything about signals, circuits and local operators, similar to our knowledge about digital electronics and computer architecture. Would we know what the brain is calculating? Would we know the "program" and the emerging cognitive processes? Of course, not. From just the knowledge of the PL of a neuronal calculus it is not possible to obtain either the algorithm or meaning of this calculus. This description needs to be completed with the descriptions corresponding to the intermediate symbolic level (the "program") and to the knowledge level, both based on elements of knowledge associated with the structure of the external environment. Evolution, culture, history and civilization do not appear explicitly in the anatomy and signals of the calculating mechanism. It is necessary to inject them to understand the meaning of the calculus.

What the CP suggests to us now is to follow the ascending path of the building of levels and domains to find a symbolic description of the neuronal calculus. In other words, we accept the conjecture that neuronal networks have accumulated enough knowledge in their structure throughout evolution to be able to accommodate a superimposed organization consisting of a set of neurophysiological symbols (NSs) and relations between these symbols ("programs"), (figure 44).

Seen from the EOD the NSs are abstractions of the cortical representations of the internal and external environment, a result of the process of biological and hereditary adaptation, selected throughout evolution and modified by learning [14, 16]. The NSs are the *sensory*, *motor* and *association* entities in the internal language from which cognition emerges and they have their *own causal reality* at an intermediate organizational level, superimposed at the PL level, but

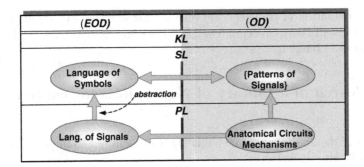

Fig. 4. Neurophysiological symbols as abstractions from neural mechanisms and specific patterns of spatio-temporal signals relevant for survival

different from it. The causality of a language of NSs is effective for all neurons systems that have the mechanisms for generating and interpreting this code. The possession of this language allows the nervous system to construct a new level of reality, an internal creation of an environment with *"objects"*, *"events"*, *"behavior modes"* and *"associations"* that are unitary and autonomous, beyond the reflex responses associated with threshold values of the individual signals that constitute these spatio-temporal patterns. Both, the patterns and the individual signals, are intervening at the same time [19]. This internal language makes it possible to handle heterogeneities in the external world in the same way by associating meanings to classes of specific experiences in a specific environment to which the animal responds in a unitary way. Subsequent instantiations of these equivalence classes are *"re-known"* by activating the same support mechanisms of these NSs and *"trigger"* the same behavior patterns. We should remember that the concept of NSs proposed here is not the one proposed by Newell and Simon [21] in their *"Physical Symbol System Hypothesis"* because in our proposal the NSs are connectionist and grounded in the mechanisms and they are not programmable in the conventional sense, but via dynamic adjustment processes of synaptic efficiency, controlled by the states of activity in the network.

Seen from the EOD these NSs *designate*: (1) Entities in the external environment relevant for survival in this environment (ranges of energies, nocireceptors, ...), (2) multimodal and temporal relations between these entities (objects and events), (3) primary conceptualizations (warning signals and homeostasis), (4) compensatory reactions and other internal strategies related to the conservation of the species (sexual, aggression or escape symbols, and descriptors of internal needs, like sleep or thirst) [14, 16]. All this semantics only exists in the observer's language (in the EOD) but not in the OD, where it is implemented in the functional architecture of the PL mechanisms [9].

The conditions necessary for the emergence of these symbols [19] are: (1) The existence of a rich and structured environment, where the possibility and convenience exists of distinguishing "objects" and "events" as autonomous entities in addition to the signals constituting them. (2) The constitutive plasticity of the biological tissue (that does not possess, for example, the silicon crystal of computers) that allows the clustering of signals by selective synaptic reinforcement and (3) the environment-system interaction at an appropriate time scale.

Considered now in their OD the NSs are active and dynamic entities associated with specific patterns of spatio-temporal signals (electrical, chemical and electronic) that are presented repeatedly in a stable, independent and autonomous way and associated with specific referents in the organisms internal and external environment [14, 16]. These patterns ("keys") are moored to the PL via neuronal mechanisms that distinguish and recognize them, giving them a unitary response ("doors" opened by these "keys"). From the first sensory processing stage to the motor stage these cortical keys act as internal representations replacing their referents in all the association processes, which are thus converted into what the external observer describes as *"associations between representations"*. These representations have developed throughout evolution as *"meaning modules"* because of their usefulness for constructing an economical and simple representation of the external environment, which in turn permits a unitary response in efficient time. The semantics of the NSs is in their syntax, it is not arbitrary, but to know it the external observer needs to know the evolutive history of the interaction with the environment from which the symbol emerged, (i.e. the goals served by this particular way of clustering signals). It seems reasonable to assume that the support mechanism of this structuring of neuronal signals in significant terms of clusterings, in classes, is the selective synaptic reinforcement and that the candidate mechanisms for controlling this selective map of reinforcements are two types: (1) *Correlational*, which are only connected with the characteristics of the cluster of signals where the selection is going to be done (spatio-temporal coincidence of states of activity, concomitance, regularities) and (2) *Reinforcement*, which are connected with comparing the characteristics of a cluster of signals with those of a prototype (filtering, tuning and resonance, feature extraction, discrimination, and classification) [28].

5   The Computational Paradigm at the Knowledge Level: Symbolic, Situated, and Connectionist Approaches

Let us assume for one moment that we have the complete repertory of *sensory*, *motor* and *association* NSs and the connectionist processes of learning that fully describe the calculus done by the nervous system at the symbol level . Let us also assume that we know the neuronal mechanisms that generate and recognize these NSs. Would we now know the meaning of what this nervous system is calculating? Would we know the purposes, goals and intentions of the behavior of an animal in a specific environment? Of course, not, because we do not have the constituent elements of cognition at the knowledge level. These elements and their relations constitute a new organization superimposed on that of the symbols and, again, with its *own causality* and characterized by the language which identify and specify this organization as a class [9]. Maturana's hypothesis is that these constituent elements of cognition at the KL reside in the observer's natural language and that, consequently, the conceptual and formal architecture of cognition is indistinguishable from the conceptual and formal architecture of natural language.

If we accept this hypothesis the work now consists of specifying a new abstraction process that enables us to link the entities and relations at the symbol level, both in the OD and EOD, with the corresponding entities and relations at the knowledge level, without any change in the underlying causal structure (figure 5).

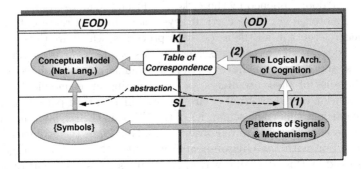

Fig. 5. The two-step abstraction process from SL to KL. (1) From patterns of signals to formal models. (2) From formal models to inferential schemes in natural language

In direct engineering the reduction process of a knowledge model to a program is done in two phases. First it passes from the conceptual model to the formal model and after the formal model is programmed [15]. If we want to proceed by analogy in the reverse process we need:

1. To link the symbols (patterns of signals) and their relations (mechanisms) with the entities and relations of a formal model.
2. To establish a table of correspondences between the entities and relations in the formal model and in a conceptual model whose constituent elements are the same as in natural language (concepts, inferential verbs and "if-then-else" control structures).

A major problem in the first step of this abstraction process is that we do not know the formal architecture of cognition. We are not sure that we have the necessary mathematics and logic to describe the mental processes exactly [16, 23]. In fact, the only complete formal architecture available is the deterministic and probabilistic modular theory of finite state automata (FSA) proposed by W.S McCulloch [13, 12] and later refined by von Neumann [29], Kleen [6] and Moreno-Díaz [23] among others. In this formalization, states of activity are associated with internal states of FSA, patterns of signals are associated with FSA "inputs", and "outputs", and support mechanisms of the dynamic activity linking the temporal evolution of these patterns of signals are associated with the FSA state transition functions.

If we start from an FSA, the second stage of the abstraction process poses fewer difficulties. Syntactically, a table of correspondence is sufficient to associate the FSA with an inferential scheme where the concepts (nouns) represent (play) input and/or output roles in the inferences (verbs) of the functional scheme resulting from analysis of the FSA. There are currently three ways of establishing this correspondence between FSAs and conceptual models, which correspond to three different approaches to the problem of knowledge modeling in AI and resulting in three computational views of cognition: (1) *representational* (symbolic), (2) *situated*, and (3) *connectionist*.

The representational approximation (symbolic) is the top-down. It starts from declarative descriptions in natural language of entities ("concepts") and relations between these entities ("inferential rules") and associates reasoning with the manipulation of these symbolic descriptions. From the first works where inference was associated with a search in a symbolic problem-space, the symbolic perspective has evolved first by separating generic components of knowledge (tasks, methods, inferences, roles and control structures) from others that are specific to each application domain [25]. After it has progressed to the construction of libraries of reusable components of modeling, essentially problem-solving methods and ontologies.

The main problem of the representational approach as a useful paradigm in neuroscience and cognition is that it does not have any direct correspondence with the lower organizational levels. Here the FSA states and their transitions are not connected to the NSs or with the mechanisms and the patterns of signals at the PL. The causal structure is not preserved when moving up one level. In other words, correspondence between causality at the KL and underlying causalities at the SL and PL is lost. This is the result of thinking of the computer as a general purpose programmable machine, instead of thinking of it as a biological machine which, if we wanted to maintain an analogy with physical machines, we would say that it is a special purpose adaptive machine, whose function is modified as a result of the interactive activity of its mechanisms with the mechanisms of the environment.

The *situated approach*, called thus to highlight that all perception and all action are structurally coupled to their environment [1, 24], is also based on natural language descriptions of the activities of an animal in its environment, but now these descriptions are of an ethological nature and are not modeled from high-level cognitive entities ("concepts" and "inferences"), but from elementary behaviors like "obstacle avoidance" or "standing-up". This approach is also therefore known as "behavior-based", or "reactive". A behavior is defined as a direct mapping between a class of "input pattern" and a predefined "pattern of motor actions", so that the correspondence with the FSA is immediate: "Behaviors" are associated with states or secuences of states, including, oscilatory loops, input patterns are associated with abstract symbols of the FSA input alphabet and patterns of actions with the elements of the FSA output alphabet. Observe however that in this approach the structural correspondence is also lost with the lower organizational levels, because we talk of *"behaviors", "patterns of perceptions"*, and *"patterns of actions"*, but we do not say anything about the correspondence between these natural language terms and the neurophysiological mechanisms supporting them. The difference with the symbolic approach is that it now suggests starting with simpler systems in less complicated environments, so that we can make some conjectures about the support mechanisms of the perceptions and actions observed.

The only alternative to storing the description of a calculus (to build up a program for a general purpose computer) is to specify the mechanism supporting this calculus, the circuit, its constituent elements and its connections (its architecture) and the set of "labeled lines" that allow signals never to lose their identity. This is the proposal of connectionism which, also at the knowledge level, continues to have a modular and distributed architecture with a large number of low granularity formal processes and programming replaced by learning. Here the causal structure of FSA is preserved. At the PL the calculus units are represented operational descriptions of dynamic systems or with automata of two states and the networks of N elements with automata of 2^N states where each element contributes with a variable of the pattern of signals. On passing to the SL, the FSA "inputs", "outputs" and "states" are associated with measurements on the activity of microstates of neural ensembles, but the resulting macroscopic FSA maintains the structure. Finally, the formalization of the step to the KL needs to be based on the structural composition of these measurements [16].

A conjecture of potential usefulness is to consider this structural composition as a *cooperative process* of almost a social nature, where the different areas of the cortex contribute with "factors" to the global functions that the external observer calls "perception", "language" or "memory", for example. The transitions in this FSA are guided by what the external observer calls purposes, intentions, motivations or desires, but we have done little to bridge the gap between neurophysiology and cognition if we are not capable of constructing correspondence tables between these EOD entities and the corresponding neuronal mechanisms that embody them in the OD. In other words, between nouns and verbs in natural language and circuits of the type suggested in Figure 3. We should however recognize that the CP at the KL still does not have the right formal tools for tackling the formalization of these concepts with the clarity and precision with which we are modeling at the PL with mathematics from physics. Formal cognition architecture has still not been discovered.

6   Conclusions

Many of the criticisms about the validity of the CP in Neuroscience and Cognition are based on arguments that misunderstand its real objectives. We are not saying that computation in general and AI in particular are equivalent to natural intelligence and human thought nor that silicon machines possess the same functionalities as the biological tissue and, consequently, that calculating is equivalent to possessing a mind. We are not saying that the brain is a Turing Machine or a network of "formal neurons" as defined by McCulloch and Pitts.

Neither are we stating that simulation in a digital computer of a neuronal model can replace the network that it is modeling, because its constituent elements are different and "all knowledge depends on the structure known". Finally, neither do we tackle the problem of semantics other than by affirming that it is a concept inherent in the EOD and that in the own domain it emerges as a result of the sensitivity of the nervous tissue that allows this kind of organization. Consequently, it is wrong to argue against the CP using these false objectives as a starting point.

What we have defended in this work is that the CP, understood in the broad and structured sense in which it has been presented here, is a useful epistemological and constructive approximation for understanding neurophysiology and cognition. In other words, the CP is a conceptual and theoretical non-exclusive framework situated above machines and living creatures (two instantiations), sufficiently solid and formalizable to allow us: (1) to ensure that the "facts" (our data) are the relevant facts to which we must pay attention and are not clouded by incorrect statements about the functioning of the nervous system. (2) To structure these facts and the available relational knowledge. (3) To interpret and use them. (4) To raise new issues and make reasonable speculations. (5) To plan new experiments and (6) To leave room for self-criticism.

We would like to stress that the capacity of calculus of the neuronal tissue, the possibility of classifying the structure of the environment into symbols and the subsequent emergence of cognition in natural language is the direct consequence of the special nature of the biological tissue that allows this kind of superimposed and nested organizations with their own causalities. Neuron, symbol and cognition phenomenologies are constitutive, ontological, consequences of the nature of the nervous system, and of the wealth of environments to which it has had to adapt.

Finally, we would like to highlight three points of our proposal. The first is the importance of the intermediate level (neurophysiological symbols) to facilitate the link between Neurophysiology and Cognition. The second is the need to prepare and formalize in greater depth the two abstraction processes that link the three organizational levels. The third is the importance of the figure of the external observer whose natural language embodies the three organizational levels, the two domains and the CP.

If these three arguments (*symbols*, *abstraction* and *observer*) are handled with precision, it is thus possible with the CP to accommodate basically very different proposals like those of Craik [2], Fodor [5], Pylyshyn [22], or Johnson-Laird on the one hand, together with those of Clancey, Searle, Dreyfus [1], Edelman [4] or Gibson on the other. This view of the CP makes it possible to combine mechanistic physicalism (PL in the OD), associated with the description of a calculus in the OD at the PL, with emergentist thought, based on two abstraction processes that complete the description of the same calculus in the two superimposed causal organizations (SL and KL, in both the OD and EOD). All we need, Maturana tells us, is to know how to do the accounting well.

References

1. W.J . Clancey. *Situated cognition. On human knowledge and computer representation*. Univ. Press, Cambridge, 1997.
2. K. Craik. *The Nature of Explanation*. Cambridge University Press, Cambridge, 1943.
3. P. Dayan and L.F. Abbott. *Theoretical Neuroscience*. The MIT Press, Cambridge, Mass, 2001.
4. G.M. Edelman. *Neural Darwinism*. Basic Books, Inc., N. York., 1987.
5. J.A. Fodor. *El Lenguaje del Pensamiento*. Alianza Editorial, Madrid, 1984.
6. S.C. Kleene. Representation of events in nerve nets and finite automata. In Shannon and J. McCarty, editors, *Automata Studies*, pages 3–42. Princeton Univ. Press., Princeton, New Jersey, 1956.
7. T.S. Kuhn. *La Estructura de las Revoluciones Científicas*. Fondo de Cultura EconómicaFondo de Cultura Económica., México, 1971.
8. D. Marr. *Vision*. Freeman, New York, 1982.
9. H. Maturana. Ontology of observing. the biological foundations of self consciousness and the physical domain existence. http://www.inteco.cl/biology/ontology/, 2002.
10. H.R. Maturana. The organization of the living: A theory of the living organization. *Int. J. Man-Machine Studies*, 7:313–332, 1975.
11. H.R. Maturana and F. Varela. *El Árbol del Conocimiento Humano*. Ed. Debate, Madrid, 1990.
12. W.S. McCulloch. *Embodiments of Mind*. The MIT Press, Cambridge, Mass., 1965.
13. W.S. McCulloch and W. Pitts. A logical calculus of the ideas immanent in nervous activity. *Bulletin of Mathematical Biophysics*, 5:115–133, 1943.
14. J. Mira. Reverse neurophysiology: The embodiments of mind revisited. In R. Moreno-Díaz and J. Mira-Mira, editors, *Brain Processes, Theories and Models*, pages 37–49. The MIT Press, Massachusetts, 1995.
15. J. Mira and A.E. Delgado. Some comments on the antropocentric viewpoint in the neurocybernetic methodology. In *Proc of the Seventh International Congress of Cybernetics and Systems*, pages 891–95, 1987.
16. J. Mira and A.E. Delgado. Neural modeling in cerebral dynamics. *BioSystems*, 71:133–144, 2003.
17. J. Mira and A.E. Delgado. Where is knowledge in robotics? some methodological issues on symbolic and connectionist perspectives of ai. In Ch. Zhou, D. Maravall, and Da Rua, editors, *Autonomous robotic systems*, page 334. Physical-Verlag. Springer, Berlin, 2003.
18. J. Mira, A.E. Delgado, J.G. Boticario, and F.J. Díez. *Aspectos básicos de la inteligencia artificial*. Sanz y Torres, SL, Madrid, 1995.
19. J. Montserrat. *La Percepción Visual*. Ed Biblioteca Nueva, Madrid, 1998.
20. A. Newell. The knowledge level. *AI Magazine*, 120, 1981.
21. A. Newell and H.A. Simon. Computer science as empirical inquiry: Symbols and search. *Communications of ACM*, 19:113–126, 1976.
22. S.W. Pylyshyn. *Computation and Cognition. Towards a Foundation of Cognitive Science*. The MIT Press, 1986.
23. Moreno-Diaz R. Deterministic and probabilistic neural nets with loops. *Mathematical Biosciences*, 11:129–136, 1971.
24. Brooks RA. Intelligence without reason. A.i. memo, MIT, N°. 1293 1991.

25. G. Schreiber, H. Akkermans, and R. de Anjo Anjewierden. *Engineering and Managing Knowledge: The CommonKADS Methodology.* The MIT Press, Cambridge, Mass, 1999.

26. G.M. Shepherd, editor. *The Synaptic Organization of the Brain.* Oxford Univ. Press., 1990.

27. F.J. Varela. *Principles of Biological Autonomy.* The North Holland Series in General Systems Research, New York, 1979.

28. H. von Foester. *Understanding Understanding.* Springer, 2003.

29. J. von Neumann. Probabilistic logics and the synthesis of reliable organisms from unreliable components. In Shannon and McCarthy, editors, *Automata Studies.* Princeton Univ. Press., Princenton, New Jersey, 1956.

30. N. Wiener. *Cybernetics.* The Technology Press. J. Wiley & Sons, New York, 1948.

Modules, Layers, Hierarchies, and Loops Where Artificial Intelligence Meets Ethology and Neuroscience – In Context of Action Selection

Pinar Öztürk

Norwegian University of Science and Technology (NTNU),
Department of Computer and Information Science,
Sem Saelandsvei 7-9, NO-7491 Trondheim, Norway
pinar.ozturk@idi.ntnu.no

Abstract. The paper presents arguments for why AI methodologies should be informed of both behavioural science and neuroscience studies, and argues why this is possible. Through identifying the resemblance points, we will discuss whether findings of ethology and neuroscience can be used in the process of design and development of non-classical AI systems. To this end, we focus on a specific example that has long been investigated in all the concerned disciplines: the action selection problem. The paper overviews action selection mechanisms in behaviour-based AI and neuroscience in order to identify the commonalities that underly the understanding of action selection in different disciplines, and that may constitute the pieces of a common language with which AI can communicate ideas and findings to and from ethology and neuroscience.

1 Introduction

This paper voices the idea that AI systems should be informed of both behavioral and neuroscience studies. David Marr's work on vision [1] points to three stages underlying the computation involved in the nervous system:'task analysis', 'algorithm and representation', and 'implementation. Our initial hypothesis is that Marr's account does align well with AI researcher Chandrasekaran's 'task-structure' approach [2] to knowledge modelling and this parallelity can best be reflected through a mapping between behavioural attitudes and neural substrates. Although Chandrasekaran's account is mostly used in 'classical' AI by knowledge modelling and reuse community, one of his main points is rather useful when considered in the context of nonclassical-AI methodologies. The point is that a task (e.g., *selection of the next motor action*) can be realized through several alternative algorithms or *methods* (in Chandrasekaran's term). However, when several alternative algorithm may realize a particular functionality, it is not an easy job to discover them all and even more difficult is to find out which one is the best. The size of the search space of the algorithms is only limited with the

J. Mira and J.R. Álvarez (Eds.): IWINAC 2005, LNCS 3561, pp. 16–26, 2005.

imagination of AI researchers, and hence, rather large. So, an informed search is needed. One rational way to take, in order to overcome the intractability of the search space, is to get inpired from neuroscience that investigates the mapping of functional behaviour to the neural substrate for describing the subject matter algorithm. Thus, a neuroscience-inspired methodology is suitable for design and development of AI systems since it brings a solution to the intractable search space problem.

Another reason for supporting the use of neuroscience-inspired methodology is to avoid the interface problem Brooks [3] mentiones when he criticizes the traditional AI approaches. This resembles a problem hidden in Minsky's idea of 'Society of minds'[4] envisaging mind as a collection of highly interacting agents. If we view agents as modules responsible for certain functionalities, we should ensure that they are interoperable in order that they collectively can manage to exhibit a larger functionality. The use of brain as a source of premises and constraints when implementing functionalities(e.g., navigation,or active perception) of agents may spare us from encountering problems at the time of putting together the functionalities designed in isolation and on different assumptions. After all, the brain works quite well as a whole!

So, we have important reasons to suggest that AI design and development methodologies should be nformed by neuroscience. The question is then whether and how this could be done. Through scanning the literature in both AI, ethology and neuroscience we try to identify the points of agreements, and will see whether these commonalities may constitute a basis for suggesting that neuroscience findings should be a justified part of AI.

The subject of section 2 is classical AI methodologies, suggesting that classical AI methodologies can be modified so that neuroscience can inform AI systems. In section 3 we review AI, ethology and neurosceince models of a concrete example: action seleection mechanisms. In section 4 we try to identify points of agreements between the architectures we considered in section 3. Finally, section 5 provides our conclusions.

2   Design Methodologies in (Classical) AI

Although never has been explicitly formulated as a methodology per se, Newell's knowledge level hypothesis [5] bears the taste of a methodology as it puts an order on the description of knowledge and its representation. Newel argues for a level called *knowledge level*(KL)where knowledge is explicitly handled independently of how it will be represented and used at the second level; the symbol (implementation) level. He neither types nor structures knowledge at KL. His successors, however, structure knowledge by emphasizing the *task* knowledge and the hierarchical structure between a task and its subtasks, at the KL level.

Marr's account of nervous system divides the analysis of computation into three levels: (i) General analysis of the task: Answering the question of what to do (e.g., grasping an object, deciding the next motor action), (ii) The question of how to accomplish this task: description of input-output relations, and the algorithms, (iii) How the actual implementation will happen: relates to the wetware, the brain areas.

Chandrasekaran, who was concerned with identification and reuse of the the generic modules in the knowledge-based systems provides some guidelines for designing KB systems [2]. His 'task structures' approach makes a distinction between *goals*, *tasks*, and *methods*. Researchers in knowledge modeling and reuse community agree on that an overarching task needs to be decomposed into smaller tasks, recursively, until reaching 'primitive actions' that are immediately executable. The 'methods' are the elements that describe how to decompose a task into its subtask. The tasks and methods upward in the decomposition tree correspond to high-level tasks and methods while methods at the lowest level of such a 'task tree' correspond to primitive actions. Chandrasekaran makes one point which is important for our argument: a task can be accomplished in different alternative ways, i.e., 'methods' (see Our interpretation of the 'task structures' approach is that, for example, the task of grasping an object may be accomplished by different methods, one for human, another for a monkey, and yet another for a lion. Fig. 1). The methods (i.e., the 'algorithm', in Marr's

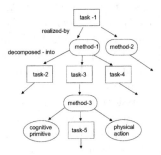

Fig. 1. An interpretation of 'task structures' - a task with two alternative methods

terms) in AI may be inspired from human, monkey or lion biologies, or designed entirely independently from biological models. So, there may be a one-to-many mapping between a particular functionality (of a task) and the corresponding algorithms and their implementations. An AI methodology should describe how this mapping takes place. A significant amount of AI research relies on behavioral sciences as an inspiration source for identification and analysis of a 'task'. Ethology, for example, has extensively been relied upon for this purpose. However, when it comes to the mapping between the task analysis in one side and algorithm and representation on the other side, existing AI methodologies do not provide clear guidelines. A key question is whether one sensible way of describing the methods/algorithms could be drawing upon neuroscience. A possible positive answer to this question is conditioned on the existence of a common currency for the interaction between AI and other disciplines.

In this paper, through elaborating the agreement points between AI, cognitive psychology, neuroscience and ethology we seek such a currency. For this purpose, we will be using the 'action selection' problem as a concrete example which has been of particular importance in both ethology, AI, neuroscience, and cognitive psychology.

3   Overview of Action Selection

Action selection defines the situation where more than one sub-system (of an agent) demand control of the scarce motor resources. The way one sub-system wants the motor unit to act may conflict with what another sub-system wants. An example is an intelligent agent that is both hungry and thirsty seeking both food and water. When the water and food resources are at a distance from each other the agent should decide which one to prioritize at a particular moment in time. A decision mechanism is, therefore, needed to coordinate the ultimate motor behaviour. Numerous action selection mechanisms have been proposed in AI so far, most being inspired from behavioural sciences, in particular ethology. In the rest of this section we give a brief overview of the most known action selection models, both in behaviour-based AI (BBAI) and in neuroscience.

3.1   Action Selection in AI

BBAI, pioneered by Brooks [3], provides a good example for the relationship between behavioral sciences and AI - without including much neuroscience. The suggested architectures are, nevertheless, inspired from physiology, although loosely - especially, the ones that have a 'connectionist' spirit (e.g.,[6]).

We start our review by Brooks' subsumption architecture [3]. His action selection algorithm rests on a number of layers that work in parallel. At the lower layer are the most primitive behaviours (e.g., avoid objects) while at the higher levels are the increasingly complex behaviours(e.g., explore environment). All layers receive input from the environment simultaneously. So, here are the layers 'vertical'. Brooks' approach to action selection is *priority-based*, i.e., higher layer modules have priority over lower layers. We can call it a loose form of hierarchy - although it has been consider as non-hierarhical by some researchers. Priorities are predefined at design time, and are enforced through inhibition of outputs and suppression of inputs to lower layers.

Maes improved Brooks' approach by making the action selection process more *distributed* and flexible. Her behaviour networks architecture is a non-hierarchical feed-forward network. The network consists of a number of behaviour nodes that have relationships to goals and external resources, and are linked to each other via three different types of links. Each behaviour node receives a set of inputs and employs spreading of activation through its links to other nodes. Behaviour nodes don't have any kind of design-time defined priorities. The degree of impact of internal (goals and drives) and external (related to the environment, e.g., food, water, etc) type of inputs are parameterized, and determined at design time. The behaviour that gets most activation at a moment decides the motor action - a distributed winner-take-all algorithm.

Rosenblatt's "Distributed Architecture for Mobile Navigation" is implemented on a mobile robot system that receives information from several different sources (e.g., video cameras, sonars) and performs diverse tasks involved in navigation, such as following roads, avoiding obstacles, and reaching goal destinations [7]. The system is implemented as a set of distributed behaviours that communicate their suggested action to a *centralized* module that takes the ultimate decision on how the motor subsystem will behave. Behaviour nodes sends their votes to the centralized unit, which in turn, combines the votes of the behaviour subsystems, so that no possibly useful information is lost. No communication exists between behaviours, contrary to Brooks and Maes' architectures. The control is, thus, distributed among behaviour nodes, but centralised at the voter.

Arkin [8] identifies two types of schemas: perception schemas and motor schemas. He states that schema theory 'helps to define brain functionality in terms of concurrent activity of interacting behavioural units' -schemas. Each schema is a generic specification of a particular task. A motor schema such as 'obstacle avoidance schema' is responsible from keeping the robot away from the obstacles. For accomplishing this, the motor schema needs to communicate with 'obstacle detection' perceptual schema. Arkin's 'divide and conquer' strategy resembles Chandrasekaran's task structures. Coordination of the schemas is realized through behaviour fusion in form of vector summation of the individual schema outputs. Schemas are implemented as independent asynchronous computational agents executing in parallel. Arkin's architecture is significatly different from Brooks' as it attaches explicit 'algorithms' to behaviours(i.e., schemas), and as such they may have much more complex computational capabilities compared to Brooks' finite state automatas. Arkin's approach to the combination of schemas and neural networks opens up integration of reactive and deliberative behaviour. The resulting design process allows a combination of a top-down approach (i.e., the task analysis level of the design) and a bottom-up approaches (the neural network implementation). Arkin's work gives us fortunate clues regarding the cooperation between AI, ethology and neuroscience.

3.2   Biological Action Selection

Similar to ethology and BBAI, functional decomposition is the main strategy used in understanding brain. In neuroscience, the action selection function is commonly attained to a brain area called basal ganglia (BG). Although there is not yet a consensus on the exact roles of basal ganglia, it has been conjectured to be involved in both motor behaviour and cognitive functions, e.g., sensory-motor associative learning, reinforcement learning, the formation of temporal sequences of actions, choosing between competing actions, and the initiation of voluntary movements.

BG include the following five nuclei: the caudeta, putamen, globus pallidus, subthalamic nucleus, and substantia nigra (see Fig. 2). Neuroscientist believe that different parts of BG have different neuronal structures and different functionalities, indicating a modularity and possibly a hierarchical organisation.

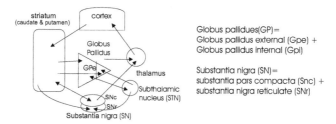

Fig. 2. Schematic representation of the anatomy of basal ganglia. Adapted from [10]

Striatum and STN (see Fig. 2)have been identified as the input nuclei while
SNc and GPi are the output neuclei of BG. Striatum receives its input from sensory and motor cortex, associational area, as well as from hippocampal formation
and amygdala. So, the type of input BG receives is both motivation-related input (e.g., reward-punishment craving and hunger-thirst-mating hormones) and
sensorimotor stimuli[9]. The output of BG is sent to frontal cortex via thalamus,
and to the brainstem.

The information flow starting from the cortex, projecting to BG, and back
to cortex is called cortico-basal ganglia loop. This loop has been anticipated to
take a role in the action selection mechanism. Yet, the precise localisation of the
action selection process in the whole BG enterprise is not so easy as there may
be partially segregated parallel processing streams (i.e., the pathways) going on
in the cortico-basal ganglia loop.

The key function of BG in action selection is to enable desired actions and
to inhibit undesired ones. The selection mechanism is a result of a combinations
of inhibitory and excitatory connections between the neurons of the BG nuclei
named above. Cortex has an excitatory connection to striatum, which inhibits
the activities of GP, which in turn inhibits the activity of thalamus. The total
effect is disinhibition of the parts of thalamus that, in turn, excites only the part
of the motor cortex that relates to the desired action. This is a clear example to
hierarchical organisation of control.

There is huge amount of literature targeting the functionality and circuitry of
the BG, with both agreements and conflicts. Our interpretation of this literature
jungle is that there are two important issues responsible from the most of the
differences between action selection models (i) how does the information flow
from one nucleus to another within BG, (ii) how is the decisions taken - and
where, (iii) how is learning realized. Regarding the *information flow* we see two
main groups of ideas: convergence model and multi-channel model. According
to the convergence model, information coming from cerebral cortex converges in
the striatum which is relatively small compared to the cortical area that project
onto it, and then striatum projects to pallidum which is smaller, which in turn
projects to the motor areas. Here the important issue is the convergence process
itself. The other group of BG models suggest multiple channels of information
processing within the cortico-basal ganglia loop [11]. This account suggests that
striatum is topographically organized. For example, each cortical area projects
to a certain part of the striatum that topologically correspond to the cortical

area from which it receives information. This topological structure is kept in GP as well, so that there are multiple parallel channels of information. Montes-Gonzales [12] embeds this model into a robot. As it comes to *decison taking* the most important aspect is distributed versus centralized control of selection of action, One of the popular approaches favors a combination of distributed and centralized decision processes: each topologically distinguished part of striatum receives competing inputs and employs a local winner-take-all process enabling one single action. As there are several such distinguished areas in striatum more than one non-conflicting actions may be enabled in this way. *Learning* is one of the most controversial subjects. However, researchers agree that substantia nigra part compacta facilitates learning through projecting into striatum and effecting dopaminergic receptors.

In the next section we try to identify the points of agreements between the basal ganglia models and BBAI models of action selection.

4   Points of Agreements

We attempt to enumerate a set of criteria through which the various action selection architectures can be compared. Our criteria will be different from the criteria used for performance measurements (i.e., energy consumption, life span, goal-orientedness, situatedness, persistence, etc. [13]) Our criteria are modularity, hierarchy, layered structure, distributed vs centralized control, cooperation vs competition, an learning.

Modularity: Both BBAI and basal ganglia models rely on the notion of modularity, both functional and architectural. The question of what qualifies as module is not easy to answer. Modules are conceptualized and realized in somewhat different ways. Architectural modularity in neuroscience is based on the types and connectivity of nerve cells as well as the underlying functional requirements. Common to all disciplines is that a modular entity has a particular role or responsibility (i.e., function) in accomplishing an overarching task. As we have seen, a module can be a sophisticated agent (e.g., as intelligent agents field), finite state automata (i.e., corresponding to 'behaviours' as in Brooks' architecture), or a brain area (e.g., striatum in neuroscience). BG models also recognize functional decomposition and functional modularity. The adopted modularity notion allows information flow between modules through specific pathways with special missions. Different forms of 'communication' is used between modules, spanning from special languages (e.g., KQML in multiagent system field), to inhibition and excitation signals (neuroscience), or inhibition and suppression links (e.g., subsumption architecture). Each module, hence, is the connection point of 'what, 'how, and 'where' questions of Marr's: each module has a responsibility (i.e., what), an algorithm (i.e., how), and an implementation of the algorithm(i.e., where).

Hierarchy: A dimension along which the action selection algorithms are often differentiated is the hierarchical organisation. Even though talking about hierarchical organisations is very common we find it difficult to decide whether a certain architecture is hierarchical or non-hierarchical. For example, Brooks' subsumption architecture is not counted as hierarchical, yet the outputs of the higher-level elements may override the suggestions of the lower-levels. We believe the best way to talk about hierarhy is to look at how the control and information flows through the modules. There isa connection between the notions of hierarchical structures and layers. In the existence of layers, hierarchical relations are more explicit. This is also closely related to which type of information is available to which layer. For example, existence of reciprocal inhibition relations between the elements within a layer imposes that only one element is allowed to pass its activity/information into the next level (e.g., as anticipated in most models of basal ganglia). In basal ganglia models striatum receives input from different parts of cortex simultaneously, and there is a more 'hierarchical' relation between striatum and GP, compared to Brooks' approach. Most of the models we have seen in this paper (both AI and neuroscience) don't encompass a uniform hierarchical structure, but involve a mixture of different types of hierarchical relations. Of the architectures we have had an insight into, the only pure non-hierarchical one is Maes' architecture.

In general, advantages of hierarhies have been articulated in connection with the reduction of he complexity of decision making.

Layers: This issue relates to the issue of hierachical structures. Architectures are often classified in AI as either non-layered, vertically layered or horizontally layered. Horisontal layering indicates a pure hierarchy where reactive behaviour is managed at the lowest layer and the deliberative planning at the highest level, and control flows from the bottom towards the top. Many early AI approaches employ such a horisontal layering for combining reactive and deliberative behaviour. 'Layer' notion in biological models is richer where horizontal and vertical layers are combined within the interactions between the modules that 'cooperate' for exhibition of a larger functionality. For example, cortical inputs project to modules in more than one layers of basal ganglia (i.e., 'layer' in a loose sense), to both the striatum and the subthalamic nucleus, while inputs from striatum converges to a smaller area in the GP. So, not all information striatum receives is passed to the GP. Also, in AI models such as DAMN, behaviours are represented at a vertical layer where all nodes receive all external stimuli which, in turn, send their decision to the cenfalized arbiter module in the next level of hierarchy which fuses the commands sent by the behaviour modules.

Distributed or centralized control: This issue also has to do with hierarchical versus nonhierarchical structures. Fully distributed architectures such as in Maes' excludes strict hierarchies while centralized approaches imply the existence of a hierarchy such as in Rosenblatt's system and many basal ganglia models. In fully distributed architectures each module may need to have a connection to all other modules which means that number of connections is much higher compared to centralized (e.g., command fuser) architectures.

Competitive versus cooperative coordination: Since the different subsystems want require the same motor elements, it is fully possible that they demand conflicting actions. Architectures differ in the way they implement conflict resolution. Some architectures apply winner-take-all as the coordination strategy which may lead to the loss of useful information. For example, a strictly competitive action selection mechanism cannot accomplish multiple non-conflicting actions such as walking and chewing. Some basal ganglia models that operationalize multiple parallel channels allow achievement of multiple goals simultaneously. For example, Gurney et al [11] anticipate several intermodules in the striatum that receives different stimuli from cortex. Within each module there is competition because stimuli concurs for the same motor areas while different modules focuse on different motor resources, and don't concure. DAMN also involves a cooperative architecture relying on command compromise.

Learning and adaptation: Learning and adaptation is almost absent from the existing AI approaches. It seems that neuroscientists have put more effort into the combination of learning and action selection. Many models of basal ganglia point to the interconnection between various roles of BG, among others action selection and reinforcement learning. Learning in BG models has been related to the notion of closed-loops, the feedback connections. In some recent models of BG actor-critic model of reinforcement learning has been integrated. Although AI has been investigating reinforcement learning quite some time, for example, in machine learning, few AI architectures combines action selection and learning; one is Blumberg's [14] system that incorporates temporal-difference learning on top of Maes' behaviour networks.

Incrementalism: Most BBAI approaches we have seen are concerned mostly with only motor actions, that is 'physical actions' and they don't address what we may call 'mental actions'. They were rather limited to reactive behaviour. Do reactive and deliberative processes rest upon the same architecture? That is, whether incrementalism applies in building a deliberative architecture based on a reactive architecture? Brooks' position was that such an incremental development is possible. Not everybody agrees with him. Rosenblatt's approach facilitates incremental development of reactive and deliberative behaviour in a special way. In neuroscience we see that vertebrate BG has attained roles related to both motor behaviour and cognitive functions. Maybe an evolutionary perspective can give us some hints on this issue, on the basis of differences between the BGs of vertebrates at different levels.

5   Conclusions

One of the main issues that will invade the research agendas in the near future is AI design and development methodologies that are biologically-informed. A lot of loose threads should be connected on the way of solving the puzzle. If our intention is to develop a robot with sophisticated functionalities then we need a holistic functional analysis including both perceptual, cognitive, and motor functionalities. Some functionality can be discussed by referring to brain regions, and some others at a more micro level, even down to the neuronal level. AI needs a uniform methodology that is operational at all levels of granularity. That is, a methodology is needed that works for all types of functionalities (e.g., perceptual, behavioural, cognitive), and at all levels of functionality (e.g., neuron level, or system level).

Ethology guides AI scientists in a top-down manner where an overall task or functionality is decomposed into simpler, lower-level tasks. We would then need to specify the 'methods' (as in Chandrasekaran's terminology) or algorithms that realize the corresponding tasks. This perspective may seem too traditional, but it is still useful if we manage to be careful not to fall into the same trap as the traditional AI did. What we need is to open up for a bottom-up approach in the stages of algorithm specification and algorithm implementation. Chandrasekaran's idea of multiple methods for a task fits very well here. The challenge is choosing among the algorithms. The algorithms can be symbolic or subsymbolic in nature. The role of neuroscience is exactly at this decision point. If AI chooses to use neuroscientific findings, then the mission of neuroscience would be putting constraints on the description of algorithms so that the researchers' search for algorithms becomes more efficient, and systems developed as such will exhibit higher performance.

Finally, the important question: does our initial hypothesis on the need of neuroscience for AI methodologies evaluates to true? We belive yes. There are enough points of agreements between AI, ethology, cognitive psychology and neuroscience, as highlighted in section 4, for building a bridge connecting these disciplines for the benefit of AI research.

References

1. Marr, D.: Vision: A computational approach. Freeman, San Francisco (1982)
2. Chandrasekaran, B.: Generic tasks involved in knowledge-based reasoning: High-level building blocks for expert system design. IEEE expert **1** (1986) 23–29
3. Brooks, R.: Intelligence without representation. Artificial Intelligence (1991) 139–159
4. Minsky, M.: The society of mind. Simon and Schuster, New York (1986)
5. Newell, A.: The knowledge level. Artificial Intelligence **18** (1982) 87–127
6. Maes, P.: Situated agents can have goals. Robotics and Autonomous Systems (1990) 49–70
7. Rosenblatt, K.J.: A distributed architecture for mobile navigation. PhD thesis (1997)
8. Arkin, R.C.: Behaviour-based Robotics. Cambridge MA, MIT press (1998)
9. Graybiel, A., Saka, E.: The basal ganglia and the control of action. In: The cognitive neuroscience III, Ed. Gazzaniga, M S, xx (2004) 495–510
10. Weedman Molavi, D.: Tutorial. http://thalamus.wustl.edu/course/cerebell.html (2005)

11. Gurney, K., Humpries, R.W., Prescott, T.J., Redgrave, P.: Testing computationa hypotheses of brain systems function: a case study with the basal ganglia. Network: Computation in neural systems (2004) 263–290
12. Montes-Gonzales, F.: A robot model of action selection in the vertebrate brain. Sheffield University (2001)
13. Tyrrell, T.: Computational mechanisms for action selection. Center for cognitive science, University of Edinburg (1993)
14. Blumberg, B., Todd, P., Maes, P.: No bad dogs: Ethological lessons for learning in hamsterdam. In: From animals to animats, MIT Press (1996) 295–304

A Unified Perspective on Explaining Dynamics by Anticipatory State Properties[*]

Jan Treur[1,2]

[1] Utrecht University, Department of Philosophy,
Padualaan 8, NL-3584 CS, Utrecht, The Netherlands
[2] Vrije Universiteit Amsterdam, Department of Artificial Intelligence,
De Boelelaan 1081a, NL-1081 HV, Amsterdam, The Netherlands
treur@cs.vu.nl
http://www.few.vu.nl/~treur

Abstract. In Cognitive Science, recently Dynamical Systems Theory (DST) has been advocated as an approach to cognitive modelling that is better suited to the dynamics of cognitive processes than the symbolic/computational approaches are. Often the differences between DST and the symbolic/computational approach are emphasized. However, if two approaches are used also their commonalities can be analysed, and a unifying framework can be sought. In this paper the possibility of such a unifying perspective on dynamics is analysed. The analysis does not only cover dynamics in the cognitive discipline, but also in other disciplines: Physics, Mathematics and Computer Science. The unifying perspective warrants the development of integrated approaches covering both DST aspects and symbolic/computational aspects.

1 Introduction

Due to dynamics the world occurs in different states, i.e., states at different points in time that differ in some of their state properties. In recent years, within Cognitive Science dynamics has been recognized and emphasized as a central issue in describing cognitive processes; for example, (Port and Gelder, 1995; Kelso, 1995). Van Gelder and Port (1995) emphasize and position dynamics as opposed to the symbolic/computational approach and the connectionist approach. The *Dynamical Systems Theory* (DST) is advocated as a new paradigm that is better suited to the dynamic aspects of cognition.

The notion of a state-determined system lies at the heart of DST. This type of system is based on the assumption that properties of a given state fully determine the properties of future states. Taking this assumption as a premise, in this paper the explanatory problem of dynamics is analysed in more detail. The analysis of four cases within different disciplines (Cognitive Science, Physics, Mathematics, Computer Science) shows how in history this perspective has led

---

[*] Extended version accepted for *Philosophical Psychology Journal*.

J. Mira and J.R. Álvarez (Eds.): IWINAC 2005, LNCS 3561, pp. 27–37, 2005.

to a number of often used concepts within these disciplines. In Cognitive Science the concepts desire and intention were introduced, and in classical mechanics the concepts momentum, energy, and force. Similarly, in Mathematics a number of concepts have been developed to formalise the state-determined system assumption. Derivatives (of different orders) of a function, and Taylor approximations are examples of such concepts. Furthermore, also transition systems, a currently (within Computer Science and related areas) popular format for specification of dynamic systems can be interpreted from this perspective. One of the main contributions of the paper is that the case studies provide a unified view on the explanation of dynamics across the chosen disciplines. All approaches to dynamics analysed in this paper share the state-determined system assumption and the (explicit or implicit) use of anticipatory state properties. Thus it places explanation and modelling of cognitive phenomena on one line with explanation and modelling of other phenomena in Nature.

Within Cognitive Science realism is one of the problems identified for the symbolic/computational approach, i.e., how do internal states described by symbols relate to the real world in a natural manner. As DST is proposed as an alternative to the symbolic/computational approach, a natural question is whether for DST the realism of the states can be better guaranteed. As a second main contribution the paper provides an evaluation of DST compared to the symbolic/computational approach, which shows that in this respect, i.e., for the realism problem, DST does not provide a better solution than the other approaches. This shows that DST and the symbolic/computational approach not only have the state-determined system assumption and the use of anticipatory state properties in common, but also the realism problem.

2   Dynamical Systems Require Anticipatory State Properties

Assuming that the world (including the mental world) occurs in different states (i.e., states at different points in time that differ in some of their state properties), and given a particular state that just changed with respect to some of its state properties, it is natural to ask for an explanation of why these new state properties occurred. In a state-based approach, as a source for such an explanation, state properties found in the previous state form a first candidate, and for dynamical systems they are assumed to form the only candidate source. Thus, to analyse the claims underlying Dynamical Systems Theory, a main question becomes how to determine for a certain state that it is going to change to a different state, and, more specifically, how to determine (on the basis of some of the state properties in the given state) those state properties for which the new state will differ from the given one. This poses the challenge to identify state properties (or combinations of state properties) occurring in a given state that in some way or the other indicate which of the (other) occurring state properties will be different in a subsequent state; by having these properties the state anticipates on the next state: *anticipatory state properties*. If such state properties (historically sometimes called *potentialities*) are given, anticipation to change is somehow encoded in a state. The existence of such properties is the crucial factor for the validity of the assumptions underlying the Dynamical Systems Theory. In Ashby (1960), a similar claim is expressed as follows:

'Because of its importance, science searches persistently for the state-determined. As a working guide, the scientist has for some centuries followed the hypothesis that, given a set of variables, he can always find a larger set that (1) includes the given variables, and (2) is state-determined. Much research work consists of trying to identify such a larger set, for when it is too small, important variables will be left out of account, and the behaviour of the set will be capricious. The assumption that such a larger set exists is implicit in almost all science, but, being fundamental, it is seldom mentioned explicitly.' (Ashby, 1960), p. 28.

Ashby refers to Temple (1942) and Laplace (1825) for support of his claims. He distinguishes phenomena at a macroscopic level for which his claim is assumed to hold from phenomena at the atomic level, for which the claim turns out not to hold.

3   Explaining Changed States by Introducing Potentialities

In this section only concepts that relate to states are included in the ontology to conceptualise reality. We focus on the possibility to include concepts in the ontology to conceptualise states that are useful to describe properties of changed states. Aristotle did introduce such a concept; he called it *potentiality* (to move), or movable. The difference between the arrow at rest and the snapshot of the moving arrow at time t at position P is that the former has no potentiality to be at P', whereas the latter has. This explains why at a next instant t' the former arrow is still where it was, at P, while the latter arrow is at a different position P':

Why is the arrow at t' at position P'?
The arrow is at position P' at t' because
at t it was at position P, and
at t it had the potentiality to be at P', and
at t nothing in the world excluded it to be at P'

Aristotle did not only consider changes of positions (due to locomotion), but also, for example, a young man becoming an old man, and a cold object becoming hot. For each of these types of changes a specific type of potentiality is considered; e.g., the potentiality to be at position P', the potentiality (of a young man) to be an old man, the potentiality (of a cold object) to be hot. In general, if the potentiality (occurring in a state S) to have state property X has led to a state S' where indeed X holds, then this state property X of state S' is called the *fulfilment* or *actualisation* of the potentiality for X occurring in state S.

4   Mental State Properties as Potentialities

In the previous section the examples refer to motion of non-living objects. Another type of motion to be explained is motion of a living being. Often used explanations of human (or animal) actions refer to internal mental states. For example, for a person B which has the capability to move:

> *Why is person B at t' at position P'?*
> *Person B is at position P' at t' because*
> *person B was at position P at t, and*
> *at t person B had the desire to be at P', and*
> *at t nothing in the world excluded B from being at P'*

The type of explanation discussed in Section 3 has some similarity to an explanation of human behaviour from internal mental state properties such as desires. In the example, the desire (which is often considered as a kind of future-directed mental state property) plays a role similar to that of the potentiality for being at P'. Indeed this similarity can be traced back in history, for example, to Aristotle. For example:

> Now we see that the living creature is moved by intellect, imagination, purpose, wish, and appetite. And all these are reducible to mind and desire. (Aristotle, 350 BC, *De Motu Animalium*), Part 6

He describes what today is often called *means-end reasoning*. He explicitly summarises that 'things in the soul' control action:

> Now there are three things in the soul which control action and truth - sensation, reason, desire. Of these sensation originates no action; this is plain from the fact that the lower animals have sensation but no share in action. (Aristotle, 350 BC, *Nicomachean Ethics*), Book VI, Part 2

Based on Aristotle's analysis the pattern to explain motion of animals or humans is as follows:

> at any point in time
> if A has a desire D
> and A has the belief that X is a (or: the best) means to achieve D
> then A will do X

This form of analysis has been called 'practical syllogism'.

5   Giving Potentialities a Place in Physics

In later times successors of Aristotle, such as René Descartes (1596-1650), Christiaan Huygens (1629-1695), Isaac Newton (1643-1727) and Gottfried Wilhelm Leibniz (1646-1716), among others, have addressed the question how to further develop the phenomenon of change or dynamics and, in particular, the concept potentiality within physics. Contributions of these will be discussed in this section and in Treur (2005); for more specific references, see there. Indeed they succeeded in giving certain types of potentialities a well-respected place in modern physics (actually in more than one way).

5.1   Potentialities in Physics

Descartes took the product of mass and velocity of an object as an appropriate foundation for its potentiality to be in a changed position, or quantity of motion. Thus he related the vague concept potentiality to other, better known concepts. Notice that this anticipatory state property 'quantity of motion' is a *relative* potentiality: the actualisation of a given quantity of motion entails being at another position as specified by this quantity relative to the current position (and not as being at some absolutely specified position). Descartes also expresses a law of conservation for this quantity of motion.

In modern physics this 'quantity of motion' concept is called *linear momentum*, or just *momentum*, and the conservation, for example, during elastic collisions, is called the 'law of momentum conservation'. Newton incorporated this notion in his approach to motion. This is one way in which a concept 'potentiality' was given a well-respected place in physics, in particular in classical mechanics.

Huygens (1629-1695), and later his student Leibniz (1646-1716), used a different way to give a concept 'potentiality' a place in physics. Leibniz called this concept *vis viva (living force)* or *motive force*, or *moving force*, or *force of motion*, or simply *force* or *power*. Leibniz claimed that the potentiaility 'motive force' was propertional not with velocity as in the case of 'quantity of motion', but with the square of velocity. In this way Leibniz put the foundation for the law of conservation of energy, in this case involving kinetic energy (which actually was later taken $1/2\ mv^2$) and potential energy, and exchange between the two. For more details, see the extended version Treur (2005).

5.2   What Kind of State Property Is a Potentiality?

Within physics, potentialities have found their place in different manners. Basic concepts in classical mechanics such as momentum, kinetic energy and force can be considered variants of potentialities. In Section 5.1 the first two of these concepts are discussed; the concept force will be discussed in Treur (2005) in the context of higher-order potentialities and exchange of potentialities by interaction. Both for momentum and kinetic energy a conservation law has been found, and both concepts can be expressed in terms of mass and velocity. Does this mean that in these two forms, potentialities have become genuine state properties, which are even definable in terms of other state properties? Even leaving relativity theory aside, this is not a simple question. A straightforward answer would be: indeed, potentialities are genuine state properties because they are defined in terms of mass and velocity which are assumed to be genuine state properties. For the sake of simplicity accepting this claim for mass, a question, however, remains what kind of state property velocity is.

A first approach is to take velocity to be distance traversed divided by time passed over some chosen time interval from t' to t; i.e.: $v = (x(t) - x(t')) / (t - t')$. This definition involves states at different points in time, so it is not based on one state at one time point. This notion of velocity actually is *velocity over the given time interval from t' to t*, so a property of a sequence of states indexed by the time points of the interval, or, to simplify it a bit, a property of a pair of states for the starting point and the end point of the time interval. This is not what one would call a genuine state property.

A second approach is to identify velocity at some point in time with what a *speedometer* displays. Indeed at a point in time t the position of the pointer of a speedometer is a genuine state property. Would this offer an appropriate solution? A first objection may be that this property is just the position of the pointer, not velocity. For every type of object and speedometer a different state concept would arise: think of speedometers for cars compared to those of airplanes, ships, rockets; and what about the velocity of a bird or an approaching meteor. Even if for a certain class of objects, such as cars, a standardisation would be reached for a speedometer, then still the position of the pointer of the speedometer is itself not velocity; at most it has a *relation* to velocity. In particular, the pointer position itself does not affect the (changed) position of the car at the next instant; velocity does affect this position. However, for such a relation between pointer position and velocity we are still in need for a genuine state property velocity; apparantly this problem was not solved by the pointer position.

As a further objection, the position of the pointer of a speedometer results from (or is affected by) the actual motion, so there is a small time delay between having some velocity and what the pointer displays. Therefore the pointer actually indicates speed over time points t' < t, which, although close to t are not equal to t. Thus the pointer position, which is a state property of the state at t, actually relates to states at t' < t , so even if it would be related to a certain velocity state property, this would be a state property of the state at some t' < t, not a state property of the state at t. This makes clear that the speedometer concept will not help to make velocity a genuine state property of the state at t.

A third approach is what is sometimes called the notion of *instantaneous velocity*. In modern physics and mathematics, for continuous processes that satisfy sufficiently strong conditions of smoothness, this is usually defined as a limit:

$$v(t) = \lim_{t' \to t} (x(t') - x(t)) / (t' - t)$$

Note that this limit is defined in terms of the whole family or sequence of states around t, i.e., in terms of the state properties $x(t')$ for all t' in a neighbourhood of t. In mathematical terms this limit can be defined as

$$\forall \varepsilon > 0 \; \exists \delta > 0 \; \forall t' \; [\, 0 < | \, t' - t \, | < \delta \Rightarrow | \, (x(t') - x(t)) / (t' - t) - v(t) \, | < \varepsilon]$$

It is clear that this statement refers to a whole sequence of states for time points t' around t. In this sense the notion of instantaneous velocity does not provide a better solution for a foundation of potentialities as genuine state properties than the other two approaches. Moreover, behind the assumption of smoothness some further assumptions are hidden, namely, to obtain such smoothness, either some inertia of speed has to be assumed (e.g., in free space), or otherwise a type of persistence of other state properties inducing velocity, which may be strongly situation-dependent.

In summary, it turns out that for all three approaches discussed the notion used to define potentiality for a state at one time point t takes into account different states, not only the one state at t. So, do we have to admit that addressing motion and change by extending the state ontology by some form of additional state properties for potentialities has failed? The answer on this question seems to be: yes and no. The answer is 'yes' in the sense that in the three possibilities considered there has not been found anything physically real in the state as a basis for a potentiality as a genuine state property. The answer is 'no' in the sense that the historical developments as discussed in this section have provided quite powerful mathematical means (calculus, differential equations) to model all kinds of problems in diverse application areas. In our daily life we all rely on artefacts constructed using classical mechanics; e.g., bridges, buildings, transportation means. Given that this conceptual machinery works quite well in predictions, makes that the question is still there: what is it that makes this machinery so successful?

6   Transition Systems and Potentialities

An often used method (in Computer Science and related areas) to specify how a state in a system may change is known as 'transition systems'. These are collections of specifications that each consist of a pair (φ, ψ), also denoted as $\varphi \rightarrow \psi$ and sometimes called a *transition rule* with antecedent φ and consequent ψ. In this specification:

- the first description φ indicates a combination of state properties for the current state
- the second description ψ indicates one or more state properties for the next state

The idea is that if the combination of properties specified in the first description holds in a (current) state, then in a next state the properties specified by the second description will hold. This is illustrated by a simple model of agent behaviour based on beliefs desires and intentions. Consider an agent walking down a street and observing an ice cream sign at the supermarket across the street he believes the supermarket sells ice cream. Based on this belief (b1) the agent generates a desire (d) for ice cream. Given this desire, and the belief (b2) that the supermarket is reachable (by crossing the street) the agent generates the intention (i) of having ice cream. Given this intention and the belief (b3) that no traffic is on the street he actually crosses the street and obtains the ice cream (e). In this case the state ontology is described by six basic state properties: b1, b2, b3, d, i, e. The simple scenario can be described in transition system format as follows:

    b1 → d
    b2 ∧ d → i
    b3 ∧ i → e

Based on such a specification a trace of subsequent states is made as follows.

- Given a current state S, take the transition rules for which the antecedent holds in the current state. This is the set of applicable rules.
- Collect the consequents of all applicable rules and obtain the next state S' by modifying S so that all these consequents hold in S' (and the rest of S is persisting).

So, for example, the subsequent states for a given initial state for which the three beliefs hold are as follows:

    0 [b1, b2, b3]
    1 [b1, b2, b3, d]
    2 [b1, b2, b3, d, i]
    3 [b1, b2, b3, d, i, e]

How can this be interpreted in terms of potentialities. For example, consider state 1. As in the next state, state 2, state property i holds, in state 1 the potentiality for i to hold has to be present. On the other hand, i occurs in state 2 because of the second transition rule. Taken together this means that this transition rule can be interpreted for state 1 as indicating that, due to the occurrence of both b2 and d in this state, also the potentiality $p(i)$ for i occurs in state 1. Similarly the other transition rules can be interpreted as indications of which potentialities occur in a given state. In general, according to this interpretation a transition system specifies for each state which potentialities occur: for each transition rule φ → ψ, if in a state S its antecedent φ holds, then in this state S also the potentialities $p(\psi)$ for ψ occur. Thus a transition rule φ → ψ can be interpreted as an implication φ → $p(\psi)$, describing a logical relationship between state properties in a given state. In the example scenario the subsequent states can be described as follows:

    0 [b1, b2, b3, p(d)]
    1 [b1, b2, b3, d, p(i)]
    2 [b1, b2, b3, d, i, p(e)]
    3 [b1, b2, b3, d, i, e]

In principle, other variants of transition systems are possible as well, for example, for nondeterministic behaviour where only one (choice of an) applicable transition rule at a time is applied. The analysis of such variants is similar.

7   Conclusion

Dynamical Systems Theory (DST) has recently been put forward in Cognitive Science as an approach to the dynamics of cognitive processes, and proposed as an alternative for the symbolic/computational approach; for example, (Port and Gelder, 1995; Kelso, 1995; Gelder and Port, 1995). The notion of a state-determined system (Ashby, 1952) is central for DST; such a system is based on the assumption that properties of a given state fully determine the properties of future states. As DST has been advocated as a better alternative to symbolic/computational approaches, the question may be raised whether this fundamental assumption is what makes it distinctive from these other approaches. The answer to this question has been shown to be 'no'. It turns out that the state-determined system assumption, and the use of anticipatory state properties it entails, unifies different approaches to cognitive processes that are often seen as mutually exclusive. Moreover, it places the explanation and modelling of cognitive phenemena on one line with the explanation and modelling of other phenomena in Nature.

As within Cognitive Science one of the problems identified for the symbolic/computational approach is the problem of realism (i.e., how do internal states described by symbols relate to the real world in a natural manner), a further relevant question is in how far DST does better in this respect than the symbolic/computational approach: in how far is the realism of the states better guaranteed by DST.

Thus for the explanatory pattern for dynamics based on anticipatory state properties that is considered here, the commonalities were summarised for applicability in the different scientific disciplines Cognitive Science, Physics, Mathematics and Computer Science. One of the main contributions of the paper is that this provides a unified view on the explanation of dynamics across these different disciplines. This unified view is at a high level of abstraction; it does not deny that at more detailed levels there are differences between these disciplines as well. For example, some of the disciplines (Physics, Mathematics) allow for an appropriate form of quantification, whereas for other disciplines (Cognitive Science, Computer Science) such a quantification is not possible, or at least debatable.

The different examples in different disciplines show that the postulation of new anticipatory state properties in addition to the basic state ontology is a fruitful approach to describe dynamics based on the state-determined system assumption, as DST does. The question in how far these additional 'synthetic' state properties are genuine or 'real' state properties was shown to be a hard question that is not simple to answer, even in Physics. Nevertheless, the fruitfulness of having such state properties added is uncontroversial; for example we all trust artefacts in our environment that were constructed based on physics and mathematics using such state properties.

Given the assumption on state-determinedness, anticipatory state properties are essential to DST. In an explanatory context the question of realism of these anticipatory state properties in DST has to be answered. The analysis made in this paper, however, answers this question negatively. It is shown that even in Physics the foundational question on whether such postulated anticipatory state properties are genuine, real state properties cannot be answered positively.

As an evaluation of DST compared to the symbolic/computational approach, this shows that not only the two types of approaches have the assumption of state-determined systems, and the use of anticipatory state properties in common, but also share the realism problem: DST does not provide a better solution than the symbolic/computational approach. However, the investigations in the history of other scientific disciplines also have shown that this shortcoming does not necessarily stand in the way of fruitfulness of any of these approaches within Cognitive Science. It may well be the case that DST will develop further, especially to address cases where the timing aspects are crucial, whereas symbolic/computational approaches will still develop further for other cases, for example, where more complex types of states are relevant, such as in reasoning or language processing.

Given that the approaches are not as exclusive as sometimes suggested, as shown by the commonalities found in this paper, integrated approaches will be developed, in which both DST aspects and symbolic aspects can be covered. Proposals for such integrated approaches can already be found, for example, in Sun (2002), or Jonker and Treur (2002, 2003) and Jonker, Treur, and de Vries (2002). In Jonker and Treur (2002) it is shown how the expressive temporal symbolic language TTL can be used to model cognitive phenomena, including their quantitative and continuous temporal aspects, and it is shown how DST techniques fit in this language. Sun (2002) claims that quantitative, continuous modelling approaches (such as DST) are suitable to address the implicit, non-representational internal agent states, whereas symbolic, discrete approaches are suitable for more explicit, representational internal agent states. He also shows the crucial role of the interaction between the two types of states, and is an advocate of hybrid modelling approaches where both types of aspects play a role.

Potentialities as postulated state properties often have relationships to other state properties of the state in which they occur (they can be said to be realised by these other state properties). But potentialities usually also relate, in a temporal manner, to state properties in other (past and future) states. These realisation relationships and temporal relationships can be exploited to obtain a further analysis of the foundational question. Such a further analysis is the focus of current research.

References

1. Aristotle (350 BC). *Physica* (translated by R.P. Hardie and R.K. Gaye)
2. Aristotle (350 BC). *De Motu Animalium* On the Motion of Animals (translated by A. S. L. Farquharson)
3. Aristotle (350 BC). *Nicomachean Ethics* (translated by W.D. Ross)

4. Ashby, R. (1952). *Design for a Brain*. Chapman & Hall, London.Burns, P.T. (2000). *The Complete History of The Discovery of Cinematography*. URL: http://precinemahistory.net/
5. Descartes, R. (1998). The World or Treatise on Light. In: Descartes, R. (1998), Descartes: The World and Other Writings (edited by Stephen Gaukroger). Cambrige University Press. (translated by M.S. Mahoney)
6. Descartes, R. (1644), *Principles of Philosophy* (translated by M.S. Mahoney)
7. Gelder, T.J. van (1999). Defending the dynamical hypothesis. In W. Tschacher & J.-P. Dauwalder ed., *Dynamics, Synergetics, Autonomous Agents: Nonlinear Systems Approaches to Cognitive Psychology and Cognitive Science*. Singapore: World Scientific, 13-28
8. Gelder, T.J. van, and Port, R.F., (1995). It's About Time: An Overview of the Dynamical Approach to Cognition. In: (Port and van Gelder, 1995), pp. 1-43.
9. Jonker, C.M., Treur, J., and Vries, W. de, (2002). Temporal Analysis of the Dynamics of Beliefs, Desires, and Intentions. *Cognitive Science Quarterly* (Special Issue on Desires, Goals, Intentions, and Values: Computational Architectures), vol. 2, 2002, pp.471-494.
10. Jonker, C.M., and Treur, J., (2002). Analysis of the Dynamics of Reasoning Using Multiple Representations. In: W.D. Gray and C.D. Schunn (eds.), *Proceedings of the 24th Annual Conference of the Cognitive Science Society, CogSci 2002*. Mahwah, NJ: Lawrence Erlbaum Associates, Inc., 2002, pp. 512-517.
11. Jonker, C.M., and Treur, J., (2003). A Temporal-Interactivist Perspective on the Dynamics of Mental States. *Cognitive Systems Research Journal*, vol. 4, 2003, pp. 137-155.
12. Kelso, J.A.S. (1995). *Dynamic Patterns: the Self-Organisation of Brain and Behaviour*. MIT Press, Cambridge, Mass.
13. Kosman, L.A., Aristotle's Definition of Motion, *Phronesis* 14(1969), 40-62
14. Laplace, P.S. (1825). *Philosophical Essays on Probabilities*. Springer-Verlag, New York, 1995. Translated by A.I. Dale from the 5th French edition of 1825.
15. Leibniz, G.W. (1686). *A memorable error of Descartes*
16. Leibniz, G.W. (1686). *Discourse on Metaphysics*
17. Leibniz, G.W. (1989). Les surprises du Phoranomus. *Les Études Philosophiques* (April-June), pp. 171-86.
18. Leibniz, G.W. von (1956). *Philosophical Papers and Letters*, LeRoy E. Loemker, ed., Chicago: University of Chicago Press, 1956.
19. Leibniz, G.W. (1991). Phoranomus seu De potentia et legibus naturæ. Dialogus II. *Physis* 28:797-885.
20. Newton, I. (1729), The Mathematical Principles of Natural Philosophy; Newton's Principles of Natural Philosophy, Dawsons of Pall Mall, 1968.
21. Newton, I. (1958), Isaac Newton's Papers and Letters on Natural Philosophy (ed. By B. Cohen). Cambridge, Mass.
22. Nussbaum, M. (ed.) (1978). *Aristotle's De Motu Animalium*. Princeton: Princeton University Press.
23. Port, R.F., Gelder, T. van (eds.) (1995). Mind as Motion: Explorations in the Dynamics of Cognition. MIT Press, Cambridge, Mass.
24. Sachs, J. (2001). Aristotle: Motion and its Place in Nature. *The Internet Encyclopedia of Philosophy*. URL: http://www.utm.edu/research/iep/a/aris-not.htm.
25. Sun, R. (2002). *Duality of the Mind*. Lawrence Erlbaum Associates, Mahwah, NJ.
26. Treur, J., (2005). States of Change: Explaining Dynamics by Anticipatory State Properties. *Philosophical Psychology Journal*. In press, 2005.

A Novel Intrinsic Wave Phenomenon in Low Excitable Biological Media

Roustem Miftahof

Korea Advanced Institute of Science and Technology, 305-701,
Taejon, Korea

Abstract. Based on a novel concept of a functional unit a mathematical model of a segment of the gut is developed. Numerical investigation into the dynamics of the electromechanical wave phenomenon reveals the fundamental principles of wave initiation, formation, and propagation along electrically anisotropic longitudinal and isotropic circular smooth muscle syncytia. A pattern of self-sustained electrical activity with the formation of spiral waves is discovered in the longitudinal syncytia and is attributed to the change in conductivity in the syncytia as a result of mechanical deformation of smooth muscle fibers. Although no direct experimental comparison to the theoretical findings is possible at this stage, the model provides new insight onto the basics of physiological processes – slow wave activity, electromechanical conjugation, and a clinical entity, gastrointestinal dysrhythmias.

1 Introduction

Electromechanical conjugation, a fundamental neuromuscular phenomenon in the gut, is responsible for the major motor functions, mixing, grinding, and propulsion of the intraluminal content. The temporal and spatial pattern of conduction of electrical activity determines to a large extent the type of contractions of the organ. Motility studies have long been dominated by single-electrode and/or intraluminal pressure recordings. Recently, however, there have been departures from this tradition. Thus, with the application of high resolution electrical mapping technique it has become possible to reconstruct a variety of modes of the spread of excitation within the smooth muscle layers of the gut [11, 12]. Multiple foci of electrical activity along with the formation of self-sustaining waves have been found. Clinically identified as spontaneous gastrointestinal dysrhythmias they are attributed to manifestations of diabetes, anorexia nervosa, overabundance of prostaglandins, and drug overdose, e.g., erythromycin or atropine [1, 6, 7]. As a possible mechanism of the initiation, wandering pacemaker activity and/or abnormal conduction – the analogous events that are responsible for trigger and reentry phenomenon in the heart – have been proposed. However, the primary pathological feature of reentry in heart muscle is either a myocardial ischemia or an ectopic beat. The first condition is incompatible with viability of the gut, while the exact role of the second proposed mechanism in a low excitable smooth muscle syncytium is uncertain.

J. Mira and J.R. Álvarez (Eds.): IWINAC 2005, LNCS 3561, pp. 38–47, 2005.

Much of the theoretical work in intestinal motility has concentrated on the separate analysis of the mechanical aspects of the phenomenon [3, 4], and only a few studies have been dedicated to model the electrical processes [2, 16]. These (and many other) models have shed light on some of the general mechanisms of the propagation of action potentials, slow wave and bursting activity, migrating motor complex, and shape change of the organ. However, they do not reveal the fundamental principles of the origination and propagation of the electromechanical wave phenomenon in the gut. One regrettable consequence of this deficit lies in the prominence of gastrointestinal dysfunction in medicine and the limited effectiveness of existing methods to treat it.

It is our intention here to formulate a mathematical model of a segment of the gut and to study the dynamics of the electromechanical wave propagation within it. We provide numerical results of normal and self-sustained spiral wave formation in the electrically anisotropic longitudinal smooth muscle layer and attempt to expose the fundamental physiological processes underlying its cause. To maintain focus on key ideas and to avoid unnecessary repetition in the discussion of general physiological facts, which support the basic scientific concepts implemented in the model, we omit the topics, which are relevant to our subject but have been analyzed in full detail in our previous publications [13]-[17].

2   Model Assumptions

The principle anatomical, morphological, electrophysiological, and mechanical properties of organization of the functional unit, a segment of the gut, are specified as the following assumptions:

A1. The functional unit is the underformed tubular locus of a given length and radius is modeled as a thin soft orthotropic shell.

A2. The wall of the unit is composed of two muscle layers, embedded into the connective tissue network. Muscle fibers in the outer layer are oriented in the longitudinal direction of the organ and muscle fibers of the inner layer are arranged in an orthogonal, circular direction. Mechanical properties are different for the two layers but are assumed to be uniform along the wall. The constitutive relationships for the biocomposite are nonlinear and are an approximation of experimental data. Deformations are finite.

A3. Both muscle layers are electrogenic, two-dimensional bisyncytia with electrical cable properties: the outer layer exhibits anisotropic and the inner layer has isotropic electrical properties.

A4. Smooth muscle cells form a continuum of spatially distributed autonomous oscillators connected weakly and homogeneously to form a syncytium. All oscillators can be divided into pools according to their natural frequencies. If two oscillators have nearly equal frequency, they weakly communicate in the sense that the phase of one of them is sensitive to the phase of the other. If two oscillators have equal frequencies then they are strongly connected; and if two oscillators have essentially different frequencies then their phases uncouple and they are disconnected. Strong connections among pools are frequency modulated. Each oscillator is in a stable (silent) state. Transition to an excitable state occurs as a result of an external input.

A5. The electrical activity of the myogenic syncytium, either slow wave or bursting, represents the integrated function of ion channels, presumably: a) Ca^{2+} channels of L- and T- types, b) Ca^{2+}-activated K^+ channels and potential sensitive K^+ channels, and c) leak Cl^- channels. Slow waves have a natural frequency; the discharge of pacemaker cells (ICC-MP) is responsible for their generation.

A6. ICC-MP modulate the properties of L-type Ca^{2+} channels via cholinergic synapses; this effect is mainly chronotropic with an increase in the time of permeability for calcium ions. The synapses are modeled as a null-dimensional three-compartment open pharmacokinetic model. All chemical reactions of neurotransmitter transformation are described by Michaelis-Menten kinetics.

A7. Electromechanical coupling in muscle layers with the generation of deformation and forces are considered a consequence of the evolution of the excitatory wave. It assumes that the "internal" Ca^{2+}, which enters the cell during the period of excitation, activates the contractile protein system.

The above basic physiological assumptions have been implemented in the mathematical formulation, numerical algorithm and the computational platform – *Gut Discovery*$^©$ (Ain Company, LLC, USA; www.aincompany.com). The following results of numerical simulations have been obtained using *Gut Discovery*$^©$.

3   Results

Under normal resting physiological conditions, the functional unit displays stable periodic electrical activity, which is a result of the dynamic interplay of the ion currents through the L- and T-type Ca^{2+}, Ca^{2+}-activated K^+, K^+, and leak Cl^- channels. As a consequence of the balanced activity of the ionic currents, slow oscillations of the membrane potential (φ) are generated. They have a frequency 0.18 Hz and constant amplitude of approximately 25 mV. The wave of depolarization achieves its maximum, exhibits a short plateau of duration, and decreases slowly to the resting value, φ= -51 mV.

Discharges of the interstitial cell of Cajal (MP) located on the left boundary of the functional unit precedes electrochemical coupling at the ICC-MP-syncytium synapse. The depolarization of the presynaptic membrane activates a short-term influx of calcium into the terminal. The concentration of cytosolic Discharges of the interstitial cell of Cajal (MP) located on the left boundary of the functional unit precedes electrochemical coupling at the ICC-MP - syncytium synapse. The depolarization of the presynaptic membrane activates a short-term influx of calcium into the terminal. The concentration of cytosolic Ca^{2+} rises to 5.55 mM and is sufficient to initiate vesicular acetylcholine (ACh) release. With the achievement of a sufficient free concentration, the release of acetylcholine into the synaptic cleft begins. The main part of acetylcholine in the cleft reaches the postsynaptic membrane and reacts with the choline receptors on the soma of the secondary sensory neuron and causes the generation of fast excitatory postsynaptic potentials (EPSP). The EPSP has the maximum, 87.1 mV. rises to 5.55 mM and is sufficient to initiate vesicular acetylcholine (ACh) release. With the achievement of a sufficient free concentration, the release of acetylcholine into the synaptic cleft begins. The main part of acetylcholine in the cleft reaches the postsynaptic membrane and reacts with the choline receptors on the soma of the secondary sensory neuron and causes the generation of fast excitatory postsynaptic potentials (EPSP). The EPSP has the maximum, 87.1 mV.

The generated EPSP exceeds the threshold value for the activation of the smooth muscle syncytia – longitudinal and circular muscle layers. The waves of depolarization, φ_l and φ_c, (subscript l refers to the longitudinal and subscript c to the circular smooth muscle layers, respectively), have an average amplitude 68-70 mV. The velocity of the wave of excitation is not constant but is associated with the mechanical deformation of the wall. Thus while the wall of the functional unit is in underformed state, the wave φ_l has a velocity 0.45 cm/s and φ_c – 0.5 cm/s. In time the velocities of propagation decrease and are: 0.37 cm/s for φ_l and 0.38 cm/s for φ_c. At this time the functional unit of the gut accommodates three waves of depolarization (fig. 1). The anisotropic electrical cable properties of the longitudinal layer make the front of the wave non-uniform, while in the electrically isotropic circular layer it remains uniform throughout.

The waves of depolarization φ_l and φ_c establish strong connections among spatially distributed oscillators and cause cyclic transitory changes in their myoelectrical patterns: a stable slow wave mode transforms to bursting chaos followed by regular bursting with generation of spikes on the crests of slow waves and finally converts back to slow wave mode. The frequency and the amplitude of slow waves remain unchanged throughout. Fast action potentials are generated at the top of the plateau of slow waves at a frequency of 19.5 Hz, have maximum amplitude of 72 mV, and oscillate at a frequency of 21 Hz.

The dynamics of the total forces in the longitudinal (T^l) and circular (T^c) syncytia demonstrates a development of the electromechanical waves of the length 0.3 cm and amplitude 15.6 g/cm (T^l) and 20.6 g/cm (T^c), respectively, close to the left boundary. Because of electrical anisotropy of the longitudinal layer the anterior front of the wave T^l has a form of an ellipse. The front of the wave T^c has the shape of a ring of contraction. A uniform wave of contraction is developed in both muscle layers close to the aboral. However, the contraction in the outer longitudinal layers exceeds the wave of contraction in the circular layers. The maximum $T^l = 16.9$ g/cm and $T^c = 43.9$ g/cm are produced. Strongly connected oscillators in both syncytia supply a homogeneous stress-strain distribution in the wall of the bioshell. The active force of contraction has average amplitude 17.2 g/cm in the longitudinal and 26.9 g/cm in the circular layers. The waves of contraction propagate in the aboral direction at a velocity 0.08 cm/s.

At $t = 13.3$ s there are two visible contractions in the smooth muscle syncytia with the fronts located to 2.7 cm and 0.3 cm in the longitudinal and 2.6 cm and 0.28 cm in the circular layers, respectively. There is a decrease in magnitude of

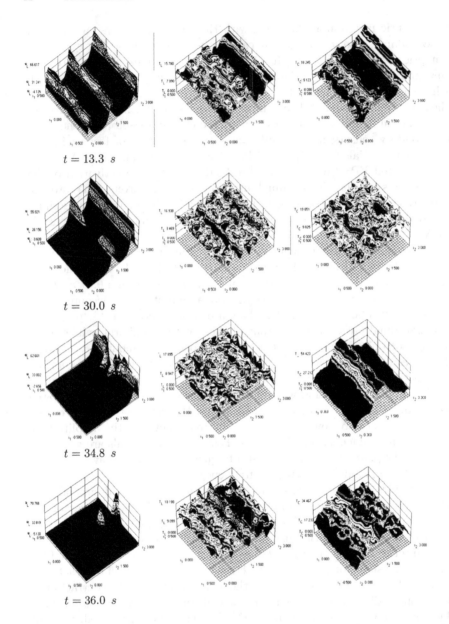

$t = 13.3\ s$

$t = 30.0\ s$

$t = 34.8\ s$

$t = 36.0\ s$

Fig. 1. Development of the wave of depolarization (left column) and the total forces in the longitudinal (middle) and circular (left column) smooth muscle syncytia

total forces T^l and T^c to 15.8 g/cm and 18.2 g/cm. At this stage the maximum amplitude of the wave φ_l and φ_c reduce to 63 mV and it affects the connectivity among oscillators.

At $t = 30$ s the wave of depolarization, φ_l, splits into two separate waves in the middle part of the functional unit. It is important to note that at this

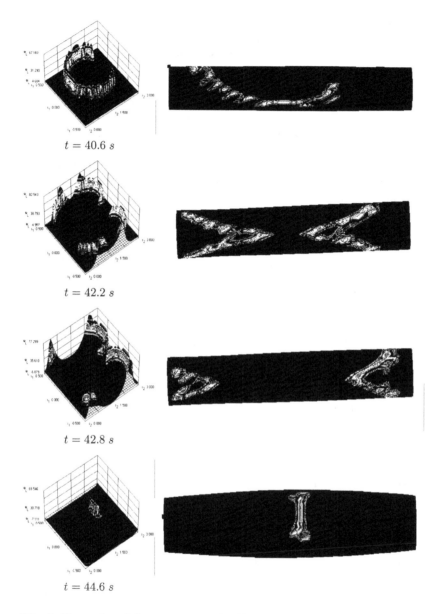

Fig. 2. Dynamics of the generation of self-oscillatory activity in smooth muscle syncytium

stage of the dynamic process the deformation of elongation is the highest in this region. The velocity of the propagation of φ_l is not equal but is slower along the stretched regions of the bioshell. As the waves of average amplitude 55.7 mV reach the right boundary they vanish while their tails collide with a significant rise in the amplitude 70.8 mV. After reflection from the boundary a spike of amplitude 65 mV is generated that propagates backwards. It reaches the middle of the segment and produces a circular wave (fig. 1). At $t = 42.2$ s the wave φ_l covers the entire surface of the gut. Again two tails are seen in the proximity of the right boundary. They again collide and another set of spiral waves is generated. The process is repeated as above. The amplitude of the spiral waves are above 65 mV and provide strong connections among the spatially-distributed oscillators. Therefore, the pattern of electrical activity causes retrograde propagation of the mechanical wave of contraction-relaxation in the longitudinal syncytium. Above changes with the formation of the self-sustained spiral waves are not produced in the electrically isotropic circular smooth muscle layer. There is a split of the wave of depolarization, φ_c, at any time of the dynamic process. The wave φ_c reach the right unexcitable boundary of the functional unit and abolish.

3.1   Effect of Lidocaine N-Ethyl Bromide Quaternary Salt QX-314

In an attempt to convert the system back to normal, i.e., the aboral spread of the electromechanical wave, we simulated the effect of a sodium channel blocker, Lidocaine. Quaternary Lidocaine derivative blocks both fast, Na^+-dependent action potentials and voltage-dependent, non-inactivating Na^+ conductance. Its effect has been achieved in the system by decreasing the parameter of permeability of sodium channels, $g_{Na} = 0$ mSm/cm^2 (normal value) to 70 and 50 (mSm/cm^2), respectively. A slight decrease in g_{Na} does not affect the dynamics of the electrochemical processes. The self-excitatory spiral wave activity persists. Further decrease in permeability of sodium channels, $g_{Na} = 0$ mSm/cm^2, abolishes the generation of the wave of excitation φ_l, and the spiral wave phenomenon. The segment of the gut returns to a silent state.

4   Discussion

Since the "rediscovery" of the migrating complex phenomenon it has been the subject of enormous scientific research and debate [18]. Much of in vivo and in vitro work has been done on the investigation into the physiology of the phenomenon and its role in the pathogenesis of gastrointestinal motor dysfunction. The scientific explanations to the questions about the origin, propagation, effects of the intrinsic and extrinsic regulatory mechanisms, neurotransmitter involvement have been proposed. It has been postulated that the complex is a result of interrelated electrical – slow wave and spiking, and mechanical-contractions of smooth muscle, processes. Based on a pure visual perception and a priori assumptions of the propagation rather than a scientific analysis of the event, the terminology and quantitative measures from physics of waves (e.g., velocity, frequency, amplitude, etc) have been adopted to describe the dynamics of migration. However, whether slow waves and the migrating complex are indeed propagating phenomena have never been questioned.

If one assumes that slow waves do really propagate, then a simple analysis of the experimental data indicates that the wavelength of the slow wave varies within 2-110 cm! The result is based on the fact that frequency of slow waves varies between 0.01-0.3 (Hz) and velocity – 0.6-1.1 (cm/s). If we disregard wide-range variability, which is not unusual in biological observations, the results imply that smooth muscle syncytia could be viewed as an infinite number of dispersed areas of myoelectrical activity, which are either connected over a long distance (110 cm) of the gut or represent spatially distributed units 2 cm in length. The convincing proof of that comes from the experiments on isolated preparations of smooth muscle syncytia [8, 11]. Using a brush of (24×10) electrodes arranged in a rectangular array the authors studied patterns of electrical activity over a large area, and the effect of a single and multiple dispersed pacemakers on the spatial conduction of the wave of excitation. However, within the frame of old concepts of gastrointestinal motility no reasonable explanation was given to their remarkable observations. Thus, a superposition of slow waves traces obtained simultaneously from 224 points (fig. 1) [10] shows that at the resting state, when no pacemaker activity is present, there are no significant phase and amplitude differences among all slow waves recorded. This simple analysis leads to suggestions that: i) smooth muscle syncytium is formed of a continuum of spatially distributed autonomous oscillators, and ii) slow waves do not propagate in the silent state.

A spontaneous discharge of a pacemaker preceded the spread of high amplitude action potentials on crests of the slow waves. Remarkably, the analysis of figs. 3 and 6 (see [10]) shows that in excited media slow waves recorded at different points of the syncytium have equal frequencies. This condition persists with every excitatory input provided by the pacemaker. This experimental evidence confirms our thought that the external excitatory input connects spatially distributed oscillators. The propagating calcium waves have been offered as a possible mechanism that sustains the spread of the electromechanical wave. However, the electrical wave propagates at higher velocity, 2.3-10.8 (cm/s) depending on the species and tissue, compared to the always delayed mechanical wave of contraction-relaxation that moves at 0.2-0.41 (cm/s) [5]. Also, the intracellular calcium waves cannot provide the extensive, ten-of-inches conduction within morphologically inhomogeneous smooth muscle syncytia.

Experiments demonstrated that action potentials in smooth muscle propagate over a short distances 1.3-12.8 (cm) [8]. Therefore, another mechanism must be involved in the efficient spread of the excitation along smooth muscle syncytia. There is compelling evidence of the crucial role of the interstitial cells of Cajal (ICC-MP) in the generation and the conduction of slow waves. They form a planar neural network that topographically lies between the elements of the enteric nervous system and smooth muscle. But the detailed analysis of the exact mechanisms of the coordination of slow wave electrical activity by ICC-MPs is beyond the scope of this investigation. With our approach we have tried to eliminate the mistakes and misinterpretations of the physiological findings that have been dominating the field for several decades.

The concepts and assumptions employed in our model comprise real anatomical, morphological, physiological and pharmacological data about the function of the longitudinal and circular smooth muscle syncytia of the gut. Despite the fact that the syncytia are inherently spatially inhomogeneous, we treat them as homogeneous low excitable media. The results of numerical simulations provide an insight into the processes of electromechanical wave formation and suggest a possible self-sustaining spiral wave phenomenon in electrically anisotropic longitudinal smooth muscle syncytium. Earlier we studied the effect of mechanical deformation on the propagation of the electrical wave of excitation in a longitudinal smooth muscle fiber and we found the condition that blocks the spread of excitation along it [17]. Extrapolating the result onto a two dimensional syncytia, which possesses cable and oscillatory electrical properties, explains the split in the wave of depolarization, φ_l.

Remarkably, the organized wave of contraction is lost in the longitudinal layer with the development of electrical spiral waves, while the pattern of contractile activity remains unchanged in the circular smooth muscle layer. The disrupted mechanical activity could be an explanation to gastrointestinal dysrhythmias with the failure of propulsive activity.

There is no direct confirmable experimental evidence to our theoretical findings, at the moment. In view of the new concepts formulated above, some existing data could be re-interpreted to support our results. Thus, what the authors refer to as a spatial distribution of multiple spike patches, (see fig. 2; [11]) could actually represent self-sustained electrical activity reproduced numerically. A mixed pattern of motor activity including "broad peristaltic, narrower propagated (ripples) and longitudinal muscle rhythmic contractions" (see fig. 2; [5]) is comparable to the total force dynamics (T^l) in the longitudinal smooth muscle syncytium. According to the experimental measurements, the velocity of propagation of the spontaneous wave of contraction is 0.41 cm/s, while the resulting simulations demonstrate its variability from 0.3 to 0.4 cm/s.

With the model we were able to study the effects of Lidocaine on the dynamics of the propagation of electromechanical waves. Lidocaine effectively blocks sodium conductance and thus abolishes the spread of the electrical wave along the syncytium. This pharmacological property has been used successfully to treat cardiac arrhythmias. The choice of Lidocaine as a pharmacological tool to suppress self-sustained spiral wave formation in the longitudinal syncytium was arbitrary. There is no data in medical literature to support the idea that it has been or can be used in treatment of gastrointestinal dysrhythmias.

5   Conclusion

A mathematical model of the gut has been developed and implemented in form of a novel computational platform *Gut Discovery*©(Ain Company, LLC, USA) that has demonstrated its effectiveness in the simulation of electromechanical wave phenomena and pharmacological interventions. A new numerically discovered pattern of self-sustained spiral wave formation in electrically anisotropic longitudinal smooth muscle syncytium could explain normal and pathological motor patterns produced by the organ and observed *in vitro*. The basis of the theoretical results needs to be established and verified experimentally. It would have enormous implication for our understanding of motor disorders of the gut.

References

1. Abell, T.L., Malagelada, J.R., Lucas, A.R., Brown, M.L., Camilleri, M., Go, W.L.M., Azpiroz, F., Callaway, C.W., Kao, P.C., Zinsmeister, A.R., Huse, D.M.: Gastric Electromechanical and Neurohormonal Function in Anorexia Nervosa. Gastroenterol. **92** (1987) 958-965
2. Aliev, R.R., Richards, W., Wikswo, J.P.: A Simple Nonlinear Model of Electrical Activity in the Intestine. J Theor. Biol. **204** (2000) 21-28
3. Amaris, M.A., Rashev, P.Z., Mintchev, M.P., Bowes, K.L.: Microprocessor Controlled Movement of Solid Colonic Content Using Sequential Neural Electrical Stimulation. Gut. **50** (2002) 475-479
4. Gao, C., Petersen, P., Liu, W., Arendt-Nelsen, L., Drewes, A.M., Gregersen, H.: Sensory Motor Responses to Volume Controlled Duodenal Distension. Neurogast. and Mot. **14** (2002) 365-374
5. D'Antona, G., Hennig, G.W., Costa, M., Humphreys, C.M., Brookes, S.J.H.: (2001) Analysis of Motor Patterns in the Isolated Guinea-pig Large Intestine by Spatio-Temporal Maps. Neurogastroenterol. and Mot. **13** (2001) 483-492
6. Holle, G.E., Steinbach, E., Forth, W.: Effects of Erythromycin in the Dog Upper Gastrointestinal Tract. Am. J. Physiol. **263** (1992) G52-G59
7. Koch, K.L., Sterm, R.M., Steward, W.R.: (1989) Gastric Emptying and Gastric Myoelectrical Activity in Patients with Diabetic Gastroparesis: Effects of Long-Term Domperidone Treatment. Am. J. Gastroenterol. **84** (1989) 1069-1076
8. Lammers, W.J.E.P.: (2000b) Propagation of Individual Spikes as "Patches" of Activation in the Isolated Feline Duodenum. Am. J. Physiol., Liver Physiol. **278** (2000) G297-G307
9. Lammers, W.J.E.P., Dhanasekaran, S., Slack, J.R., Stephen, B.: Two-Dimensional High-Resolution Motility Mapping in the Isolated Feline Duodenum: Methodology and Initial Results. Neurogastroenterol. and Mot. **13** (2001) 309-323
10. Lammers, W.J.E.P., Kais, A., Singh, S., Arafat, K., El-Sharkawy, T.Y.: Multielectrode Mapping of Slow-Wave Activity in the Isolated Rabbit Duodenum.. J. Appl. Physiol. **74(3)** (1993) 1454-1461
11. Lammers, W.J.E.P., Slack, J.R.: Of Slow Waves and Spike Patches. News in Physiol. Sci. **16** (2001) 138 - 144
12. Lammers, W.J.E.P., Slack, J.R., Stephen, B., Pozzan, O.: The Spatial Behavior of Spike Patches in the Feline Gastroduodenal Junction in Vitro. Neurogastroent. and Motil. **12** (2000) 467-473
13. Miftakhov, R.N., Abdusheva, G.R., Christensen, J.: Numerical Simulation of Motility Patterns of the Small Bowel. II. Comparative Pharmacological Validation of a Mathematical Model. J. Theor. Biol. **200** (1999) 261-290
14. Miftakhov, R., Christensen, J.: A model of the Enteral Sympathetic Communication. NC'2000 Symposium, Berlin, Germany, (2000) 454-462
15. Miftakhov, R., Christensen, J.: A Physicochemical Basis of Synaptic Transmission in the Myenteric Nervous Plexus. Biophysical Neural Networks, Mary Ann Liebert Inc., New York, (2001) 147-176
16. Miftakhov, R., Vannier, M.V.: Nonlinear Dynamic Waves in Electromechanical Excitable Biological Media. Advances in Fluid Mechanics, WIT Press, Southampton, Boston. (2002) 725-735
17. Miftakhov, R.N., Wingate, D.L.: Numerical Simulation of the Gradual Reflex of the Small Bowel. 14th Int Conf IEEE/EMBS, Paris, France. (1992)1634 -1637
18. Wingate, D.L.: Backwards and Forwards with the Migrating Complex. Dig. Dis. Sci. **26** (1981) 641-666

Conceptual Idea of Natural Mechanisms of Recognition, Purposeful Thinking and Potential of Its Technical Application

Zinoviy L. Rabinovich[1] and Yuriy A. Belov[2]

[1] Institute of Cybernetics,
Academician Glushkov prospectus, 40, Kyiv, 03680, Ukraine
http://www.icyb.kiev.ua
[2] Kyiv National Taras Shevchenko University,
Academician Glushkov prospectus, 2, building 6, Kyiv, 03680, Ukraine
belov@ukrnet.net
http://www.unicyb.kiev.ua

Abstract. In general, this paper concerns the topic "From Artificial to Natural", namely the subject "Computational Neuroscience" (with emphasis on the subsubject "Plasticity"), as well as the subject "Neuro-informatics" (with emphasis on the subsection "Neurostimulators"). The paper also includes some basic considerations and real examples related to the topic "From Natural to Artificial", namely the subject "Bio-inspired Circuits and Mechanisms" ("Electronics" and "Sensory and motor prostheses").

1 Introduction

The conceptual level of modeling natural mechanisms of psyche means penetrating into it from the top to the bottom - from certain psychical functions to information principles of its physical realization, i.e. from the psychical result to the information mechanism of its obtaining [1]-[4].

Thus, the conceptual model determines (using top-theme) "the understanding of the relationships between structure and function in biology".

Information processes taking place in the physical substance of nervous system consist of two main classes: purely combinational processes (as unrelated to memorizing) which occur in sensory organs, organs moving control etc., and combinational- accumulating processes which occur in the memory.

One should consider that thinking processes as a starting postulate belong to these processes (including processing of both the input and output information coming into and out of the memory, respectively); i.e. the memory is also a thinking medium encircled into the total neural net of the whole nervous system.

Where is the limit of the memory and how is it organized?

We shall consider below the conceptual model of the memory and processes which occur in it, and direct our attention both to the recognition function relating to the memory concept as itself and the function of solving problems already

J. Mira and J.R. Álvarez (Eds.): IWINAC 2005, LNCS 3561, pp. 48–57, 2005.

relating to purposeful thinking as the most important one for metabolism. These functions are dominating for artificial intelligence problems and the conceptual model (CM) will, therefore, be of interest within the subject "From Nature to Artificial".

2   CM – Memory and Recognition

The memory functions include such concepts as "recognizing", "remembering", "imagining" etc. Unlike perceiving the information from environment, it is suitable to represent all these actions as demonstrating the so-called "mental sight", i.e. a look which is initialized within the memory and acted as exciting the certain sensory memorized structures in the memory network, their combination etc. So, what is this structure?

To answer this question and understand the global defining principle of the memory organization (that is required for the conceptual model), it is necessary to start with the basic hypothetical prerequisite which ensues, as it were, from "common sense" (but doesn't coordinate with experimental data which it doesn't follow since it is impossible to observe sufficiently in general and in details).

And such a prerequisite as the main basic hypothesis is the following one:

"The pattern reconstruction in the memory (imagination, mental sight) is determined by exciting all its elementary components which took part in the pattern perception".

It would be considered as a law of nature, as an immutable but inexplicable fact. Really, how does it happen that fluctuations of the potential of the neural net components are transformed to, visible inside (also audible, tangible, palpable) patterns? But it happens, isn't it? So, this fact should be the basis for the further constructions.

These constructions should already result in capability to form the signals inside the memory. These signals are able to excite the components of the pattern perceived and fixed in the memory, and there is already no information about this pattern at the memory input. Concerning this initial information, the internal exciting signals are the information flow, which is reverse with regard to the information formed in the memory. Fig. 1 illustrates the aforementioned fact: memory is represented with its main fields (see below) and the arrows denote delimitation between the direct and inverse information. So, we conclude that the "memory" begins, in fact, where the reverse bonds finish. They finish just at the level of "c" conceptors, which repeat "r" receptors. Reverse bonds shouldn't extend over "r" receptors since the illusions occur during the mental sight, i.e. one sees, hears and touches that it isn't sensed at present time. So, the memory limit is just positioned between "r" receptors and "c" conceptors; therefore, duplicating the first ones by the last ones is necessary for it.

The latter ones, "c" conceptors, are actually the smallest sensory components of the perceived patterns and the most elementary sub-patterns regarding to the Pattern. These sub-patterns hierarchically and naturally group together into larger sub-patterns, then the latter ones group in even larger sub-patterns, etc.,

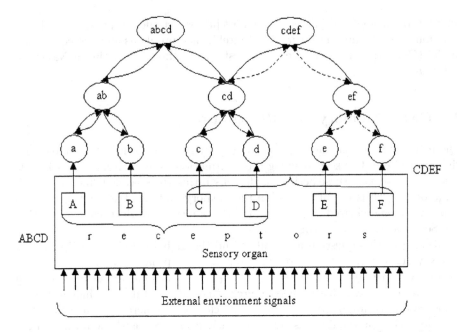

Fig. 1. Elementary structures of pattern perception, memorizing and recognition

up to the sensory concentration of the whole Pattern in the single structure unit which denotes only its symbol.

Thus, when constructed information signals enter the memory from outside, the nonlinear pyramidal model of the pattern expressed by this information will be constructed in the memory. The down-going reverse bonds from the top of the pattern pyramid up to its base (consisting of conceptors which repeat input receptors) are necessary for memorizing the pattern, i.e. for making possible the pattern reconstruction by "mental sight" according to the aforementioned main hypothesis. So, the model of the concrete pattern, both as an object and as a structure in the memory, is the pyramid hierarchical construction with up-going convergent inductive (from partial to general) bonds and down-going divergent deductive (from general to partial) bonds.

The memory as a whole consists of the set of such models as the memory loops associatively bound by the general components. The memory is a structural realization of the semantic network as a system with knowledge fixed in it.

The aforementioned fact can be illustrated with the network, which is enough simple for visualization (Fig. 1). The network consists of fully constructed (i.e. fixed in the memory) ABCD pattern and CDEF pattern entered into the memory but not fixed in it.

It is caused by the fact that the pyramid (i.e. the model) of the ABCD pattern already has reverse bonds and pyramid of the CDEF pattern isn't still constructed in full.

These patterns have one general CD sub-pattern which is the associative element of both pyramids. Such using of the general parts of the pattern provides:

first, economy of constructing the semantic structures in the memory; second, spontaneous (i.e. self-originated) parallelism during processing the pattern information in it that is essentially favorable for effectiveness of this processing (see below). Finishing the construction of the CDEF pattern means creating the reverse bonds in it (shown via dotted line).

Finishing the construction, i.e. changing the model of demonstrated pattern into the memorized one, can be performed by several successive demonstrations of the same pattern that results in beating the genetically innate reverse bonds or even in their appearance.

The creation of the pattern structures in the memory (models) is scientifically valid. The aforementioned conception on constructing these structures, nevertheless, requires the following plausible hypothesis: constructing the direct up-going bonds for perceiving the concrete patterns by the memory precedes an appearance of the reverse down-going bonds which ensure their memorization (see the main hypothesis) by learning.

Thus, a brain memory medium of a new-born child (or other individuals with enough developed nervous system) is already saturated by bonds needed to obtain the information from organs of sense. The memory itself is originated in full by constructing the already reverse bonds in metabolic process begun from the birthday (and while preparing the relevant capabilities even before it). Therefore, the brain as the thinking mechanism creates itself but under the environment impact. Let rather look aside and notice that so-called innate abilities are apparently caused by peculiarities of the genetically innate direct bonds. The demonstration of these abilities (as the construction of the relevant full pattern structures in the memory) already occurs as a result of learning; i.e. phenomenal talents appear as a result of coincidence of two factors: the genetically constructed relevant structures in the memory Medium and its subsequent learning as an impact of external factors.

As a consequence, the brain's memory in its conceptual representation is a hierarchical semantic network restricted from above with ending of ascending connections (straight lines coming from sensory organs) and from bottom - with ending of backward connections to structure units, which directly perceive information entering the memory (formed with specific sensory organs from specific signals). And whole thinking from simple recognition of pattern information to its analysis and synthesis and subsequent actions on sets of patterns, connected with creation and transformation of different situations, are performed in the closed restricted space, connected with dual-side informational links with external (toward it) environment. The first link is intended for information perceiving from the environment, the second - for information (directing, informative and other, created directly in memory) production into the environment.

Now we consider recognition as the simplest psychical function, which determines the presence of the model of the recognized pattern in the memory as its definite structure.

If such structure exists (e.g. abcd model described in Fig. 2) then the secondary spike of exciting this structure already passed by the reverse bonds will

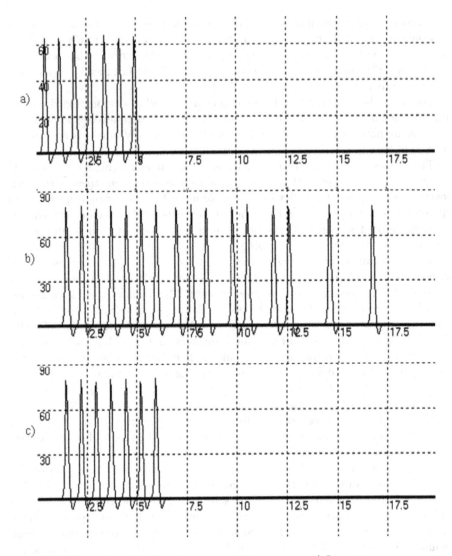

Fig. 2. Model's work result of pattern recognition process: a) Inputs to memory onto elementary conceptores (receptors exits) b) Lower lever elementary conceptors' output with presence of backward links c) Lower lever elementary conceptors' output with presence of broken backward links

take place after presenting the prototype of the structure and its switching-off,. It means implementing the "mental sight", i.e. recognizing the presented pattern in accordance with the main hypothesis. If the fully reverse bonds are absent, i.e. the pattern isn't filled in (such as cdef in Fig. 2) the secondary spike will not take place.

Thus, it is hypothetically claimed that recognition of the demonstrated pattern is determined with the secondary spike of exciting the relevant structure as its remembrance.

This hypothesis in supported by modeling of the components of the olfactory sensory system (performed by neurophysiologist G.S. Voronkov (Moscow State University, Russia). Results of computational experiment represented on Fig. 2 were obtained by PhD students (Tkachuk. S, Yamborak R., Yamborak V.), with assistance of biological network modeling tools developed by them (the secondary spike doesn't occur if the existing reverse bonds in the olfactory sensory system are interrupted and it obviously occurs if the reverse bonds exist, i.e. there is the full model of the demonstrated pattern in the memory). Since exciting processes can be both probabilistic and possibilistic, the results are not defined in full and misoperations (e.g. due to associative bonds (cdef) between the models of two patterns) are certainly possible in a recognizing process. But this indeed takes place in fact.

"Recognizing" in the presented above interpretation is the function implemented quite automatically in any living organism, which has the relevant nervous system. The implementation of this function in the broad sense of this term already including "comprehension" should relate to the thinking process in which described actions are implemented only as its first stage. During the construction of mathematical models of memory patterns the apparatus of growing pyramidal networks [5] can be effectively used, as a type of semantic networks.

3   CM – Memory and Purposeful Thinking

Introduced conception of pattern structures in the memory as their models is general for all thinking functions since it is a basis for organizing all the memory.

It is very important for modeling the human thinking processes to consider the memory as a whole system consisting of two subsystems: sensory one, lingual one [2] and higher associative subsystem, which store the patterns and their lingual notations and notions. The structures of these subsystems are interrelated by direct and reverse bonds, which define a correspondence between them and their mutual effect by transfer of the excitation. Note, that language variety is realized by the additional lingual subsystems, whose structures wouldn't bind with the sensory system structures but interact with it only by the structure of subsystem of single language. The type of bonds between sensory and one or another lingual subsystems defines capability and the level of recognized thinking in one or another language.

The process of human thinking, as specified above, is determined by interaction between the sensory and lingual subsystems of the Medium at the cognitive and intuitive (as uncognitive) levels.

The cognitive thinking is organically bound with the lingual expression of thoughts, i.e. individual talks, as it were, with himself or herself. Thus, the cognitive thinking is called verbal although the language wouldn't be a speech. For example, the relevant signals even enter the organs of speech in the first dominating case. The principally sequential character of cognitive thinking results from it, because it is impossible to pronounce more than one thought at the same time.

In general, the cognitive thoughts are represented by the so-called "complete" dynamic structures, which integrate the relevant excited structures of the sensory and lingual subsystems at different levels of their hierarchy.

Excitement of incomplete structures is related to uncognitive thinking, which isn't restricted by strong interaction of structures of the sensory and lingual subsystems.

Therefore, such dynamic structures can arise at once at different levels of these subsystems not resulted in "pronouncing" the excited "sense".

So, the amount of information processed (even spontaneously) can be much greater here than during cognitive thinking.

Furthermore, such combinations can arise at the intuitive thinking level, which do not have lingual equivalents and so they don't rise to the consciousness level (e.g. savage thinking).

Therefore, the component of uncognitive intuitive thinking is of great importance besides the cognitive component of the purposeful process of solving the problems (man does think!).

The following hypothesis will be quite natural in accordance with aforementioned facts [1].

The Problem to be solved is specified in the memory by the models of original and goal situations, and its solution is an activated chain of cause-effect bonds that results in transforming the first one to the second one. The process of constructing the chain consists of two interrelated processes operating at the same time: sequential cognitive process (as reasoning) and spontaneous activation of the structures in the memory by their associative bonds with the models of the original and goal situations. We shall ignore below the "model" term and terminologically equate the structure in the memory to the situation itself.

Realizing the problem to be solved (the Problem situation) creates, some "tension" in the memory, and so the special "problem generator" (PG) [1] term is very useful for obvious illustrating and considering the process specified in the hypothesis, which poles are the original and goal situations and its "tension" maintains the existence of the problem situation.

Creating the activated chain, which locks these poles and means solving the Problem, liquidates this "tension", i.e. terminates the PG existence. The links of the specified chain are intermediate situations between the original and goal situations (Fig. 3) and they can be defined not only by single-sided transformation but also by cross-transformation of these situations. However, the locking chain wouldn't be created in the continuous process (e.g. if there isn't enough knowledge in the memory) that promotes creating the new intermediate PG which defines the break in the constructed chain of its poles locking, i.e. a new pair of original and goal situations.

The Problem solution can result in creating the new structure (i.e. the new knowledge) in the memory due to beating the new bonds between its components. If there are enough intension and lifetime of the dynamic structure, this process is similar to conversion of the information dynamically stored (i.e. as a short-term memorizing in the statistically fixed memory).

As the "distance" between the original and goal situations decreases due to creating the links of the required chain, which locks them, increase of the activity of the second process and the process as a whole can result in avalanche sudden locking of the PG poles, i.e. solving the Problem as a result of enlightenment. Furthermore, it can occur quite unexpectedly and accidentally just as a result of the second process, only if the first process which is the cognitive reasoning doesn't exist. The second process occurs nevertheless because PG is already excited.

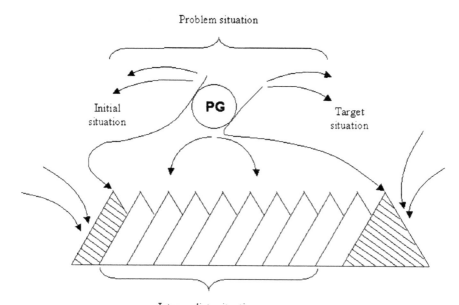

Problem situation

Initial situation

PG

Target situation

Intermediate situations

Fig. 3. Chain of the problems' solution as a pattern situations transmission

In general, enlightenment is a property of creative processes, which can be schematically considered as a sequence with step-by-step domination of cognitive and intuitive thinking [6].

In the first case, the obtained result is comprehended and the new intermediate chain (sub-chain) is proposed. In the second case (i.e. at the next stage), this sub-chain is already reached and changed, if possible, up to obtaining the final result. Thus, the total process of solving the problem is probabilistic (or possibilistic) with a wide range of its quantitative characteristics. So, the rate and the time of its proceeding depend on the degree of the PG excitation and the complexity of the solved problem, respectively. The fact how man occupies himself with this problem determines the first factor. The length of the chain of interrelated structures, which connect the original and goal situations (i.e. the distance between them), determines the second one.

In general, the above mentioned conceptual model explains a number of psychological phenomena of the thinking process and illustrates the material substance of its mechanisms including human talents, erudition, quickness of wit, inspiration etc.

4   Possibilities of CM Bionic Use

CM, which is constructed using the idea "From Artificial to Natural", has already essential bionic value in addition to the cognitive value in accordance with the idea "From Natural to Artificial". It makes no sense to apply nature as a whole to machines (e.g. legs and wheels). For example, when implementing the analogues of CM characteristics in computer architecture, it is necessary to remember its destination first of all.

For developing the universal high-performance and high-intellectual computers (i.e. computers which have enough advanced internal intelligence), it is quite reasonable to reflect the following characteristics in their architectures (as some analogues of the main CM principles):

- Distributed processing the information and its operational storage (i.e. processing part of the machine is to be some memory-processor medium).

- The two-component machine computational process: first, the successive one which perceives the user's tasks, initiates, organizes and controls the process of their performing; and second, parallel one which is the component of the total computational process and is responsible for performing the tasks in the corresponding branch.

- Machine tasks and knowledge representation as semantic and associative networks realized by graphs and their hierarchical processing; knowledge is represented by complex date structures at its upper level and in details at the lower level.

- Possibility to adapt to tasks and to organize the total computational process step-by-step with its dynamic scheduling and controlling the results obtained at each step.

The universal computer with the architecture, built on the basis of specified principles must promote the effective realization of different information technologies of alternative classes: a symbolism and connectualism including neural-computer ones. But, in the second case the technologies are realized at the program models, which can also have sufficiently high characteristics (e.g. large number of neuron-like elements) and can be structurally realized in part by paralleling processing in memory-processor medium. Furthermore, such computer must promote the effective technology realizing and integrating where different technology processes would alternate step-by-step including logic processing and learning (e.g. realizing the sequence of "rational" and "intuitive" conclusions [6]-[7]).

A new class of multimicroprocessor cluster computers with specified properties (the so-called intellectual solving machines (ISM)) and, in particular, its models for broad using were developed at the Institute of Cybernetics of NASc of Ukraine with the assistance of one of the authors of this paper, Rabinovich Z. (this work was supported by the grant from the USA, chief of the project - Prof. V. Koval).

ISMs, in accordance with aforementioned fact, combine the distributed information processing with the internal higher-level language (which has the developed means of knowledge presenting and processing) and dynamic centralized-decentralized (successive and parallel, respectively) control. Exactly this set of characteristics stipulates the belonging of ISMs to a new class [8]-[10].

References

1. Rabinovich Z. L.: A particular bionic approach to structural modeling of purposeful thinking. // Cybernetics, No 2.(1979) 115–118
2. Voronkov G. S., Rabinovich Z. L.: Sensor and lingual system - two forms of knowledge representation. // News of artificial intelligence. (1993) 116–124
3. Rabinovich Z. L.: About Thinking Mechanisms and Intelligent Computers. // Cybernetics and system analysis, No 3. (1993) 69–78
4. Haken H., Haken-Krel M.: Mystery of perception. // Translation from German, Moscow: Institute of computer researches (2002)
5. Gladun V.P.: Decision planning. //Kyiv: Nauk. Dumka (1987) 167p.
6. Glushkov V.M., Rabinovich Z. L.: Problems of deductive constructions automation. // Information management and intelligence. Moscow (1976) part 4, chapter 2, 300 – 326.
7. Geoffrey E. Hinton mapping part-whole hierarchies into connectionist networks. //Artif. Intellig. **46** (1990) No 1/2, 47–75
8. Rabinovich Z.L.: About artificial intelligence concept and its development. // Cybernetics and system analysis, No 2 (1995) 163–173
9. Koval V.N., Bulavenko O.N., Rabinovich Z.L.: Intellectual solving machines as a basis of highly productive computational systems. // Controlling systems and machines. **36** (1998) 43–52.
10. Koval V., Bulavenko O., Rabinovich Z.: Parallel Architectures and Their Development on the Basis of Intelligent Solving Machines. // Proc. of the Intern. Conf. of Parallel Computing in Electrical Engineering. - Warsaw, Poland, September 22-25 (2002) 21–26.

Simulation of Orientation Contrast Sensitive Cell Behavior in TiViPE

Tino Lourens[1] and Emilia Barakova[2]

[1] Honda Research Institute Japan Co., Ltd.,
8-1 Honcho, Wako-shi, Saitama, 351-0114, Japan
tino@jp.honda-ri.com
[2] Brain Science Institute, Riken,
2-1 Hirosawa, Wako-shi, Saitama, 351-0198, Japan
emilia@brain.riken.jp

Abstract. Many cells in the primary visual cortex respond differently when a stimulus is placed outside their classical receptive field (CRF) compared to the stimulus within the CRF alone, permitting integration of information at early levels in the visual processing stream that may play a key role in intermediate-level visual tasks, such a perceptual pop-out [11], contextual modulation [7, 3, 4], and junction detection [13, 3, 5]. In this paper we construct a computational model in programming environment TiViPE [9] of orientation contrast type of cells and demonstrate that the model closely resembles the functional behavior of the neuronal responses of non orientation (within the CRF) sensitive $4C\beta$ cells [5], and give an explanation of the indirect information flow in V1 that explains the behavior of orientation contrast sensitivity.

1 Introduction

Neurons in the primary visual cortex (V1) respond in well defined ways to stimuli within their classical receptive field (CRF), but these responses can be modified by additional peripheral stimuli. The size of the periphery (non classical surround) provides input from a larger portion of the visual scene than originally thought, permitting integration of information at early levels in the visual processing stream. Recent works indicate that neuronal surround modulation at cross-orientation, an orientation orthogonal to the preferred orientation of the classical receptive field, might play a key role in intermediate level visual tasks, such as perceptual pop-out [11], contrast facilitation [2, 15], and contextual modulation [7, 3, 4]. The strength of this contextual influence on a neuron can be predicted from a model of local connection based on simple overlap with particular features, which indicates that local intra cortical circuitry could endow neurons with a graded specialization for processing angular visual features such as corners and junctions [13, 3, 5].

Depending on the orientation of an inner and outer grating pattern, these neuronal cells have the tendency to respond strongly to a center orientation

J. Mira and J.R. Álvarez (Eds.): IWINAC 2005, LNCS 3561, pp. 58–67, 2005.

preference or orientation contrast[1] between inner and outer pattern. Neuronal output activity was enhanced in both cat and macaque primary visual cortex (V1) when, a surrounding field at a significantly different orientation (30 degrees or more) was added to the preferred orientation of the classical receptive field [13]. Cells in layer 4Cβ, which are non-orientation sensitive within their CRF, also show these response profiles indicating that there must be a strong feedback from other areas (within V1) that create these more complex profiles. We assume that these cells obtain feedback from complex cells in layers 2, 3, 5, and 6 of V1. The aim of this paper is to setup a computational model of this type of cells which we will term *orientation contrast cells*, and to simulate these cells in visual programming environment TiViPE [9].

The paper is organized as follows: Section 2 elaborates on the properties of non orientation tuned cells with respect to orientation contrast, their pathway in early vision, and provides a computational model. Section 3 gives a TiViPE simulation that provides the results of this model when applied to the stimuli given by Jones et al [5]. The paper finishes with a discussion.

2 Non Orientation Tuned Cells

In primate V1 cells 94 percent had a response to orientation contrast stimuli that exceeded the response to the inner stimulus alone, independent from the diameter of the surround patch, while the responses were somewhat inhibitory when the orientation of the inner and outer stimuli were the same, compared to the response to the inner stimulus alone [5]. They found that the responses of 4Cβ cells could be modulated by varying both orientation of a center grating patch (inside the CRF) and a surround grating patch (outside the CRF), despite the cell's lack of orientation tuning within the CRF. Its response output was extremely sensitive to orientation differences between center and surround patches.

The LGN parvo cellular cells (P) have center-surround shaped receptive field profiles which optimally respond to a spot of light. In a feed-forward processing stream one could expect a similar receptive field type in layer 4Cβ. For instance, a set of center-surround profiles that are aligned in a certain way, may respond strongly to a line or bar of a specific orientation. However, such profile does not provide center orientation preference nor is it able to provide a measure for center-surround orientation contrast. The modulation of its response behavior must be caused by an indirect (feedback loop) information stream, as illustrated in Figure 1.

2.1 Organization of the Primary Visual Cortex

The primary visual cortex (V1) consists of six layers (1-6) between the pial surface and the underlying white matter. The principal layer for inputs from the

---

[1] Orientation contrast is the difference between preferred orientation of a center patch (which roughly covers the CRF) and preferred orientation of a surround patch (outside the CRF). This contrast is strongest when the center and surround orientations are orthogonal and weakest when both are the same.

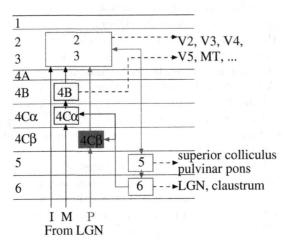

Fig. 1. Information flow in the primary visual cortex (V1) based on anatomical connections [6]

lateral geniculate nucleus (LGN) is layer 4, which is subdivided into four sub layers (4A, 4B, 4Cα, and 4Cβ), see also Figure 1. This flow can be described by means of input, intra cortical, and output connections [6]:

- **Inputs.** Axons from magno cellular (M) and parvo cellular (P) cells in the lateral geniculate nucleus (LGN) end on spiny stellate cells in the sub layers of 4C, and these cells project axons to layers 2, 3, or 4B. Axons from cell in the intra laminar (I) zones of the LGN project directly to layers 2 and 3.
- **Intra cortical connections.** Axon collaterals of the pyramidal cell in layers 2 and 3 project to layer 5 pyramidal cells, whose axon collaterals project both to layer 6 pyramidal cells and back to cells in layers 2 and 3. Axon collaterals of layer 6 pyramidal cells then make a loop back to layer 4C onto smooth stellate cells.
- **Outputs.** Each layer, except for 4C, has outputs and each is different. Cells in layers 2, 3, and 4B project to extra striate visual cortical areas. Cells in layer 5 project to the superior colliculus, the pons, and pulvinar. Cells in layer 6 project to claustrum and back to the LGN.

The assumption that a 4Cβ cell receives input from simple (layer 2) or complex cells (layer 3) through layers 5 and 6 makes it plausible that these cells have a far more complex receptive field profile than one can expect from a feed-forward mechanism alone.

2.2 Orientation Sensitive Input Responses

In order to model the profiles suggested by [5] we assume that layer 4Cβ receives complex cell (indirect) input from layers 2, 3, 5, and 6. A computational model of simple and complex cells [8, 14] is used to form the input of the orientation contrast cells and is introduced only briefly.

The receptive fields of simple cells can be modeled by complex valued Gabor functions:

$$\widehat{G}_{\sigma,\theta}(x,y) = \exp\left(i\frac{\pi x_1}{\sqrt{2}\sigma\lambda}\right)\exp\left(-\frac{x_1^2 + \gamma^2 y_1^2}{2\sigma^2}\right) , \tag{1}$$

where $i = \sqrt{-1}$, $x_1 = x\cos\theta + y\sin\theta$ and $y_1 = y\cos\theta - x\sin\theta$. Parameters σ, λ, γ, and θ represent scale, wavelength, spatial aspect ratio, and orientation, respectively. These Gabor functions have been modified, such that their integral vanishes and their one-norm (the integral over the absolute value) becomes independent of σ, resulting in $G_{\sigma,\theta}(x,y) = \eta\widehat{G}_{\sigma,\theta}(x,y)$, where $\eta = \eta_{\mathrm{Re}}^+$ for the positive valued real part of $\widehat{G}$, $\eta = \eta_{\mathrm{Re}}^-$ for the negative valued real part of $\widehat{G}$, and $\eta = \eta_{\mathrm{Im}}$ for the imaginary part of $\widehat{G}$. For details about these constants see [8]. A spatial convolution was used to transform input image $I(x,y)$ by these operators to yield the simple cell operator, and the amplitude of the complex values [10]

$$\mathcal{C}_{\sigma,\theta} = ||I * G_{\sigma,\theta}|| \tag{2}$$

was taken to obtain the complex cell operator.[2] This operator forms the basis of the orientation contrast cell operator $\mathcal{O}$ to be described later in this paper. A high value at a certain combination of (x,y) and θ represents evidence for a contour element (bar or edge) oriented orthogonally to θ. Orientations are sampled linearly $\theta_j = \pi/N, j = 0,\ldots,N-1$, and the scales are sampled $\sigma_k = \sigma_{k-2} + \sigma_{k-1}$, for $k = 2\ldots S-1$, where σ_0 and σ_1 represent constants.

2.3   Orientation Contrast and Center Orientation Preference

Neuronal cells in area V1 respond to both orientation contrast and center orientation. Depending on the size and orientation of the peripheral patch compared to the preferred orientation of the center patch (which covers the CRF) the response is inhibitory or excitatory. When the patch is similar in size compared to its center patch the cell tends to respond strongly to orientation contrast, while a patch that has a diameter of four times the diameter of the central patch tends to respond strongly to the preferred orientation of the central patch [5]. These findings suggest a varying gain value that depends on the size of the surround patch. This is modeled as follows:

$$G_x(s) = \left[-\frac{2s^2}{30} - \frac{s}{10} + \frac{2}{3}\right]^{\geq 0} , \tag{3}$$

where s denotes the surround patch diameter in degrees, and $[x]^{\geq 0} = x$ if $x \geq 0$ and 0 otherwise. The curve obtained by varying the surround patch diameter is illustrated in Figure 2a.

The normalized response profile (weight matrix) is modeled as a blend between orientation contrast preference and preferred center orientation:

$$W(c_p, c_o, c_s, s_o, s_s) = G_x(s_s/c_s)X(c_o, s_o) + G_c C(c_p, c_o), \tag{4}$$

---

[2] The preferred orientation $\theta \in [0,\pi)$, since $\mathcal{C}_{\sigma,\theta} = \mathcal{C}_{\sigma,\theta+\pi}$.

where c_p, c_o, and c_s denote the preferred orientation, used orientation, and diameter of the center patch, all between 0 and 360 degrees. Likewise s_o, and s_s denote the used orientation and diameter of the surround patch. The normalized orientation contrast profile is as follows:

$$X(c_o, s_o) = \begin{cases} 0.5 - 0.5 \cos\left(\frac{|c_o - s_o|\pi}{W_x}\right) & \text{if } \alpha_X(c_o, s_o) \leq W_x \\ 1 & \text{otherwise} \end{cases}, \qquad (5)$$

where $W_x = 90$ degrees is a constant, and $\alpha_X(c_o, s_o) = \min(|c_o - s_o|, |360 + c_o - s_o|, |360 + s_o - c_o|)$. The normalized preferred center orientation is

$$C(c_p, c_o) = \begin{cases} 0.5 \cos\left(\frac{|c_p - c_o|\pi}{W_c}\right) + 0.5 & \text{if } \alpha_C(c_p, c_o) \leq W_c \\ 0 & \text{otherwise} \end{cases}, \qquad (6)$$

where $W_c = 90$ degrees is a constant, and $\alpha_C(c_p, c_o) = \min(|c_p - c_o|, |360 + c_p - c_o|)$.

The response of the $4C\beta$ cell as measured by [5] in Figure 6 shows a maximum response of around 70 while the minimum response is around 15. To obtain the response profile as given in Figure 2b-d the following response was used:

$$R_W = 70(W + 0.2). \qquad (7)$$

2.4   Orientation Contrast Cell Operator

The response of a center patch which covers the CRF is obtained as follows:

$$\mathbf{C}_{\sigma, c_s} = \mathbf{C}_{\sigma, \theta_i} * g_{c_s/6}, \qquad (8)$$

where $\theta_i = i\pi/N$, $i = 0, \ldots, N$, and $g_\sigma(x, y) = 1/(2\pi\sigma^2) \exp(-(x^2 + y^2)/2\sigma^2)$ is a 2D Gaussian function.

The response of a surround patch is obtained by taking the maximum response of differently sized surround patches

$$\mathbf{S}_{\sigma, c_s, s_{\min}, s_{\max}, Q} = \max_{(x1, y1)} (W(c_p, c_o, c_s, s_o, s_s) - W_s) \max_j \left(\widehat{\mathcal{C}}_{\sigma, \theta_i}(x1, y1)\right), \qquad (9)$$

where $\widehat{\mathcal{C}}_{\sigma, \theta_i} = \mathcal{C}_{\sigma, \theta_i} * g_{s_{s_j}}/6$, $s_{s_j} = j(s_{\max} - s_{\min})/(Q - 1) + s_{\min}$ has a linearly increasing patch size between $s_{\min}$ and $s_{\max}$, $j = 0, \ldots, Q - 1$, and Q is the number of surround patch sizes. Let $j_{\max}$ denote the index j for which holds $\widehat{\mathcal{C}}_{\sigma, \theta_i}$ is maximal. Weight W from (4) is in the 0 to 90 degree range, since we assume that the grating pattern is static rather than moving in a specific direction, W_s is an inhibitive weight, and $(x1, y1)$ are the spatial positions of the outer stimulus. Since these patches largely overlap resampling is used to reduce computational time. Preferred center orientation c_p, center orientation c_o, surround orientation s_o, and surround patch size s_s are as follows:

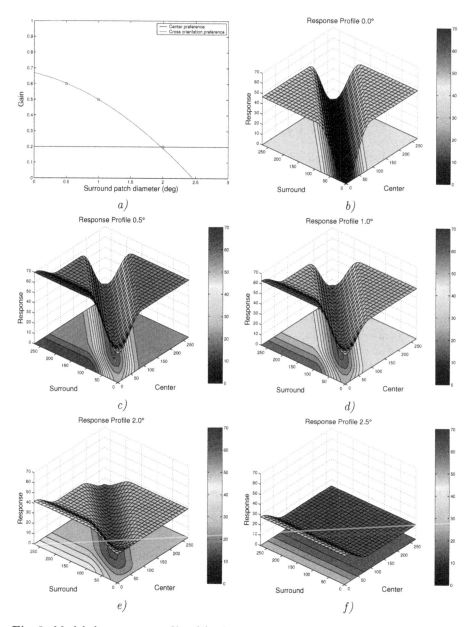

Fig. 2. Modeled response profiles (7) of a non orientation tuned layer 4Cβ cells to varying the orientation of both center and surround patch, for a comparison with the measured responses, see Fig. 6 of [5]. *a)* Blending curve between orientation contrast and center orientation preference. *b)* Modeled profile for $s_s = 0$, which gives solely a preference to orientation contrast. *c-e)* Profiles for $s_s = 0.5$, 1.0, and 2.0 degrees, respectively. *f)* Modeled profile for $s_s \geq 2.5$, which solely prefers the center orientation. Parameters used are preferred center orientation $c_p = 0$ degrees, and center radius $c_s = 0.5$ degrees

$$c_p = c_o = \begin{cases} 180i/N & \text{if } i \le N/2 \\ 180(N-i)/N & \text{otherwise} \end{cases} \tag{10}$$

$$s_s = s_{\min} + j_{\max}\frac{s_{\max} - s_{\min}}{Q-1} \tag{11}$$

$$s_o = \begin{cases} 180j_{\max}/N & \text{if } j_{\max} \le N/2 \\ 180(N - j_{\max})/N & \text{otherwise} \end{cases} \tag{12}$$

The cross-orientation operator which comprises a center response and a surround response that depends on the center response is as follows:

$$\mathcal{O}_{\sigma,c_s,s_{\min},s_{\max},Q} = (\mathbf{C}_{\sigma,c_s} + w\mathbf{S}_{\sigma,c_s,s_{\min},s_{\max},Q}) * g_{c_s}/6, \tag{13}$$

where weight $w = \mathbf{C}_{\sigma,c_s}/R$ is a weight that is dependent on the center response $\mathbf{C}$. In all simulations constant $R = 255$ was used to bound w between 0 and 1.

3   Responses to Test Patterns

The input stimuli used in the simulation have a center radius of 24 pixels and surround radii of 24 (Figure 3a), 48, 72 (Figure 3b), or 96 pixels. The block gratings consist of alternating black and white bars which are both 8 pixels wide. A complex cell operator $C_{\sigma,\theta}$ with $\sigma = 4\sqrt{2}$ and an orientation θ corresponding to the preferred orientation of the grating pattern yields an optimal response (255), see also Figure 3a, for the complex cell operator C in the center of the input stimuli of Figure 3a and b. When the center-only input stimulus is applied to orientation contrast operator ($\mathcal{O}$) for the preferred horizontal center orientation the $\mathcal{O}$-operator has a very similar response profile compared to the C-operator, but where the results of C-operator remain the same, the $\mathcal{O}$-operator is influenced by its surround as illustrated in Figure 3b ("Orientation contrast fixed center 1:3"). The profile is very similar to the one given by [5].

The orientation contrast cell operator $\mathcal{O}$ from (13) has been implemented in visual programming environment TiViPE [9]. The orientation contrast simulation that is represented by a network of connected icons consists of a "Read-Image" icon which generates the input stimulus, its connected "Display" icon yields the images provided in Figure 3a and b. The "ComplexAndEndstopppedResponse" produced the responses of the C-operator (2). Its output forms the input of the "ComplexCrossOrientationResponses" and gives the responses of the $\mathcal{O}$-operator (13). The values at the center of the two other "Display" icons have been used to construct Figure 3c.

4   Discussion

Many neurons in primary visual cortex (V1) respond differently to a simple visual element presented in isolation compared to when it is embedded in a more

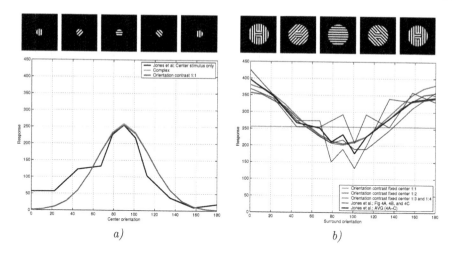

a)   b)

Fig. 3. Response characteristics of orientation contrast sensitive cells, see Figure 4A-F from [5]. *a)* Input stimuli with preferred orientations of 0, 45, 90, 135, and 180 degrees, and below the response profiles to these stimuli of the measured V1 cells, complex cells ($\mathcal{C}$-operator) and orientation contrast type of cell ($\mathcal{O}$-operator). *b)* Input stimuli with surround, with preferred center orientation of 90 degrees and varying surround orientation from left to right from 0 to 180 degrees. The ratio between center and surround of these stimuli is 1:3 (top). Response profiles for measured cells and center-only (1:1), and center-surround (1:2, 1:3, and 1:4) stimuli. Responses have been normalized to the maximum response of the modeled complex cells (255). The inhibitory weight $W_s = 0.45$ which yields similar response profiles as the measured V1 cells

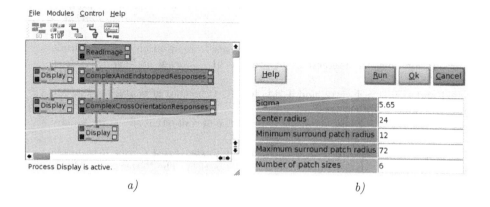

a)   b)

Fig. 4. *a)* TiViPE network. *b)* Parameters used for cross orientation type of cells

complex stimulus. Typically the surround influence was suppressive when the surround grating was at the neuron's preferred orientation [2], but when the orientation in the surround was perpendicular to the preferred orientation facilitation became evident [13, 12, 2, 5]. The difference is in the modulation by surrounding elements, hence it could provide neurons with a graded specialization for processing junctions [13, 3]. These neurons also respond to a grating or a single bar of a preferred orientation and are in that respect too general to be purely responding to junctions. In the monkey the majority of cells showed response suppression with increasing grating patch diameter [1, 13] therefore it is likely that a group of these neurons responds to junctions and facilitates pop-out patterns [11].

The proposed model for orientation contrast cells uses complex cell input that is provided by the indirect pathway from layers 2, 3, 5, and 6 of V1 and yields appropriate characteristics to test patterns as used by Jones et al [5]. Future work will involve patterns for junction detection, pop-out, and will be applied to natural images. The model itself will be integrated into a highly parallel vision system that will be used in a humanoid robot.

References

1. R. T. Born and R. B. H. Tootell. Single unit and 2-deoxyglucose studies of side inhibition in macaque striate cortex. *Proc. Natl. Acad. Sci. USA*, 88:7071–7075, 1991.
2. J. R. Cavanaugh, W. Bair, and J. A. Movshon. Selectivity and spatial distribution of signals from the receptive field surround in macaque v1 neurons. *Journal of Neurophysiology*, 88:2547–2556, 2002.
3. A. Das and C. D. Gilbert. Topography of contextual modulations mediated by short-range interactions in primary visual cortex. *Nature*, 399:655–661, June 1999.
4. V. Dragoi and M. Sur. Dynamic properties of recurrent inhibition in primary visual cortex: Contrast and orientation dependece of contextual effects. *Journal of Neurophysiology*, 83:1019–1030, 2000.
5. H. E. Jones, W. Wang, and A. M. Sillito. Spatial organization and magnitude of orientation contrast interactions in primate v1. *Journal of Neurophysiology*, 88:2797–2808, 2002.
6. E. R. Kandel, J. H. Schwartz, and T. M. Jessell. *Principles of neural science*. Mc Graw Hill, fourth edition, 2000.
7. J. B. Levitt and J. S. Lund. Contrast dependence of contextual effects in primate visual cortex. *Nature*, 1997.
8. T. Lourens. *A Biologically Plausible Model for Corner-based Object Recognition from Color Images*. Shaker Publishing B.V., Maastricht, The Netherlands, March 1998.
9. T. Lourens. Tivipe –tino's visual programming environment. In *The 28th Annual International Computer Software & Applications Conference, IEEE COMPSAC 2004*, pages 10–15, 2004.
10. M. C. Morrone and D. C. Burr. Feature detection in human vision: A phase-dependent energy model. *Proc. of the Royal Society of London*, 235:335–354, 1988.
11. H. C. Nothdurft, J. L. Gallant, and D. C. Van Essen. Respons e modulation by texture surround in primate area v1: Correlates of "popout" under anesthesia. *Visual Neuroscience*, 16:15–34, 1999.

12. I. A. Shevelev, N. A. Lazareva, G. A. Sharaev, R. V. Novikova, and A. S. Tikhomirov. Selective and invariant sensitivity to crosses and corners in cat striate neurons. *Neuroscience*, 84(3):713–721, 1998.
13. A. M. Sillito, K. L. Grieve, H. E. Jones, J. Cudiero, and J. Davis. Visual cortical mechanisms detecting focal orientation discontinuities. *Nature*, 378:492–496, November 1995.
14. R. P. Würtz and T. Lourens. Corner detection in color images through a multiscale combination of end-stopped cortical cells. *Image and Vision Computing*, 18(6-7):531–541, April 2000.
15. C. Yu, S. A. Klein, and D. M. Levi. Facilitation of contrast detection by cross-oriented surround stimuli and its psychophysical mechanisms. *Journal of Vision*, 2:243–255, 2002.

Formulation and Validation of a Method for Classifying Neurons from Multielectrode Recordings

M.P. Bonomini[1,*], J.M. Ferrandez[2], J.A. Bolea[1], and E. Fernandez[1]

[1] Instituto de Bioingeniería, Univ. Miguel Hernández, Alicante
[2] Dept. Electrónica y Tecnología de Computadores, Univ. Politécnica de Cartagena

Abstract. The issue of classification has long been a central topic in the analysis of multielectrode data, either for spike sorting or for getting insight into interactions among ensembles of neurons. Related to coding, many multivariate statistical techniques such as linear discriminant analysis (LDA) or artificial neural networks (ANN) have been used for dealing with the classification problem providing very similar performances. This is, there is no method that stands out from others and the right decision about which one to use is mainly depending on the particular cases demands. Therefore, we developed and validated a simple method for classification based on two different behaviours: periodicity and latency response. The method consists of creating sets of relatives by defining an initial set of templates based on the autocorrelograms or peristimulus time histograms (PSTHs) of the units and grouping them according to a minimal Euclidian distance among the units in a class and maximizing it among different classes. It is shown here the efficiency of the method for identifying coherent subpopulations within multineuron populations.

1 Introduction

Understanding how information is coded in different sensory systems is one of the most interesting challenges in neuroscience today. Technology is now available that allows to acquire data with more accuracy both in the temporal and the spatial domains. However this process produces a huge neural database which requires new tools to extract the relevant information embedded in neural recordings.

While the neural code is partially understood in the auditory [1] and olfactory systems [2], the visual system still presents a challenge to neuroscientists due to its intrinsic complexity. Different studies have used different analysis tools to approach the decoding objective. Fitzhugh [3] applied a statistical analyzer to the neural data in order to estimate the characteristics of the stimulus, whereas Warland used linear filters [4] for the decoding task. Other approaches used

---

* Corresponding author: p.bonomini@umh.es

J. Mira and J.R. Álvarez (Eds.): IWINAC 2005, LNCS 3561, pp. 68–76, 2005.

to get insights into the coding process are discriminant analysis [5], principal component analysis [6] and supervised and non-supervised neural networks [7].

The nearest neighbour clustering decomposes the whole population in disjoint groups, minimizing the distance betwween the elements of a group and maximizing the avaraged squared distance between the centroids of different clusters. This division will achieve temporal boundaries for classificating the neural responses according to their periodicity or PSTH schemes.

In this paper, it is developed and validated a simple method for classification based on two different parameters: periodicity and latency response. The process consists of creating sets of relatives by defining an initial set of templates based on the autocorrelograms or PSTHs of the neural population response and grouping them according to a minimal Euclidian distance among the units in a class and maximizing it among different classes.

The method was implemented using the K-means algorithm [8]. This clustering method was found to be slightly more robust to large amounts of noise in neuronal data than others [9].

It will be shown here the efficiency and robustness of the method for identifying coherent subpopulations within multineuron populations.

2   Materials and Methods

Experimental data was obtained from rabbit retina recordings using a one hundred multielectrode array (Cyberkinetics Inc.)[10]. Visual stimulus consisted of full-field flashes at 0.5 Hz with a 300 ms ON period. Units were sorted using a free open-source program, Nev2lkit, (source code and documentation is freely available at http://nev2lkit.sourceforge.net) by means of principal component analysis (PCA), specifically the correlation method.

Afterwards, the autocorrelograms and PSTHs were calculated on each of the cells in the dataset at two different bin sizes, 50 ms and 100 ms respectively. For the autocorrelation, a maximum lag of 200 was used such that the complete occurrence of a flash transition at any of the former bin widths were included. Accordingly, a 2 second window was chosen for the PSTH computation. For each bin size, either the autocorrelograms or the PSTHs fed a partitional clustering method for the creation of a number of clusters which varied between two and five. For the clustering method, the nearest-neighbour or k-means approach was used. This approach decomposes the dataset into a set of disjoint clusters and then minimizes the average squared distance from a cluster centroid among the elements within a cluster, while maximizes this distance when regarding the centroids of the different clusters. This defines a set of implicit decision boundaries that separate the clusters or classes of units according to their periodicity when clustering their autocorrelograms or to its response delay when clustering their PSTHs. In this way, we end up with groups of relatives that are a subset of the entire array.

In order to validate the efficiency of the method, a first graphical arrangement of the units was accomplished for a fast visualization of the coherence in the classification among different number of classes and bin widths. Then, a quantitative analysis based on population responses was done. On each class, a population response was build up by summing up all the single response trials from each item within the class at a 5 ms bin size. On these signals, the autocorrelation and correlation with stimulus was calculated.

3   Results

In order to test the accuracy of the method either for an autocorrelation-based or PSTH-based classification, we first arranged the classified units in a graphical way for an appropriate first visual inspection. For both of the two bin sizes used for the analysis, a diagram as the one in Fig. 1 showing the classification performance, was built up for each of the autocorrelation and PSTH approach as well as for each number of clusters.

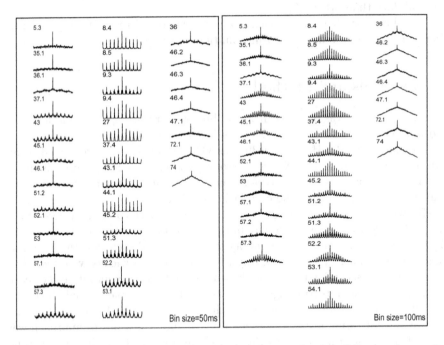

Fig. 1. Graphical representation of the k-means output for five different classes. On each column it is shown the autocorrelograms of the classified units belonging to a same class. Left panel: autocorrelation calculated at a bin size of 50 ms. Right panel: autocorrelation calculated at a bin size of 50 ms. Maximum lag used: 200. Legends on top of the autocorrelograms describe unit identifiers. For clarity reasons not all the units of the original dataset are presented

Specifically, Fig. 1 shows in a graphical fashion the output of the k-means method for the autocorrelograms of a number of units at two bin sizes; 50 and 100 ms respectively, both of them with a maximum lag of 200. All the units belonging to the same class are aligned in the same column. Legends on top of each plot represent the unit identifiers. In the case shown in Fig.1, the dataset was disjoint into three different clusters while in Fig. 2 original data were split into five sets of relatives. Not all of the units are shown for clarity reasons. Notice that units belonging to the class represented by the middle column in both panels present a strong periodicity, following the periodic stimulation while units grouped on the third column do not respond consequently to, at least, the applied stimulation. The same holds when increasing the number of classes, as seen in Fig. 2, with the advantage that the cell periodicities can be clustered in a more graded fashion.

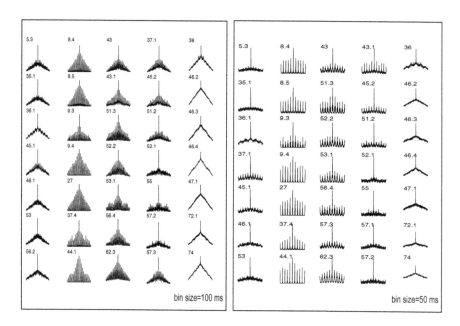

Fig. 2. Graphical representation of the k-means output for five different classes. On each column it is shown the autocorrelograms of the classified units belonging to a same class. Left panel: autocorrelation calculated at a bin size of 50 ms. Right panel: autocorrelation calculated at a bin size of 50 ms. Maximum lag used: 200. Legends on top of the autocorrelograms describe unit identifiers. For clarity reasons not all the units of the original dataset are presented

Once the qualitative assessment of the performance for different bin sizes and number of classes was done as explained above, a more quantitative analysis of how efficiently the method separated units was carried out. Firstly, it was calculated as many subpopulation responses as number of classes were chosen in the K-means algorithm for the classification task. For each class, the activity of all the units belonging to that class were summed up in order to create a population response for that class. Population responses were computed as follows: a bin width of 5 ms was defined and the activity of all the cells in the subpopulation was summed up at each bin. Afterwards, the signals were smoothed by convolving them with a Gaussian function with a standard deviation of 0.5. Then, both the autocorrelation and correlation to stimulus of these signals were calculated. For an appropriate comparison with the original entire population, Fig. 3a shows the raster of the original population while Fig. 3b shows its smoothed population response. Accordingly, Fig. 3c shows the autocorrelation of the population response normalized so that at zero lag the autocorrelations are identically one, and Fig. 3d shows the population response correlated to stimulus without any scale applied, this is, non normalized. For both the autocorrelation and cross-correlation computation, the entire signal was lagged. Continuous trace on top of Fig. 3a represents the temporal evolution of the stimulus, a 2 second full-field flash with a 300 ms ON period.

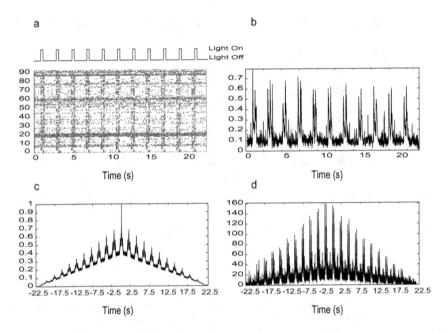

Fig. 3. Representation of the original data. a) Raster plot of the original dataset. Stimulus evolution is represented in the continuous top trace. b) response population resulting from summing up the activity across all the units in the recording for every bin (5 ms). c) normalized population signal autocorrelogram and d) population signal correlated to stimulus. For both correlations the entire signal was lagged

The same analysis as the one done to the original population was applied to every subpopulation in order to highlight the properties that were kept or removed from the original dataset in each of the subpopulations created. As an example, Fig. 4 displays the rasters, population responses, autocorrelation and correlation to stimulus (Fig .4a, 4b, 4c and 4d respectively) of three different

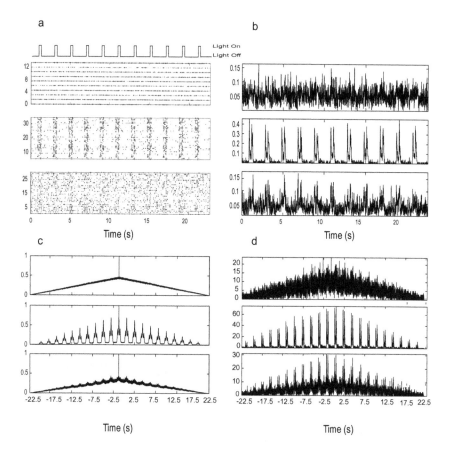

Fig. 4. Representation of the classified data by means of the autocorrelation approach. a) Raster plots of the three subpopulations found by defining the number of classes three. Stimulus evolution is represented in the continuous top trace. b) response populations resulting from summing up the activity across all the units in each subpopulation for every bin (5 ms). c) normalized population signal autocorrelograms and d) population signals correlated to stimulus for each filtered population. For both correlations the entire signal was lagged

subpopulations found by means of the autocorrelation approach using a bin size of 100 ms and a maximum lag of 200.

Consistently, the number of classes set in the classification task was three. The processes were calculated as explained above for the original data. Here, each subpopulation is represented in a subplot across every panel respecting the order through out the different panels for a clear visualization. Notice the clean temporal pattern of the subpopulation represented as a raster (Fig. 4a, middle panel), which is also reflected on its response population (Fig. 4b, middle panel) as well as on both of the correlograms (Fig. 4c and Fig. 4d, middle panels), where either of them turned out improved with respect to the correlograms from the original entire population (Fig. 3c and Fig. 3d), which display certain amount of noise.

In the same direction, Fig. 5 shows other three different subpopulations obtained from the same original data but using the PSTH approach.

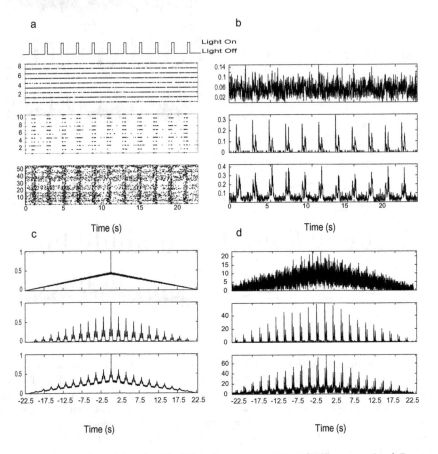

Fig. 5. Representation of the classified data by means of the PSTH approach. a) Raster plots of the three subpopulations found by defining the number of classes three. Stimulus evolution is represented in the continuous top trace. b) response populations resulting from summing up the activity across all the units in each subpopulation for every bin (5 ms). c) normalized population signal autocorrelograms and d) population signals correlated to stimulus for each filtered population

Notice that the purity of the filtered populations depends strongly on the number of classes defined. For instance, in this example three classes are not enough for getting pure ON or pure OFF subpopulations but is enough for getting predominant ON (Fig. 5b, middle subplot) and predominant OFF (Fig. 5b, lower subplot), since one class is reserved for noisy units. The ON and OFF predominance is also noticed in Fig. 4c, middle and lower subplots respectively when showing correlated activity to stimulus.

4   Conclusions

Two different approaches for an automatic classification of electrophysiological multineuron data were developed and validated. It was established that both approaches worked satisfactorily at splitting original datasets although temporal dynamics of filtered populations vary from one approach to another. When using the autocorrelation approach, units will cluster according to its periodicity while when using the PSTH approach, they will do according to its latency response. The number of classes defined by the experimenter also contribute to the purity within each class, since the more clusters are defined, the more specificity will be achieved. The method turns out to be robust and effective whereas simple and versatile. Even though this is an automatic method, the experimenter has to keep track of the parameters applied in the processing stage as well as the number of classes that would expect to be generated at every particular case.

Acknowledgements

This research is being funded by E.C. QLK-CT-2001-00279, and Ministerio de Ciencia y Tecnología TIC 2003-09557-C02-02.

References

1. Secker, H. and Searle, C.: Time Domain Analysis of Auditory-Nerve Fibers Firing Rates. J. Acoust. Soc. Am. 88 (1990) 1427-1436
2. Buck, L.B.: Information Coding in the vertebrate olfactory system. Ann. Rev. Neurosci. 19 (1996) 517-544
3. Fitzhugh, R. A.: A Statistical Analyzer for Optic Nerve Messages. J. Gen. Physiology 41 (1958) 675-692
4. Warland, D., Reinagel, P., Meister, M.: Decoding Visual Information from a Population of Retinal Ganglion Cells. J. Neurophysiology 78 (1997) 2336-2350
5. Fernández, E., Ferrández, J.M., Ammermüller, J., Normann, R.A.: Population Coding in spike trains of sinultaneosly recorded retinal ganglion cells Information. Brain Research 887 (2000) 222-229
6. Tovee, M.J., Rolls, E.T., Treves, A. and Bellis, R.P.: Encoding and the Responses of Single Neurons in the Primate Temporal Visual Cortex. Journal of Neurophysiology, 70 (1993) 640-654
7. Ferrández, J.M., Bolea, J.A., Ammermüller, J. , Normann, R.A. , Fernández E.: A Neural Network Approach for the Analysis of Multineural Recordings in Retinal Ganglion Cells: Towards Population Encoding. IWANN 99, Lecture Notes on Computer Science 1607 (1999) 289-298

8. MacQueen, J.: Some methods for classification and analysis of multivariate observations. Le Cam, L. M. and Neyman, J., editors, Proceedings of the Fifth Berkeley Symposium on Mathematical Statistics and Probability. University of California Press, Berkeley and Los Angeles, CA (1967)

9. Sim A. W. K., Jin C., Chan L. W., and Leong P. H. W.: A comparison of methods for clustering of electrophysiological multineuron recordings. IEEE Engineering in Medicine and Biology Society (1998) 1381-1384

10. Jones, K.E., Campbell, P.K. and Normann, R.A.: A glass/silicon composite intracortical electrode array. Annals of Biomedical Engineering 20 (1992) 423-437

Gap-Junctions Promote Synchrony in a Network of Inhibitory Interneurons in the Presence of Heterogeneities and Noise

Santi Chillemi, Alessandro Panarese, Michele Barbi, and Angelo Di Garbo

Istituto di Biofisica CNR, Sezione di Pisa,
Via G. Moruzzi 1, 56124 Pisa, Italy
{santi.chillemi, alessandro.panarese, michele.barbi,
angelo.digarbo}@pi.ibf.cnr.it
http://www.pi.ibf.cnr.it

Abstract. Recent experiments revealed that inhibitory interneurons networks are coupled by both electrical and inhibitory synapses. Moreover these findings suggest that a population of interneurons operate as a clockwork affecting the processing of neural information. In this paper we determine, in the weak coupling limit, the parameter values leading to the emergence of synchronous regime in a pair of Fast Spiking interneurons coupled by chemical and electrical synapses. Then, our results will be compared with those obtained recently in [1] for a pair of coupled Integrate & Fire neural models. Next, the effects of heterogeneities and noise on the coherence properties of the network (containing two or more coupled units) will be investigated numerically.

1   Introduction

Synchronization phenomena occur between interneurons and recent experimental findings reveal that networks of GABAergic inhibitory cells contribute to the generation of Gamma rhythms (30-100 Hz) in neocortex [2]. Moreover, both experimental and theoretical results relate the synchronous discharge of a population of inhibitory interneurons to the features of their coupling [3, 5, 6, 7, 8, 9]. Recently, it was found that Fast Spiking (FS) neocortical inhibitory interneurons are interconnected by electrical synapses too [10, 11]. Thus, at present, an open problem is to understand how these two types of synaptic coupling determine the network dynamical behavior. In a recent paper this topic was addressed for a pair of identical Leaky Integrate & Fire (LIF) models in absence heterogeneities and noise [1]. In this paper we show that the results found in [1] do not hold in general for more realistic biophysical models of FS interneurons. In particular we will show that both the after-hyperpolarization amplitude and the rise time constant of the inhibitory postsynaptic current (IPSC) affect the synchronization properties of a network of coupled cells. Moreover the effects of noise and heterogeneities will be investigated too also for a network containing more than two cells.

J. Mira and J.R. Álvarez (Eds.): IWINAC 2005, LNCS 3561, pp. 77–85, 2005.

2   Methods

FS interneurons are not capable of generating repetitive firing of arbitrary low frequency when injected with constant currents [12, 13], thereby they have type II excitability property [14].

Recent experiments carried out on in vitro FS cells reveal that they have high firing rates (up to $\approx 200 \ Hz$), average resting membrane potential of -72 mV and input resistance $\approx 89 \ M\Omega$; their action potential has a mean half-width $\approx 0.35 \ ms$, average amplitude $\approx 61 \ mV$ and after-hyperpolarization amplitude $\approx 25 \ mV$ [10, 11, 13].

2.1   Model Description

A single compartmental biophysical model of a FS interneuron well accounting for the features above reads:

$$C\frac{dV}{dt} = I_E - g_{Na}m^3h(V - V_{Na}) - g_K n(V - V_K) - g_L(V - V_L) \qquad (1)$$

$$\frac{dx}{dt} = \frac{x_\infty - x}{\tau_x}, \quad x_\infty = \frac{\alpha_\infty}{\alpha_\infty + \beta_\infty}, \quad \tau_x = \frac{1}{\alpha_\infty + \beta_\infty}, \quad (x = m, h, n) \qquad (2)$$

where $C = 1 \ \mu F/cm^2$, I_E is the external stimulation current. The maximal specific conductances and the reversal potentials are respectively: $g_{Na} = 85$ mS/cm^2, $g_K = 60 \ mS/cm^2$, $g_L = 0.15 \ mS/cm^2$ and $V_{Na} = 65 \ mV$, $V_K = -95 \ mV$, $V_L = -72 \ mV$. In order to model the kinetic of the Na^+ current we estimated theoretical curves for the voltage dependence of the activation and deactivation rate variables $\alpha_m(V)$, $\beta_m(V)$. This was carried out by using the experimental curve of the steady-state activation and the data on the activation/deactivation time constants obtained from hippocampal FS interneurons [15]. A similar work was done for $\alpha_h(V)$, $\beta_h(V)$, leading to the following expressions: $\alpha_m(V) = 3.0exp[(V + 25)/20]$, $\beta_m(V) = 3.0exp[-(V + 25)/27]$, $\alpha_h(V) = 0.026exp[-(V + 58)/15]$, $\beta_h(V) = 0.026exp[(V + 58)/12]$.

Voltage-gated K^+ channels are assembled from subunits of four major subfamilies, designated as Kv1, Kv2, Kv3, Kv4 [16]. FS inhibitory interneurons express Kv3 subunits at a very high level and this has been found necessary for their phenotype [17]. A model for the gating of Kv3 channels in hippocampal FS interneurons has yet been proposed [17]. In this paper we will use this model with a different value of the parameter vAHP, shaping the amplitude of the after-hyperpolarization phase of the action potential, in order to reproduce the corresponding experimental average value for FS cells. This leads to the rate variable expressions : $\alpha_n(V) = [-0.019(V-4.2)]/exp[-(V-4.2)/6.4]-1$, $\beta_n(V) =$ $0.016exp(-V/vAHP)$, where vAHP $= 13 \ mV$. In this model the onset of periodic firing occurs through a subcritical Hopf bifurcation for $I_E \approx 1.47 \ \mu A/cm^2$ with a well defined frequency ($\approx$16 Hz), according to the type II excitability property [14](data not shown).

2.2   Synaptic Coupling

The electrical and chemical synapses are modeled as follows. The postsynaptic current for the GABAergic synapse at time $t > t_N$ is given by

$$I_{Sy}(t) = g_{Sy}s_T(t)(V_{Post}(t) - V_{Rev}) = g_{Sy}\sum_j s(t - t_j)(V_{Post}(t) - V_{Rev}), \quad (3)$$

where g_{Sy} is the specific maximal conductance of the inhibitory synapse (in mS/cm^2 unit), $s(t) = A[exp(-t/\tau_{Decay}) - exp(-t/\tau_{Rise})]$, t_j $(j = 1, 2, , N)$ are the times at wich the presynaptic neuron generated spikes, τ_{Decay} and τ_{Rise} are the decay and rise time constants of the IPSC. For FS interneurons the IPSC is characterized by a reversal potential V_{Rev}= -80 mV, a mean rise time constant $< \tau_{Rise} > = 0.25$ ms and a mean decay time constant $< \tau_{Decay} > = 2.6$ ms [13]. The electrical synapse is modeled as

$$I_{El} = g_{El}(V_{Post} - V_{Pre}), \quad (4)$$

where g_{El} is the maximal conductance of the gap junction (in mS/cm^2 unit). For all simulations the values of parameters $g_{Sy}, g_{El}, \tau_{Rise}, \tau_{Decay}$ were all within their physiological ranges [10, 11, 13].

2.3   Analysis Method

Let us consider the case of two identical coupled oscillators and assume that the coupling intensity ϵ is weak ($\epsilon << 1$). Moreover, let us assume that for $\epsilon = 0$ each oscillator possesses a stable limit cycle $X_0(t)$ of period T. Then the state of each oscillator is defined by its phase θ_i $(i = 1, 2)$ and the corresponding dynamical behavior is determined by the following equations:

$$\frac{d\theta_1}{dt} = 1 + \epsilon H_1(\phi) \quad (5)$$

$$\frac{d\theta_2}{dt} = 1 + \epsilon H_2(-\phi), \quad (6)$$

where $\phi = \theta_2 - \theta_1$ and $H_1(\phi)$, $H_2(-\phi)$ are T-periodic functions which take into account the effect of coupling. The details for the computation of $H_1(\phi)$, $H_2(-\phi)$ can be found in [18, 4].

The time evolution of ϕ is determined by

$$\frac{d\phi}{dt} = [H_2(-\phi) - H_1(\phi)] = -\epsilon D(\phi) \quad (7)$$

($\epsilon = g_{Sy}$ for inhibitory coupling, g_{El} for electrical coupling) and the phase locked states can be determined by searching the solutions of the equation $D(\phi^*) = 0$. A phase locked state is characterized by a constant phase difference ϕ^* between the two oscillators and it will be stable (or unstable) according as $dD/d\phi > 0$ ($dD/d\phi < 0$). For mixed coupling $\epsilon D(\phi) = [g_{Sy}D_s(\phi) + g_{El}D_e(\phi)]$.

In the numerical simulations the degree of coherence between two spikes trains is quantified by the normalized cross-correlation

$$C_{m,n} = \sum_{j=1}^{n_{bin}} [x(j)y(j)] / \sqrt{\sum_{j=1}^{n_{bin}} x(j) \sum_{j=1}^{n_{bin}} y(j)} \qquad (8)$$

where n_{bin} is the total number of bins, each one $0.1T$ msec of duration, and T is the average network discharge period. Moreover, for each spike train, the variable $x(j)$ (or $y(j)$) is defined to be 1 or 0 according to whether a spike is fired or not in the $j - th$ bin.

3   Results

First of all we study the pattern of phase-locked states of a pair of identical biophysical models of FS cells against the stimulation current.

Panel a) of figure 1 shows that, in absence of electrical coupling, the increase of I_E affects the synchronization properties of the cells in two way. For I_E values lower than $\approx 7\mu A/cm^2$ the increase of the discharge frequency promotes synchronization, while for $I_E > 7\mu A/cm^2$ antisynchrony is promoted. These results contrast with the corresponding findings for a pair of coupled LIF model (for which

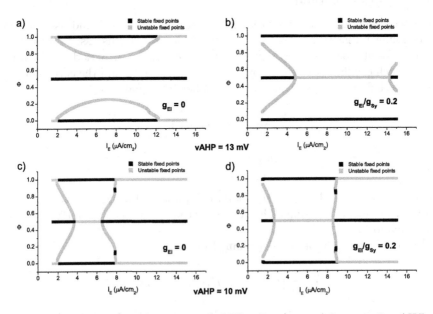

Fig. 1. Phase locking states for two coupled FS cells: a) $g_{Sy} \neq 0$, $g_{El} = 0$, $vAHP = 13mV$; b) $g_{El}/g_{Sy} = 0.2$, $vAHP = 13mV$; c) $g_{Sy} \neq 0$, $g_{El} = 0$, $vAHP = 10mV$; d) $g_{El}/g_{Sy} = 0.2$, $vAHP = 10mV$. For all panels it is $\tau_{Rise} = 0.25 \ ms$ and $\tau_{Decay} = 2.6$ ms. Stable fixed points (black squares), unstable fixed points (gray squares)

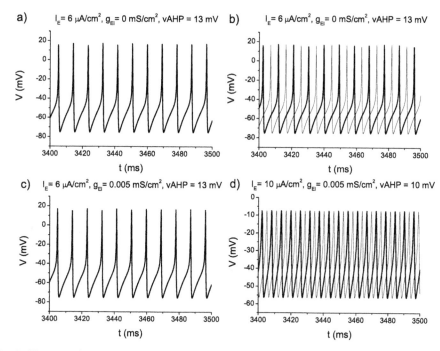

Fig. 2. Numerical simulation of two coupled FS cells: a) and b) different initial conditions leads to synchrony (left) or antisynchrony (right); c) electrical coupling is on; d) a smaller $vAHP$ leads to antisynchrony. For all panels it is $g_{Sy} = 0.02 mS/cm^2$

IPSC time course was modeled by α–functions). By setting on gap-junctions there is a deep change of the pattern of the phase-locked states (see panel b) of figure 1). Now there is range of I_E values $(4.8\mu A/cm^2 < I_E < 14.5\mu A/cm^2)$ where only synchronous states are stable. In panels c) and d) the results obtained for $vAHP = 10$ are shown. In this case the amplitude of the action potential AHP decreases and the synchronization properties of the network change. In particular by comparing the results plotted in panel c) and d) it follows that the effects of the electrical coupling to promote higher network coherence levels are reduced. To test the above theoretical findings, obtained in the weak coupling limit, we performed numerical simulations of a pair of coupled FS cells, and the corresponding results are shown in figure 2.

In panels a) and b) are plotted the results obtained by using different initial conditions for the case $g_{El} = 0$ and $I_E = 6\mu A/cm^2$. They show that there is bistability as predicted theoretically (see panel a) of figure 1). Panel c) of figure 2 shows the corresponding results when the electrical coupling is set on. Instead, in panel d) the results obtained for $I_E = 10\mu A/cm^2$, $vAHP = 10$ and with $g_{El} \neq 0$ are reported. In this case the only stable state is the antiphase one. Next the effects of heterogeneities and noise on the synchronization properties of two coupled cells were investigated. By comparing panels a) and b) of figure 3 it follows that the presence of a small amount of heterogeneity $(I_E(1) \neq I_E(2))$

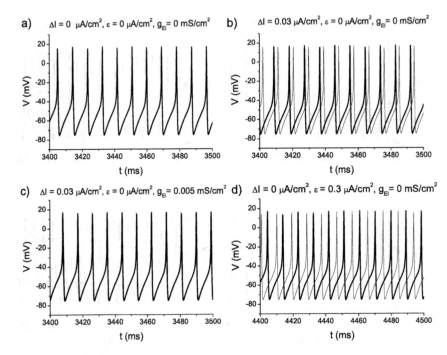

Fig. 3. Heterogeneities and noise effects: a)synchronous firing in a noiseless ($\epsilon = 0$)and homogeneous ($\Delta I = I_E(1) - I_E(2) = 0$) network with $g_{El} = 0$; b)$\Delta I = 0.03 \mu A/cm^2$; c)$\Delta I = 0.03 \mu A/cm^2$ and $g_{El} = 0.005 mS/cm^2$; d)$\Delta I = 0$, $g_{El} = 0$ and $\epsilon = 0.3 \mu A/cm^2$

breaks the synchronous firing of the cells. However by setting on the electrical coupling the initial network synchrony is recovered (see panel c). In panel d) of figure 3 the results obtained when each cell receive independent gaussian stochastic inputs alone are shown ($I_E(1) = I_E(2) = 6\mu A/cm^2$). Also in this case synchrony is lost, but recovers when adding the electrical coupling (data not shown). When heterogeneity and noise are introduced simultaneously, the presence of the electrical coupling is capable, as in the previous cases, to promotes synchronization(data not shown).

We investigated more systematically the effects of heterogeneity and noise on the network coherence levels and the results are shown in figure 4. It is worth noticing that the data reported in this figure (and in the next one) were obtained by averaging the results over five randomly chosen initial conditions. By comparing the left and right panels it follows that the presence of the electrical coupling is a key factor to prevent asynchrony for small amounts of heterogeneity ($\Delta I = I_E(1) - I_E(2)$) and noise amplitude ($\epsilon$). Now we come to study how the population size affects the network coherence level when heterogeneities and noise are present. The corresponding results are reported in figure 5. Panels a) and b) show the results obtained when heterogeneities are introduced by using stimulation current values from a Gaussian distribution ($< I_E > = 6\mu A/cm^2$, $\sigma(I_E) = 0.1\mu A/cm^2$). For both panels the synchronization levels increase as the

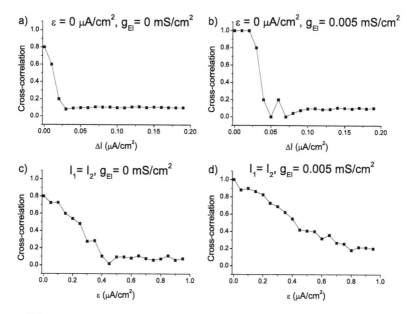

Fig. 4. Coherence levels of the spike trains of two coupled cells: panels a) and b) against the heterogeneity amplitude ΔI; c) and d) against the noise amplitude

number of units in the network grows. Moreover the presence of the electrical coupling leads to a relevant increase of the coherence level. Panels c) and d) show the results obtained in presence of noise, and they look qualitatively similar to the previous ones.

4   Conclusions

In this paper the synchronization properties of a network of FS interneurons coupled by inhibitory and electrical synapses were studied theoretically and numerically. We used a biophysical model of each FS cell able to reproduce the main experimental features of these interneurons. First, to understand how the electrical and inhibitory coupling affects the synchronization properties of a pair of FS interneurons, we determined the corresponding patterns of phase-locking states in the weak coupling limit. We found that our results are contrasting with those obtained recently for a pair LIF models [1]. In particular we found that for realistic time course of the IPSC ($\tau_{Rise} \ll \tau_{Decay}$) the increase of the discharge frequency of the cells promotes synchronization in well defined range of I_E values (see figure 1). Moreover, we shown that the amplitude of AHP play a relevant role in shaping the pattern of phase-locked states. Next the synchronization properties of the network, in the presence of heterogeneities and noise, were investigated numerically. We shown that, in absence of electrical coupling, small amounts of heterogeneities (or noise) are sufficient to break synchrony. However, our simulations shown that the addition of the electrical coupling (with intensity

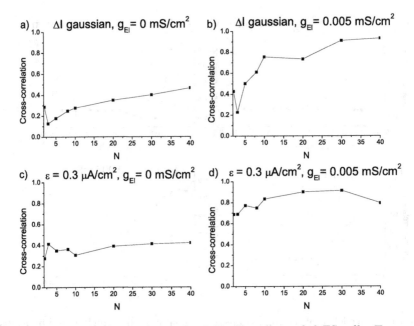

Fig. 5. Coherence levels in a network of N all-to-all coupled FS cells. Top panels: $\Delta I \neq 0$ and $\epsilon = 0$; bottom panels $\Delta I = 0$ and $\epsilon \neq 0$

within the physiological range) is able to contrast these decoherence effects. We studied a larger network of coupled FS cells (in the presence of heterogeneities and noise) and we shown that the electrical coupling plays a relevant role in promoting synchrony.

References

1. Lewis T., Rinzel J.: *Dynamics of spiking neurons connected by both inhibitory and electrical coupling.* J. Comp. Neuroscience **14** (2003) 283-309
2. Galarreta M., Hestrin S.: *Electrical synapses between GABA-releasing interneurons.* Nat. Neurosci. **2** (2001) 425 - 433
3. Di Garbo A., Barbi M., Chillemi S.: *Synchronization in a network of fast-spiking interneurons.* BioSystems **67** (2002) 45 - 53
4. Di Garbo A., Panarese A., Chillemi S.: *Gap-junctions promote synchronous activities in a network of inhibitory interneurons.* BioSystems **79** (2005) 91 - 99
5. Van Vreeswijk C. A., Abbott L. F., Ermentrout G. B.: *Inhibition, not excitation, synchronizes coupled neurons.* J. Comp. Neuroscience **1** (1995) 303 - 313
6. Wang X. J., Rinzel J.: *Alternating and synchronous rhythms in reciprocally inhibitory model neurons.* Neural Comput. **4** (1992) 84 - 97
7. Wang X. J., Buzsaki G.: *Gamma oscillations by synaptic inhibition in an interneuronal network model.* J. Neurosci. **16** (1996) 6402 - 6413
8. Whittington M. A., Traub R. D., Jefferys J. G. R.: *Synchronized oscillations in interneuron networks driven by metabotropic glutamate receptor activation.* Nature, **373** (1995) 612-615.

9. Whittington M. A., Traub R. D., Kopell N., Ermentrout B., Buhl E. H.: *Inhibition-based rhythms: experimental and mathematical observations on network dynamics.* Int. J. Psychophysio. **38** (2001) 315 - 336

10. Galarreta M., Hestrin S.: *A network of fast-spiking cells in the cortex connected by electrical synapses.* Nature **402** (1999) 72-75

11. Gibson J.R., Beierlein M., Connors B.W.: *Two networks of electrically coupled inhibitory neurons in neocortex.* Nature **402** (1999) 75-79

12. Erisir A., Lau D., Rudy B., Leonard C. S.: *Function of specific K^+ channels in sustained high-frequency firing of fast-spiking neocortical interneurons.* J. Neurophysiology **82** (1999) 2476-2489

13. Galarreta M., Hestrin S.: *Electrical and chemical Synapses among parvalbumin fast-spiking GABAergic interneurons in adult mouse neocortex.* PNAS USA **99** (2002) 12438-12443

14. Rinzel J., Ermentrout B.: *Analysis of neural excitability and oscillations.* Eds. Koch and Segev, Methods in neural modelling (1989), The MIT Press, Cambridge

15. Martina M, Jonas P.: *Functional differences in Na^+ channel gating between fast spiking interneurons and principal neurons of rat hippocampus.* J. of Physiol. **505.3** (1997) 593-603

16. Coetzee W. A. et al.: *Molecular diversity of K^+ channels.* Annals of the New York Academy of Sciences **868** (1999), 233 - 285

17. Lien C. C., Jonas P.: *Kv3 potassium conductance is necessary and kinetically optimized for high-frequency action potential generation in hippocampal interneurons.* J. of Neurosci. **23** (2003) 2058-2068

18. Ermentrout B.: *Neural networks as spatio-temporal pattern-forming systems.* Rep. Prog. Phys. **61** (1998) 353-430.

A Conceptual Model of Amphibian's Tectum Opticum with Probabilistic Coded Outputs

Arminda Moreno-Díaz[1], Gabriel de Blasio[2], and Roberto Moreno-Díaz[2]

[1] School of Computer Science, Madrid Technical University
amoreno@fi.upm.es
[2] Instituto Universitario de Ciencias y Tecnologías Cibernéticas,
Universidad de Las Palmas de Gran Canaria
gdeblasio@dis.ulpgc.es
rmoreno@ciber.ulpgc.es

1   Introduction

Systems theoretical methods used to describe lower vertebrate retinal processing [1] can be extended to include models of the amphibian tectum. Figure 1 is a diagram of the frog's visual system with the eyes enlarged so that the ganglion cells can be illustrated collecting information from the retina and conveying it to both the optic tectum and the geniculate nucleus. Connections to the tectum are the principal ones. Experiments indicate that only the presence of blue light in the field of view is reported to the geniculate.

Figure 2 represents schematically the retina and optic tectum for amphibians. The photoreceptors, which respond to light, stimulate the bipolar cells, which in turn stimulate the ganglion cells, whose axons comprise the optic nerve. The axons of each of four kinds of ganglion cells terminate on four separate layers of the tectum. Each type of ganglion cells performs a different analysis of the retinal image and reports, through the optic nerve, to the tectum, in the following accepted manner for *Rana Pipiens* [2]:

1. Group 1 ganglion cells respond to edges, i. e., to differences in illumination on the receptive field.
2. Group 2 ganglion cells respond to the image of a small dark object, moving centripetally into their visual field.
3. Group 3 ganglion cells respond to any change, spatial or temporal, in the luminance of the scene.
4. Group 4 ganglion cells respond to dimming of light.

Lettvin, et al. [3] were the first to find correlations between the form and function of each of these cells. Their description of the tectum has not been as thorough, however, since they were concerned only with the response of tectal cells in an animal acquiring food. Tectal responses related to procuring food and to scape reactions have been described for toads [4], [5].

J. Mira and J.R. Álvarez (Eds.): IWINAC 2005, LNCS 3561, pp. 86–94, 2005.

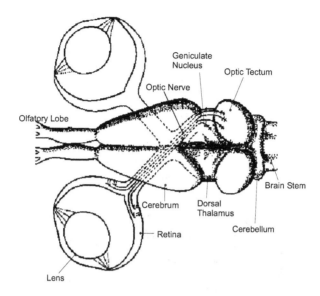

Fig. 1. Diagram of frog's visual system

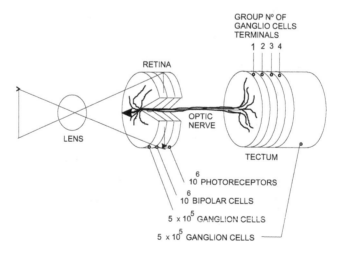

Fig. 2. Schematic of amphibian visual system

2 Neurophysiological Basis of the Model

Pedro Ramon y Cajal's classical drawing of the frog's tectum is shown in Fig 3. Although the drawing is not correct in detail, there are four main features that appear clearly. First, the cell bodies lie in the deeper layers (layers 1 to 6 in Fig 3).

Second, the neurons seem to have a restricted dendritic tree. Third, axons leave the tectum branching from a main ascending dendrite. Fourth, interconnections among cells lying at the same depth is strongly suggested.

Figure 4 is deducted from the discussions in [2]. It shows the essential anatomical features of the tectum. Axons from the four major retinal ganglion cell groups enter the tectum as the optic tract and map onto the four superior tectal layers called the superficial neuropile. Axons from ganglion cells which respond to adjacent retinal areas map onto adjacent points in the neuropile. Thus, each small area in the retina corresponds to four small areas, one on each layer, in the neuropile. Within the neuropile there also exist a few cells - not represented in the figure - which appear to be lateral connectives.

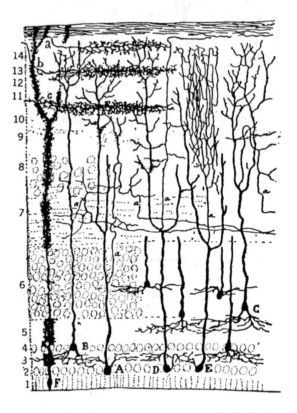

Fig. 3. Drawing of the optic tectum in the frog

All other tectal cell bodies lie in the granular layer, beneath the superficial neuropile 3. Their dendrites, which have the appearance of columns, extend up through the four neuropile layers covering a small area in each layer. Their axons leave the tectum by branching from the main dendrites, at the lower part of the neuropile.

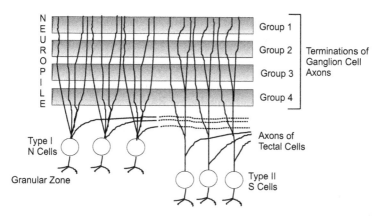

Fig. 4. Essential anatomical features of the tectum

For toads (*Bufo Bufo*), the most extensive neurophysiological recordings, and the correlation of these with behavioral patterns have been made by Ewert [4],[5] also in prey-catching situations.

In the toad retina there are similarly three types of ganglion cells that send their fibers by way of the optic nerve to the optic tectum. However, Group 1 is missing. The three ganglion-cell types in the toad (as in the frog) differ in several characteristics, including in particular the diameter of their excitatory receptive fields: about four degrees for the so-called Group 2 ganglion cells, about eight degrees for Group 3 cells and from 12 to 15 degrees for Group 4 cells.

Lettvin, et al. have distinguished from several varieties of tectal neurons two extreme types: "Newness" cells, N, sensitive to new objects in the visual field, and the "Sameness" cells, S, sensitive to the same object for a specific period of time. The bodies of Sameness cells lie deep in the granular zone. The properties of Newness and Sameness cells are shown by the following outline:

1. Newness Cell
   a. Its receptive field is approximately 30° in diameter, with considerable overlap.
   b. It yields small responses to sharp changes in illumination.
   c. Its response frequency increases if the motion of an object is irregular. Its response also depends on the velocity and size of the object and on the direction of motion.
   d. It has no enduring response.
   e. Its adaptation time is less than 1 second.
   f. Adaptation is erased with a step to darkness.
2. Sameness Cell
   a. Its receptive field is almost all the visual field, but includes a null region.
   b. It does not respond to changes in illumination.
   c. It responds to the presence of an object in its field by generating a low frequency pulse train. Response is maximum to objects about 3° in diameter.

    d. It "notices" with a "burst" of pulses all motion discontinuities. It discriminates among objects, fixing its "attention" on the one with the most irregular movement.

Lettvin, et al. [3] comment that N and S cells may be considered two extremes of several types of tectal cells. From Pedro Ramon y Cajal's drawings, tectal cells do not seem to be significantly different, anatomically, although S cells could receive excitation from N cells. A general tectal model should be one in which N and S cells would be differentiated only by values of parameters, such as adaptation time, width of the receptive field, etc. This is the point of view adopted here. While the receptive field is relatively wide for both types of cells, the dendritic tree is very restricted. Interaction between tectal cells would account for this observation.

For toads, Ewert also distinguishes two essential types, I and II, which may correspond to Lettvin et al. [3] N and S tectal cells.

3   Conceptual Model

In the conceptual model of tectal-cell behavior, object velocity and size dependant is determined by signals assumed to originate on the retina. For convenience these signals are referred to simply as "activity". Assume that each layer of the superficial neuropile is divided into small areas, such as area K in layer i, see Fig 5) Some ganglion-cell axons of Group i map into this area and emit their signals there. These are spike trains signals providing for multiple 'meaning' (multiplexing) probably by using white or gaussian noise. It is natural to assume a decoding process of spike coded modulated signals which produces analog values for them. This process must occur at the tectal cell dendrites, because neuropile recordings are of retinal ganglia spike outputs. Let $V_{iK}(t)$ be the amplitude signal indicating the response of the ganglion cells which map into area K of layer i.

Assume further that the dendrites of each tectal neuron are restricted to an area K, and there is little overlapping. The first process which is suggested is an interaction process among signals $V_{Ki}(t)$ (vertical line K layer i) providing for a total signal V_K in line K, which speed of change is:

$$\dot{V}_K(t) = f_K[V_K(t), V_{K1}(t), V_{K2}(t), V_{K3}(t), V_{K4}(t), t]$$

where f_K is an arbitrary function (having, in general, discontinuities and being variable in time).

Notice that signal $V_{Ki}(t)$ can either be excitatory (increasing $\dot{V}_K(t)$) or inhibitory (decreasing), also existing thresholds and saturations for the activity of vertical lines V_K.

As it is seen, the structure of these four layers where the retinal cells project is that of a layered computer (like the retina): the 'vertical' processing lines are laterally interconnected by lateral interaction (linear, non linear, algorithmic).

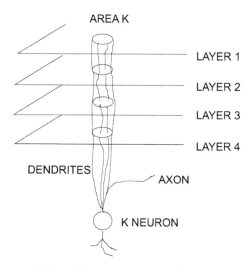

Fig. 5. Diagram of a tectal neuron

The simplest situation is that V_K is time independent, does not take into account history and is a linear function of $V_{Ki}(t)$. That is

$$V_K(t) = \sum_{i=1}^{4} \alpha_{iK} \cdot V_{iK}(t)$$

In this case, the coefficient α_{iK} may be either positive or negative and are assumed to be constant. If it is positive, the corresponding $V_{iK}(t)$ is said to be excitatory; if it is negative, $V_{iK}(t)$ is said to be inhibitory. For a qualitative approximation, α_{1K}, α_{2K} and α_{3K} are positive, whereas α_{4K} is negative. By summation, a series of "vertical" lines are obtained, which transmit signals $V_K(t)$ through the superficial neuropile. These signals interact (conserving the labels of geometrical situation) due to the interactivity of the medium through which they pass.

Four interaction processes may account for tectal cell properties, (See Fig. 6). Process A at the level of the dendrites, and Process D, at the level of the axons, are of the same nature, i.e., a type of lateral interaction. Process B,which accounts for adaptation, may be produced by the existence of a distributed analog retard which, once charged, damps the transmission of the signals. Process C is a 'maximum selector' which requires non-symmetric elements that could be provided by the cellular membrane. This selector must preserve spatial coordinates, if it is to be useful in the overall system behavior.

3.1 Interaction Process A (Facilitation)

Lateral spreading of activity increases the response to an irregularly moving object. $V_K(t)$ is referred to as the initial activity of line k, i.e., the signal before interaction. Its effect on another line j may depend upon the distance d_{kj}

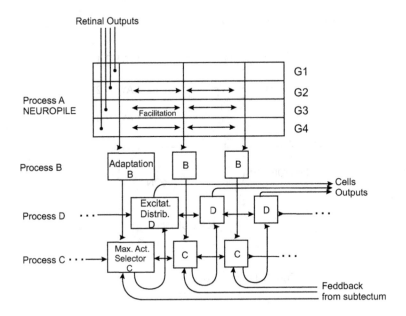

Fig. 6. Diagram of tectal cell inputs and interaction processes

according, for example to a decreasing function $f(d_{kj})$; $f(d_{kj})$ is such that as $d_{kj} \to \infty$, $f \to 0$,

The locus of maximal activity propagates like a wave (see [6],[7]) with some constant or variable speed. Notice that the labels corresponding to time and space must always be preserved.

Processes involving wave propagation of localized points of activity provide for a maximal activity if there is a 'jerky' motion restricted to areas type K, mostly if wave processes are non linear and increase almost like a 'scape time' process.

Lateral coupling is assumed for Type N neurons over a zone covering 30° of the visual field; for Type S, over the entire visual field.

3.2   Interaction Process B (Adaptation)

Proper adaptive saturation coupling will cause a line that has been excited to adapt and damp any further transmission for certain time (some type of refractoriness).

3.3   Interaction Processes C and D (Maximum-Activity Selection and Distribution)

Process C blocks all paths except the one having maximum activity, and Process D distributes this activity to the other lines, in a manner similar to Process A. In these processes, subunits (tectal cells) must know (preserve) the address where the maximal activity occurs and they must respond with a coded signal which

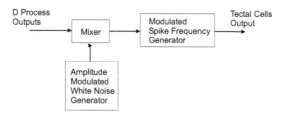

Fig. 7. Block diagram for spike frequency coded output mimics of tectal cells

is inversely proportional to the geometrical distance to the location of the maximum. Each type of these two processes are a result of competitive/cooperative lateral interaction (coupling) in each level. Maximum activity selection is similar to that used to model modal cooperative decision [8].

3.4   Output Coding Process

The final coder receives the results of process D as an input and goes through a noisy pulse-frequency conversion [9]. The basic diagram of this encoder is shown in figure 7. An information-carrying noise generator is mixed (linearly, in principle) with the results of Process D, to frequency modulate a spike generator which would mimic the output of tectal cells [10].

Possible feedback from subtectum is a potencial candidate to account for part of the so called 'multiple meaning' (multiple coding) in visual pathways [2].

4   Parameters of Types N and S Neurons

With lateral coupling for Type N neurons assumed for a zone covering 30° of visual field, necessary time constants for adaptation are such that damping of signals occur in 1 second and persist for 50 to 70 seconds. The difference between the time required for adaptation and the time for erasure of adaptation is a consequence of the nonlinear coupling (maximum selection) that cause Process C.

For Type S neurons, $V_K(t)$ is made higher by the cooperation of the signals coming from the retina of the output of some Type N neurons. For Type S cells lateral coupling occurs over almost all the visual field. Adaptation is negligible, implying a low value for the adaptation constant.

5   Conclusion

The main properties of tectal cells may be conceptually accounted for in a relatively simple manner by the model. Additional neurophysiological data and refinements of the model will permit further approximations. As a recapitulation, the main hypotheses of the model are as follows:

The same basic mechanism can in fact be assumed for both types N and S cells. To produce the desired response for the cells, this mechanism should include the four processes described above, namely, lateral facilitation, adaptation, selection of maximum activity, and distribution of it. The output of the conceptual model is obtained through a probabilistic (noisy) spike-frequency encoder.

References

1. Moreno-Díaz, R., de Blasio, G.: Systems Methods in Visual Modelling. Systems Analysis Modelling Simulation **43(9)** (2003) 1159–1171
2. Lettvin, J.Y. et al: What the Frog's Eye Tells the Frog's Brain. Proc. I.R.E., **47** (1959) 1940–1951
3. Lettvin, J.Y. et al: Two remarks on the Visual System of the Frog. in Sensory Communication. W. A. Rosemblith ed. MIT Press, (1961) 757–776
4. Ewert, J.P.: Neuroethology of Releasing Mechanism: Prey-Catching in Toads. Behavioral and Brain Sciences, **10** (1987) 337–405
5. Ewert, J.P.: The Neural Basis of Visually Guided Behavior. Scientific American, **230(3)** (1974) 34–42
6. Moreno-Díaz, R., Mira-Mira, J., Delgado, A.: Hacia una Teoría Social de Redes Neuronales. Porc. II Simposium Nacional de Bioingeniería (1983) 215–210
7. Delgado, A., Moreno-Díaz, R., Mira-Mira, J.: Qualitative Computation with Neural Nets: Differential Equations like Examples. In Lecture Notes in Computer Science, **769**. Pichler, Moreno-Díaz, eds. Springer, (1994) 366–379
8. Delgado, A., Mira-Mira, J., Moreno-Díaz, R.: A Neurocybernetic Model of Modal Cooperative Decision in the Kilmer-McCulloch Space. Kybernetes, **18 (3)** (1989) 48–57
9. Moreno-Díaz, R., Mira-Mira, J., Roy-Yarza, A.: Realización de Redes No-Determinísticas. Automática, **11 (V)** (1976) 5–14
10. Moreno-Díaz, R.: Deterministic and Probabilistic Neural Nets with Loops. Math. Biosciences, **11** (1971) 129–131

Realistic Stimulation Through Advanced Dynamic-Clamp Protocols

Carlos Muñiz[1], Sara Arganda[2], Francisco de Borja Rodríguez[1], and Gonzalo G. de Polavieja[2,†]

[1] Grupo de Neurocomputación Biológica (GNB),
Dpto. de Ingeniería Informática,
Escuela Politécnica Superior,
Universidad Autónoma de Madrid,
28049 Madrid, Spain
[2] Laboratorio de Procesamiento Neuronal,
Dpto. de Física Teórica, C-XI and
Instituto 'Nicolás Cabrera', C-XVI, planta 4,
Facultad de Ciencias,
Universidad Autónoma de Madrid,
28049 Madrid, Spain
carlos.muniz@uam.es

Abstract. Traditional techniques to stimulate neurons in Neuroscience include current injection using several protocols. In most cases, although neurons are able to react to any stimulus in the physiological range, it is difficult to assess to what extent the response is a natural output to the processing of the input or just an awkward reaction to a foreign signal. In experiments that try to study the precise temporal relationships between the stimulus and the output pattern, it is crucial to use realistic stimulation protocols. Dynamic-clamp is a relatively recent method in electrophysiology to mimic the presence of ionic or synaptic conductances in a cell membrane through the injection of a controlled current waveform. Here we present a set of advanced dynamic-clamp protocols for realistic stimulation of cells that allow from the addition of single and multiple ionic or synaptic conductances, to the reconfiguration of circuits and bidirectional communication of living cells with model neurons including plasticity mechanisms.

1 Introduction

Traditionally, neurophysiologists have used current and voltage clamp protocols to assess the electrical properties of neurons. In the current clamp technique, a current (typically a pulse) is injected into the neuron while the membrane potential is being recorded. In voltage clamp, the membrane potential is kept at a controlled value while the transmembrane current is being recorded. Both

---

[†] Additional author: Pablo Varona[1]. The regulations of this conference allow only four authors in the title page. However, this paper has five fully contributing authors.

J. Mira and J.R. Álvarez (Eds.): IWINAC 2005, LNCS 3561, pp. 95–105, 2005.

techniques have contributed to the understanding of the biophysical properties
of excitable cells and allowed the design of conductance-based models of neurons.

A decade ago, a new technique in neuron electrophysiology known as dynamic-
clamp was introduced [1, 2, 3]. This technique can simulate in a living neuron
the addition of new ionic and synaptic conductances. In these ten years, mod-
elers and experimentalists have used dynamic-clamp to design new experiments
about excitable cell properties which were impossible or difficult to carry out
with classical techniques. For example, dynamic-clamp have been used to simulate the effects of introducing or removing conductances [3], simulating the effects
of pharmacological conductance blocks [4], increasing or decreasing motoneuron
activity [5], simulating in vivo conditions [6] and building or modifying neural
circuits with artificial synapses and artificial neurons in hybrid circuits [7,9,8,10].

The dynamic-clamp technique operates in a cyclic way. An electrode is inserted
into a neuron and the membrane potential is recorded into a computer that calculates the current to inject in a postsynaptic neuron, which can be the same or
a different cell. Usually the current is calculated after solving a set of differential
equations that describe artificial models of ionic or synaptic conductances, or even
models of artificial neurons and networks. This process is repeated indefinitely
with a fixed frequency. The time between two consecutive membrane potential
acquisitions is known as the *update rate*. The correct election of the update rate is
critical for the well functioning of the application. The maximum update rate it is
usually determined by the data acquisition board. A slow update rate avoids the
correct simulation of conductances and a realistic stimulation. On the other hand,
a fast update rate requires a very fast computer, depending on the computational
load, to solve all the differential equations in time.

In this paper, we describe a set of advanced dynamic-clamp protocols that we
are developing for the realistic stimulation of neurons to investigate neural in-
put/output relations, the effect of intracellular transient memory and synaptic or
intracellular plasticity mechanisms. The software tries to satisfy all the requirements that an experimentalist will expect from an ideal dynamic-clamp system
and extend the existing protocols and techniques to achieve a more natural stimulation. Some of the features that our system accomplish are the following: high
resolution time between updates, real time monitoring of all time series and parameters, flexible control of model parameters in real-time, an extended library
of ionic, synaptic and neural models, the ability to change online the models
used in hybrid configurations, easy implementation of new models in the library,
an easy to use Graphical User Interface (GUI), easy installation, and lastly the
software is intended to be a general-purpose dynamic-clamp for vertebrate and
invertebrate preparations. We intend this software can be used in any configuration of dynamic-clamp, from the simulation of ionic or synaptic currents, to
the implementation of pattern clamp protocols [11] and hybrid circuits.

2   Existing Dynamic-Clamp Options

Today and according to [12], there are more than 20 different dynamic-clamp
systems in the electrophysiology community. We can classify them into soft-

ware or hardware implementations. Hardware implementations can also be divided basically into two groups. Firstly, there are very fast commercial hardware implementations in built-in analog devices that can operate up to 50 kHz in real-time (see [6] and http://www.instrutech.com). This approach is appropriate only to carry out easy experiments that simulate constant conductances or when they use a computer system to deal with more complex conductances. Secondly, analog devices can be replaced by quick embedded-processor or DSP systems [13, 17, 14]. However this solution is quite expensive.

Increase in speed of data acquisition boards (DAQ) and personal computers allows the implementation of dynamic-clamp in software. These implementations are more suitable than hardware implementations as they are easier to program, generally inexpensive, and more flexible to modify and customize to a particular experiment. As we mentioned in the previous section, the dynamic-clamp protocols rely critically on the update rate. This update rate must be strictly accomplish, requiring precise timing and no jitter. General-purpose OS are good in the overall performance, but are not suited to deal with applications that require deterministic control and timing. Here we describe briefly some of the most popular dynamic-clamp software, indicating their advantages and disadvantages and how they solve the update problem mentioned above (see also [12]).

1. **Extended Dynamic Clamp** by Pinto et al. [8]. This a Windows-based program. The authors have solved the problem of working with a non real time operating system by reading the time clock in the DAQ, so they know exactly how long it takes between two successive updates. This solution limits the speed of the application and it works perfectly as long as the computational load is low. The implementation of models in this system can also be complex as the increment in time in every update is different. An additional demultiplexing circuit has been built to control up to four neurons. This can be useful for electrophysiologists as they can control more than two neurons or two spatially separated sites in the same neuron, e.g. the soma and the dendrites. This program can simulate up to 8 Hodgkin-Huxley conductances and up to 18 chemical or electrical synapses. It provides a Graphical User Interface but does not display or save the time series, and other application is required for this task. The user can modify the parameters of the conductances on-line but can not build his own models.

2. **RTLDC (Real-Time Linux Dynamic Clamp)** by Dorval et al. [15]. This software runs in Real- Time Linux, it is flexible, easy to use (provides a GUI), with high-speed, low cost and good performance. It allows the loading of neuronal models and the modification of parameters on-line. It lacks on-line and off-line analysis tools. It also lacks a log to record what has happened to the variables and parameters in the model during the experiment.

3. **MRCI (Model Reference Current Injection)** by Railkov et al. [16]. This dynamic-clamp also operates in Real-Time Linux at a high-speed. There is no GUI, instead it has a shell where the electrophysiologist writes down the commands. Among other features, provides a model specification language,

scripts to implement repeatable protocols, the ability to perform data logging of variables, on-line modification of parameters and easy installation.

4. **LabVIEW-RT Dynamic Clamp** by Kullmann et al. [17]. This dynamic-clamp runs under a proprietary OS, the labVIEW-RT from National Instruments [http://www.ni.com]. LabVIEW-RT dynamic-clamp operates with two computers. One computer contains the GUI and runs Windows while the other computer, which they call embedded controller, runs the dynamic-clamp engine in labVIEW-RT OS. It is a fast, flexible implementation with a GUI and facilitates the building of conductances to the user.

3   Advanced Dynamic-Clamp Description

The dynamic-clamp protocols that we are developing run in a hard real-time extension to the Linux Kernel (http://www.kernel.org) called Real-Time Application Interface (RTAI, http://www.aero.polimi.it/~rtai/). This extension ensures that the operations specified in the software are executed in real time, while the general performance of the operating system is kept up. To control the data acquisition board, the software uses an open-source project known as COntrol and MEasurement Device Interface (http://www.comedi.org) that provides drivers, tools and libraries to control a wide variety of common data acquisition plug-in boards. Most of the DAQ boards used by electrophysiologists are supported by these libraries.

Our software intends to be user-friendly, thus it includes a customizable GUI, which is programmed with two well-supported C++ graphical libraries: Qt (http://trolltech.com/products/qt) version 3.3 and Qt Widgets for Technical Applications (QWT) (http://qwt.sourceforge.net/) version 0.4.1. QWT is especially useful for displaying autoscaled and non-autoscaled time series.

The RTAI Linux can manage real-time processes and non real-time processes. The user can work with Linux ordinary applications like any general-purpose OS and at the same time work with a real-time application. In our system two processes are running simultaneously: a RT Process (the core of the system) written in C and the GUI, which is a non real time process. The real-time process lives in kernel-space while the GUI resides in the user-space. This means that they are two independent processes working in parallel that need to interchange data and communicate to each other. We have implemented two FIFO (First-in First-out) queues. In one of them, the *data* FIFO, the RT process writes the voltage obtained by the DAQ or the output produced by the models. The GUI needs a timer to read from this FIFO. This timer can be modified by the user to adapt the performance of the application. The other FIFO, the *command* FIFO is used to interchange commands between the RT process and the GUI, and it allows the GUI to modify the RT behavior. The model parameters and the network topology is stored in a shared memory that can be modified by both the RT process and the GUI.

Figure 1 shows a schematic representation of our system's architecture. Firstly, the RT-Process initializes its internal structures, internal variables, FIFOs, shared

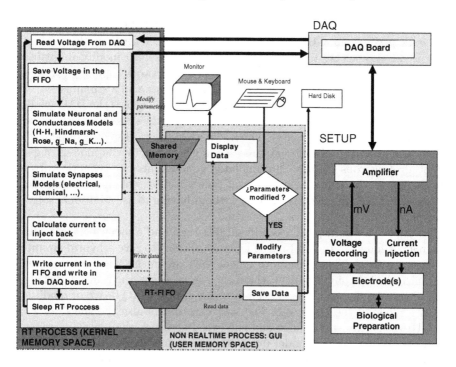

Fig. 1. Schematic software architecture. The membrane potential is recorded by electrode(s) and processed by the intracellular amplifier. The output of the amplifier is converted by the DAQ board to digital data. In the computer a RT process periodically checks the arrival of new data from the DAQ. When this happens the RT process calculates the current to inject to the neuron as the result of a model simulation. The data generated during the simulation as well as the voltage of the cell and the current are stored in a FIFO queue. Meanwhile, the GUI reads from time to time the time series stored in the FIFO to display it, and the GUI can also save it at the same time into a file. The user can interact with the program by changing online all parameters and models. In this case, the GUI informs of the changes asynchronously to the RT process through a shared memory between both processes. The RT FIFO shown in this figure refers to the data FIFO. The command FIFO is not shown for the sake of clarity

memory and the DAQ board. The shared memory contains among other information a graph representation of the ionic channel, synapse and neuron models loaded (*network topology*). After the initialization, the RT-process enters a hard RT thread which is suspended and awakened with a fixed periodicity. Every time the DAQ reports the arrival of a new voltage the selected gain and offset is adjusted and introduced in the FIFO. Secondly, the RT-process calculates the response of the models to this voltage and the selected user variables and parameters are inserted into the FIFO. Thirdly, the RT-process calculates the current to be injected to the neuron and the selected parameters and variables are inserted into the FIFO. In the last step, the current is inserted into the neuron.

The GUI reads periodically the data inserted in the FIFO by the RT-Process. The workspace of our GUI is divided in three parts. On top of the workspace there is an easy-access tool bar, that allows the user to modify in real-time the model parameters and variables, create a new network topology (an hybrid circuit), start or stop the data recording or the generation of a stimulus, etc. The rest of the workspace is splitted into two parts. One of them is the DAQ window, a multiplot where the user can monitor the biological system and the current being injected into the neuron(s). The last part of the workspace can be used to display the evolution of different time series generated by the models, e.g. the potential of neurons models, the simulated conductances or synaptic currents, etc. Therefore, the experimentalists can take advantage of the dynamic-clamp protocols and the data acquisition tool without having to use additional software or displays.

On top of the workspace, experimentalists can find a menu bar with all the features not found in the tool bar. There are options to save a hybrid circuit for posterior retrieval. This is useful to repeat the experiment. There are options to customize the DAQ input output channels, e.g. to adjust the range and the gain of the channels, change the operation frequency of the DAQ board or modify the measure mode of the DAQ, etc. The user can control also the speed at which the GUI reads the FIFO queue and the speed at which the plots are updated. The user can change the model integration step, modify the number of points drawn in the plots, and personalize the windows, or the zooming options. The windows that display the time series also have their own properties: autoscaling in the screen, trace color, gain, offset, etc.

To build hybrid circuits the program offers a wizard where users can select the topology of the network. Firstly, the user chooses the different neuron models from the library. The user also has to specify whether he/she wants to interact with elements outside the computer such as the biological neurons or other devices. Secondly, the user can choose the topology of connections between the elements selected in the previous step, indicating the kind of synapses and its number. In the last step, the user can change the default values of the state variables and parameters of the models.

4   Example Experimental Protocols

As an example of the dynamic-clamp protocols we show here a set of experiments on the mechanoreceptor touch cell (T-cell) of the leech *Hirudo medicinalis*. We followed standard preparation techniques [18]. In brief, hungry leeches were obtained from a German supplier (Zaug GmbH) and maintained in artificial pond water at 15°C in natural light. Leeches were anesthetized in cold saline for few minutes. Ganglia 7th-9th were dissected out and pinned ventral side up in a Petri dish with a Sylgard base and filled with Ringer solution containing 115 mM $NaCl$, 4 mM KCl, 1.8 mM $CaCl_2 \cdot 2H_2O$, 1.5 mM $MgCl_2 \cdot 6H_2O$, 10 mM Glucose, 4.6 mM Tris Maleate and 5.4 mM Tris base, driven to pH 7.4. Neuron T was identified from its position in the ganglia and firing characteristics. Intracellular recordings were made using a single quartz microelectrode filled with 4 M potassium acetate and pulled to resistances 40-70 MΩ (P2000, Sutter Instruments). Signals were amplified using an Axoclamp 1A amplifier and acquired with a National Instruments data acquisition card.

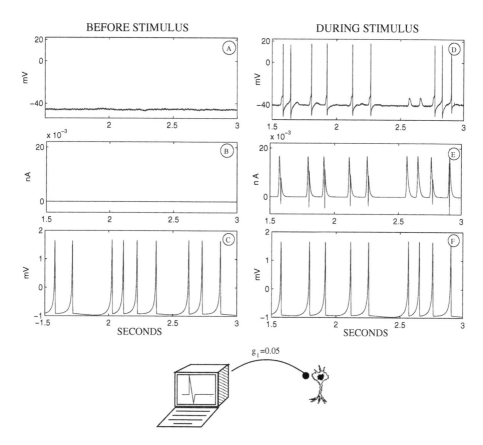

Fig. 2. Stimulation of a T-cell of the leech *Hirudo medicinalis* through a realistic neural model: Left panels show the control experiment. Panel A shows the state of T-cell without external stimuli: no current is flowing through the artificial synapse (panel B). Panel C shows the activity of the artificial neuron (HR model) in the ADC. The parameters were chosen to place the model in the spiking regime. Right panels show the activity when the artificial neuron is connected to the living cell through the graded synapse implemented in the dynamic-clamp. Panel D shows the activity of the T-cell under the stimulus. Panel E shows the artificial current injected to the T-cell through the dynamic-clamp synapse. Panel F shows the activity of the model during the stimulation period

We have built two different circuits, connecting a T-Cell through artificial synapses to a Hindmarsh-Rose model neuron [19]. The artificial chemical graded synapse we implemented is described in [8]. In this model, a unique differential equation describes the postsynaptic conductance. The current injected by this synapse is the following:

$$I_{inj}(t) = g \cdot S(t) \cdot (E_{syn} - V_{post}(t)),$$

where g is the maximum synaptic conductance, $S(t)$ is the synaptic activation variable, E_{syn} is the synaptic reversal potential and V_{post} is the postsynaptic potential. The synaptic activation is described by the following differential equation:

$$dS(t)/dt = (S_\infty(V_{pre}) - S(t))/(\tau_{syn} \cdot (1 - S_\infty(V_{pre})),$$

$$S_\infty(V_{pre}) = \tanh[(V_{pre}(t) - V_{th})/V_{slope}], \quad \text{when} \quad V_{pre} > V_{th},$$

$$\text{else } S_\infty(V_{pre}) = 0,$$

where V_{th} is the synaptic threshold voltage, τ_{syn} is the synaptic characteristic time constant, S_∞ is the steady synaptic activation, V_{pre} is the presynaptic potential and V_{slope} controls the slope of the function.

The presynaptic neural model does not need to work in the electrophysiological range of the biological neurons. In fact, the Hindmarsh-Rose model used here works in a range from -1.5 mV to 2 mV. This model is very suitable to use in hybrid circuits as it has a very rich individual dynamics and many realistic bifurcations in the behavior [7,9]. To increase the width and reduce the frequency of the model in this experiment, all equations were multiplied by 0.25.

In our first configuration, the artificial model was connected to a T-cell unidirectionally through a fast excitatory chemical synapse, as the one described above. The parameters of the synapse were the following: $g = 0.05$ μSiemens, $\tau_{syn} = 10$ ms, $E_{syn} = 0$ mV, $V_{th} = -0.5$ mV. Panel A in Fig. 2 shows the membrane resting potential of the T-cell before being connected to the model. Panel D shows the action potentials induced by the stimulation from the Hindmarsh-Rose. Every time the Hindmarsh-Rose potential overpasses -0.5 mV, which is the synaptic threshold voltage, the dynamic-clamp synapse injects current into the T-Cell depolarizing its membrane.

We repeated the experiment with a bidirectional connection between the model and the real neuron implemented through two excitatory graded synapses. The parameters of the fast synapse between the model and the neuron were the followings: $g_1 = 0.2$ μSiemens, $\tau_{1syn} = 10$ ms, $E_{1syn} = 0$ mV, $V_{1th} = -0.5$ mV. The parameters of the slow synapse between the neuron and the model were: $g_2 = 2$ μSiemens, $\tau_{2syn} = 100$ ms, $E_{2syn} = 0$ mV, $V_{2th} = 0$ mV. The connection of the neuron with the model is stronger because the neuron spiking rate is much slower. As can be seen in Fig. 3, the model induces bursts of action potentials in the postsynaptic neuron. When the neuron fires, current is injected back to the model. As can be seen in the figure, this produces a change in the model regime, from tonic spiking to an irregular bursting mode.

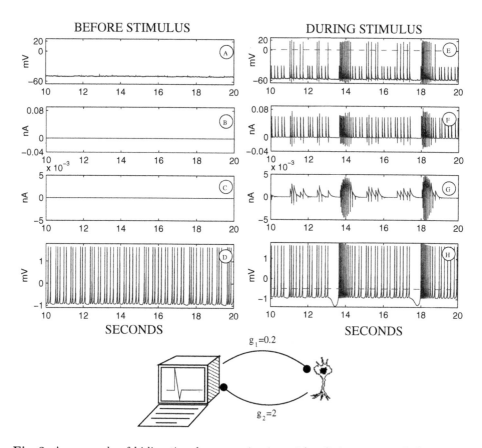

Fig. 3. An example of bidirectional communication with a living neuron. Left panels in the graph show the control experiment. Panel A shows the state of T-cell without external stimuli: no current through the artificial synapses (panels B and C). Panel D shows the voltage value that exhibits the model neuron implemented in the dynamic-clamp. The parameters were chosen to place the model in the spiking regime. Right panels show the activity when the artificial neuron is connected to the T-cell through the graded synapses. In this case the interaction is bidirectional. Panel F shows the artificial current injected into the T-cell. We can see that due to this interaction the T-cell begins to fire bursts of action potentials (see panel E). Panel G shows the artificial current injected to the model neuron through the other synapse. Note in panel H the robust bursting behavior in the artificial neuron as reaction to this bidirectional interaction

5   Discussion

Advanced dynamic-clamp software can contribute to a realistic stimulation of neurons by allowing pattern clamp protocols, the construction of hybrid circuits of interacting artificial and living cells, and the implementation of artificial synaptic and intracellular plasticity. Existing dynamic-clamp software is usually too oriented to a particular experimental setup or neural system, difficult to use, lacking extensive libraries of ionic channels, synapses and neuron models, limited in the set of parameters that can be changed on real time and without a customizable graphical interface. We are developing an advanced dynamic-clamp software that tries to solve these problems and provides additional features. In the near future, this software will have the capability to control microinjectors of neuromodulators and neurotransmitters, so that neurons can be stimulated in a less invasive manner than electrode injection. The kernel of this software will also be adapted to control more advanced protocols that involve multi-photon microscopy and laser stimulation. Dynamic-clamp has provided a long list of successful experiments in the last ten year. Hopefully its more remarkable results are yet to come.

Acknowledgments. This work was supported by Fundación BBVA and MEC (BFI 2003-07276, TIN 2004-04363-C03-03).

References

1. Robinson HPC. 1991. Kinetics of synaptic conductances in mammalian central neurons. Neurosci. Res. 16:VI.
2. Robinson HPC, Kawai N. 1993. Injection of digitally synthesized synaptic conductance transients to measure the integrative properties of neurons. J. Neurosci. Methods 49: 157-1-65.
3. Sharp AA, O'Neal MB, Abbott LF and Marder E. 1993. Dynamic clamp: computer-generated conductances in real neurons. J. Neurophysiol. 69. 992-995.
4. Ma M and Koester J. 1996. The role of potassium currents in frequency-dependent spike broadening in Aplysia R20 neurons: a dynamic clamp analysis. J. Neuroscience. 16: 4089–4101.
5. Kiehn O, Kjaerulff O, Tresch MC, Harris-Warrick. 2000. Contributions of intrinsic motor neurons properties to the production of rhythmic motor output in the mammalian spinal cord. Brain Res. Bull. 53: 649–659.
6. Chance FS, Abbott LF and Reyes AD. 2002. Gain modulation from background synaptic input. Neuron 35: 773–782.
7. A. Szucs, P. Varona, A.R. Volkovskii, H. D. I. Abarbanel, M.I. Rabinovich, A.I. Selverston. 2000. Interacting Biological and Electronic Neurons Generate Realistic Oscillatory Rhythms. NeuroReport, 11 (3): 563–569.
8. Pinto R.D. et al (2001). Extended Dynamic Clamp: controlling up to four neurons using a single desktop computer and interface. J. Neuroscience Methods 108: 39–48.
9. R.D. Pinto, P. Varona, A.R. Volkovskii, A. Szucs, H.D.I. Abarbanel and M.I. Rabinovich. 2000. Synchronous behavior of two coupled electronic neurons. Physical Review E, 62(2): 2644-56.
10. Nowotny T, Zhigulin VP, Selverston AI, Abarbanel HD, Rabinovich MI. 2003. Enhancement of synchronization in a hybrid neural circuit by spike-timing dependent plasticity. J Neurosci. 23(30): 9776–85.
11. Szucs A, Rozsa KS, Salanki J. 1998. Presynaptic modulation of Lymnaea neurons evoked by computer-generated spike trains. Neuroreport 9(12): 2737–42.
12. Prinz AA, Abbott LF and Marder E. 2004. The dynamic clamp comes of age. Trends in Neurosciences 27: 218–224.

13. LeMasson G, LeMasson S, Moulins M. 1995. From conductances to neural network properties: analysis of simple circuits using the hybrid network method. Prog. Biophys. Molec. Biol. 64: 201–220.
14. Sorensen M, DeWeerth S, Cymbalyuk G, Calabrese RL. 2004. Using a Hybrid Neural System to Reveal Regulation of Neuronal Network Activity by an Intrinsic Current. Journal of Neuroscience 24: 5427–5438.
15. Dorval AD, Christini DJ and White JA. 2001. Real-time linux dynamic clamp: a fast and flexible way to construct virtual ion channels in living cells. Annals of Biomedical Engineering 29: 897–907.
16. Raikov I, Preyer A, Butera RJ. 2004. MRCI: a flexible real-time dynamic clamp system for electrophysiology experiments. Journal of Neuroscience Methods 132: 109–123.
17. Kullmann PH, Wheeler DW, Beacom J, Horn JP.2004. Implementation of a fast 16-Bit dynamic clamp using LabVIEW-RT. J Neurophysiol. 2004 91(1): 542–54.
18. Muller K, Nicholls J, and Stent G. 1981. Neurobiology of the Leech. Cold Spring Harbor Laboratory: New York.
19. Hindmarsh JL and Rose RM, 1984. A model of neuronal bursting using three coupled first order differential equations. Philos, Trans. R Soc. London. B221: 87–102.

Interacting Slow and Fast Dynamics in Precise Spiking-Bursting Neurons

Fabiano Baroni[1], Joaquin J. Torres[2], and Pablo Varona[1]

[1] Grupo de Neurocomputación Biológica (GNB),
Dpto. de Ingeniería Informática,
Escuela Politécnica Superior,
Universidad Autónoma de Madrid,
28049 Madrid, Spain
{fabiano.baroni, pablo.varona}@uam.es

[2] Departamento de Electromagnetismo y Física de la Materia,
Universidad de Granada, 18071 Granada, Spain
Institute Carlos I for Theoretical and Computational Physics,
Universidad de Granada 18071 Granada, Spain
jtorres@onsager.ugr.es

Abstract. We have explored the role of the interaction of slow and fast intracellular dynamics in generating precise spiking-bursting activity in a model of the heartbeat central pattern generator of the leech. In particular we study the effect of calcium-dependent currents on the neural signatures generated in the circuit. These neural signatures are cell-specific interspike intervals in the spiking-bursting activity of each neuron. Our results show that the slow dynamics of intracelullar calcium concentration can regulate the precision and shape of the neural signatures.

1 Introduction

There are many different intracellular dynamics that directly or indirectly influence the electrical activity of neurons [1, 2, 3]. These different dynamics can act in several timescales. The combination of ionic currents with different kinetics and time constants are responsible, for example, of the characteristic spiking-bursting activity of many central pattern generator neurons. Some of these channels are fast and affect the action potential generating mechanisms of the cells. Other channels are slow and regulate the burst depolarization waves. Calcium dynamics is often considered to affect mainly the slow depolarization waves. However, the interaction of fast and slow dynamics gives rise to phenomena on both time scales.

A Central Pattern Generator (CPG) is a neural network that, acting alone or together with other CPGs, drives a motor movement that must be repeated in time. CPG neurons are typically spiking-bursting cells [4, 5, 6, 7] that generate a characteristic rhythmical pattern of activity. The signal produced by a CPG consists of a sequence of rhythmic bursts of action potentials. Typically, it is

J. Mira and J.R. Álvarez (Eds.): IWINAC 2005, LNCS 3561, pp. 106–115, 2005.

thought that the intraburst spike structure is not important to characterize the CPG behavior and its communication with other external neurons and systems. However, recent studies using in vitro preparations [8] have revealed that each neuron of the pyloric CPG of the lobster has a specific individual neural signature that coexists with its characteristic slow rhythm. These signatures consist of a cell-specific interspike interval (ISI) distribution. The origin of these signatures has been studied using model networks [9, 10, 11]. In these models the synaptic configuration of the network is the main factor shaping the signatures, although the intrinsic dynamics of each cell can also contribute to the specific ISI distributions.

In this paper we assess the influence of slow calcium dynamics on the generation and precision of neuronal signatures. We study this using a model of a well-known CPG: the circuit pacing the heartbeat of the medicinal leech, *Hirudo medicinalis*. The main neurons of this circuit are modeled with a previously developed Hodgkin-Huxley type formalism in which we include dependence upon calcium concentration in an outward potassium current.

2   CPG Model

The rhythmic activity of the leech heartbeat CPG originates from two segmental oscillators located in the third and fourth midbody ganglia, each one formed by a couple of reciprocally inhibiting heart interneurons, called oscillator interneurons [12]. The interneurons located in the first and second ganglion act to coordinate the activity of the two segmental oscillators, and are then referred to as coordinating interneurons [13, 14, 15]. The system exhibits a symmetry between the third and the fourth ganglion, so that only a couple of heart oscillator interneurons are depicted in the schematic representation of the circuit shown in Fig. 1A. Leech heart interneurons are commonly referred to with a nomenclature indicating the emibody and the ganglion where they are located, and we will use this nomenclature throughout the paper.

We depart from a conductance-based model developed by the Calabrese group [16]. The general equation that describes the membrane potential of each model neuron is the following:

$$C_m \frac{dV}{dt} = -(I_l + I_{SynS} + I_{SynG} + I_{ion}) \tag{1}$$

where C_m is the total membrane capacitance, I_l is a leakage current, I_{SynS} is the spike-mediated synaptic current, I_{SynG} is the graded synaptic current, and

$$I_{ion} = I_{Na} + I_P + I_{CaF} + I_{CaS} + I_h + I_{K1} + I_{K2} + I_{KA} \tag{2}$$

is the total ionic current for oscillator heart interneurons. For the coordinating heart interneurons:

$$I_{ion} = I_{Na} + I_{K1} + I_{K2} \tag{3}$$

Five inward currents are included in the oscillator interneuron model: a fast Na^+ current I_{Na}, a persistent Na^+ current I_P, a fast, low-threshold Ca^{2+} current I_{CaF}, a slow, low-threshold Ca^{2+} current I_{CaS}, and a hyperpolarization-activated cation current I_h [17, 18, 19, 20]. Three outward currents are also included: a delayed rectifier like K^+ current I_{K1}, a persistent K^+ current I_{K2}, and a fast K^+ transient current I_{KA}. Each of these currents have Hodgkin-Huxley type kinetics. Coordinating model interneurons implement only a subset of the ionic currents described for oscillator heart interneurons. For a complete description of the model see [16].

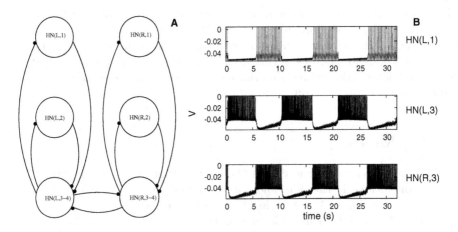

Fig. 1. A schematic representation of the circuit and electrical activity of the CPG model. A: the timing network contains four pairs of bilaterally symmetric interneurons that have cell bodies in the first four midbody ganglia (G1 to G4). The representations of oscillator heart interneurons in the 3rd and 4th ganglia are combined. Open circles represent cell bodies and lines ending in small filled circles represent inhibitory synapses. B: from top to bottom: voltage traces of heart interneurons HN(L,1), HN(L,3), HN(R,3)

This model produces rhythmic oscillations (Fig. 1B) with a mean period of $10.4 \pm 0.2s$. Several studies have pointed out the role of presynaptic calcium concentration in graded [21] and spike-mediated [22] synaptic transmission among oscillator interneurons coupled in a half center oscillator, while no existing model takes into account homeostatic regulation of ionic conductances based upon calcium concentration in this circuit. Nevertheless, calcium dependent potassium conductances have been described in several other cells in the leech nervous system, such as Retzius [23], anterior pagoda [24], and anterior lateral giant [25], and might be ubiquitous in invertebrate neurons. The modifications that we perform on the original model [16] (described below) are meant to incorporate calcium dependence in the activation variable of the persistent (non-inactivating) potassium current I_{K2}. We chose this among the three potassium currents described in the original model because its relatively slow kinetics makes it vary on a time scale closer to the burst time scale than to the individual spike time scale, albeit incorporation of the same kind of calcium dependent regulation on the kinetics of I_{K1} activation variable leads to similar results (data not shown). Nevertheless this study is meant to be more a prove of principle of what might be the influence of slow homeostatic regulation of ionic conductances on burst temporal structure than a biophysically realistic description of the subcellular mechanism underlying this regulation. We implemented calcium dependence upon the activation variable of I_{K2} as a calcium dependent voltage shift V_{shift} in the steady state and time constant of the activation function:

$$m_{\infty,K2} = \frac{1}{1 + \exp(-83(V + V_{shift} + 0.02))} \tag{4}$$

$$\tau_{m,K2} = 0.057 + \frac{0.043}{1 + \exp(200(V + V_{shift} + 0.035))} \tag{5}$$

where V_{shift} is calculated according to the equation:

$$V_{shift} = k \cdot \ln\left(\frac{[Ca^{2+}]_{eff}}{[Ca^{2+}]_{eff_0}}\right) \tag{6}$$

We chose $k = 0.002V$ and $[Ca]_{eff_0} = 1.36 \cdot 10^{-12}$. An additional equation describes the dynamics of intracellular effective calcium concentration,

$$\frac{d[Ca^{2+}]_{eff}}{dt} = \frac{I_{Ca} - B[Ca^{2+}]_{eff}}{\tau_C} \tag{7}$$

where

$$I_{Ca} = \max(0, -I_{CaF} - I_{CaS}) \tag{8}$$

In Eq. 7 parameter B is a buffering rate constant and τ_C is a lumped time constant describing both calcium diffusion and the kinetics of binding/unbinding of calcium to receptors modulating the gating of I_{K2} ionic channels. It is important to note that $[Ca^{2+}]_{eff}$ is not necessarily meant to reproduce the actual calcium concentration in a concrete intracellular region, even if intracellular calcium concentration in the proximity of the cell membrane is surely rate limiting for $[Ca^{2+}]_{eff}$. This model will from now on be referred to as the calcium model, as opposed to the original model described in [16]. Note that the implementation of the coordinating interneurons and of the synaptic coupling is the same in the two models. The calcium model also produces a robust oscillatory rhythm (Fig. 2A) with a period of $6.16 \pm 0.07s$.

The models were implemented in C and equations were integrated with a variable step Runge-Kutta 6(5) integration routine for at least 950 burst in each cell. The first 100 seconds of simulated data were not included in the analysis. Spike times were collected and the corresponding ISIs were calculated as difference of successive spike times. ISI distributions in all cases were clearly bimodal, so it was possible to unambiguously define a burst threshold. Cycle period was calculated considering the semi-sum of the first and last spike time of each burst.

Two different forms of visualizing and quantifying burst structures were used: ISI return maps and ISI versus ISI index plots. ISI return maps were constructed by plotting each ISI against the previous ISI. For neurons with a robust bursting behavior as the one considered in this study, is it possible to uniquely determine an intraburst region in the ISI return map (corresponding to triplets of spikes belonging to the same burst) and two interburst regions (corresponding to triplets belonging to different bursts). ISI versus ISI index plots represent the mean and standard deviation of the ISI with the same index inside a burst, plotted against the ISI index. Indexes with less than three ISIs were not included in the analysis. Due to the symmetry of the model only results from one neuron of each kind are presented.

3   Results

Figure 2B shows the membrane potential of an oscillator interneuron together with the time course of $[Ca^{2+}]_{eff}$ and V_{shift}. As the cell depolarizes, Ca^{2+} enters through voltage dependent calcium channels and $[Ca^{2+}]_{eff}$ correspondingly rises until approximately one third of the whole duration of the burst. As $[Ca^{2+}]_{eff}$ rises, V_{shift} produces a shift towards a more hyperpolarized potential of the steady state and time constant curves of the I_{K2} activation function, resulting in a greater activation of this current. This corresponds to a $[Ca^{2+}]_{eff}$ dependent increase in the total repolarizing current, enhancing firing frequency adaptation along a burst (as we will discuss in Fig. 4C) and eventually causing a precocious escape from inhibition of the contralateral neuron and an overall increase in cycle frequency. This is consistent with the general intuition of calcium dynamics as a mechanism providing a delayed negative feedback to the dynamics of neuron models.

The calcium model generates bursts with a more precise temporal structure than the original model. Figure 3 shows the ISI return maps for an oscillator

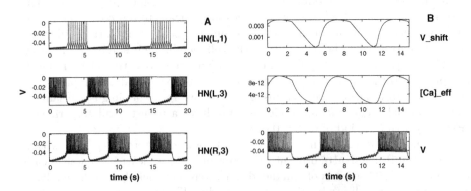

Fig. 2. Dynamics of the calcium model. Panel A, from top to bottom: voltage traces of heart interneurons HN(L,1), HN(L,3), HN(R,3). Panel B, from top to bottom: $[Ca^{2+}]_{eff}$, V_{shift} and voltage time course for oscillator interneuron HN(L,3)

and a coordinating interneuron in the two models. It is remarkable how the sparse triplets present in the intraburst region of the ISI return map of the coordinating interneuron HN(L,1) in the original model disappear in the calcium model, for which all intraburst triplets fall in a small interval. More detailed analysis revealed that those sparse triplets in the original model all share the feature of being the last triplets of a burst. The interburst regions for the two kinds of heart interneurons under study are both fairly more precise in the calcium model than in the original model (only interburst regions for oscillator interneurons are shown). This is consistent with the lower standard deviation in the cycle period for the calcium model.

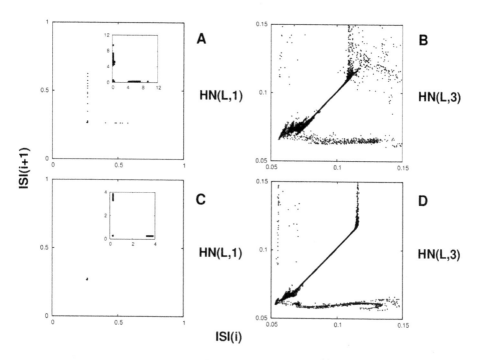

Fig. 3. Intraburst region of ISI return maps for the coordinating interneuron HN(L,1) (left) and the oscillator interneuron HN(L,3) (right) in the original (top) and calcium (bottom) models. Insets in A,C show full ISI return maps for coordinating interneuron HN(L,1) in the two models

The ISI return maps are useful to identify the correlation between successive ISIs. However, they do not provide detailed information about the distribution of each ISI index inside the burst. Thus, we also plot the mean and standard deviation of the ISIs against the ISI index inside a burst (Fig. 4A,C).

Besides the shorter cycle period in the calcium model, reflected in a smaller number of spikes per burst, two main features qualitatively differentiate the burst structure in the two models: the precision of the burst structure and the

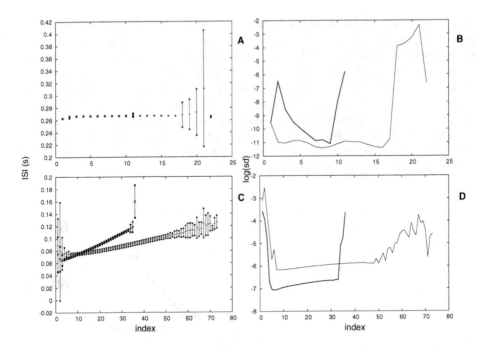

Fig. 4. Left panels: mean and standard deviation of ISI of the same index plotted against the index of the ISI inside a burst for HN(L,1) (top) and HN(L,3) (bottom) for the two models. Original model: mean ISI plotted with bars, errorbars ending with filled circles. Calcium model: mean ISI plotted with crosses, errorbars ending with filled squares. Right panels: standard deviation of ISI of the same index plotted in logarithmic scale against the index of ISI inside a burst for HN(L,1) (top) and HN(L,3) (bottom) for the two models. Original model: thin line. Calcium model: thick line

profile of the mean ISI along a burst. The coordinating interneurons do not differ in the two models, so that any change in the dynamics of these neurons in the two models is attributable only to network activity and not to intrinsic neuronal properties. Nevertheless, while the mean ISI for each index inside a burst is almost unchanged in the two models, ISI precision is slightly corrupted in the calcium model over all indexes except for the last indexes of the burst, for which the calcium model achieves enhanced precision (Fig. 4B). This is consistent with the previous observation that all intraburst triplets in the ISI return map for HN(L,1) in the calcium model are included in a small interval. Oscillator interneurons are the neurons where calcium dynamics was effectively added, so that any change in the dynamics of these neurons in the two models is attributable to both network activity and intrinsic neuronal properties. ISI precision for these neurons in the calcium model is greatly enhanced with respect to the original model for all possible ISI indexes inside a burst, except for the last ISI ending a burst. This improvement in ISI precision is particularly evident in Fig. 4D, where standard deviation of ISI with a given index is plotted against ISI index inside a burst in logarithmic scale. Note that there is almost one order of magnitude of improved precision with respect to the original model. The mean of ISI with a given index increases along a burst with a higher slope in the calcium model than in the original model, reflecting the added spike time adaptation mechanism provided by the dependence of non-inactivating potassium channels upon calcium concentration.

4   Discussion

Bursts are traditionally considered as unitary events. The intraburst spike distribution of CPG neurons has not been analyzed in great detail since, typically, it is thought that the slow wave dynamics is the main factor shaping the rhythmic behavior of the system. However, several recent experimental and modeling results indicate that the temporal structure of the bursts can be important for CPG neurons [8, 9, 10, 11]. In particular, CPG cells could use the specific temporal structure of their fast dynamics, in addition to the phase and frequency of the slow wave, as an information encoding mechanism. In this study we have demonstrated how the modulation of an outward current by intracellular calcium can provide a model CPG with enhanced precision in its burst structure.

An interesting question about oscillator heart interneurons is whether they can present endogenous spiking-bursting activity when they are pharmacologically isolated. It has been reported that this behavior can be achieved in the original model neuron only in a narrow region of the parameter space (E_{leak}, g_{leak}), i.e. the reversal potential and maximal conductance of the leak current, respectively [28]. Endogenous bursting has remarkable consequences on the activity of the timing network, ensuring robust bursting characteristics such as period, phase, and duty cycles in face of weakening of mutual inhibition or random or imposed changes in membrane parameters [28]. In fact, robust pacing rhythm with little sensitivity to parameter change has been observed in the living system but not in the original model.

A key question in this context is whether the consideration of an intracellular calcium dynamics modulating the conductances of outward currents can induce spiking-bursting activity in the single neuron model for a broader range of parameters, thus ensuring robust dynamics in the model network. Preliminary results in our neuron model show that coupling of intracellular calcium dynamics with the activation of I_{K2} potassium current can induce robust spiking-bursting behavior similar to that reported in the experiments. Nevertheless it has been reported that the addition of a FMRFamide-activated potassium current [27] with Hodgkin-Huxley like kinetics also produces an expansion of the bursting region for the single neuron model. Thus, future analysis will aim at assessing if the enhanced robustness observed in the calcium model can be attributable only to a net increase in total outward currents during the burst phase of a neuron's activity or if the modulation of outward currents by calcium concentration along a burst plays a key role in ensuring correct bursting activity over a wide range of parameters.

Further work will assess how the improved precision of the spiking-bursting activity depends on parameter values and on the specific implementation of calcium dependence on potassium channel conductances, how it relates to single neuron's activity, and to which extent it is functionally relevant to the dynamics of the network. Comparison with experimental data will test the physiological plausibility of our modeling choices and guide future research.

Acknowledgments. This work was supported by Fundación BBVA and Spanish MEC (BFI-2003-07276). J.J. Torres acknowledges financial support from MCyT and FEDER (project No. BFM2001- 2841 and *Ramón y Cajal* contract).

References

1. Berridge M.J. 1998. Neuronal calcium signaling. Neuron, 21: 13–26.
2. Varona P., Torres J.J., Abarbanel H.D.I., Rabinovich M.I., Elson R.C. 2001a. Dynamics of two electrically coupled chaotic neurons: Experimental observations and model analysis. Biological Cybernetics, 84 (2): 91–101.
3. Varona P., Torres J.J., Huerta R., Abarbanel H.D.I., Rabinovich M.I. 2001b. Regularization mechanisms of spiking-bursting neurons. Neural Networks, 14: 865–875.
4. Hartline, D. K. and Maynard, D. M. 1976. Motor patterns in the stomatogastric ganglion of the lobster panulirus argus. J Exp Biol, 62(2): 405–420.
5. Russell, D. F. and Hartline, D. K. 1978. Bursting neural networks: a reexamination. Science, 200(4340): 453–456.
6. Marder E. and Calabrese R.L. 1996. Principles of rhythmic motor pattern generation. Physiol Rev, 76: 687–717.
7. Selverston, A. I., Elson, R. C., Rabinovich, M. I., Huerta, R., and Abarbanel, H. D. I. 1998. Basic principles for generating motor output in the stomatogastric ganglion. Ann. N.Y. Acad. Sci, 860(1): 35–50.
8. Szucs, A., Pinto, R. D., Rabinovich, M. I., Abarbanel, H. D. I., and Selverston, A. I. 2003. Synaptic modulation of the interspike interval signatures of bursting pyloric neurons. J Neurophysiol, 89: 1363–1377.
9. Latorre, R., Rodriguez, F. B., and Varona, P. 2002. Characterization of triphasic rhythms in central pattern generators (i): Interspike interval analysis. Lect. Notes Comput. Sc., 2415: 160–166.
10. Rodriguez, F. B., Latorre, R., and Varona, P. 2002. Characterization of triphasic rhythms in central pattern generators (ii): Burst information analysis. Lect. Notes Comput. Sc., 2415: 167–173.
11. Latorre, R., Rodriguez, F. B., and Varona, P. 2004. Effect of individual spiking activity on rhythm generation of central pattern generators. Neurocomputing, 58-60: 535–540.
12. Peterson EL. 1983a Generation and coordination of heartbeat timing oscillation in the medicinal leech. I. Oscillation in isolated ganglia. J. Neurophysiol. 49: 611–626.
13. Peterson EL. 1983b Generation and coordination of heartbeat timing oscillation in the medicinal leech. II. Intersegmental coordination. J. Neurophysiol. 49: 627–638.
14. Hill AA, Masino MA, Calabrese RL. 2002. Model of intersegmental coordination in the leech heartbeat neuronal network.J Neurophysiol. 87(3):1586–602.
15. Jezzini SH, Hill AAV, Kuzyk P, Calabrese RL. 2004. Detailed model of intersegmental coordination in the timing network of the leech heartbeat central pattern generator. J. Neurophysiol. 91: 958–977.

16. Hill A.A.V., Lu J., Masino M.A., Olsen O.H., Calabrese R.L. 2001. A model of a segmental oscillator in the leech heartbeat neuronal network. Journal of Computational Neuroscience 10: 281–302.
17. Angstadt JD, Calabrese RL. 1989. A hyperpolarization-activated inward current in heart interneurons of the medicinal leech. J. Neurosci. 9: 2846–2857.
18. Angstadt JD, Calabrese RL. 1991. Calcium currents and graded synaptic transmission between heart interneurons of the leech. J. Neurosci. 11: 746–759.
19. Olsen OH, Calabrese RL. 1996. Activation of intrinsic and synaptic currents in leech heart interneurons by realistic waveforms. J. Neurosci. 16: 4958–4970.
20. Opdyke CA, Calabrese RL. 1994. A persistent sodium current contributes to oscillatory activity in heart interneurons of the medicinal leech. J. Comp. Physiol. A 175: 781–789.
21. Ivanov AI, Calabrese RL. 2000. Intracellular Ca2+ dynamics during spontaneous and evoked activity of leech heart interneurons: low-threshold Ca currents and graded synaptic transmission. J Neurosci. 20(13): 4930–43.
22. Ivanov AI, Calabrese RL. 2003. Modulation of spike-mediated synaptic transmission by presynaptic background Ca2+ in leech heart interneurons. J Neurosci. 23(4): 1206–18.
23. Beck A, Lohr C, Nett W, Deitmer JW. 2001. Bursting activity in leech Retzius neurons induced by low external chloride. Pflugers Arch. 442(2): 263–72.
24. Wessel R, Kristan WB Jr, Kleinfeld D. 1999. Dendritic Ca(2+)-activated K(+) conductances regulate electrical signal propagation in an invertebrate neuron. J Neurosci. 19(19): 8319–26.
25. Johansen J, Yang J, Kleinhaus AL. 1987. Voltage-clamp analysis of the ionic conductances in a leech neuron with a purely calcium-dependent action potential. J Neurophysiol. 58(6): 1468–84.
26. Calabrese RL, Nadim F, and Olsen OH. 1995. Heartbeat control in the medicinal leech: a model system for understanding the origin, coordination, and modulation of rhythmic motor patterns. J Neurobiol 27: 390–402.
27. Nadim F, Calabrese RL. 1997. A slow outward current activated by FMRFamide in heart interneurons of the medicinal leech. J. Neurosci. 17:4461–4472.
28. Cymbalyuk GS, Gaudry Q, Masino MA, and Calabrese RL. 2002. Bursting in leech heart interneurons: cell-autonomous and network-based mechanisms. J. Neurosci. 22(24): 10580–10592.

An Integral Model of Spreading Depression: From Neuron Channels to Field Potentials

Ioulia Makarova[1,2], Iria R. Cepeda[1,2], Fivos Panetsos[2], and Oscar Herreras[1]

[1] Dept. Investigaciøn Hospital Ramón y Cajal, 28034 Madrid, Spain
[2] Dept. Matemática Aplicada, Fac. Biología,
Universidad Complutense de Madrid, 28040 Madrid, Spain
oscar.herreras@hrc.es

Abstract. Spreading depression is an inactivating wave of massive cell depolarization that moves slowly through the brain associated to large negative DC potentials. Its biophysical nature is unclear. Based on our recent finding that longitudinal depolarizing gradients arise on individual neurons affected by SD, we have initiated the construction of an integral model to reproduce large macroscopic potentials in the extracellular space from channel activity on individual neurons. The model includes a simulation of the extracellular space and features V_o feedback to calculate transmembrane currents and potentials.

1   Introduction

Spreading depression (SD) [1] is a propagating mass event that spreads through nervous tissue associated to large ion changes and negative DC potentials (V_o). The biophysical nature of SD is poorly understood [2] The large negative V_o signal is considered an extracellular index of cell depolarization. Earlier hypotheses attributed the negative V_o to sinks of current created during the washout of elevated interstitial potassium into glial cells [3] but we ruled out a major contribution of this mechanism [4-5]. On the other side, neurons have rarely been regarded as a likely candidate, since they become silent during SD and recordings from individual cells show near complete depolarization, extreme shunt of membrane input resistance (R_{in}), and loss of electrical responsiveness. Although the inactivated electrical status of somata was presumed for the entire somatodendritic membranes, direct confirmation is missing. In fact, core conductor theory predicts that homogeneously depolarized cells do not generate currents across their membranes. We earlier described sharp V_o and interstitial ion gradients along the pyramidal cell axis in the hippocampal CA1 [6] that suggest spatially heterogeneous events. In a recent subcellular study of membrane excitability [7] during SD in the CA1 region we demonstrated that depolarization and membrane shunt is only restricted to discrete subregions of the dendritic tree matching the negative V_o. To bring into harmony the macroscopic and unitary results, we developed a biophysical model to study how subcellular changes are causally related to the macroscopic variables during SD. The model is designed to render macroscopic field potentials and extracellular ion concentrations

J. Mira and J.R. Álvarez (Eds.): IWINAC 2005, LNCS 3561, pp. 116–122, 2005.

from the compartmental transmembrane currents in a realistic CA1 ensemble architecture within a volume conductor, and features V_o feedback to calculate true V_m.

2   The CA1 Aggregate Biophysical Model for the Calculation of the Vo During SD Conditions

The compartmental model reproduced in detail the average pyramidal cell morphology. This was simulated using 265 compartments, distributed in soma, apical and basal dendritic trees and axon (a 2-D projection of the model neuron is shown in Fig. 2). Compartment length was always >0.01 and <0.2 λ. A detailed description of the cell morphology can be found in the http://navier.ucsd.edu/ ca1ps address. Total effective area of the neuron was 66,800 μm^2 (including spine area). The electrotonic parameters for soma and dendrites were R_m=70,000 Ω/cm^2, R_i=75$\Omega \cdot cm$ and C_m=0.75 $\mu F/cm^2$, while R_m was 100 and 0.5 Ω/cm^2 for non-myelinated and myelinated axon compartments, respectively. The later had a C_m of 0.04 $\mu F/cm^2$. The input resistance measured at the soma was 140 $M\Omega$ and τ was 25 ms. The ion channels, their kinetics and distribution along the morphology of the model neuron are as described in Ibarz and Herreras [8]. Conductance variables were described with Hodgkin-Huxley type formalism. The details can be obtained from http://navier.ucsd.edu/ca1ps. We used a generic Na^+/K^+ mixed conductance to reproduce the increased membrane conductance during SD (termed here I_{SD}). The component Na^+ and K^+ conductances were modeled as separate voltage-independent non-inactivating currents. Their spatial distribution and maximum densities were optimized to reproduce the experimental spatial profiles of R_{in} during SD. For the specific purposes of modeling SD conditions we aimed the reproduction of the spatial profile of V_m and Rin along the entire neuron morphology during its main phase. This was defined by the near-zero steady depolarization and R_{in} below 10 percent of control. Only some of the extraneuronal variables are relevant, such as the spatial distribution of interstitial ion concentrations and the V_o. This was calculated as previously [9] and fed back into the calculation of ionic currents. The equilibrium potential for each ionic subspecies was calculated directly from their varying concentrations in the extracellular space. See http://navier.ucsd.edu/depres for details on channel kinetics and spatial distribution. We used here a best 1:4 Na:K ratio selected from a screening of 1:10 to 10:1. In control, E_{Na} and E_K were +50 and -90 mV, respectively, while SD-affected compartments had +10 and -20 mV, corresponding to $[Na^+]_o$ and $[K^+]_o$ of 80 mM and 32 mM, and $[Na^+]_i$ and $[K^+]_i$ of 40 mM and 92 mM, respectively. Except for Ca^{++}, we used fixed intracellular concentrations.

3   Experimental Recordings

Anesthetized female rats were used. Surgery and stereotaxic procedures were as previously described [6]. Glass micropipettes filled with the appropriate electrolytes were used for intra and extracellular recordings (DC and evoked potentials), and concentric bipolar electrodes used for electrical stimulation. SD waves were elicited by ejection of KCl microdrops into either the apical dendritic tree or the basal tree.

4   Results

Typically, SD waves run through the CA1 layer affecting the entire somato-dendritic extent of pyramidal cells except for the distal most parts. During its passage, a large negative DC signal arises, that is taken as an extracellular correlate of massive cell depolarization. These negative potentials, however, have a peculiar spatiotemporal heterogeneous pattern, lasting much longer in the apical than the somatobasal dendritic band (Fig. 1,B). Intracellularly, a similar heterogeneous distribution of depolarization is recorded, so that only the subcellular elements embedded in the extracellular negative area reached a near-zero potential (Fig. 1,A). The result is an internal gradient of depolarization from SD-affected subcellular regions down to rest in far distal dendrites. During the SD span all evoked activity is abolished (Fig. 1,B), suggesting the total inactivation of the whole CA1 pyramidal cell. However, when SD was forced to initiate in the basal tree, so that the apical tree was unaffected for some time before the SD finally invaded, evoked activity can still be recorded [7,10]. These re-

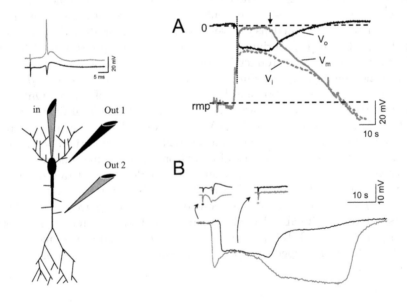

Fig. 1. Experimental recordings showing the major features of an SD wave. A: intra and extracellular recordings at the soma level. B: extracellular records at the soma and apical levels. Note that evoked activity is abolished during SD

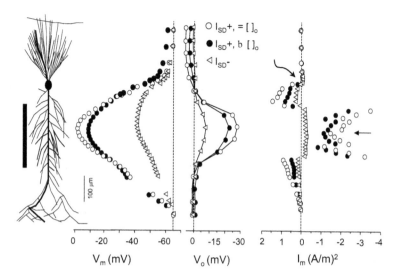

Fig. 2. Model of SD in the apical tree. A discrete band of dendrites is set on SD conditions. A comparison is made between a model neuron using the normal battery of V- dependent channels (triangles) and with an additional Na/K mixed current ($I_{SD}+$), with fixed (open circles) or free (black circles) ion concentrations in the extracellular space

sults indicate that the basal and apical trees can be inactivated independently from each other, and that the non-seized dendritic tree must maintain a Vm close enough to resting as to develop normal electrogenesis. We used the experimentally recorded heterogeneous depolarization as the fundamental premise of our model of SD. The first step was to check whether the standard battery of V-dependent channels alone was enough to reproduce the large negative potential when SD conditions were set for a strip of dendrites (Fig. 2). By setting the ENa and EK at SD values for the selected compartments a mild stationary depolarization of (20 mV was achieved, but Vo barely reached a few milivolts, in correspondence with a small inward current. Besides, the R_{in} dropped to about 25 percent, far from the undetectable values in actual SD. Therefore, an additional conductance is required. We used a V-independent Na/K mixed conductance (I_{SD}) and studied the density required to cause the near-zero potential and the R_{in} experimental drop. Figure 2 shows the fittings for two different approaches, using fixed ion equilibrium potentials (open circles) or allowing ion to change freely in the simulated extracellular space to equilibrium (filled circles). In both cases, near-zero potential was achieved matching the I_{SD} implemented area and a strong 30 mV negative potential was created in the extracellular space. This DC potential was caused by a stationary sink (inward current) of (2-4 A/m^2. The sink had a concave spatial waveform due to central cancellation of inward and outward currents (arrow in Fig. 2). Net outward currents develop at the flanks of the central sink. In the basal direction, they stopped at the soma (curved arrow), meaning that almost no current can be drawn from basal dendrites and the axon because the strong resistance and impedance mismatch at the soma junction.Figure 3 shows the share of currents through the different channels for the same three approaches in Fig. 2. Also, a comparison with an homogeneous distribution of I_{SD} along the entire anatomy allows the study of the effect of zonal versus homogeneous activation. When no I_{SD} is present, the small inward current is carried through K^+ channels, due to the strong change of their E_K. The use of SD channels enabled the necessary carrier, while Na and Ca conductances become inactivated. However, when depolarization is homogeneous, I_m was negligible. If I_{SD} is present, most current enters through it and leaves across all other K channels.Simultaneous records in the apical tree and the soma layer in actual experiments show that the beginning of negative signals in one region is typically mirrored by positive humps on other subregions already under SD (Fig. 4,A). This mirror behavior was reproduced in a simulation in which a somatic SD was initiated over an earlier apical one (arrow in Fig. 4,B). If the source-sink dipoles affecting only some parts of individual neurons are the cause of the macroscopic negative DC potential it should be possible to modulate these by acting on adjacent non-participating regions. We used extracellular microinjections of KCl on subregions where SD had terminated and simultaneously recorded in adjacent regions where it did not. In the injecting location potassium drops always caused negative potentials, but in the adjacent SD affected regions mirror positive potentials developed (Fig. 4,C).

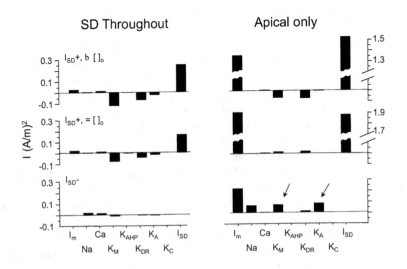

Fig. 3. Disclosure of currents through the different ion channels for different SD approaches. A comparison of a whole neuron and a discrete SD-affected region is presented

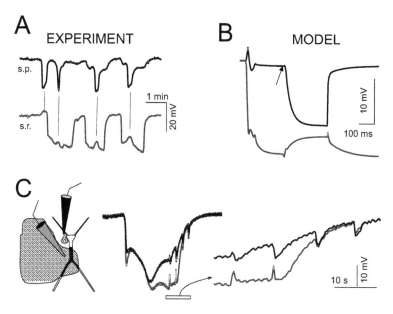

Fig. 4. Mirror behavior of positive and negative potentials associated to SD. A: spontaneous experimental occurrences in several SD waves. B: Reproduction of mirror potentials by model computation. C: experimental assessment by depolarizing subregions where SD had terminated and recording on adjacent ones where SD still went on

5   Discussion

The present results demonstrate that neurons are not inactivated along their entire anatomy during the massive cell depolarization featured on SD. In fact, neurons maintain their integrity and electrical function except on discrete membrane domains where the opening of a large membrane conductance creates a sustained constant depolarization flanked by longitudinal gradients of depolarization. We propose that a classical source-sink dipole transmembrane current distribution is the main electrical event responsible for the uneven subcellular depolarization of SD-seized neurons. The present simulations indicate that such uneven depolarization on individual neurons is the electromotive force generating axial currents toward less depolarized regions, as found on experiments [5]. As a consequence, inward current flows through the electrically shunted membranes causing a large negative potential. For decades, it has prevailed the idea that neurons were electrically impaired during SD[2,3,11]. This concept arose from two long known facts: neurons become electrically unresponsive, and potassium floods the ES, depolarizing cells. The idea of a global membrane breakdown and failure of ionic homeostatic mechanisms prevailed, a chaotic scenario in which all cells were thought depolarized and neuron membranes leaky. The present results refute such notion as far as it concerns to the electrical state of neurons and question the idea of breakdown of ionic homeostasis. In fact, neurons are the active players that initiate and maintain the process by the opening of an unknown conductance in a specific portion of their membranes. It was evident that unless the depolarization is heterogeneous, there would not be large negative potentials. Previous model [12] failed to reproduce this major feature because they did not incorporate any additional conductance. Without our SD-conductance, neither V_m reaches zero, nor R_{in} drops to the experimental value, nor large negative V_o is generated. We showed here the complete causal line from channels on individual model neurons to large macroscopic fields affecting large portions of the tissue.

Acknowledgment

We thank M.J. Yagüe for technical assistance. This work has been supported by grants 8.5/15/98 of the Comunidad Autónoma de Madrid and PB97/1448 and BEFI 2004/1767 of the Spanish DGICYT.

References

1. Leão A.A.P. Spreading depression of activity in the cerebral cortex. J Neurophysiol (1944) 7:359-390.
2. Somjen G.G. Mechanisms of spreading depression and hypoxic spreading depression-like depolarization. Physiol Rev (2001) 81:1065-96.
3. Sugaya E., Takato M., and Noda Y. Neuronal and glial activity during spreading depression in cerebral cortex of cat. J Neurophysiol (1975) 38:822-841.
4. Largo C., Cuevas P., Somjen G.G., Martín del Río R., and Herreras O. The effect of depressing glial function on rat brain in situ on ion homeostasis, synaptic transmission and neuronal survival. J Neurosci (1996) 16:1219-1229.
5. Largo C., Ibarz J.M., and Herreras O. Effects of the gliotoxin fluorocitrate on spreading depression and glial membrane potential in rat brain in situ. J Neurophysiol (1997) 78:295-307.
6. Herreras O. and Somjen G.G. Analysis of potentials shifts associated with recurrent spreading depression and prolonged unstable SD induced by microdialysis of elevated K_+ in hippocampus of anesthetized rats. Brain Res (1993) 610:283-294.
7. Canals S., Makarova I., López-Aguado L., Largo C., Ibarz J.M. and Herreras O. Longitudinal depolarization gradients along the somatodendritic axis of CA1 pyramidal cells: a novel feature of spreading depression. J. Neurophysiol. 2005, in press.
8. Ibarz J.M. and Herreras O. A study of the action potential initiation site along the axosomatodendritic axis of neurons using compartmental models. Lecture Notes in Computer Science (2003) 2626:9-15.
9. Varona P, Ibarz JM, López-Aguado L, and Herreras O. Macroscopic and subcellular factors shaping CA1 population spikes. J Neurophysiol (2000) 83:2192-2208.
10. Makarova I., Ibarz J.M., López-Aguado L., and Herreras O. Selective Inactivation of Neuronal Dendritic Domains: Computational Approach to Steady Potential Gradients. Lecture Notes in Computer Science (2003) 2626:40-45.
11. Czéh G., Aitken P.G. and Somjen G.G. Membrane currents in CA1 pyramidal cells during spreading depression (SD) and SD-like hypoxic depolarization. Brain Res (1993) 632:195-208.
12. Kager H., Wadman W.J., and Somjen G.G. Simulated seizures and spreading depression in a model incorporating interstitial space and ion concentrations. J Neurophysiol (2000) 84:495-512.

Separation of Extracellular Spikes: When Wavelet Based Methods Outperform the Principle Component Analysis

Alexey Pavlov[1], Valeri A. Makarov[2], Ioulia Makarova[2,3], and Fivos Panetsos[2]

[1] Nonlinear Dynamics Laboratory, Department of Physics,
Saratov State University, Astrakhanskaya St. 83, 410026 Saratov, Russia
[2] Neuroscience Laboratory, Department of Applied Mathematics, School of Optics,
Universidad Complutense de Madrid,
Avda. Arcos de Jalon s/n, 28037 Madrid, Spain
[3] Dept. Investigación, Hospital Ramón y Cajal, 28034 Madrid, Spain

Abstract. Spike separation is a basic prerequisite for analyzing of the cooperative neural behavior and neural code when registering extracellularly. Final performance of any spike sorting method is basically defined by the quality of the discriminative features extracted from the spike waveforms. Here we discuss two features extraction approaches: the Principal Component Analysis (PCA), and methods based on the Wavelet Transform (WT). We show that the WT based methods outperform the PCA only when properly tuned to the data, otherwise their results may be comparable or even worse. Then we present a novel method of spike features extraction based on a combination of the PCA and continuous WT. Our approach allows automatic tuning of the wavelet part of the method by the use of knowledge obtained from the PCA. To illustrate the methods strength and weakness we provide comparative examples of their performances using simulated and experimental data.

1 Introduction

Current extracellular experiments provide recordings of multi-unitary activity, where several neurons nearby to the electrode tip produce short lasting electrical pulses, spikes, of different amplitudes and shapes (see for details e.g. [1]). Consequently, extracting useful information from these measurements relies on the ability of separating the recorded firing events into groups or clusters. Ideally each cluster should contain all spikes emitted by only one neuron. Errors occur when spikes belonging to different neurons are grouped together (false positive) or when some spikes emitted by a single neuron are not included into the group (false negative). The performance of this procedure defines the final quality and reliability of any bio-physical results obtained upon the analysis of spike timings. However, the quality of the spike separation by a human operator is significantly below the estimated optimum [2]. Besides, amount of the data generated by modern experimental setups is really huge, thus there is a big demand for automatic separation techniques.

J. Mira and J.R. Álvarez (Eds.): IWINAC 2005, LNCS 3561, pp. 123–132, 2005.
© Springer-Verlag Berlin Heidelberg 2005

Nowadays there exist a number of numerical techniques targeting classification of the extracellular action potentials (see e.g. [1, 3] and references therein). Any method for sorting of spikes relays on two basic steps: i) Extracting the important (most discriminative) features of the spikes and thus lowering the dimension of the parametric set representing the spikes, and also reducing noise influence; and ii) Clustering of the parametric sets into groups, i.e. identifying the number of different spike types (neurons) and the membership of spikes in these groups. Also there are many clustering algorithms (see e.g. [4, 5]) showing different performances on different data sets, as a mater of fact, the final performance of the spike separation is mostly defined by the quality of the extracted spike features, i.e. the quality of the first step. Currently available features extraction methods may be divided into three groups: 1) "naive", threshold based; 2) based on the Principal Component Analysis (PCA); and 3) based on the Wavelet Transform (WT). First two methods are the most widely used now, while the third method becomes more popular and has been demonstrated to have advantages [6, 7, 8]. Although these methods show a good performance, the best representation of the spike feature is still a challenging problem. Here we analyze strength and weakness of the methods and present our novel approach that combines the PCA and Continous Wavelet Transform (CWT).

2   Spike Features Extraction Methods: General Possibilities and Limitations

The simplest approach to the problem of spike separation is high-pass filtering following by the amplitude thresholding. Obvious disadvantage of this approach is that the amplitude is not the only feature of a spike waveform, and separation of spikes close enough in amplitudes degenerates drastically the method performance.

Another simple but significantly more powerful tool for spike sorting is the PCA. Within this framework a set of orthogonal vectors is estimated being the eigenvectors of the covariance matrix constructed from the data. Each spike is completely represented by a sum of the principal component vectors with the corresponding weights or scale factors, so called scores. The latter ones are considered as spike features for sorting. In practice the use of first two or three components is optimal, since they account for the most important information about the shapes of action potentials, while higher components are usually very noisy and decreases the algorithm performance.

A problem occurs when among some number of different waveforms there are two types with similar shapes and clearly expressed distinctions appearing only on small time scales. Such distinctions are usually not reflected in the first principal components, and consequently the method fails to separate such spikes. To illustrate this we generated a test data set consisting of 500 spikes of five different waveforms (Fig. 1A) corrupted by a noise.

Application of the PCA to the data set reveals four different clusters. First three clusters correspond to spikes of the WFs 1–3, so demonstrating the poten-

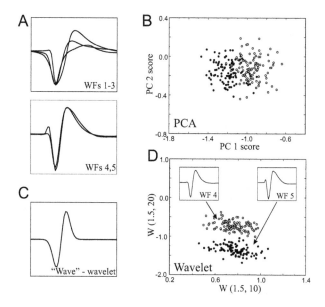

Fig. 1. An example where the wavelet-based approach outperforms the spike separation by the PCA. A) Original spike waveforms used for generation of the data set. We use 3 clearly different waveforms (WF 1-3) and 2 similar waveforms (WF 4 and 5). The difference between two similar WFs appears on small time scales. B) Feature space of the first two principal components. A zoomed region corresponding to the fourth cluster is shown. Spikes of two waveforms (open and solid circles for WFs 4 and 5, respectively) are mixed, and their acceptable separation is impossible. C) The Wave - function chosen for the wavelet-analysis. D) Zoomed region corresponding to fourth and fifth clusters (WF4 and 5) in the wavelet space. Two clearly distinct clouds are formed, and separation with high fidelity is possible

tial of the PCA approach. However, fourth cluster contains a mixture of spikes of two similar waveforms: WFs 4 and 5 (Fig. 1B). Analysis of the Principle Components confirms that the difference between WFs 4 and 5 is not reflected in the first of them. Thus PCA based methods fail to separate spikes with differences appearing on small scales.

Recently a new approach for spike sorting based on the WT has been developed [6, 7, 8]. This approach is claimed to have advantages in comparison with the techniques traditionally used for classification of action potentials. The Continuous Wavelet Transform (CWT) of a one-dimensional signal $f(t)$ involves its decomposition (somehow similar to the Fourier transform):

$$W(a,b) = \frac{1}{\sqrt{a}} \int\limits_{-\infty}^{+\infty} f(t)\psi_{a,b}(t)\, dt, \quad \psi_{a,b}(t) = \psi\left(\frac{t-b}{a}\right), \qquad (1)$$

where $\psi_{a,b}(t)$ is a translated and scaled mother wavelet, $\psi(t)$, with b and a defining the time location and scale. Instead of the continuous transform (1), its

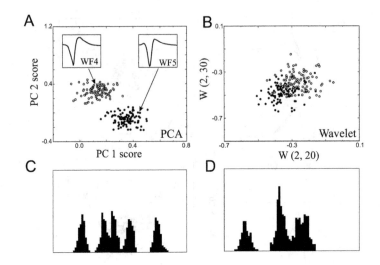

Fig. 2. A case where the PCA provides better separation. Like in Fig. 1, we use a data set with spikes of 3 clearly different and 2 similar waveforms. However, now the difference between similar spikes is not so pronounced and is not in small scales. A) Principle components show a good separation of spikes of WF4 and WF5 (open and solid circles, respectively). B) Wavelet classification. The chosen wavelets-coefficients demonstrate multi-modal distributions allowing separation of clearly different spikes. However, separation of WF4 and 5 is not achieved. C,D) Histogram of spike density along the first component score (C) and one of the wavelet-coefficients (D). The wavelet-coefficient demonstrates a multi-modal distribution however the number of peaks (four in (D)) corresponding to clusters is less than in the PCA case (five in (C))

discrete counterpart (DWT) is usually used. In the DWT the scale takes only some fixed values (usually $a = 2^i$).

Several methods for spike separation based on the DWT have been proposed [6, 7, 8]. They use the fact that the WT of a signal (spike) can be considered as a set of filters with different bandwidth controlled by the scale parameter a. Then the values of the energy found in specific frequency bands during each spike profile are considered as quantities for spike classification within the framework of the Wavelet-based Spike Classifier (WSC) [6].

In the case where spike waveforms have a multi-scale structure with any significant characteristics appearing on small scales, like in the data set used in Fig. 1, the wavelets are able to resolve these features. Indeed, application of the wavelet technique to the data set of Fig. 1 shows that this approach finds all five clusters. Figure 1D illustrates a good separation of WF 4 and 5 into two clusters, where the PCA had difficulties (Fig. 1B).

Although the WT is potentially more powerful there are a number of problems restricting its considerable application for spike separation. Here we discuss main of them:

i) An arbitrary choice of the mother wavelet
ii) Complicated selection of the best wavelet-coefficients

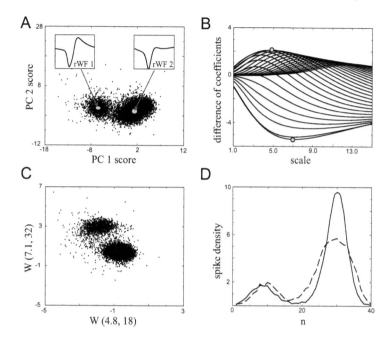

Fig. 3. Working principle of the WSAC method. A) Two overlapping clouds correspond to spikes of different types on the PCA plane. Insets show representative spike waveforms obtained by averaging over neighborhoods of the cloud centers. B) The difference between the wavelet-coefficients for the representative spikes as a function of scale. Circles mark the coefficient pairs ($a = 4.8$, $b = 18$ and $a = 7.1$, $b = 32$) that correspond to the most prominent distinctions between rWF1 and rWF2. C) New spike features space. The found coefficients are used. D) Spike density along the clouds. Peaks correspond to the centers of the clouds. Dashed line corresponds to the PCA space, and solid line shows the results obtained in the wavelet space. The later distribution shows better separated and prominent peaks resulting in a better localization in feature space of spike of different waveforms (compare clouds in (A) and (C))

i) Apparently, the results of the analysis, e.g. the wavelet-coefficients, depend on the mother wavelet ψ. Generally, there is no standard answer on how to choose the mother wavelet in a particular case. Thus the performance of the method may strongly vary from one case to another. For spike separation different mother wavelets have been advocated: Daubechies [6], Coiflet [7], Haar [8]. Possible advantages of one or another depend on the particular spike waveforms of the analyzed data set, and no a-priori assumption which mother wavelet will perform better can be given. In our experience, success of the classification can be often achieved by selection of the mother wavelet similar to the shape of spike waveforms. For instance, in the example shown in Fig. 1, to obtain a good separation we used the so-called "Wave" – wavelet (Fig. 1C). Visually, this wavelet is very similar to the WFs 4 and 5 (Fig. 1A), and a good separation of all waveforms, including WF4 and WF5, has been obtained.

ii) Let us assume that the mother wavelet has been somehow selected. Then the WT of spike waveforms is performed, thus obtaining a number (usually 64 in the case of the DWT and even more for the CWT) of different wavelet coefficients. The right choice of some of them for spike classification is also crucial. Different authors suggested different procedures for coefficient selection, e.g. large standard deviation, large average, multi-modal distribution [6]. There is a more complicated but at the same time mathematically better justified method based on the information theory [7]. However, there is no one universal approach for the choice of the WT-features capable to provide all the time the best classification and a counterexample can be always found. The difficulties especially occur when the analyzed data contains spiking activity of many neurons, and among which there are both, clearly different and rather similar types of spike waveforms.

To illustrate a kind of problem that can be found we again generated a test data set similar to that used in Fig. 1, however now the difference between the WF 4 and 5 is more pronounced, and no clear differences on small scales exist. This helps the PCA to separate all spike groups including those of similar waveforms (Fig. 2A). According to one of the wavelet coefficient selection procedure [6, 8] the features used for classification should show multi-modal distribution. However in many practical cases multi-modal distribution is obtained for many wavelet-coefficients and there is no clue on how to perform their automatic comparison in order to select the most informative ones. An example of such quasi arbitrary (unsuccessful) choice of the coefficients is illustrated in Fig. 2B. Although the chosen wavelets coefficients have multi-modal distributions (Fig. 2D) allowing separation of the first three clearly different spike waveforms, the wavelet approach gives worse classification of two similar waveforms than that provided by the PCA (Fig. 2AC).

3   Our Novel Approach for Spike Features Extraction

Let us start with a typical situation frequently appearing when processing real electrophysiological data. We assume that a conventional method of spike features extraction (e.g. the PCA) gives two overlapping clouds. For the sake of simplicity we suppose that these clouds consist of spikes of two types (or we may aim at separation of spikes of a given type, say WF1, from the rest, possibly noisy spike-like pulses). Figure 3A shows such an example with the PCA of spikes from a real electrophysiological recording. Let us now sketch our three-steps approach based on a combination of the PCA-method and the CWT method that we shall refer to as Wavelet Shape-Accounting Classifier (WSAC).

First step: Calculation of the representative WaveForms (rWFs).

The usual PCA is performed on the all available waveforms. Then we average spike waveforms in a small neighborhood of each cloud center. As a result two representative (mean) spike waveforms are obtained (insets in Fig. 3A). Since these waveforms are related to the centers of the corresponding clouds we can suppose that they represent "real" spike waveforms with lowest noise impact.

Second step: Wavelet transform of the rWFs and selection of the coefficients that optimally depict the differences between them.

Two obtained representative spike waveforms are analyzed in the wavelet space. We seek for those wavelet coefficients that maximize the difference between the rWFs. Thus the differences between corresponding wavelet coefficients are estimated and coefficients showing maximal dissimilarity are selected. Figure 3B shows examples of the differences between the obtained wavelet coefficients (for rWFs 1 and 2) as a function of scale. Circles mark two points where the differences between the wavelet-coefficients are maximized. Note that there may be more than two extrema for different scales, so increasing the number of features (i.e., the wavelet-coefficients) that may be used for classification. Because the given procedure performs a search of the most prominent distinctions for the representative spike waveforms, the estimated features can provide a better resolution between the clusters than that obtained with the PCA.

Third step: Estimation of the selected coefficients for all available spike waveforms and their use as new features for classification.

The coefficients selected at the second step are estimated for all available spike waveforms and then considered as spike features for classification (Fig. 3C). Figure 3D shows densities of spikes in the PCA and Wavelet feature spaces. Main peak in the wavelet space becomes narrower and more pronounced in comparison with the distribution obtained for the PCA method. This means that we can now better separate clouds into clusters and reduce classification errors that mostly originate from a misclassification of spikes in the intermediate (common) part of the clouds.

4   Results

We test the proposed approach on three different data sets (S1, S2, and S3). Each data set has been obtained in the following way. We take two experimental electrophysiological recordings. One of the recordings is selected in the way that spikes of one type can be easily separated from the rest by the conventional thresholding method. Then these spikes are mixed with another experimental recording demonstrating complex spiking activity. This procedure, from one side, allows keeping all characteristics (level and type of noise, spike waveform variation etc.) essential to a real electrophysiological experiment, and from the other hand we possess the a-priori information about the membership of spikes for one target cluster formed by the "additional" spikes. Hence we can estimate the classification error for the given cluster.

The generated data sets have been used as an input to four feature extraction algorithms above discussed. Then clustering by the superparamagnetic method [9] has been performed and the number of misclassified spikes has been estimated.

Figure 4 illustrates results obtained for the data set S1 consisting of 16568 spike waveforms including 3069 "additional" spikes. The PCA gives 2 clusters (Fig. 4A) shown in black and gray corresponding to the additional (targeting) and the original action potentials, respectively. Squares mark unclassified spikes

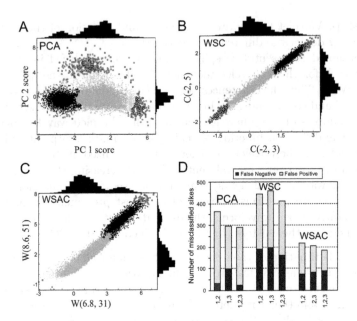

Fig. 4. Results of spike separation by different methods for the data set S1. A) Projection of the feature space for the PCA into first two components, and corresponding histograms of spike densities. Black points correspond to spikes classified to be belonging to the targeting cluster. B) The same as in (A) but for the WSC method. C) The same as in (A) but for the WSAC method. D) Number of misclassified spikes for different methods and for different spike features subsets used for classification

being not related to either of the clusters. Classification of spikes by three first PCs gives 290 misclassified spikes: 24 false negative and 266 false positive, i.e. 0.8% and 8.6% from the total number of spikes in this cluster. The histograms of spike densities for each coordinate in the features space show a bimodal distribution for the PC1, and uni-modal distribution for the PC2. The former allows separation of different waveforms into two clusters, while the later does not actually provide additional information for spike classification.

Figure 4B illustrates the results of spike sorting performed by the WSC method [6]. Following the author recommendations, we have chosen for classification the wavelet-coefficients showing the largest standard deviations, the largest values and the bimodal distributions. Note, that unlike to the PCA, the histograms in Fig. 4B are both bi-modal, and therefore they actually provide useful information for spike sorting. However, for the considered example we obtain higher classification error: 410 misclassified spikes (5.2% of false negative and 8.1% of false positive). Thus a quasi arbitrary choice of the wavelet-coefficients satisfying the mentioned recommendations did not allowed improving classification in comparison to the PCA method.

Figure 4C shows results of spike classification obtained by our WSAC method. We found that three pairs of coefficients: (6.8, 31), (8.6, 51), and (6.2, 20) max-

Table 1. Classification error rates for all data sets and different methods (percentage of the misclassified spikes to the total number of spikes in the cluster). FN and FP denote False Negative and False Positive errors

	S1		S2		S3	
	FN/FP	Sum	FN/FP	Sum	FN/FP	Sum
PCA	0.8/8.6	9.5	41.6/11.8	53.4	0.1/2.6	2.7
WSC	5.2/8.1	13.3	34.2/13.8	48.0	6.7/2.9	9.6
WMMC	7.5/8.9	16.4	28.7/0.8	29.5	9.5/4.4	13.9
WSAC	2.8/3.1	5.9	26.4/8.2	34.6	1.8/0.3	2.1

imize the difference between the characteristic spike shapes, and used them for classification. This method provides the best results: 185 errors or 2.8% of false negative and 3.1% of false positive.

Figure 4D shows results of spike classification done by these three methods for different combinations of features used in each particular technique. For instance, classification performed by the use of first two principle components gives 364 errors (first bar in Fig. 4D), while the same done with PC1 and PC3 results in 296 errors. This means that in this case PC3 describes better the variation in the data set than PC2. The use of all three components improves a bit the classification resulting in 290 errors. Considering WSC we note that each coefficient improves the results of classifications, but the overall performance is the worst. The WSAC approach is the winner giving in average the minimal classification error for any combination of the spike features.

Table 1 summarizes results obtain for all data sets. We also included in the table classification errors obtained by the WMMC method based on the approach proposed in [8]. This approach performs considerably better for the set S2, while showing poor performance in S1 and S3.

5   Conclusions

Addressing the question: when the wavelet-based methods outperform the PCA, we have shown that the main advantage of the WT techniques reveals when dealing with the detailed structure of experimental signals in a wide range of scales. Considering the WT-approach as a "mathematical microscope", the following interpretation can be given: the wavelets can resolve fine details of a signal structure, but we need to choose appropriately the focusing point and the resolution of this "microscope". From the mathematical viewpoint the latter means that the selection of the parameters a and b in (1) responsible for the resolution and focusing is of a crucial importance. In the case of successful selection, the "microscope" elucidates the differences in spike waveforms. That is why the problem of selection of the optimal wavelet-coefficients is an important trend in the problem of spike separation. Unlike the PCA-based methods where the first principal component scores are used as spike features due to their natural order, optimal selection of features within the framework of the WT techniques is significantly more complicated procedure.

In order to eliminate arbitrariness in the selection of the spike features here we have proposed a novel technique, the WSAC method. It is based on the choice of the wavelet-coefficients tuned to the spikes shapes. The main idea of the method is to find such features of the WT that maximize the differences between two or more kinds of representative waveforms selected from the experimental recordings, and then to use them for classification of all spikes. Using different data set we have shown that the proposed method of features selection outperforms the PCA and the other wavelet-based techniques.

Summarizing, we emphasize that there are at least two cases when the wavelet-based techniques potentially are preferable than the Principal Component Analysis: (i) the presence of small-scale structure in waveforms that is not reflected in the first principal components, and (ii) the presence of strong enough low-frequency noise that strongly reduces the PCA-method performance whereas this noise statistics is less essential for the wavelets. In other situations the WT-based approaches show results comparable with the classical technique.

Acknowledgments

This work was supported by the European Project ROSANA (EU-IST-2001-34892), and by the programm Ramon y Cajal (V.A.M.).

References

1. Lewicki, M.: A review of methods for spike sorting: the detection and classification of neural action potentials. Net. Com. Neu. Sys. **9** (1998) R53–78.
2. Harris, K.D., Henze, D.A., Csicsvari, J., Hirase, H., Buzsaki, G.: Accuracy of tetrode spike separation as determined by simultaneous intracellular and extracellular measurements. J. Neurophysiol. **84** (2000) 401–14.
3. Wheeler, B.: *Automatic Discrimination of Single Units.* (CRC Press, Boca Raton, FL. 1999).
4. Kaufman, L., Rousseeuw, P.J.: *Finding Groups in Data: An Introduction to Cluster Analysis.* (Wiley-Interscience. 1990).
5. Downs, G.M., Barnard, J.M.: Clustering methods and their uses in computational chemistry. Rev. Comput. Chem. **18** (2002) 1–40.
6. Letelier, J., Weber, P.: Spike sorting based on discrete wavelet transform coefficients. J. Neurosc. Methods **101** (2000) 93–106.
7. Hulata, E., Segev, R., Ben-Jacob, E.: A method for spike sorting and detection based on wavelet packets and Shannon's mutual information. J. Neurosc. Methods **117** (2002) 1–12.
8. Quian Quiroga, R., Nadasdy, Z., Ben-Shaul, Y.: Unsupervised spike detection and sorting with wavelets and superparamagnetic clustering. Neur. Comput. **16** (2004) 1661–1687.
9. Blatt, M., Wiseman, S., Domany, E.: Superparamagnetic clustering of data. Phys. Rev. Lett. **76** (1996) 3251–3254.

Structural Statistical Properties of the Connectivity Could Underlie the Difference in Activity Propagation Velocities in Visual and Olfactory Cortices

Mavi Sanchez-Vives and Albert Compte

Instituto de Neurociencias de Alicante, Universidad Miguel Hernández - Consejo Superior de Investigaciones Científicas, 03550 Sant Joan, Alicante, Spain
`mavi.sanchez@umh.es`

Abstract. We show experimentally that the properties of the propagation of activity in cortical slices depend critically on the cortical area explored. Thus, olfactory cortex slices present a much faster speed of propagation than neocortical slices. In order to explore the possibility that this reveals different statistical properties of the underlying synaptic connectivity, we study the small-world properties of a computational network model of slow oscillatory activity that we have previously shown to replicate closely the activity in the slice. We show that for the Gaussian probability connectivity used, progressive reduction of the Gaussian spread makes the network transition from a random, to a small-world and to an ordered network. We then relate the small-world parameters of the connectivity to the velocity of activity propagation in the model. We conclude that the locality parameter C, and not the mean path length L, determines primarily the velocity of propagation.

1   Introduction

The cerebral cortical network performs different functions that range all the way from sensory processing and motor functions up to higher or cognitive functions. Strikingly, these functional diversity is processed and supported by a network with a high degree of similarity, with a basic columnar modularity that repeats itself across the cortex [1, 2, 3]. Still, this apparent uniformity coexists with certain cytoarchitectonic diversity that allows the differentiation of cortical areas, together with local differences in cell types, neurotransmitters, distribution of the thalamic inputs or gene expression.

This cortical network generates ongoing spontaneous activity. During slow wave sleep, spontaneous activity is organized in a slow rhythm (< 1 Hz) of cortical origin [4]. This rhythm consists on epochs of discharge or up states interleaved with silent periods or down states that propagate across the cortex. While *in vitro*, the cortical network also generates an almost identical activity if maintained in an environment that closely mimics the conditions *in situ* [5].

J. Mira and J.R. Álvarez (Eds.): IWINAC 2005, LNCS 3561, pp. 133–142, 2005.

The rhythmic activity that is generated by the cortical network integrates the synaptic and intrinsic properties of the network, and the main characteristics of this activity can be succesfully reproduced in a realistic cortical model contaning excitatory and inhibitory neurons that are reciprocally connected [6]. However, the cytoarchitectonic and functional diversity across areas mentioned above is also reflected in the resulting emergent activity, evidenced when different cortical areas are compared.

In previous studies we have analyzed the properties of the activity generated by the visual and the olfactory cortex [5, 9]. Both networks give rise to activity consisting of alternating up and down states that propagate along the cortex. However, some differences exist between the activity arising from both areas. In order to analyze the underlying properties that induce differences in generation or propagation of rhythmic ongoing activity, we have found useful a combined experimental-computational approach. We described previously the role of inhibition and $GABA_A$ reversal potential on the speed of activity propagation [9]. Here we explore in detail the impact of different network connectivity patterns on the propagation of emergent activity.

2   Methods

2.1   Experiments

Cortical slices were prepared from the brain of ferrets (2-12 month old, either sex) as we have previously described [5]. The animals were anesthetized with sodium pentobarbital (40 mg/kg) and decapitated. Horizontal slices (0.4 mm thick) from the ventral side of the temporal lobe were prepared on a vibratome, the first 4-5 slices containing the piriform cortex and endopiriform nucleus. Additional slices from the primary and secondary visual cortical areas (areas 17, 18, and 19) were also placed in the bath and recordings were made for comparison. After preparation, slices were placed in an interface style recording chamber. The normal bathing medium contained (in mM): NaCl, 126; KCl, 2.5; $MgSO_4$, 2; NaH_2PO_4, 1.25; $CaCl_2$, 22; $NaHCO_3$, 26; dextrose, 10, and was aerated with 95% O_2, 5% CO_2 to a final pH of 7.4. The modified solution in which the rhythmic activity was evoked had the same ionic composition except for different levels of (in mM) KCl, 3.5; $MgSO_4$, 1 and $CaCl_2$, 1-1.2 [5]. Electrophysiological recordings started after allowing at least 2 hours recovery. Extracellular multiunit recordings were obtained with 2-4 $M\Omega$ tungsten electrodes. Horizontal (same layer) propagation was measured by means of simultaneous recordings from 2 or 3 tungsten electrodes placed on the surface of the slice. To analyze the time structure of the multiunit extracellular recordings, they were converted into spikes or events. Upward swings of the multiunit recording (120 seconds acquired at 10 KHz) that crossed the threshold counted as events. The propagation, synchrony and duration of the oscillations were measured in auto or crosscorrelograms. The temporal relationship between the neuronal activities at varying recording sites was examined with auto and crosscorrelograms. The speed of propagation was calculated from the offset of the central peak in crosscorrelograms of the activity recorded by electrodes located at a known distance, which was measured with a calibrated grid in the microscope.

2.2   Computational Modelling

We use the network model of slow oscillatory activity in visual cortical slices of the ferret that has been presented elsewhere [6]. Briefly, the network model consists of a population of 1024 excitatory cells and 256 inhibitory neurons (2048 and 512, respectively, in 'network size 2' of Fig. 3C) equidistantly distributed on a line and interconnected through biologically plausible synaptic dynamics. Some of the intrinsic parameters of the cells are randomly distributed (Gaussian probability distributions), so that the populations are heterogeneous. This and the random connectivity (determined by Gaussian synaptic probability distributions; see Figure 3A) are the only sources of noise in the network. The neurons in the network are sparsely connected to each other through a fixed number of connections that are set at the beginning of the simulation. Neurons make 20 5 contacts (mean standard deviation) to their postsynaptic partners (multiple contacts onto the same target, but no autapses, are allowed). For each pair of neurons separated by a distance x in the network (see Fig. 3A), the probability that they are connected in each direction is decided from a Gaussian probability distribution centered at 0 and with a prescribed standard deviation σ_E, which we call connectivity footprint. For inhibitory connections a Gaussian probability distribution is also used but with a smaller, fixed standard deviation $\sigma_I = 0.025$. All values of σ_E or σ_I reported here are expressed as a fraction of the standard network size ('network size 1': 1024 e-cells and 256 i-cells). Notice that because of this, both 'network size 1' and 'network size 2' have the same neuronal density in the spatial scale of the footprint, so that they are suited to check for the appearance of free-boundary effects in our measures. In Figs. 3 and 4 we also use the absolute footprint $\sigma\prime_E = 1024\sigma_E$. Our model neurons are implemented in the Hodgkin-Huxley formalism. All details about the exact parameter values and model implementations can be found in [6].

3   Results

The cortical network generates a rhythmic pattern of spontaneous activity while maintained *in vitro* in conditions that are similar to the ones *in vivo*. Our recordings from neocortex (visual) and piriform cortex reveal that both networks generate rhythmic activity that recruits a large number of neurons in the network, both excitatory and inhibitory, and that alternates periods of discharge (up states) and silence (down states). However, we have reported significant differences in some aspects of this activity [9], probably reflecting underlying structural and functional disparities. Two of the most prominent differences are illustrated in Fig. 1, that includes representative examples of recordings in both cortices. These differences include the frequency of oscillations (higher in olfactory cortex, Fig. 1C) and the speed of propagation, which is faster in olfactory cortex [9]. This is noticeable in the recordings shown in Fig. 1, where speeds of 11.2 mm/s for the visual cortex and 245 mm/s for the olfactory cortex are displayed. We have been interested in the quantitative exploration of the basis for those differences and with that aim we have done a combined experimental-computational effort. In [6] we showed that a network model featuring strong recurrent excitation, feedback inhibition, and slow cellular spike frequency adaptation could reproduce very closely the main features of slow oscillatory activity in visual cortical slices [5]. This model has also been used in [9] to explore *in silico* the potential mechanisms involved in the much faster activity propagation in olfactory cortex as compared to visual cortex that we have discussed above. There, we cite as plausible mechanisms the broader spread of excitatory connections, reduced inhibition, and/or more depolarized reversal potential through $GABA_A$ channels in olfactory cortex. Here, we go one step further by exploring what structural changes are induced in the model network of excitatory neurons by the progressive enlargement of the Gaussian probability distribution of synaptic contacts between excitatory cells, and how these changes relate to the velocity of propagation of slow oscillatory activity. To this end, we characterize the structural properties of a particular realization of the connectivity within the framework of small-world networks, as proposed in [10].

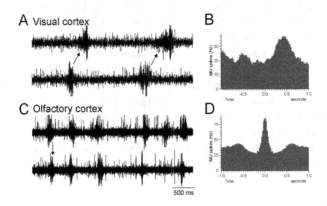

Fig. 1. Multiunit recordings of activity propagation in neocortical (visual) and piriform (olfactory) slices. A. Recordings obtained with two tungsten electrodes placed 6 mm apart on layer 4 of the visual cortex. Rhythmic activity consisting on up states (periods of discharge) and down states was recorded from both locations. The occurrence of multiunit activity indicates that is a population phenomenon. The arrows point out at the propagation of the activity. B. Cross-correlogram of the activity recorded with the two electrodes in the visual cortex (A). The average time of propagation between electrodes was estimated by the offset of the peak, resulting in a speed of propagation of 11.2 mm/s in this particular case. C. Recordings obtained with two tungsten electrodes placed 4.5 mm apart at the border of layers 2 and 3 of the piriform cortex. Observe that the frequency of occurrence of the upstates is higher in piriform than in visual cortex. D. Cross-correlogram of the activity recorded by the two electrodes in the piriform cortex (C). The average time of propagation between electrodes was estimated at the peak, resulting in a speed of propagation of 245 mm/s in this particular case

We use two complementary structural measures: the characteristic path length L and the clustering coefficient C. The characteristic path length is defined as the average number of synaptic contacts that separate two neurons picked at random in the network. Intuitively, L will be larger the narrower the connectivity footprint is, because neurons at the two ends of the network will need to go through many intermediate relay connections to reach each other. The clustering coefficient is calculated as the average probability that one neuron connects via its postsynaptic neurons and their postsynaptic partners back to itself, thus constituting a 3-synapse loop. This is a particular example (loop pattern) in directed networks of the measure termed transitivity (especially in sociology), and it gives a measure of the locality, or cliquishness, of the network [10]. Networks with only local connectivity are characterized by a high L and a high C, whereas in the other end lie networks with totally random connectivity (lowest L and C). There is, however, a significant range of intermediate cases where C is high and L is low, and these are called small-world networks [10]. In these networks, despite the fact that individual nodes feel locally very clustered (high C), as in a locally connected network, they can reach practically the whole network through just a few synapses (low L), as in a random network. We have applied these measures to our network model, and we have followed their evolution as we changed the footprint of excitatory connectivity to excitatory cells σ_E (see Figs. 3 and 3). We see that as σ_E is increased, both L and C decrease, as expected, but there is a sharp slope change in the decrease of the locality parameter C that is not contingent upon network size, as it would if it corresponded to some free-boundary effect of this finite network (Fig. 3 panels B.2, B.3, C.2, C.3). This is suggestive of some structural change in the network connectivity at the point of the slope change, which occurs at the local, not the global, level. Our calculations further show that an increase in the width σ_E of the Gaussian probability of connection alone can induce a transition from ordered, to small-world to random networks (see Fig. 3), without the need of short-cut connections extraneous to the Gaussian probability distribution.

For each of the various synaptic connectivities characterized above (except for those corresponding to 'network size 2' in Fig. 3C), we let the full simulation run, which displayed the characteristic slow oscillatory activity and traveling waves of this biophysical network model [6]. Subsequently, we measured the propagation velocity of these wave fronts as σ_E was changed (see caption to Fig. 3), in order to study the relationship between the structural properties of the connectivity and an experimentally accessible measure like propagation velocity. We repeated this for two different initializations of our random number generator (rows B and C in Fig. 3). The measures of propagation velocity are noisy and, because we are interested in evaluating the relation of the propagation velocity to structural changes that might depend on the precise realization of the connectivity (see for instance that the sharp slope change in Fig. 3B.2 and C.2 occurs for different value of σ_E), averaging over different noise realizations of one given average connectivity shape is not a good variance-reduction mechanism. Despite the noisiness of this measure, several aspects can be highlighted: 1) as expected, the

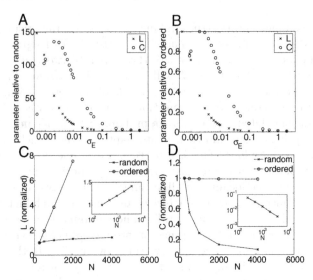

Fig. 2. Connectivity parameters L and C for 'network size 2' (some of the data as in Fig. 3C.2 and Fig. 3C.3) but now plotted: (A) normalized to the minimum values $L_1 = 3.06$ and $C_1 = 0.004$ that would correspond to a randomly connected network; (B) normalized to the maximum values $L_0 = 453$ and $C_0 = 0.527$, thus comparing with the ordered network. In both cases, the characteristic path length L (crosses) is always small, whereas the clustering coefficient C (circles) climbs up first as the footprint gets smaller (the redecrease of C for very small σ_E is due to the effective reduction of the number of links when most neurons make multisynaptic contacts onto just a few postsynaptic cells). This shows that the network for footprints below ~ 0.05 is approaching the small-world network regime, whereas above this value it approaches the domain of random networks. This is further demonstrated by studying the dependence of the parameters C and L with the number of cells in the network: (C) L increases linearly with network size N for an ordered network ($\sigma_E = 0.01$, normalizing constant $L_{norm} = 8.423$), whereas it increases only logarythmically (see semilog inset) for a random network ($\sigma_E = 0.3N/1024$, normalizing constant $L_{norm} = 2.3849$). (D) C is unaffected by network size in an ordered network ($\sigma_E = 0.01$, normalizing constant $C_{norm} = 0.459$), while it decays as $1/N$ (see log-log inset) as one increases the size of a random network ($\sigma_E = 0.3N/1024$, normalizing constant $C_{norm} = 0.063$). These dependencies with N are characteristic of the extreme cases around small-world networks [10]

propagation velocity increases as both L and C decrease; 2) it is apparent from Fig. 3, panels B.1 and B.2, that the slope change in the locality parameter C (Fig. 3, B.2 and C.2) corresponds in the velocity graphs (Fig. 3, B.1 and B.2) to a sudden speed increase in activity propagation; and 3) from the previous point and also looking at the shapes of the L vs. σ_E and C vs. σ_E graphs, it appears like the velocity of propagation depends primarily on the local properties of the network.

To further test this hypothesis we identified three network parameter regimes, where the connectivity properties between any two of them differ in just one of

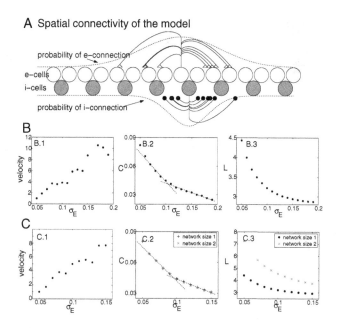

Fig. 3. Simulations in the network model of slow oscillatory activity show that increasing the width σ_E of the excitatory connectivity probability of connection (connectivity footprint, see illustration in panel A) results in increased propagation velocity (panels B.1 and C.1) with some apparent discontinuities, which are related to a change in the structural properties of the graphs of excitatory connections among excitatory cells, as assessed through their characteristic path length L (panels B.3 and C.3) and their clustering coefficient C (panels B.2 and C.2). Rows B and C correspond to runs of the same network simulation starting off a different random seed. For each value of excitatory connectivity footprint σ_E, 30 seconds of network activity were simulated, from which the velocity of propagation was evaluated by fitting a Gaussian to the cross-correlogram between the spikes of two cells found on the propagation front. The center of the Gaussian was taken to be the travel time between these two cells, from which a velocity could be calculated. Velocities are plotted normalized to the velocity found when σ_E=0.05. Footprints (σ_E) are indicated as the fraction of the size of the network ('network size 1'). In panels C.2 and C.3 simulations were carried out for an additional network size (twice as big as network 1, see Methods) to ensure that the effect was not due to the network free boundaries

the parameters L, C and N (network size) (see Fig. 4). We let the simulation run and develop traveling waves and we measured the velocity of propagation of these waves. We then evaluated which one of the parameters L, C and N had a more direct influence in the velocity of propagation (Fig. 4E). As previously conjectured, the parameter C, scaled with N, was a good predictor of the speed of propagation, while the L parameter did not make correct predictions. Notice the very unintuitive result that the velocity of propagation in the case C is much slower than that in the case A despite the fact that σ'_E is 4 times larger in C than in A and they have the same number of neurons N.

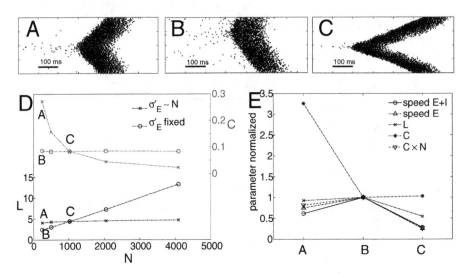

Fig. 4. Relation between the propagation velocity and the small-world parameters C and L. Three different network simulations are run (sample rastergrams in panels A-C, x-axis is time, y-axis is the neuronal network), corresponding to different situations: networks A and B share a similar value of L, while networks B and C have virtually the same C and networks A and C have the same number of neurons N (see panel D, lower curves for L and upper curves for C as N is varied for two scaling regimes of $\sigma'_E = 1024\sigma_E$: σ'_E scales with N or $\sigma'_E = 51$ remains fixed). Velocities in A-C are measured as indicated in the caption to Fig. 3 for two cases of the network: full case with functional excitation and inhibition (labeled 'speed E+I'), and case with inhibition removed (labeled 'speed E', these correspond to the fromts depicted in panels A-C). (E) The velocity of propagation for each situation A-C (left, middle and right points, respectively), whether in the disinhibited or normally inhibited case, does not correlate with the parameter L but it is directly predicted by the parameter C times the size of the network N

4   Discussion

We have shown here that the connectivity of biophysically realistic network models of spontaneous activity in cortical slices can be studied through the measures of small-world networks. Based on the usual assumption of a Gaussian distribution of synaptic contacts, and a low number of synaptic partners per neuron, we show that a transition from the ordered network, to the small-world regime to the random network regime can be operated by progressively changing the width of the excitatory connectivity footprint σ_E. Thus, it is not necessary in this system to introduce long-range, short-cut connections in addition to the local Gaussian patches of synaptic contacts in order to induce this regime change. This type of patchy long-range connections, though, are very prevalent in the visual cortex (horizontal patchy connections), and they presumably enhance the small-world character of this network, although their exact influence on the structural properties of the connectivity remains to be carefully checked (for their effect on the velocity of propagation see [6]).

As for the functional effects of such structural conditions we have started here to look at how these relate to the velocity of propagation of activity fronts in the network model, with the intention, eventually, of generating hypothesis and predictions to be tested experimentally on the slowly oscillating slice [5]. We have found that the measure of velocity of front propagation relates inversely, but monotonously, to the parameters L and C of the connectivity, as it was expected. In addition, sudden jumps in the propagation velocity have a suggestive relation to sudden slope changes in the locality parameter C. This remains to be thoroughly checked, but it suggests that a sharp increase in propagation velocity occurs when there is a breakdown in the compactness of local connectivity, and maybe this characterizes the circuitry of olfactory cortex. Finally, we provide evidence that propagation of slow oscillatory activity in the cortex relies importantly on the compactness of local cortical connectivity (Fig. 4). This last point is consistent with the fact that activity both in the network model and in the *in vitro* preparation is generated through reverberation in the local circuit [5, 6]. This suggests a principal role for the local, transitivity parameter C, which accounts for the incidence of tri-synaptic closed loops apt for reverberation. However, effective path length L also has a critical effect on propagation speed, as shown through the addition of patchy long-range connections [6]. Further investigations are needed to establish the exact relationship of the parameters C and L and experimentally measurable quantities like the velocity of propagation of activity.

Our results suggest that visual cortex might operate in a small world regime, olfactory cortex in a random connectivity regime, and the instrumental difference at the structural level would fall on the spatial extent of excitatory connectivity, but with constant number of synaptic contacts per neuron. A complete proof and understanding of the implications of these structural properties requires at this point further intensive investigations combining biophysical modeling, mathematical analysis, electrophysiological experiments, and anatomical measures. Our preliminary analyses suggest that the velocity of propagation of activity fronts in a slice preparation might reveal some of these structural differences experimentally.

Acknowledgments

We thank Dr R Gallego, VF Descalzo and R Reig for their contribution to the experiments. This work was sponsored by Human Frontiers Science Program Organization and *Ministerio de Educación y Ciencia* to MVSV and *Ministerio de Educación y Ciencia* to AC.

References

1. Lorente de No, R.: Cerebral cortex: architecture, intracortical connections, motor projections. In: Physiology of the nervous system 3rd edn, J Fulton ed. (Oxford, Oxford University Press, 1949), pp. Chapter 15: 288-330

2. Edelman, G. M., Mountcastle, V. B.: The mindful brain : cortical organization and the group-selective theory of higher brain function Cambridge, MIT Press (1978)
3. Szentagothai, J.: The neuron network of the cerebral cortex: a functional interpretation. The Ferrier Lecture. Proc R Soc Lond B Biol Sci (1978) 201: 219-48
4. Steriade, M., Núñez, A., Amzica, F.: A novel slow (¡ 1 Hz) oscillation of neocortical neurons in vivo: depolarizing and hyperpolarizing components. J Neurosci (1993) 13:3252-3265
5. Sanchez-Vives, M.V., McCormick, D.A.: Cellular and network mechanisms of rhythmic recurrent activity in neocortex. Nat. Neurosci. (2000) 10:1027-34
6. Compte, A., Sanchez-Vives, M.V., McCormick, D.A., Wang, X.-J.: Cellular and network mechanisms of slow oscillatory activity (¡1 Hz) and wave propagations in a cortical network model. J. Neurophysiol. (2003) 89:2707-25
7. Wang, X.-J.: Synaptic basis of cortical persistent activity: the importance of NMDA receptors to working memory. J. Neurosci. (1999) 19:9587-9603
8. Wang, X.-J., Buzsáki, G.: Gamma oscillation by synaptic inhibition in a hippocampal interneuronal network model. J. Neurosci. (1996) 16: 6402 6413
9. Sanchez-Vives, M.V., Descalzo, V. F., Figueroa, A., Reig, R., Compte, A., Gallego, R.,: Rhythmic spontaneous activity in piriform cortex. under revision in Cerebral Cortex
10. Watts, D. J., Strogatz, S H.: Collective dynamics of 'small-world' networks. Nature (1998) 393:440

Rules and Roles of Dendritic Spikes in CA1 Pyramidal Cells: A Computational Study

José M. Ibarz[1], Ioulia Makarova[1,2], Iria R. Cepeda[1,2], and Oscar Herreras[1]

[1] Dept. Investigación Hospital Ramón y Cajal, 28034 Madrid, Spain
[2] Dept. Matemática Aplicada, Fac. Biología, Universidad Complutense de Madrid, 28040 Madrid, Spain
jose.m.ibarz@hrc.es

Abstract. The presence of dendritic spikes in pyramidal cells modifies the classical concepts of dendritic integration but the strong non-linear behaviour makes difficult the experimental study. A number of factors control their initiation and spread. We explore here with an experimentally calibrated model the relation of these spikes to cell output during synchronous input and spontaneous firing. A small number of firing dendrites within a reduced timescale is enough to cause cell output. We show that dendritic inhibition and modulation of the excitability at the main apical shaft offer a wide variety of input-output relations.

1 Introduction

Voltage-dependent currents in dendrites and inhibition modulate the information flow between synaptic and decision areas. During the electrotonic spread of synaptic potentials there is an unavoidable recruitment of intrinsic dendritic currents that may either maintain or modulate their impact at the action potential (AP) triggering zones [1-3]. A complicating factor for the study of input-output relations is that these currents may also initiate APs (dendritic spikes) in multiple loci [4,5] which may remain local or propagate to adjacent regions or the soma and axon initial segment (AIS), in which case dendrites become alternative zones for output decision [6,7]. The later possibility supports the notion of parallel dendritic processing in dendritic subregions. Their experimental study is subject to error because V-dependent currents have a strong non-linear behaviour that makes them highly sensitive to many factors hard to control during experiments. In fact, we have reported some differences between the gross behaviour of dendritic spikes and their relation to cell output between the in vitro slice preparation and the intact animal [3,6]. Up or down regulation of dendritic excitability depends on intrinsic and network factors [8], many of which are likely to be modified during ongoing activity, by tissue manipulation and recording, or by the choice of the preparation. In fact, available data are highly variable and often contradictory. In this work we used a realistic CA1 pyramidal cell model of average dendritic morphology calibrated with in vivo and in vitro results [9,10] to analyse the influence of some well known factors that govern the synaptic initiation of dendritic APs and the capability of the apical shaft to decide cell output.

J. Mira and J.R. Álvarez (Eds.): IWINAC 2005, LNCS 3561, pp. 143–149, 2005.

2   The Model

We used a previously developed compartmental model [9] that reproduced in detail the average pyramidal cell morphology. This was simulated using 265 compartments, distributed in soma, apical and basal dendritic trees and axon (a 2-D projection of the model neuron is shown in Fig. 2). A detailed description of the cell morphology can be found in http://navier.ucsd.edu/calps. In this study we chose a pyramidal morphology without major apical dendritic bifurcation, that represents about half of the CA1 pyramidal cells. The electrotonic parameters for soma and dendrites were R_m=70,000 Ω/cm^2, $R_i = 75$ $\Omega \cdot cm$ and C_m=0.75 $\mu F/cm^2$, while R_m was 100 and 0.5 Ω/cm^2 for non-myelinated and myelinated axon compartments, respectively. The later had a C_m of 0.04 $\mu F/cm^2$. The input resistance measured at the soma was 140 $M\Omega$, and τ was 25 ms. The ion channels, their kinetics and distribution along the morphology of the model neuron are as described in previous work [9,10]. Conductance variables were described with Hodgkin-Huxley type formalism. The details can be obtained from http://navier.ucsd.edu/calps. Simulation of antidromic stimuli was made by 1 nA, 0.1 ms long pulses in a Ranvier node. Alpha functions were used to simulate synaptic activation. GABAA mediated currents had a τ of 7 ms and reversal potential of -75 mV. These were distributed in the soma and proximal apical shaft, and initiated 1.5 ms after Glu activation. The non-NMDA type of Glu-r mediated excitatory activation was simulated with a τ of 2 ms and reversal potential of 0 mV. The number of activated synapses was simulated by varying the total excitatory conductance. This was evenly distributed throughout the entire surface of all individual dendritic branches making up the band of dendritic activation. Antidromic activation was used to fit the individual AP waveforms in different dendritic loci without the interference of previous synaptic activity. The simultaneous fitting along the entire somatodendritic axis was obtained by large-scale reproduction of the spatial field potential map for the antidromic population spike, using a realistic spatial arrangement of thousands pyramidal cells [9,10]. The computed population spike is extremely sensitive to small variations in unitary parameters. The so calculated channel distributions yielded AP thresholds of 7 to 9 mV and 30 mV at the soma and apical dendritic loci, respectively, which is well within reported experimental values [3]. The detailed channel distributions can be obtained from http://navier.ucsd.edu/calps. Briefly, the above AP features were obtained using a moderate homogeneous Na$^+$ channel density along the somatodendritic compartments (500 pS/m^2) and a slightly higher axon density (1200 pS/m^2), a somatofugally increasing density of the KA current [11], and a slow recovery from inactivation of the dendritic Na$^+$ channels [12]).

3   Experimental Recordings

Tissue slices of the hippocampus were obtained as described elsewhere [3]. Glass micropipettes filled with the appropriate electrolytes were used for intra and extracellular recordings, and monopolar tungsten electrodes used for electrical stimulation. Some drugs (TTX) were microejected locally by pressure ejection from the pipette to affect a small subcellular region, while others (PTX) were directly applied to the bath.

4   Results

In actual experiments, the synchronous activation of the Schaffer collaterals caused cell firing of the CA1 pyramidal population, giving rise to the ortho-dromic population spike (PS) that can be recorded at the soma and proximal apical layers. Stimulation of the CA1 axons caused the antidromic invasion of the AP that backpropagated into the soma and dendrites sequentially (Fig. 1,A). The local blockade of dendritic Na channels by TTX application (grey tracings) caused the blockade of the orthodromic PS and AP in dendrites and somata but not the antidromic PS (Fig. 1,A, upper). However, when the TTX is applied in the axonal region (lower) the orthodromic PS and AP remained intact, and the antidromic PS was reduced. These results indicate that the AP is initiated in the apical dendrites during synchronous activation and propagates in a contin-uous manner toward the soma and axon. When inhibition is blocked by bath application of the GABA antagonist picrotoxin (PTX), the local blockade of Na channels at the proximal apical shaft does not avoid firing, although it was

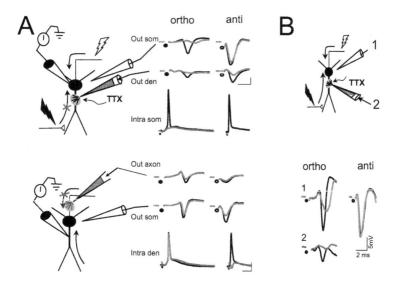

Fig. 1. Dendritic initiation of AP firing in the CA1 pyramidal cells in vitro. A: Local blockade of Na$^+$ channels in the proximal apical shaft (upper) or the axon (bottom) blocked the orthodromic or the antidromic firing, respectively. B: Removal of inhibition restored the Schaffer capability to fire cells even in absence of proximal (blocked) Na channels

delayed (Fig. 1,B), likely originated at the axon initial segment. These results indicate that the Schaffer input alone is not enough to fire the pyramidal cells, and the level of inhibition may control when cell output occurs and the spike initiation zone. Figure 2 shows the model results for the study of the initiation

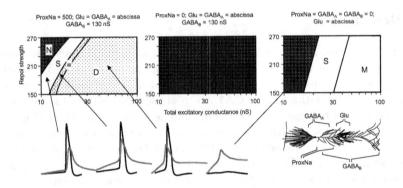

Fig. 2. Parametric study of the initial AP sites. The input strength is co-analyzed with the repolarizing ability of the apical dendrites. Left panel shows the control results. Middle panel is in absence of proximal Na channels. Right panel shows the firing modes after removal of inhibition and under proximal Na channel blockade. S: soma-first; D: dendrite first; M: multiple firing; N: no firing

site as a function of the input strength and the repolarizing strength of the apical dendrites (slope of density of KA channels). At low intensities in control, the AP either was not generated or it was in dendrites (left panel). Strong inputs biased the initial locus to the axon-soma regions. When the proximal Na currents were blocked, there was no firing at any input strength (middle panel). Removal of inhibition restores the firing ability, even during proximal Na channel blockade (right panel). These results reproduce the experimental results faithfully. We next studied the mechanisms of dendritic integration that result on AP output. Figure 3 show the model results for different firing modalities. During low intensity, even if no AP is generated at the axon, some lateral dendrites did fire a local AP. Increasing the number of lateral dendrites that fire a local AP managed to set an AP at the distal portion of the apical shaft. This failed to conduct the AP, which began to fade off (curved arrow). However, the AP was regenerated at the axon initial segment in a pseudosaltatory manner. When a sufficient number of lateral dendrites fired an AP within a short time, the AP generated in the apical shaft propagated in a continuous non-decremental manner (full forward). The examples shown are not directly related to the strength of the input. In fact, it was lower for the full-forward modality. The difference is that for the higher excitatory input, there was also a stronger inhibition, which scattered the firing of the lateral dendrites so they could not overlap their contributions to the apical shaft. It is usually thought that spontaneous firing is always initiated

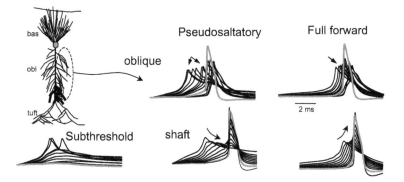

Fig. 3. The number and temporal scatter of local spikes determine the mode of firing and dendrite-to-axon coupling. Low intensity does not fire a somatic AP (gray tracings), although ha few local spikes were initiated. When more local spikes fired, an AP is initiated in the distal part of the apical shaft that failed to propagate actively to the axon. However, a full AP is regenerated there (thin tracing). Decreasing the temporal scatter of local spikes causes the apical shaft AP to conduct nondecrementally all the way to the axon

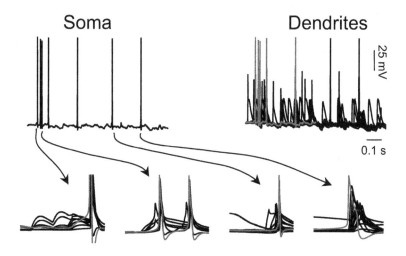

Fig. 4. Spontaneous firing is also preceded by dendritic spiking. One second of spontaneous firing in a model neuron in which random activation of excitatory and inhibitory inputs was set on their apical dendrites. Each AP has a different history. Four out of six had at least one dendritic spike ahead of the axon AP

in the axon. There is no experimental assessment for such claim, however. We computed a few cases of spontaneous firing by entering excitatory inputs in the apical compartment at a random frequency (below $5Hz$). Figure 4 shows an example of 1 second during which the model neuron fired six APs. Each AP had quite a different history, and in 4 out of 6 there was some local dendritic spike ahead of the AP. In a strict sense, these should be considered pseudosaltatory firing cases, i.e., the output decision was already made on the apical dendrites.

5   Discussion

Several mechanisms have been found by which distal inputs amplify their impact on cell output, such as the somatofugal increase of synaptic conductance [13] or the recruitment of dendritic boosting current [2]. In predominant single cell in vitro research, the shift of the AP initial locus from the axon/soma to the apical trunk is achieved by increasing the stimulus intensity. This test is considered evidence for the axon/soma locus being the only subregion deciding the final output [14]. However, using the same intensity different modalities of firing and dendrite-to-axon coupling can be achieved [3] that highlight the complex processing within the dendritic tree. Among other variables, we showed that inhibition and modulation of Na channels at the proximal apical shaft causes major shifts in the amount of excitatory input required to fire a cell. The possible differences between the EPSP required to initiate a soma-first spike or a dendrite-first spike could hardly be large enough as to imply that the axon/soma is the preferred AP initiation site during ongoing activity. In fact, these cells fire at a very low rate and they do so upon bursting activity in the CA3. In our preliminary study with ongoing activity we found that the axonal output was preceded in many cases by at least one dendritic spike combined with other slower inputs. In the end, we might say that dendrites can take output decision in most cases, although this is hardly visualized by conventional recordings that have a difficult access to the small lateral dendrites. The computational approach can at least describe the principles on which this spike-based dendritic output decision is taken. Knowing which combinations of inputs and in which circumstances will cause the cell to fire is the next challenge of unitary studies.

Acknowledgement

This work has been supported by grants 8.5/15/98 of the Comunidad Autónoma de Madrid and PB97/1448 and BEFI 2004/1767 of the Spanish DGICYT.

References

1. Stuart G and Sakmann B. Amplification of EPSPs by axosomatic sodium channels in neocortical pyramidal neurons. Neuron 15:1065-76, 1995.
2. Lipowsky R, Gillessen T, and Alzheimer C. Dentritic Na^+ channels amplify EPSPs in hippocampal CA1 pyramidal cells. J. Neurophysiol 76:2181-2191, 1996.
3. Canals S, López-Aguado L, and Herreras O. Synaptically-recruited apical currents are required to initiate axonal and apical spikes in hippocampal pyramidal cells: modulation by inhibition. J Neurophysiol 93: 909-918, 2005.
4. Masukawa LM and Prince DA. Synaptic control of excitability in isolated dendrites of hippocampal neurons. J Neurosci 4: 217-227, 1984.

5. Wong R, and Stewart M. Different firing patterns generated in dendrites and somata of CA1 pyramidal neurones in guinea-pig hippocampus. J Physiol 457:675-687, 1992.

6. Herreras O. Propagating dendritic action potential mediates synaptic transmission in CA1 pyramidal cells in situ. J. Neurophysiol 64: 1429-1441, 1990.

7. Chen WR, Mitgaard M, and Shepherd GM. Forward and backward propagation of dendritic impulses and their synaptic control in mitral cells. Science 278:463-468, 1997.

8. Tsubokawa H and Ross WN. IPSPs modulate spike backpropagation and associated $[Ca^{2+}]i$ changes in the dendrites of hippocampal CA1 pyramidal neurons. J Neurophysiol 76: 2896-2906, 1996.

9. Varona P, Ibarz JM, López-Aguado L, and Herreras O. Macroscopic and subcellular factors shaping CA1 population spikes. J Neurophysiol 83:2192-2208, 2000.

10. López-Aguado L, Ibarz JM, Varona P, and Herreras O. Structural inhomogeneities differentially modulate action currents and population spikes initiated in the axon or dendrites. J Neurophysiol 88:2809-2820, 2002.

11. Hoffman, D. A., Magee, J. C., Colbert, C. M. and Johnston, D. K^+ channel regulation of signal propagation in dendrites of hippocampal pyramidal neurons. Nature 387: 869-875, 1997.

12. Colbert, C.M., Magee, J.C., Hoffman, D.A., and Johnston, D. Slow recovery from inactivation of Na^+ channels underlies the activity-dependent attenuation of dendritic action potentials in hippocampal CA1 pyramidal neurons. J. Neurosci. 17:6512-6521, 1997.

13. Magee JC and Cook EP. Somatic EPSP amplitude is independent of synapse location in hippocampal pyramidal neurons. Nat Neurosci 3: 895-903, 2000.

14. Turner RW, Meyers DER, Richardson TL, and Barker JL. The site for initiation of action potential discharge over the somatodendritic axis of rat hippocampal CA1 pyramidal neurons. J Neurosci 11:2270-2280, 1991.

Slow Conductances Encode Stimulus History into Spike Shapes

Gonzalo G. de Polavieja[1], Annette Harsch[2], Hugh Robinson[2],
and Mikko Juusola[2]

[1] Neural Processing Laboratory, Department of Theoretical Physics,
Universidad Autonoma de Madrid, Madrid 28049, Spain
[2] Physiological Laboratory, University of Cambridge,
Cambridge CB2 3EJ

Abstract. The shape of action potentials plays an important role in synaptic integration. Action potentials of different shapes shunt excitatory potentials differentially and consequently correspond to different probabilities of generating the next spike. Thus two neurons producing different spikes shapes, say Purkinje and pyramidal cells, integrate differently the same excitatory potentials. More interestingly, there is variability in the spike shape of a single neuron when stimulated dynamically, that can then also dynamically affect the synaptic integration. Our recent experiments have shown that this variability in the spike shape is not random but depends on stimulus history. Here we analyze a simple model of cortical neuron to understand the origin of this encoding of stimulus history into spike shape. We find that slow conductances, for example calcium conductances, can be responsible for this rich encoding.

1   Introduction

The integrate-and-fire model of a neuron gives a simple intuitive picture of neuronal processing [1]. Excitatory potentials arriving at the neuron are integrated and, when reaching a threshold value, the neuron fires an action potential. The model assumes that the potential of the neuron resets to its resting value after firing. The reason for this resetting is that the action potential shunts the incoming excitatory potentials. The integrate-and-fire model is known to misrepresent important features of the real dynamics, including the resetting. Recent experiments in pyramidal and Purkinje neurons have shown that the resetting is not complete [2]. Three factors influence the degree of resetting: the action potential shape, the shape of the excitatory potential and the time interval between these two events. The same authors show that this interaction has consequences for synaptic integration. As the excitatory potentials are shunted to a degree that depends on the three factors mentioned, a train of EPSPs may reach the threshold for firing depending on them.

Different neurons then integrate the same excitatory potentials differently depending on the shape of the action potential they produce. In *in vivo* conditions, however, a more interesting picture can take place as the same neuron

J. Mira and J.R. Álvarez (Eds.): IWINAC 2005, LNCS 3561, pp. 150–155, 2005.

can produce spikes of different shapes. In dynamical conditions, there in no time for the recovery of the conductances between two action potentials. Even simple models like the Hodgkin-Huxley model for the squid axon [1] show different spike shapes when stimulated by a Poisson train of excitatory potentials.

Our recent experiments show that the shape of the action potentials depends on the past stimulus history [3]. After reviewing the results of these experiments, we analyze the behaviour of a simple model of cortical neuron when using the same stimulus than in our experiments. We find that the slow conductances are responsible for the encoding of stimulus into spike shapes.

2   Spike Shapes in Naturalistic Conditions

Details of our experimental procedures can be found in [3]. Briefly, whole-cell recordings from the somas of neurons in cortical layers 2/3 and 5 were carried out in traverse slices from occipital cortex of $13 - 23$ days old Wistar rats. To mimic natural input arriving at the spike producing site, we used the dynamic clamp technique [4, 5, 6]. While more standard experiments simply inject a current profile fixed in advance, this technique consists in injecting the current that obeys the equations of natural synapses. The equation for the dynamics of the voltage difference between the inside of the neuron and the outside, $V(t)$, is of the form

$$C\frac{dV}{dt} = I_{\text{spike}} + I_{\text{synaptic}}, \tag{1}$$

with C the membrane capacitance. The current I_{spike} contains the spike-producing conductances with dynamics described with additional differential equations that also depend on the voltage $V(t)$. The synaptic current I_{synaptic} has the form

$$I_{\text{synaptic}} = g_{\text{AMPA}}(t) \left[V - E_{\text{AMPA}}\right] + g_{\text{GABA}}(t) \left[V - E_{\text{GABA}}\right]$$
$$+ \frac{g_{\text{NMDA}}}{1 + k_1 \exp(-k_2 V(t))} \left[V - E_{\text{NMDA}}\right] \tag{2}$$

The technique of dynamic clamp consists in (1) measuring the value of voltage $V(t)$, (2) calculating I_{synaptic} and (3) injecting this current into the neuron. Note that (1)-(3) are done for every time t, in practice in the microsecond regime. To mimic naturalistic input, the conductances $g_{\text{AMPA}}(t)$, $g_{\text{GABA}}(t)$ and $g_{\text{NMDA}}(t)$ are a sum of elementary events. For example, $g_{\text{AMPA}}(t) = \sum_i g_i(t)$ with $g_i = \bar{g}\left[\exp(-(t - t_i)/\tau_d - \exp(-(t - t_i)/\tau_r\right]$ and $\bar{g} = 1000$, $\tau_d = 2$ and $\tau_r = 0.5$. Values for the GABA and NMDA cases and the rest of parameters can be found in [3]. The times $\{t_i\}$ are determined by a random process with time intervals obeying the Poisson distribution $p(T) = \lambda \exp(-\lambda T)$ or from a non-stationary process whose rate is given by $\lambda(t) = \bar{R}\sum_i \exp(-(t - T_i)/\tau_b$ for $t \geq T_i$, with $\bar{R}$ the initial peak rate, τ_b the burst decay time constant and $\{T_i\}$ withdrawn from a Poisson distribution. $\bar{R}$ and τ_b were adjusted until the neuron fired vigorously.

Figure 1(a) shows the typical statistics found for the heights and widths of action potentials during approximately 30 minutes long recordings. 56 neurons from different layers and temperatures showed different slopes for the distribution (here layer 2/3 and 23 degrees) but always an approximate linear relationship between the two variables. To examine whether action potentials of different shapes were produced by different stimuli, we divided the data by their width into the 7 groups, indicated in Figure 1(a) by vertical bands and numbers. Figure 1(b) gives the average stimulus before a spike for each of the 7 groups. It is clear that the stimulus amplitude is coded into different spike widths. In ([3]) we further analyzed this coding using information theory. We found that the spike shapes increase the information rate by 3 − 4 times.

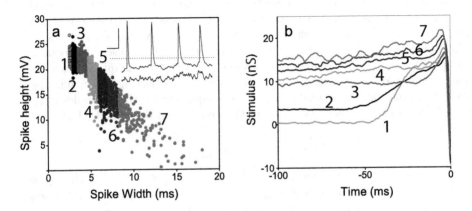

Fig. 1. Encoding of stimulus history into spike shapes in of cortical neurons. (a) Spike height versus spike width. We divide the widths into 7 groups for analysis. (b) Average stimulus before the spikes for each of the 7 groups in (a)

3   Slow Conductance Encoding Stimulus Features into Spike Shape

To understand the origin of the encoding of stimulus history into spike shapes, we have turned to modeling. Our purpose is to find the key element responsible for the encoding of stimulus history into different spike shapes. We first used a Hodgkin-Huxley model of a squid axon and found no encoding into spike waveforms even if spikes showed different shapes. We then analyzed a simple model of a cortical cell used by Wilson [7]. The Wilson model takes into account the Na^+ and K^+ spike-producing currents, the Ca^{2+} current I_T and the hyperpolarizing current I_{AHP}. The equations for the dynamics of this model are

$$C\frac{dV}{dt} = -m_\infty(V - 0.5) - 26R(V + 0.95) - 1.2T(V - 1.2)$$
$$- 3.4H(V + 0.95) + I_{external} \tag{3}$$

$$\frac{dR}{dt} = (-R + R_\infty)/4.2 \tag{4}$$

$$\frac{dT}{dt} = (-T + T_\infty)/14 \tag{5}$$

$$\frac{dH}{dt} = (-H + 3T)/45. \tag{6}$$

with $m_\infty = 17.8 + 47.6V + 33.8V^2$, $R_\infty = 1.24 + 3.7V + 3.2V^2$ and $T_\infty = 8(V + 0.725)^2$. This model has made many simplifications. For example, it takes into account very few of the dozen conductances of cortical neurons, the dynamics of sodium channel is considered instantaneous and the dependence of the on voltage in all equations has a simple polynomial form. Despite these simplifications, Wilson has shown that the model correctly describes the spike shape in response to a current step and reproduces the major classes of dynamics by changing the values of the parameters [7]. In our calculations we used the expression $I_{synaptic}$ in (2) for the external current $I_{external}$ to simulate our dynamic clamp experiments. The results were analyzed as before for the experimental data.

Figure 2 gives an analysis of the model analogous to the one given in Figure 1 for the experimental data. Figure 2(a) gives the height versus width of the numerically obtained action potentials. As in the experimental data, there is a distribution and not a single point and the relationship between the two variables is approximately linear. Note that the action potentials in the model are thinner.

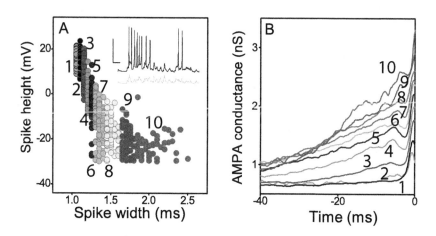

Fig. 2. Encoding of stimulus history into spike shapes in the Wilson model of cortical neuron. (a) Spike height versus spike width. We divide the widths into 10 groups for analysis. (b) Average stimulus before the spikes for each of the 10 groups in (a)

This is so because the parameters in the model are for a temperature of 34°C. Our experimental results for similar temperatures are also thin $(1-6$ ms) [3]. We classified the action potentials into 10 groups for analysis, indicated by vertical divisions in Figure 2(a). Figure 2(b) gives the average stimulus before each of the 10 groups chosen for analysis. Models without the slow conductances do not have different stimuli before spikes of different shapes. It is then clear that the calcium conductance is responsible for the encoding of stimulus amplitude into spike shape.

It is however instructive to consider the differences between experimental and modeling results. The model shows an encoding of 15 ms of stimulus history whereas experimentally this value is approximately 50 ms. Longer time scales in the slow conductances are needed. However, simply changing the time constant in the calcium dynamics of the Wilson model does not have the desired result. This is so because the ability of the slow conductance to integrate stimulus features has to be matched to an added flexibility of the action potential to vary the shape accordingly.

4   Local Processing with Information-Rich Signals

We have seen that a simple model of cortical neuron already shows that slow conductances are responsible for the encoding of stimulus features into spike shapes. What is the role of this encoding? The interaction between excitatory potentials and action potentials is an operation that compares information at two different times. This is so because the action potential shape depends on the stimulus history. For concreteness, let us consider a possible, yet speculative, scenario. Simple models of neuronal integration are based on a threshold idea: when the sum of excitatory potentials reaches a threshold value, the neuron fires an action potential. In terms of a 'voting' rule, this simple model implies that a spike is produced when there are enough votes. However, our results imply that different shapes distinguish the amplitude of the stimulus. The shape thus identifies the actual number of voters and not only that the number of votes is above a threshold. More interestingly, the different spike shapes are able to code the number of voters in the 50 ms before spiking. So if a spike has been produced by many voters then it is wider and probably shunts excitatory potentials to a larger degree. That is, the number of voters producing a spike influences the importance of future votes. Therefore the 'politics' of synaptic integration does probably allow for a rich processing.

References

1. Koch, C.: Biophysics of computation. Oxford Univ. Press: New York (1999)
2. Häusser, M. Major, G. and Stuart, G.J.: Differential shunting of EPSPs by action potentials. Science **291** (2001) 138–141
3. de Polavieja, G.G., Harsch, A., Kleppe, I., Robinson H.P. and Juusola, M.: Stimulus history reliably shapes action potential waveforms of cortical neurons (to appear)

4. Robinson, H.P.C.: Kinetics of synaptic conductances. Neurosci. Res. **16** (1991) S6
5. Robinson, H.P.C., Kawai, N.: Injection of digitally synthesized synaptic conductance transients to measure the integrative properties of neurons. J. Neurosci. Methods **49** (1993) 157–165
6. Sharp, A.A., O'Neil, M.B., Abbott L.F. and Marder, E.: Dynamic clamp-computer generated conductances in real neurons. J. Neurophysiol. **69** (1993) 992–995
7. Wilson, HR: Simplified dynamics of human and mammalian neocortical neurons. J. Theor. Biol. **200** (1999) 375–388

Comparison of Plasticity of Self-optimizing Neural Networks and Natural Neural Networks

Adrian Horzyk and Ryszard Tadeusiewicz

University of Science and Technology, Department of Automatics,
Mickiewicza Av. 30, 30-059 Cracow, Poland
horzyk@agh.edu.pl
rtad@ia.agh.edu.pl

Abstract. The paper interplays between plasticity processes of natural neural networks [9] and Self-Optimizing Neural Networks (SONNs) [7]. The natural neural networks (NNNs) have great possibility in adaptation to environment. The possibility to adapt is based on the chemical processes changing synaptic plasticity and adapting neural network topology during life. The described SONNs are able to adapt their topology to the given problem (i.e. artificial neural network environment) in the functionally similar way the natural neural networks do. The SONNs as well as the NNNs solve together the two essential problems: neural networks topology optimization and weights parameters computation for the given environment. The ontogenic SONNs development gradually adapts network topology specializing the network to the given problem. The fully automatic deterministic self-adapting mechanism of SONNs does not use any *a priori* configuration parameters and is free from different training problems.

1 Introduction

Artificial neural networks (ANNs) are very modern computational tool used to solve many difficult problems. However, we have many types of ANNs today it is still necessary to develop new better adaptable and adjustable neural networks. Today, an important group of neural networks that are able to adapt their structure and parameters to different problems establishes ontogenic neural networks. The presented ontogenic Self-Optimizing Neural Networks (SONNs) develop the network topology and adjust the weights in the self-adapting process. The SONNs gradually add neurons and interconnections between different layers in the similar way the natural neural networks (NNNs) behave during the embryological development of some parts of brain and plasticity processes of interconnections between neurons [1, 3, 9]. It is well known that natural brains can adapt to the environment by rebuilding some parts of the networks structure in plasticity processes of neurons and interconnections. This feature helps many people come into health after some damages and injuries of central nervous systems. The topology of NNNs are partially programmed by the genetic code of an individual and partially by the environment of the growing brain. There are

J. Mira and J.R. Álvarez (Eds.): IWINAC 2005, LNCS 3561, pp. 156–165, 2005.

many psycho-somatic, physical, chemical and electrical factors that can affect the brain development from its embryo to its death [9].

The NNNs are able to specialize in solving different problems sensitizing its behavior and render its intelligence into previously solved problems. The SONN also adapts its structure and adjusts the weights to the given problem it should solve. The topology of the SONN is always build up and adjusted for the given training data. In such a way, the given adapted SONN is specialized in solving the given problem. The environment and the NNN always interplay between each other changing parameters both the given environment and the NNN. In such a way, the development of ANNs can lead to produce specialized ANNs better or more effective solving some problems than NNNs do because ANNs are created in their structure and adjusted in their parameters to those given problems, i.e. ANNs environment [2, 12, 13, 14]. At the beginning, the SONNs - the classification networks - as well as NNNs project the most discriminative and well-differentiating features of the training data into the structure and parameters (weights) of the neural network development process. In the following steps the network adapts its structure and parameters to still more and more details of the training data necessary to correctly classify all training data. This way of SONNs adapting is very similar the way of remembering and classification learning people do. The SONNs stop the process of adaptation when all training data are correctly classified constructing the specific network structure and adjusting weight precisely to the given problem. If training data changes, the structure and parameters of the SONN have to change as well. This feature is also similar the natural processes. If a human specializes in solving some kind of problems, the part of brain concerned with solving of these kind of problems expands and also partially rebuilds its parameters in varied plasticity processes.

If training data are corrupted or not univocally qualify training data to the given classes, the SONN as well as NNNs notice at last that there is no possibility to find out any difference between such data using all accessible features. In such cases the adapting process of SONN stops. The result of classification for corrupted data is very natural - the SONN answers that such data represent two or more classes as precisely as they have been defined in the training data set.

Moreover, the SONNs automatically reduces the set of input features used to discriminate training data of different classes. Such processes are also natural in human learning process omitting features that do not help to differentiate data.

Furthermore, the SONN automatically recognizes inversion of the input training data, classifying the inverted data correctly to the classes and returning information about the inversion of the input data.

Finally, the SONN as well as NNNs can increase the knowledge about training data details. The adaptation process of SONNs can be continued even if all training data are correctly and univocally classified.

2   SONN – Mathematical Background

The ontogenic Self-Optimizing Neural Networks (SONNs) develops irregular multilayer partially connected neural network topology with rare interlayer and supralayer conncections (Fig. 1). The NNNs are also partially connected in contrast to many classical adaptation on ANNs, e.i. fully connected structures of multilayer perceptron. The SONN adaptation process is deterministic and free from many training problems that are characteristic for many other neural networks algorithms: initial topology and parameters (weights) establishment, redundant or insufficient network structure, local minima, stopping condition, overfitting [1, 3, 5, 6, 7] etc.

The SONN is created in complex optimizing process after the statistical analysis of the training data and probabilistic estimation of all discrimination features properties in view of all classes, whole data set and taking into account the quantitative differences of class representation [7].

Neither initial structure nor weights parameters of SONN does need to be matched experimentally, because the whole SONN structure is created from the whole beginning reflecting the most discriminant features of the training data.

The training data $U = \{(u^1, C^{m_1}), \ldots, (u^N, C^{m_N})\}$ usually consist of the pairs: the input vector $u^n = [u_1^n, \ldots, u_K^n]$ (defined for SONNs as $u_k^n \in \{-1, 0, +1\}$ with following interpretation: false (-1), don't known or undefined (0), true $(+1)$)) and the adequate class $C^{m_n} \in \{C^1, \ldots, C^M\}$. The relevance (importance) of the features $1, \ldots, K$ of any input vector $u^n \in C^{m_n}$ can be estimated in view of the classification process. In order to extract the most important features of the patterns that characterize each class and discriminate it from others the definition of the following three coefficients will be useful:

1. Coefficient $p_k^n \in [0; 1]$ is a measure of representativeness of feature k of pattern n in its class m and is computed as the Laplace probability [4] of correct classification of the given training pattern n after the feature k:

$$\forall_{n \in \{1, \ldots, N\}} \forall_{k \in \{1, \ldots, K\}} \; p_k^n = \begin{cases} \frac{P_k^m}{P_k^m + N_k^m} & \text{if } u_k^n > 0 \;\& \; u^n \in C^m \\ 0 & \text{if } u_k^n = 0 \\ \frac{N_k^m}{P_k^m + N_k^m} & \text{if } u_k^n < 0 \;\& \; u^n \in C^m \end{cases} \tag{1}$$

where $\forall_{m \in \{1, \ldots, M\}} \forall_{k \in \{1, \ldots, K\}} \; P_k^m = \sum\limits_{u_k^n \in \mathcal{P}^m} u_k^n \; \wedge \; N_k^m = \sum\limits_{u_k^n \in \mathcal{N}^m} -u_k^n,$

$\mathcal{P}^m = \{u_k^n : u_k^n > 0 \& u^n \in U \cup C^m \& n \in \{1, \ldots, N\}\},$
$\mathcal{N}^m = \{u_k^n : u_k^n < 0 \& u^n \in U \cup C^m \& n \in \{1, \ldots, N\}\}.$

2. Coefficient $q_k^n \in [0; 1]$ measures how much is the value of the feature k of the training pattern n representative for class m in view of the whole data set. This coefficient makes the SONN insensitive for quantity differences of training patterns representing individual classes in the training data set. This coefficient is computed as the Laplace probability [4] of correct classification of the training pattern n from class m after the feature k:

$$\forall_{n\in\{1,...,N\}}\forall_{k\in\{1,...,K\}} \quad q_k^n = \begin{cases} \frac{P_k^m}{P_k} & \text{if } u_k^n > 0 \ \& \ u^n \in C^m \\ 0 & \text{if } u_k^n = 0 \\ \frac{N_k^m}{N_k} & \text{if } u_k^n < 0 \ \& \ u^n \in C^m \end{cases} \quad (2)$$

where $\forall_{k\in\{1,...,K\}} \ P_k = \sum\limits_{u_k^n \in \mathcal{P}} u_k^n, \mathcal{P} = \{u_k^n : u_k^n > 0 \& u^n \in U \& n \in \{1, ..., N\}\}$ and

$\forall_{k\in\{1,...,K\}} \ N_k = \sum\limits_{u_k^n \in \mathcal{N}} -u_k^n, \mathcal{N} = \{u_k^n : u_k^n < 0 \& u^n \in U \& n \in \{1, ..., N\}\}$

3. Coefficient $r_k^n \in [0; 1]$ measures how much is the value of the feature k of the training pattern n rare in view of the whole data set and is computed as follows:

$$\forall_{N>1}\forall_{n\in\{1,...,N\}}\forall_{k\in\{1,...,K\}} \quad r_k^n = \begin{cases} \frac{N_k}{P_k+N_k-1} & \text{if } u_k^n > 0 \\ 0 & \text{if } u_k^n = 0 \\ \frac{P_k}{P_k+N_k-1} & \text{if } u_k^n < 0 \end{cases} \quad (3)$$

The coefficients (1,2,3) are used to define the fundamental coefficient of discrimination $d_k^n \in [0; 1]$ that defines how well the feature k discriminate between the training pattern n of the class m and the training patterns of the other classes:

$$\forall_{n\in\{1,...,N\}}\forall_{k\in\{1,...,K\}} \atop d_k^n = p_k^n \cdot q_k^n \cdot r_k^n = \begin{cases} \frac{(P_k^m)^2 \cdot N_k}{(P_k^m+N_k^m) \cdot P_k \cdot (P_k+N_k-1)} & \text{if } u_k^n > 0 \ \& \ u^n \in C^m \\ 0 & \text{if } u_k^n = 0 \\ \frac{(N_k^m)^2 \cdot P_k}{(P_k^m+N_k^m) \cdot N_k \cdot (P_k+N_k-1)} & \text{if } u_k^n < 0 \ \& \ u^n \in C^m \end{cases} \quad (4)$$

The discrimination coefficient is insensitive for quantitative representation of classes and concerns statistical differences of features quantity after their values in classes and the training data set. The bigger discrimination coefficient the more important and discriminative is the feature in view of classification and *vice versa*. The following architecture optimization process starts from the selection of the most important and discriminative features (called major features in the Fig. 1) for each training pattern.

The major features are selected after the maximal values of the corresponding discrimination coefficients and used to construct an ontogenic NN. The construction process of the NN aggregates the features of the training patterns of the same class after the sign of their values and the quantity of their same values in such a way to minimize the quantity of connections in the constructed SONN (Fig. 2). The fundamental connections of the NN structure form a tree for each class (Fig. 1) completed with the input connections. After each construction period the NN is evaluated for all training patterns and the answers of the NN are checked up on univocal classification. If the answers are not univocal the incorrectly classified training patterns are marked and represented by more major features in the next development period in order to achieve their discrimination from other patterns of different classes. The ontogenic construction process stops just after all training patterns are univocally classified. The correctness of training patterns classification is built in the construction process and is always fully performed. The resultant NN correctly discriminates all training patterns of

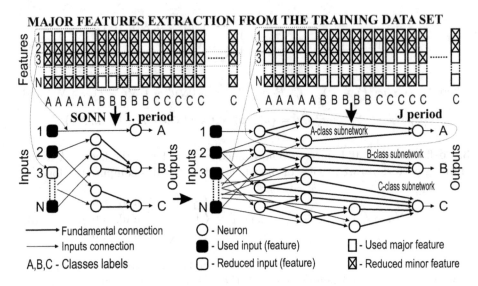

Fig. 1. The SONN development process based on the most discriminative features

different classes. The SONNs adapt simultaneously structure and weights. The weights values are computed for all major features $i \in J$ (Fig. 1) as follows:

$$\forall_{i \in J} \; w_i = \frac{u_i^n \cdot d_i}{\sum\limits_{j \in J} d_j} \tag{5}$$

where $\forall_{n \in \{1,...,N\}} u_0^n = 1$ and $d_0 = \sum\limits_{l \in L} d_l$ for each neuron except the input neurons, $\forall_{n \in \{1,...,N\}} u_0^n = 0$ and $d_0 = 0$ for each input neuron, $J = A \cup \{0\}$ is the set of features that are transformed to connections of the neuron, $A \subset \{1, 2, ..., K\}$ is the set of accordant already not used (in previous layers) features related with training patterns concerned with the neuron, $L \subset \{1, 2, ..., K\} \cup \{0\}$ is the set of features of preceding neuron that are already transformed into connections.

The activation function of neurons except the output ones is defined as

$$y = f(u) = \sum\limits_{j \in J} w_j \cdot u_j \tag{6}$$

The activation function of output neuron m is defined as

$$y^m = f_{SGMMAX}(Y) = \begin{cases} y_{MIN} & if \; |y_{MAX}| < |y_{MIN}| \\ y_{MAX} & if \; |y_{MAX}| \geq |y_{MIN}| \end{cases} \tag{7}$$

where $Y = \{y^1, \cdot, y^T\}$, y^t is the output of some neuron of a previous layer, $y_{MIN} = \min\{y_1, \cdot, y_T\}$, $y_{MAX} = \max\{y_1, \cdot, y_T\}$. The output neurons return the values in the rage $[-1; +1]$. The SONN outputs give information about the similarities of input vector to each class. The maximum output value appoints the class to which the input vector is the most similar.

Final classification is a simple process of SONN evaluation. The input vector can consist of bipolar $\{-1; +1\}$, trivalent $\{-1; 0; +1\}$ or continuous values in the rage $[-1; +1]$.

The described SONN development resembles the process of NNNs embryological and next development. The training data inputs (input data dimension) are automatically reduced as a side-effect of the described process. The minor inputs are reduced if the corresponding features are not included into the neural network construction process. The curse of dimensionality problem is automatically solved in view of SONN methodology. The maximum number of SONN configuration periods is equal the input dimension of given training data.

The development process of the SONN can be continued even after the adapting algorithm stops and the neural network correctly and univocally classifies all training data. The next development of the SONN focuses on details of the training data. The details development of SONN is sometimes necessary to correctly classify testing data that are very different from the training data. Such development increases network size and sometimes generalization property of the SONN.

The described SONN development process is very fast even for big data bases because discrimination coefficients are ones computed for all training patterns. The major features are simply selected after the maximum discrimination coefficients.

Even not exactly defined input vectors can be used as training patterns with no essential harm to the described configuration process. For the present, the described self-optimization process of NNs configuration works only on input training vectors consisting of trivalent values $\{-1, 0, +1\}$. In order to use SONN to continuous data, the data have to be quantized. The quantization of inputs of the training data can be performed by intervals or any other method.

There are sometimes more then one output with maximal value than the classification result is not univocal. Such result mean that the given input can pertain one from the classes it has been classified but the discrimination properties of the created SONN (after the given training data) can not univocally differentiate it. In some cases such answer is an advantage that can help human expert to decide or to perform other actions or research.

The negative value of the output neuron suggests that the inversion of the given input vector is more similar to the appropriate class. The classification of inverse inputs can be very interesting when working with images. There is a possibility to automatically classify negative images as well as in the recurrent Hopfield networks [1, 3]. In the other cases the negative output value means hardly any similarity to the considered class.

3   Topology Optimization Process

The SONN topology optimization algorithm minimizes the quantity of interconnections necessary to represent the major features of the training data. The minimization of interconnections leads to optimization on neurons as well as layers of the network. Use of major (the mostly relevant, discriminative and well-differentiating) features of the training data leads to omitting some of the input data features making the input data dimension sometimes smaller than the original one. The neurons quantity, layout and interconnections change from period to period. The similar features for different training samples of the same class are aggregated and transformed into single interconnections. In order to find out the minimum number of interconnections necessary to represent the data in each period of SONN development some so called divided features (DFs) have to be found in the input feature space. The DF is the specially selected feature that enables maximal saving of interconnections when representing a given group of training samples when developing SONN in the given period. (Fig. 2). The given DF enables to disjoint the given set of patterns to two subsets consisting of training patterns of the same value of the DF. The same features for all patterns of the same subset can be represented by single interconnections in the SONN. The described optimization process finds out the DFs in such a way to maximize the number of the same features in the maximally big subsets of the given set of input training patterns. There is used a special kind of table to compute DF (9) for any given group of training patterns (Tab. 1).

The columns "Sums of F" and "Sums of T" in the table 1 mean the sum of all F or T accordingly for the feature in the considered row. The columns "Same for all F" and "Same for all T" in the table 1 mean the quantity of the same signs features for other features that considered one but only for the subset of patterns for which the considered feature is F or T accordingly.

$$Saved = SameF \cdot (SumsF - 1) + SameT \cdot (SumsT - 1) \qquad (8)$$

$$DF = \max_{k=1,...,K} \{Saved_k\} \qquad (9)$$

At the beginning of the interconnection optimization process, the input samples of same classes are aggregated. The initial neural network consists only of single output neuron for each trained class and the inputs referring to avail-

Table 1. The computation of the devided feature DF for the exemplar set of 23 patterns and 6 feature that are true (T,+1) or false (F,-1)

Ft	1	2	3	4	5	6	7	8	9	10	11	12	13	14	15	16	17	18	19	20	21	22	23	Sums of F	T	Same for all F	T	Saved
1	F	T	F	F	F	F	F	F	F	F	T	F	F	F	F	F	F	F	F	F	F	F	F	21	2	1	3	23
2	T	T	T	T	T	T	T	T	T	T	T	T	T	T	T	T	T	F	F	F	F	F	F	6	17	4	1	36
3	F	F	F	F	F	T	F	F	F	F	F	F	T	T	T	T	T	T	T	T	T	T	T	11	12	2	3	**53**
4	F	F	F	F	F	F	F	T	T	T	T	T	T	T	T	T	T	T	T	T	F	F	F	10	13	1	2	33
5	F	T	F	T	F	F	F	F	F	F	F	F	F	F	T	T	F	F	T	F	F	F	T	17	6	1	2	26
6	F	F	F	F	F	F	T	F	F	F	F	F	F	F	F	F	F	F	F	F	F	F	F	22	1	1	6	21

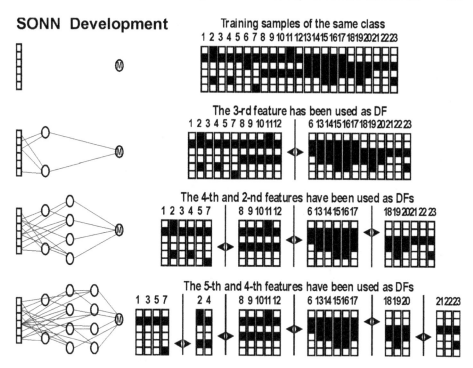

Fig. 2. The exemplar neurons and interconnections development process for the selected major features and for the training patterns of the exemplar class with partitioning of the training data using DFs

able features of the training data. The described optimization process develops a separate neural network for each trained class using k-classificator strategy. Next periods builds up the next layer (Fig. 1,2) accordingly to the computed DFs (9).

The SONN does not require any *a priori* coefficients. All necessary computations are fully automatic for given training data. The SONN topology increases only if necessary. Sometimes the quantities of some NN elements (neurons, connections, layers) can even be smaller than in previous periods. This is a result of optimization process that takes place in each development period of the SONN.

4   Experiments

The SONNs have been successfully implemented and adapted to many classification and recognition tasks: OCR, images recognition and classification, voice signal classification, medical diagnosing, searching for new pharmacological microemulsions, forecasting of credit ability, biocybernetics, pharmacoeconomy, the other economical data and the classical benchmarks [8, 10, 11]. The results of classification are very satisfying in comparison to the other classical well-known methods.

5 Conclusions

The described SONNs methodology for simultaneous configuration and adaptation of NNs builds up the topology optimized for the given training data and computes optimal weighs for them. It does not need to use any *a priori* given configuration parameters and is free from different training problems. The SONN always gradually adapts its topology and weights for the given training data in the deterministic way. Small changes of training data almost always result in non-trivial changes of optimized SONN topology and weights parameters. The methodology of SONN building can be compared to the processes that take place during natural brain embryo and next development. Moreover, the SONN can increase its discrimination ability if necessary. Such extra discrimination improvement influences generalization properties of the NN even after finishing of the fundamental SONN development. Such an additional development produces larger topologies. The created SONN always correctly classifies all training data. The SONNs offer very good generalization properties, because it is created after the most representative and discriminating features of the training data producing the model of any given classification problem. The SONN can also automatically recognize and classify inverse patterns of defined classes. Finally, the computational time of the described SONN is incomparable faster then in many other classifying or NNs training methods.

Recapitulating, the adaptation process of the SONN is very similar to the way of human learning (NNNs) in many features:

- The SONN adapts gradually beginning with more discriminative features.
- If SONN can not univocally classify training data, the process of details detection can be continued.
- The process of details detection affects only the data that have been not univocally classified.
- The process stops immediately after all training data are correctly and univocally classified.

Moreover, the SONNs have many other interesting and very similar to NNNs features:

- The SONNs automatically reduce the input features space using only the mostly discriminative and well-differentiating features.
- The SONNs automatically recognize the inversion of the input data and correctly classify these data informing about inversion of them.
- The SONNs can increase their sensitiveness for details of the training data continuing adaptation process.

Acknowledgements. Support from research funds 10.10.120.493 and the Polish Committee for Scientific Research is gratefully acknowledged.

References

1. Duch, W., Korbicz, J., Rutkowski, L., Tadeusiewicz, R. (eds): Biocybernetics and Biomedical Engineering. Neural Networks, EXIT Warsaw **6** (2000) 257–566
2. Dudek-Dyduch, E., Horzyk, A.: Analytical Synthesis of Neural Networks for Selected Classes of Problems. Knowlege Engineering and Experts Systems, Bubnicki Z., Grzech A. (eds), OWPN Wroclaw (2003) 194–206
3. Fiesler, E., Beale, R. (eds): Handbook of Neural Computation. IOP Publishing Ltd and Oxford University Press, Bristol & New York (1997) B3–C1
4. Hellwig, Z.: Elements of Calculus of Probability and Mathematical Statistics PWN Warsaw (1993) 40–50
5. Horzyk, A.: New Efficient Ontogenic Neural Networks for Classification Tasks. Advanced Computer Systems, Soldek J., Drobiazbiewicz L. (eds) INFORMA Szczecin (2003) 466–473
6. Horzyk, A.: New Ontogenic Neural Network Classificator Based on Probabilistic Reasoning. Advances in Soft Computing. Neural Networks and Soft Computing, Rutkowski L., Kacprzyk J. (eds) Physica Verlag, Springer-Verlag Company, Heidelberg (2003) 188–193
7. Horzyk, A., Tadeusiewicz, R.: Self-Optimizing Neural Networks. Advances in Neural Networks - ISNN 2004, Yin F., Wang J., Guo C. (eds), Springer-Verlag Berlin Heidelberg (2004) 150–155
8. Horzyk, A.: Self-Optimizing Neural Networks as a New Computational Tool in Biomedicine. Proc. of SIIB 2004, preTEXt Cracow (2004)
9. Konturek, S.: Neurofizjology. Human Fizjology, AGAT PRINT Cracow (1992) 9–126
10. Mendyk, A., Horzyk, A., Jachowicz, R., Polak S.: Development of new microemulsions systems using ontogenic neural networks. Proc. of SIIB 2004, preTEXt Cracow (2004)
11. Polak, S., Horzyk, A., Mendyk, A., Skowron, A., Brandys, J.: Perspective of use self-optimizing neural networks in pharmacy - modeling survival time of III-B and IV-th stage Non-Small Cell Lung Cancer patients. Proc. of SIIB 2004, preTEXt Cracow (2004)
12. Tadeusiewicz, R., Wszolek, W., Izworski, A., Wszolek, T.: Utilization of Artificial Intelligence Methods for Assistance in Interpretation of Acoustic Signals, Brambilla G., Ianniello C., Maffei L. (eds), Proc. of the 5-th European Conference on Noise Control Naples (2003) 276–282.
13. Tadeusiewicz, R., Ogiela, M.R.: Machine Perception and Automatic Understanding of Medical Visualization. Automatic Image Processing in Production Process, Damczyk M. (eds), Second Polish-German Seminar, CAMT, Wroclaw (2003) 39–48.
14. Tadeusiewicz, R., Mikrut, Z.: Neural-Based Object Recognition Support - From Classical Preprocessing to Space-Variant Sensing, Heiss M. (eds), Proc. of the International ICSC/IFAC Symposium on Neural Computation NC'98, ICSC Academic Press Canada/Switzerland (1998) 463–468.

Evaluation of Neuronal Firing Densities
via Simulation of a Jump-Diffusion Process[*]

A. Di Crescenzo[1], E. Di Nardo[2], and L.M. Ricciardi[3]

[1] Dipartimento di Matematica e Informatica, Università di Salerno,
Via Ponte don Melillo, I-84084 Fisciano (SA), Italy
adicrescenzo@unisa.it
[2] Dipartimento di Matematica, Università della Basilicata,
Campus Macchia Romana, I-85100 Potenza, Italy
dinardo@unibas.it
[3] Dipartimento di Matematica e Applicazioni, Università di Napoli Federico II,
Via Cintia, I-80126 Napoli, Italy
luigi.ricciardi@unina.it

Abstract. We consider a stochastic neuronal model in which the time evolution of the membrane potential is described by a Wiener process perturbed by random jumps driven by a counting process. We consider the first-crossing-time problem through a constant boundary for such a process, in order to describe the firing activity of the model neuron. We build up a new simulation procedure for the construction of firing densities estimates.

1   Introduction

The study of single neuron's activity is often performed by means of stochastic models based on diffusion processes. They are invoked in order to describe the time-course of the neuronal membrane potential. The formal description of the firing process is thus viewed as a first-crossing-time problem (FCT) through a suitable firing threshold (see, for instance, Ricciardi and Lánský [12] or Ricciardi et al. [11] and references therein). In the celebrated paper by Gerstein and Mandelbrot [4] interspike intervals histograms were fitted by FCT probability density function (pdf) of a Wiener process. Since then the Wiener process has been used as the starting point for the modeling of neuronal activity. Subsequently, other stochastic models have been proposed in the literature, aiming at refinements and embodiments of other neurophysiological features. Recently, the occurrence of jumps has been added to the diffusion process describing the neuronal membrane potential in order to include the contributions of more effective excitatory and inhibitory stimuli coming from synapses located in the trigger zone (see Giraudo and Sacerdote [6] and Giraudo et al.[8]). Along this line, in this paper we

---

[*] This work has been performed under partial support by MIUR (PRIN 2003) and by G.N.C.S. (INdAM).

J. Mira and J.R.Álvarez (Eds.): IWINAC 2005, LNCS 3561, pp. 166–175, 2005.

consider a jump-diffusion model for the description of the neuronal membrane potential, where the diffusion component is the Wiener process, and where the jumps have random amplitude and occur at random times driven by a suitable renewal process. Unfortunately few analytical results are available on the FCT pdf of this kind of jump-diffusion process, even in the case of constant thresholds. The available results are mainly focused on equations involving the FCT moments (see, for instance, Abundo [1], Giraudo and Sacerdote [5], and Tuckwell [15]), that however seem to be hardly manageable for practical purposes. Analytical results on FCT pdf's are fragmentary also in the case of regular diffusion processes, so that efficient algorithms have been devised in order to evaluate FCT densities (cf., for instance, Ricciardi *et al.* [11], Di Nardo *et al.* [2], [3] and references therein).

Being hard solving equations involving FCT densities, simulation techniques appear to be the most efficient methods to obtain reliable estimates of the FCT pdf's of jump-diffusion processes. The main aim of this paper is to develop a simulation procedure, different from the ones given in Giraudo *et al.* [9] and Musila and Lánský [10]. Such procedure generates FCT values through a constant boundary for jump-diffusion processes, the underlying diffusion being a Wiener process. It is strictly based on some specific features of the underlying Wiener process, such as the spatial homogeneity, the time homogeneity and the availability in closed form of the FCT density through a constant boundary. Our aim is to use the simulated FCTs in order to construct accurate estimates of neuronal firing densities. The results obtained by means of the proposed procedure will be useful to solve problems related to the identifications of neurophysiological parameters, by comparisons of estimates of neuronal firing densities with recorded interspike intervals histograms.

In Section 2 we describe the jump-Wiener model. The FCT problem of such process through a constant boundary is then addressed. A detailed description of the simulation procedure is given in Section 3. Finally, in Section 4 we show some results obtained in the special case when upward and downward jumps have constant amplitude and occur according to a Poisson process.

2   FCT Problem for a Jump-Diffusion Process

Let us consider a stochastic neuronal model, in which the dynamics of the membrane potential is described by a jump-diffusion process $\{X(t)\}_{t \geq 0}$ given by

$$X(t) = W(t) + Y(t) \tag{1}$$

where $\{W(t)\}_{t \geq 0}$ and $\{Y(t)\}_{t \geq 0}$ are independent stochastic processes and

(i) $\{W(t)\}_{t \geq 0}$ is a Wiener process with drift $\mu \in \mathbf{R}$, and variance $\sigma^2 \in (0, +\infty)$ per unit time, starting at $W(0) = x_0$;

(ii) $\{Y(t)\}_{t \geq 0}$ is a jump process such that $Y(0) = 0$ and $Y(t) = \sum_{i=1}^{N(t)} J_i, t > 0$, with $Y(t) = 0$ when $N(t) = 0$; $J_1, J_2, \ldots$ are real-valued i.i.d. r.v.'s not degenerating at zero and possessing cumulative distribution function (cdf)

$F_J(x)$ and survival function $\overline{F}_J(x) = 1 - F_J(x)$; $\{N(t)\}_{t\geq 0}$ is a counting process independent from $\{J_1, J_2, \ldots\}$ and characterized by i.i.d. absolutely continuous positive renewals $R_1, R_2, \ldots$ having pdf $f_R(x)$, cdf $F_R(x)$ and survival function $\overline{F}_R(x) = 1 - F_R(x)$.

We note that $X(t)$ possesses a conditional pdf, i.e.

$$f(x,t\,|\,x_0) = \frac{\partial}{\partial x}\,P\{X(t) \leq x \,|\, X(0) = x_0\}, \qquad t > 0.$$

Being

$$P\left\{W(t) + \sum_{i=1}^{k} J_i \in dx \,\middle|\, W(0) = x_0\right\} = \int_{\mathbf{R}} f_W(x - u, t\,|\,x_0)\,dF^{[k]}(u)\,dx,$$

where

$$f_W(x,t\,|\,x_0) := \frac{1}{\sqrt{2\pi\sigma^2 t}}\exp\left\{-\frac{(x - x_0 - \mu t)^2}{2\sigma^2 t}\right\}$$

is the pdf of $W(t)$ and where $F^{[k]}(u)$ denotes the cdf of $\sum_{i=1}^{k} J_i$, from the given assumptions on $\{X(t)\}$ we have:

$$f(x,t\,|\,x_0) = \sum_{k=0}^{\infty} P[N(t) = k]\int_{\mathbf{R}} f_W(x - u, t\,|\,x_0)\,dF^{[k]}(u). \qquad (2)$$

It can be shown that

$$E[X(t)\,|\,X(0) = x_0] = x_0 + \mu t + E(J_1)\,E[N(t)], \qquad (3)$$
$$\mathrm{Var}[X(t)\,|\,X(0) = x_0] = \sigma^2 t + \mathrm{Var}(J_1)\,E[N(t)] + E^2(J_1)\,\mathrm{Var}[N(t)]. \qquad (4)$$

Let us now discuss the FCT problem through a constant boundary S for the jump-diffusion process defined in (1). Without loss of generality we assume that $x_0 < S$. Let

$$T_X = \inf\{t \geq 0 : X(t) \geq S\}, \qquad X(0) = x_0 < S \qquad (5)$$

be the FCT of $\{X(t)\}$ through S from below, and let

$$g_X(S,t\,|\,x_0) = \frac{\partial}{\partial t}\,P\{T_X \leq t\,|\,X(0) = x_0\}, \qquad t > 0 \qquad (6)$$

be its pdf. The task of determining the FCT probability and the FCT moments can be accomplished by solving appropriate integro-differential equations (see the approach in Abundo [1] and Tuckwell [15]). A harder problem is to evaluate the FCT pdf, which identifies with the firing density of the model neuron. It seems that no analytical methods are available to get closed-form expressions for density (6).

In the following the r.v. T_W will denote the FCT from below of Wiener process $\{W(t)\}$ from $x_0 < S$ to the constant boundary S. Let us now recall some well-known results on T_W:

- the FCT cdf of $\{W(t)\}$ through S:

$$G_W(S,t\,|\,x_0) = P\{T_W \le t \,|\, W(0) = x_0\}$$
$$= \Phi\left(-\frac{S - x_0 - \mu t}{\sqrt{\sigma^2 t}}\right) + \exp\left(2\mu\frac{S - x_0}{\sigma^2}\right)\Phi\left(-\frac{S - x_0 + \mu t}{\sqrt{\sigma^2 t}}\right), \quad (7)$$

where

$$\Phi(z) = \frac{1}{\sqrt{2\pi}}\int_{-\infty}^{z} e^{-x^2/2}\mathrm{d}x, \qquad z \in \mathbf{R}; \qquad (8)$$

- the FCT pdf of $\{W(t)\}$ through S:

$$g_W(S,t\,|\,x_0) = \frac{S - x_0}{\sqrt{2\pi\sigma^2 t^3}}\exp\left\{-\frac{(S - x_0 - \mu t)^2}{2\sigma^2 t}\right\}; \qquad (9)$$

- the S-avoiding transition pdf of $\{W(t)\}$:

$$\alpha_W(x,t\,|\,x_0) = \frac{\partial}{\partial x} P\{W(t) \le x, T_W > t \,|\, W(0) = x_0\}$$
$$= \frac{1}{\sqrt{2\pi\sigma^2 t}}\exp\left\{-\frac{(x - x_0 - \mu t)^2}{2\sigma^2 t}\right\}\left[1 - \exp\left\{-\frac{2(S - x)(S - x_0)}{\sigma^2 t}\right\}\right]. \quad (10)$$

3   Simulation of First-Crossing Times

In order to build up suitable estimates for the firing density $g_X(S,t\,|\,x_0)$, in this section we propose a reliable Monte-Carlo method that simulates first-crossing times of the jump-diffusion process (1). In other terms, such numerical procedure gives as output a sampled value for the FCT r.v. T_X, defined in (5).

The procedure is based on the fact that the jump instants of $\{X(t)\}$ are regenerative, due to assumptions specified in the previous section. Process $\{X(t)\}$ is thus simulated only at the jump instants $R_1, R_1 + R_2, \ldots$. Formally, the procedure starts simulating the first renewal time R_1; a simulated Bernoulli trial thus establishes if the first crossing occurs in $(0, R_1)$. If this is the case then T_X is generated. Otherwise, the value $X(R_1^-)$ attained by process $\{X(t)\}$ immediately before time R_1 and the size J_1 of the first jump are simulated. The first crossing thus occurs at time R_1 if $X(R_1^-) + J_1 \ge S$. If this inequality is not fulfilled, then the simulation continues starting afresh with the generation of the second renewal time, and so on until the first crossing occurs or a preassigned final time $tmax$ is reached.

It should be pinpointed that the present approach is quite different from those adopted by other authors for the simulation of diffusion processes. Indeed, the common procedures are essentially based on the time-discretization of stochastic differential equations (see Giraudo and Sacerdote [7] and references therein). The two main sources of errors in such procedures are related to the adopted discretization scheme and to the undetected crossings of the boundary occurring in each discretization interval. Both such sources of errors do not appear in our approach. However, in the proposed procedure some approximations must be done in order to avoid certain computational brawbacks. These involve three tolerance parameters ε_1, ε_2 and ε_3, that are to be chosen close to zero.

Hereafter we give a detailed description of the simulation procedure.

- Current state x and current time t are initialized as $x = x_0$ and $t = 0$.
- A pseudo-random jump instant τ is generated according to the cdf $F_R(t)$.
- A Bernoulli pseudo-random number r is generated and compared with FCT probability $G_W(S, \tau \,|\, x)$, given in (7).
- The first-crossing probability during a renewal interval $G_W(S, \tau \,|\, x)$ is treated as zero if it is smaller than a tolerance parameter ε_1 in order to avoid drawbacks related to the numerical evaluation of pdf (11). If $r \leq G_W(S, \tau \,|\, x)$ then the first-crossing of process $\{X(t)\}$ through boundary S occurs at instant $\theta \in (0, \tau)$ and is due to the diffusive component of the process. Being not affected by the first jump, the first-crossing time θ is simulated according to the FCT pdf of $\{W(t)\}$ conditioned by $T_W \leq \tau$ and $W(0) = x$, that is

$$f_1(\theta) = \frac{g_W(S, \theta \,|\, x)}{G_W(S, \tau \,|\, x)}, \qquad 0 < \theta < \tau, \tag{11}$$

with g_W and G_W given in (9) and (7), respectively. The first-crossing instant θ is then given as output, and the procedure stops.
- If $r > G_W(S, \tau \,|\, x)$, the first crossing has not occurred before instant τ. The state $z \in (-\infty, S)$ attained by $\{X(t)\}$ immediately before τ is generated. Such a state is obtained according to the pdf of $W(\tau)$ conditioned by $T_W > \tau$ and $W(0) = x$, i.e.

$$f_2(z) = \frac{\alpha_W(z, \tau \,|\, x)}{1 - G_W(S, \tau \,|\, x)}, \qquad -\infty < z < S, \tag{12}$$

with α_W and G_W given respectively in (10) and (7).
- The size j of the jump performed at time τ is generated according to the probability law of r.v. J, so that $z + j$ is the new state attained by $\{X(t)\}$. If $z + j \geq S$ then τ is a first-crossing time, the procedure outputs τ and then stops. Otherwise the values of x and t are restored: the current state is set to $z + j$, the current time is set to $t + \tau$ and the procedure restarts until a preassigned final time $tmax$ is reached.

Making use of the Von Neumann acceptance-rejection method (see Rubistein [13]) a pseudo-random number from the r.v. with pdf $f_1(\theta)$ could be generated. Recalling (11), such density is bounded by its maximum $m = f_1(\theta_m)$, which is located at the abscissa

$$\theta_m = \min\{\tau, c_m\}, \quad \text{where} \quad c_m = \begin{cases} \dfrac{-3\sigma^2 + \sqrt{9\sigma^4 + 4\mu^2(S-x)^2}}{2\mu^2} & \text{if } \mu \neq 0, \\[4mm] \dfrac{(S-x)^2}{3\sigma^2} & \text{if } \mu = 0. \end{cases}$$

Hence, being $m \geq f_1(\theta)$ for all $\theta \in (0, \tau)$, it follows $m\tau \geq \int_0^\tau f_1(\theta) \, d\theta = 1$. Thus θ_m is a pseudo-random number from the r.v. with pdf $f_1(\theta)$, provided that $1/(m\tau) \leq \varepsilon_2$, where ε_2 is a prefixed tolerance parameter whose role is to avoid a loop.

A classical acceptance-rejection Von Neumann method is implemented in order to construct a pseudo-random number from the r.v. having pdf $f_2(z)$ (12). By virtue of (7) and (10) such function can be decomposed as follows:

$$f_2(z) = C \, f_3(z) \, \ell(z), \qquad -\infty < z < S, \tag{13}$$

where $C = [1 - G_W(S, \tau \,|\, x)]^{-1}$,

$$f_3(z) = \frac{1}{\sqrt{2\pi\sigma^2\tau}} \exp\left\{-\frac{(z - x - \mu\tau)^2}{2\sigma^2\tau}\right\} \left[\Phi\left(\frac{S - x - \mu\tau}{\sigma\sqrt{\tau}}\right)\right]^{-1}, \quad -\infty < z < S, \tag{14}$$

and

$$\ell(z) = \Phi\left(\frac{S - x - \mu\tau}{\sigma\sqrt{\tau}}\right) \left[1 - \exp\left\{-\frac{2(S - z)(S - x)}{\sigma^2\tau}\right\}\right], \qquad -\infty < z < S.$$

We note that $C \geq 1$, $f_3(z)$ is the truncation over $(-\infty, S)$ of a Gaussian density with mean $x + \mu\tau$ and variance $\sigma^2\tau$, while $\ell(z) \in [0, 1]$ for all $z \in (-\infty, S)$. The implemented acceptance-rejection method requires the generation of a pseudo-random number from the r.v. Z having pdf $f_3(z)$ (14). If $U \leq \ell(Z)$, where U is an uniform pseudo-random number in $(0, 1)$, then Z is a pseudo-random number with pdf $f_3(z)$ (14).

Let us now finally discuss the simulation of a pseudo-random number from the r.v. characterized by pdf (14). First of all let us observe that if Z has pdf (14) then

$$V = \frac{Z - x - \mu\tau}{\sigma\sqrt{\tau}}$$

is a truncated standard normal r.v. possessing pdf

$$f_V(z) = \frac{1}{\sqrt{2\pi}} \exp\left\{-\frac{z^2}{2}\right\} \left[\Phi\left(\frac{S - x - \mu\tau}{\sigma\sqrt{\tau}}\right)\right]^{-1}, \quad z < \frac{S - x - \mu\tau}{\sigma\sqrt{\tau}}. \tag{15}$$

So a pseudo-random number from the r.v. Z is generated by sampling a pseudo-random number from the r.v. V. To this last purpose, if

$$1 - \Phi\left(\frac{S - x - \mu\tau}{\sigma\sqrt{\tau}}\right) < \varepsilon_3, \tag{16}$$

where ε_3 is a preassigned tolerance parameter, then the distribution of V is approximated by a standard normal distribution. The role of ε_3 is to speed up the simulation procedure. If (16) is not fulfilled, the following classical acceptance-rejection Von Neumann method is employed. We consider the r.v. Y, having pdf

$$f_4(y) = \begin{cases} \exp\left\{y - \dfrac{S - x - \mu\tau}{\sigma\sqrt{\tau}}\right\} & \text{if } y < \dfrac{S - x - \mu\tau}{\sigma\sqrt{\tau}}, \\ 0 & \text{otherwise,} \end{cases} \tag{17}$$

with $x, S \in \mathbf{R}$, $x \le S$ and $\sigma, \tau > 0$. Then, the pdf given in (15) can be expressed as

$$f_V(z) = D\, f_4(z)\, \xi(z), \qquad z < \frac{S - x - \mu\tau}{\sigma\sqrt{\tau}},$$

where

$$D = \left[\Phi\left(\frac{S - x - \mu\tau}{\sigma\sqrt{\tau}} \right) \right]^{-1} \exp\left\{ \frac{S - x - \mu\tau}{\sigma\sqrt{\tau}} + 4 \right\} \ge 1$$

and

$$\xi(z) = \frac{1}{\sqrt{2\pi}} \exp\left\{ -\frac{z^2 + 2z + 8}{2} \right\}, \tag{18}$$

with $0 < \xi(z) < 1$. A pseudo-random number from the r.v. Y is generated by using the inversion method of cdf's. We thus have

$$Y = \log U_2 + \frac{S - x - \mu\tau}{\sigma\sqrt{\tau}} \quad \Longleftrightarrow \quad U_2 = \exp\left\{ Y - \frac{S - x - \mu\tau}{\sigma\sqrt{\tau}} \right\} = F_4(Y),$$

where F_4 is the cdf of Y.

To determine the complexity of the above simulation procedure is a difficult task because the number of renewals to be generated before the boundary crossing depends in a complex way on the involved parameters, such as drift μ, jump distribution, renewal distribution and distance $S - x_0$.

4   Firing Activity for a Wiener Neuronal Model with Constant Jumps

In this section we consider model (1) in a simpler case. We now assume that the jumps are separated by i.i.d. exponentially distributed random times, i.e. stimuli of high intensity arrive according to a Poisson process $\{N(t)\}$ with parameter λ. Moreover, let us assume that the jumps have positive or negative constant amplitude. In other terms, jumps J_i are distributed as

$$J_i = \begin{cases} \alpha & \text{w.p. } \eta \\ -\beta & \text{w.p. } 1\text{-}\eta, \end{cases} \tag{19}$$

with $0 < \eta < 1$, $\alpha > 0$ and $\beta > 0$. Hence, $Y(t)$ can be expressed as

$$Y(t) = \alpha\, N_1(t) - \beta\, N_2(t), \qquad t > 0$$

where $\{N_1(t)\}_{t \ge 0}$ and $\{N_2(t)\}_{t \ge 0}$ are independent Poisson processes with rate $\eta\lambda$ and $(1 - \eta)\lambda$, respectively. From (3) and (4) we have

$$E[X(t) \,|\, X(0) = x_0] = x_0 + \mu t + (\eta\alpha - (1 - \eta)\beta)\lambda t,$$
$$\mathrm{Var}[X(t) \,|\, X(0) = x_0] = \sigma^2 t + (\eta\alpha^2 + (1 - \eta)\beta^2)\lambda t.$$

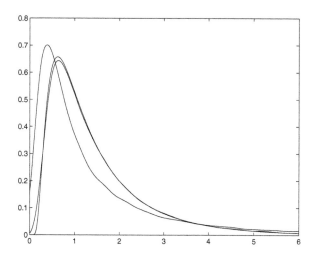

Fig. 1. Estimated FCT pdf $\hat{g}(t)$ of the Wiener process with constant Poisson-paced jumps for $\alpha = \beta = 1, 0.5$ (left to right), and FCT density $g(S, t \mid x_0)$ (last on the right) of the Wiener process, with $x_0 = 0$, $S = 1.5$, $\lambda = 2$, $\eta = 0.5$ and $\mu = \sigma^2 = 1$

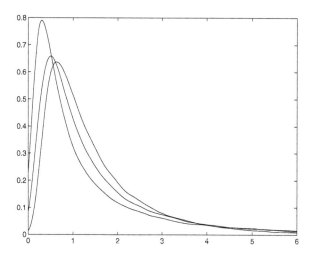

Fig. 2. Estimated FCT pdf $\hat{g}(t)$ for the Wiener process with constant Poisson-paced jumps with $\alpha = \beta = 1$, $\eta = 0.5$ and $\lambda = 0.1, 1, 3$ (from bottom to top near the origin) for $x_0 = 0$, $S = 1.5$, $\mu = \sigma^2 = 1$

The infinitesimal moments of order 1 and 2 are given respectively by

$$M_1(x_0) = \lim_{h \to 0^+} \frac{E[X(t + h) - X(t) \mid X(0) = x_0]}{h} = \mu + [\alpha\eta - \beta(1 - \eta)]\lambda,$$

$$M_2(x_0) = \lim_{h \to 0^+} \frac{E[(X(t + h) - X(t))^2 \mid X(0) = x_0]}{h} = \sigma^2 + [\alpha^2\eta + \beta^2(1 - \eta)]\lambda.$$

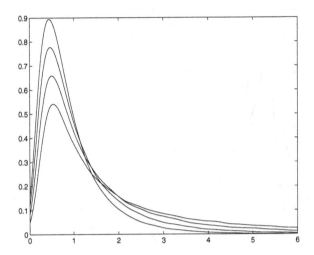

Fig. 3. Same as Fig. 2, with $\lambda = 1$ and $\eta = 0.3, 0.5, 0.7, 0.9$ (from bottom to top near the origin)

As pointed out in Giraudo *et al.* [8], for the model considered in this section if $M_1(x_0)$ is strictly positive then $P(T_X < \infty) = 1$ and

$$E[T_X] \geq \frac{S}{M_1(x_0)}, \qquad \text{Var}[T_X] \geq S \frac{M_2(x_0)}{[M_1(x_0)]^3}.$$

Note that the latter relations are satisfied as equality when $\{X(t)\}$ reduces to the Wiener process, being indeed $E[T_W] = S/\mu$ and $\text{Var}[T_W] = S\sigma^2/\mu^3$ if $\mu > 0$.

Since r.v.'s R_i now are exponentially distributed, they are simulated by means of the inversion method (cf. Rubistein [13]). Moreover, recalling (19), r.v.'s J_i are now simulated by means of a simple transformation of a Bernoulli variate.

In order to construct estimates of the firing density $g(S, t \mid x_0)$ a random sample of first-crossing instants can be obtained by means of the given procedure. To this purpose, the following kernel estimator has been considered:

$$\hat{g}(t) = \frac{1}{n\,h} \sum_{i=1}^{n} K\left(\frac{t - T_i}{h}\right), \quad t > 0, \tag{20}$$

where h is the bandwidth of the kernel $K(\cdot)$, $\{T_i; \; i = 1, 2, \ldots, n\}$ is the random sample observable by repeated use of the simulation procedure, and $K(\cdot)$ is the Epanechnikov kernel (see Silverman [14]).

We conclude with various examples, in which the initial state is set as $x_0 = 0$, the bandwidth is $h = 0.1$ and the sample size is $n = 10^5$. Moreover, the following choices are made: $\varepsilon_1 = 10^{-6}$, $\varepsilon_2 = 10^{-8}$ and $\varepsilon_3 = 10^{-4}$. In Figure 1, plots of $\hat{g}(t)$ are shown for the jump sizes given by $\alpha = \beta = 1, 0.5, 0.1$, when the boundary is $S = 1.5$. The estimated densities are compared with the FCT pdf of the Wiener process given in (9). In Figures 2 and 3, plots of $\hat{g}(t)$ are given for $\mu = \sigma^2 = 1$ and $\alpha = \beta = 1$, with different choices of λ and η.

References

1. Abundo M.: On First-passage times for one-dimensional jump-diffusion processes. Probability and Mathematical Statistics **20** (2000) 399–423.
2. Di Nardo E., Nobile A.G., Pirozzi E., Ricciardi L.M.: A computational approach to first-passage-time problems for Gauss-Markov processes. Advances in Applied Probability **33** (2001) 53–482.
3. Di Nardo E., Nobile A.G., Pirozzi E., Ricciardi L.M., Rinaldi S.: Simulation of Gaussian processes and first passage time densities evaluation. In F. Pichler, R. Moreno-Díaz & P. Kopacek (eds.), *Computer Aided Systems Theory – EURO-CAST'99*. Berlin: Springer-Verlag, (2000) 319–333.
4. Gerstein G.L., Mandelbrot B.: Random walk models for the spike activity of a single neuron. Biophys. J. **4** (1964) 41–68.
5. Giraudo M.T., Sacerdote L.: Some remarks on first-passage-time for jump-diffusion processes. In R. Trappl (ed.), *Cybernetics and Systems '96* Vienna: Austrian Society for Cybernetic Studies, (1996) 518–523.
6. Giraudo M.T., Sacerdote L.: Jump-diffusion processes as models for neuronal activity. BioSystems **40** (1997) 75–82.
7. Giraudo M.T., Sacerdote L.: Simulation methods in neuronal modelling. BioSystems **48** (1998) 77–83.
8. Giraudo M.T., Sacerdote L., Sirovich R.: Effects of random jumps on a very simple neuronal diffusion model. BioSystems **67** (2002) 75–83.
9. Giraudo M.T., Sacerdote L., Zucca C.: A Monte-Carlo method for the simulation of first passage times of diffusion processes. Methodology and Computing in Applied Probability **3** (2001) 215–231.
10. Musila M., Lánský P.: Generalized Stein's model for anatomically complex neurons. BioSystems **25** (1991) 179–181.
11. Ricciardi L.M., Di Crescenzo A., Giorno V., Nobile A.G.: An outline of theoretical and algorithmic approaches to first passage time problems with applications to biological modeling. Mathematica Japonica **50** (1999) 247–322.
12. Ricciardi L.M., Lánský P.: Diffusion models of neuron activity. In The Handbook of Brain Theory and Neural Networks (M.A. Arbib, ed.), (2002) 343–348. The MIT Press, Cambridge, ISBN 0-262-01197-2.
13. Rubistein R.Y.: Simulation and the Monte Carlo Method. New York: John Wiley & Sons (1981).
14. Silverman B.W.: Density Estimation for Statistics and Data Analysis. London: Chapman and Hall (1986).
15. Tuckwell H.C.: On the first-exit time problem for temporally homogeneous Markov processes. Journal of Applied Probability **13** (1976) 39–48.

Gaussian Processes and Neuronal Modeling

Elvira Di Nardo[1], Amelia G. Nobile[2], Enrica Pirozzi[3], and Luigi M. Ricciardi[3]

[1] Dipartimento di Matematica, Università della Basilicata,
Contrada Macchia Romana, Potenza, Italy
dinardo@unibas.it
[2] Dipartimento di Matematica e Informatica, Università di Salerno,
Via Ponte don Melillo, Fisciano (SA), Italy
nobile@unisa.it
[3] Dipartimento di Matematica e Applicazioni, Università di Napoli Federico II,
Via Cintia, Napoli, Italy
{enrica.pirozzi, luigi.ricciardi}@unina.it

Abstract. The research work outlined in the present note highlights the essential role played by the simulation procedures implemented by us on CINECA supercomputers to complement the mathematical investigations concerning neuronal activity modeling, carried within our group over the past several years. The ultimate target of our research is the understanding of certain crucial features of the information processing and transmission by single neurons embedded in complex networks. More specifically, here we provide a bird's eye look of some analytical, numerical and simulation results on the asymptotic behavior of first passage time densities for Gaussian processes, both of a Markov and of a non-Markov type. Significant similarities or diversities between computational and simulated results are pointed out.

1 Modeling Neuronal Firing

The aim of the research work outlined here is to describe the dynamics of neuronal firing by modeling it via a stochastic process representing the change in the neuron membrane potential between each pair of consecutive spikes (cf., for instance, [20]). In our approach, the threshold voltage is viewed as a deterministic function, and the instant when the membrane potential reaches it (i.e. when a spike occurs) as a first passage time (FPT) random variable. We shall focus our attention on neuronal models rooted on Gaussian processes, partially motivated by the generally accepted hypothesis that in numerous instances the neuronal firing is caused by the superposition of a very large number of synaptic inputs which is suggestive of the generation of Gaussian distributions by virtue of some sort of central limit theorems. Although the present paper is centered on mathematical modeling and simulations aspects, it opens the way to a number of comparisons with features of natural systems, i.e. neurons and neural networks.

Let us first consider a one-dimensional non-singular Gaussian stochastic process $\{X(t), t \geq t_0\}$ and a boundary $S(t) \in C^1[t_0, +\infty)$. We assume $P\{X(t_0) =$

J. Mira and J.R. Álvarez (Eds.): IWINAC 2005, LNCS 3561, pp. 176–185, 2005.

$x_0\} = 1$, with $x_0 < S(t_0)$, i.e. we focus our attention on the subset of sample paths of $X(t)$ that originate at a preassigned state x_0 at the initial time t_0. Then,

$$T_{x_0} = \inf_{t \geq t_0} \{t : X(t) > S(t)\}, \qquad x_0 < S(t_0)$$

is the FPT of $X(t)$ through $S(t)$, and

$$g(t|x_0, t_0) = \frac{\partial}{\partial t} P(T_{x_0} < t) \tag{1}$$

is its probability density function (pdf).

Henceforth, the FPT pdf $g(t|x_0, t_0)$ will be identified with the firing pdf of a neuron whose membrane potential is modeled by $X(t)$ and whose firing threshold is $S(t)$.

In order to include more physiologically significant features – such as a finite decay constant of the membrane potential, the presence of reversal potentials, time-dependent firing thresholds – and to refer to wider classes of inputs as responsible for the observed sequences of output signals released by the neuron, we also define the FPT upcrossing model. This is viewed as an FPT problem to a threshold, or boundary, $S(t)$ for the subset of sample paths of the one-dimensional non-singular Gaussian process $\{X(t), t \geq t_0\}$ originating at a state X_0. Such initial state, in turn, is viewed as a random variable with pdf

$$\gamma_\varepsilon(x_0, t_0) \equiv \begin{cases} \dfrac{f(x_0)}{P\{X(t_0) < S(t_0) - \varepsilon\}}, & x_0 < S(t_0) - \varepsilon \\ 0, & x_0 \geq S(t_0) - \varepsilon. \end{cases}$$

Here, $\varepsilon > 0$ is a fixed real number and $f(x_0)$ denotes the Gaussian pdf of $X(t_0)$. Then,

$$T_{X_0}^{(\varepsilon)} = \inf_{t \geq t_0} \{t : X(t) > S(t)\},$$

is the ε-upcrossing FPT of $X(t)$ through $S(t)$ and the related pdf is given by

$$g_u^{(\varepsilon)}(t|t_0) = \frac{\partial}{\partial t} P(T_{X_0}^{(\varepsilon)} < t) = \int_{-\infty}^{S(t_0)-\varepsilon} g(t|x_0, t_0)\, \gamma_\varepsilon(x_0, t_0)\, dx_0 \qquad (t \geq t_0),$$

where $g(t|x_0, t_0)$ is defined in (1). Without loss of generality, we set $t_0 = 0$ and $x_0 = 0$ and for this case we write $g(t) \equiv g(t|0, 0)$ and $g_u(t) \equiv g_u^{(\varepsilon)}(t|0)$, for fixed values of ϵ.

The selection of one of the various methods available to compute the firing pdf's $g(t)$ and $g_u(t)$ depends on the assumptions made on $X(t)$. For diffusion processes (cf. [1], [2]) and for Gauss-Markov processes (cf. [8]) we have proved that the firing pdf is solution of a second kind integral Volterra equation. For generally regular thresholds we have designed, and successfully implemented, a fast and accurate numerical procedure for solving such integral equation, and compared our approximations with those stemming out of standard numerical methods. Furthermore, in [8], by adopting a symmetry-based approach, we have determined the exact firing pdf for thresholds of a suitable analytical form.

Mathematical models based on non-Markov stochastic processes better describe the correlated firing activity, even though their analytical treatment is more complicated and only rare and fragmentary results appear to be available in literature.

By using a variant of the method proposed in [19], for a zero-mean non-singular stationary Gaussian process differentiable in the mean square sense, a cumbersome series expansion for the conditioned FPT pdf density and for the upcrossing FPT pdf density has been obtained ([10], [11]). In both cases we have succeeded to obtain numerically a reliable evaluation of the first term of the series expansion. By comparisons of these results with the simulated firing densities ([6], [10]), we have been led to conclude that this first term is a good approximation of firing pdf's only for small times.

The simulation procedures provide valuable alternative investigation tools especially if they can be implemented on parallel computers, (see [7]). We point out that our simulation originates from Franklin's algorithm [15]. We have implemented it in both vector and parallel modalities after suitably modifying it for our computational needs, for instance to obtain reliable approximations of upcrossing densities (cf. [5], [6]). Thus doing, reliable histograms of FPT densities of stationary Gaussian processes with rational spectral densities can be obtained in the presence of various types of boundaries.

For a thorough description of the simulation algorithm we refer the reader to [3], [4], [6].

We wish to stress that our endeavors strive to improve the simulation techniques, particularly within the context of stationary Gaussian processes, particularly relevant within the neuronal modeling context, for which alternative simulation procedures are being implemented and tested. Besides the use of algorithms based on special properties of the spectral density, such as its being of a rational type, methods based on the spectral representation of the process and circulant embedding methods are presently under investigation. The aim is to design efficient algorithms to simulate Gaussian processes of a more general type. It must also be pointed out that within our research project the role of the simulation procedure is threefold:

(i) to provide an investigation tool to explore the behavior of the firing density in a variety of different conditions;

(ii) to permit us to evaluate reliability and precision of the results obtained via numerical and analytic approximations;

(iii) to represent the only possible alternative whenever analytic and computational methods are failing.

2   An Analysis by Simulations

In order to analyze how the lack of memory affects the shape of the FPT densities, in [13] we have compared the behavior of such densities for Gauss-Markov processes versus Gauss non-Markov processes.

Let us consider a zero-mean stationary Gaussian process $X(t)$ with correlation function

$$\gamma(t) = e^{-\beta\,|t|}\,\cos(\alpha\,t), \qquad \alpha, \beta \in {I\!\!R}^{+} \tag{2}$$

which is the simplest type of correlation of concrete interest for applications [22]. Furthermore, let us assume that the boundary is of the following type:

$$S(t) = d\,e^{-\beta t}\left\{1 - \frac{e^{2\beta t}-1}{2\,d^2}\,\ln\left[\frac{1}{4} + \frac{1}{4}\sqrt{1 + 8\exp\left(-\frac{4\,d^2}{e^{2\beta t}-1}\right)}\,\right]\right\}, \tag{3}$$

with $d > 0$. Due to the assumed correlation (2), $X(t)$ is not mean square differentiable (see [13] for the details). Thus, the afore-mentioned series expansion (see [11]) does not hold. However, specific assumptions on the parameter α help us to characterize the shape of the FPT pdf.

We start assuming $\alpha = 0$, so that the correlation function (2) factorizes as

$$\gamma(t) = e^{-\beta\,\tau}\,e^{-\beta\,(t-\tau)} \qquad \beta \in {I\!\!R}^{+}, \, 0 < \tau < t.$$

In such case $X(t)$ becomes Gauss-Markov. As proved in [8], for the Gauss-Markov process $X(t)$ with covariance (2), the FPT pdf $g(t)$ in the presence of boundaries of type (3) can be evaluated in closed form. Alternatively, $g(t)$ can be numerically obtained by solving a non-singular Volterra integral equation (see [8]). In Figure 1(a) such density is plotted for $\beta = 0.5$ and $d = 0.25$.

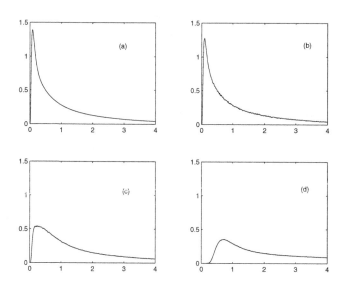

Fig. 1. Plots refer to FPT pdf $g(t)$ with $\beta = 0.5$ for a zero-mean Gaussian process characterized by correlation function (2) in the presence of the boundary (3) with $d = 0.25$. In Figure 1(a), the function $g(t)$ has been plotted. The estimated FPT pdf $\tilde{g}(t)$ with $\alpha = 10^{-10}$ is shown in Figure 1(b), with $\alpha = 0.25$ in Figure 1(c) and with $\alpha = 0.5$ in Figure 1(d)

Setting $\alpha \neq 0$ in (2), the Gaussian process $X(t)$ is no longer Markov and its spectral density is given by

$$\Gamma(\omega) = \frac{2\beta\left(\omega^2 + \alpha^2 + \beta^2\right)}{\omega^4 + 2\omega^2\left(\beta^2 - \alpha^2\right) + \left(\beta^2 + \alpha^2\right)^2}, \tag{4}$$

thus being of a rational type. Since in (4) the degree of the numerator is less than the degree of the denominator, it is possible to apply the simulation algorithm described in [9] in order to estimate the FPT pdf $\tilde{g}(t)$ of the process.

The simulation procedure has been implemented by a parallel FORTRAN 90 code on a 128-processor IBM SP4 supercomputer, based on MPI language for parallel processing, made available to us by CINECA. The number of simulated sample paths has been set equal to 10^7. The estimated FPT pdf's $\tilde{g}(t)$ through the specified boundary are plotted in Figures 1(b)$\div$1(d) for Gaussian processes with correlation function (2) in which we have taken $\alpha = 10^{-10}, 0.25, 0.5$, respectively. Note that as α increases, the shape of the FPT pdf $\tilde{g}(t)$ becomes progressively flatter, while its mode increases. Furthermore, as Figures 1(a)-1(b) indicate, $\tilde{g}(t)$ is very close to $g(t)$ for small values of α.

3   Asymptotic Results

The asymptotic behavior of the FPT densities for Gaussian processes as boundaries or time grow larger has been studied in [9], [10] and [11]. Our analysis is a natural extension of some investigations performed for the Ornstein-Uhlenbeck (OU) process [18] and successively extended to the class of one-dimensional diffusion processes admitting steady state densities in the presence of single asymptotically constant boundaries or of single asymptotically periodic boundaries (see [14] and [17]). There, computational as well as analytical results have indicated that the conditioned FPT pdf is susceptible of an excellent non-homogeneous exponential approximation for large boundaries, even though these boundaries need not be very distant from the initial state of the process. To this aim, we have estimated such a density by generating the sample paths of the Gaussian process through the parallel simulation algorithm implemented on the super-computer CRAY T3E of CINECA in order to overcome the outrageous complexity offered by the numerical evaluation of the involved partial sums of the conditioned FPT pdf series expansion. Specifically, we have considered the class of zero-mean stationary Gaussian processes characterized by damped oscillatory covariances (see [9]):

$$\gamma(t) = e^{-\beta|t|}\left[\cos(\alpha t) + \sin(\alpha|t|)\right], \tag{5}$$

where α and β are positive real numbers.

The results of our computations have shown that for certain periodic boundaries of the form

$$S(t) = S_0 + A\sin(2\pi t/Q), \quad S_0, A, Q > 0, \tag{6}$$

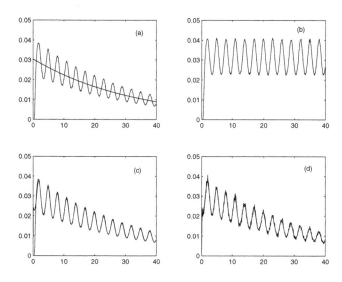

Fig. 2. For a zero-mean stationary Gaussian process with covariance (5) and $\alpha = \beta = 1$ in the presence of the periodic boundary $S(t) = 2 + 0.1 \sin(2\pi t/3)$, the simulated FPT density $\tilde{g}(t)$ is compared with the exponential density $\widehat{\lambda} e^{-\widehat{\lambda} t}$, with $\widehat{\lambda} = 0.030386$ in (a). The function $\tilde{Z}(t) = \tilde{g}(t) \exp\{\widehat{\lambda} t\}$ is plotted in (b). The asymptotic exponential approximation is plotted together with the simulated FPT density $\tilde{g}(t)$ in (c). The asymptotic exponential approximation is plotted together with the simulated upcrossing FPT density $\tilde{g}_u(t)$ in (d)

not very distant from the initial value of the process, the FPT pdf soon exhibits damped oscillations having the same period of the boundary. Furthermore, starting from rather small times, the estimated FPT densities $\tilde{g}(t)$ appears to be representable in the form

$$\tilde{g}(t) \simeq \tilde{Z}(t)\, e^{-\widehat{\lambda} t}, \tag{7}$$

where $\widehat{\lambda}$ is a constant that can be estimated by the least squares methods, and where $\tilde{Z}(t)$ is a periodic boundary of period T (see, for example, Figure 2(a) and Figure 2(b)). The goodness of the exponential approximation increases as the boundary is progressively moved farther apart from the starting point of the process. The more the periodic boundary is far from the starting point of the process, the more the exponential approximation improves.

In [11] we have shown by rigorous mathematical arguments that as boundary (6) moves away from the initial state of the process, the FPT pdf approaches a non-homogeneous exponential density of the type

$$\tilde{g}(t) \sim \tilde{h}(t) \exp\left\{ -\int_0^t \tilde{h}(\tau) d\tau \right\}, \tag{8}$$

where

$$\tilde{h}(t) = \frac{\sqrt{\alpha^2 + \beta^2}}{2\pi} \exp\left\{-\frac{S^2(t)}{2}\right\} \left[\exp\left\{-\frac{[\dot{\rho}(t)]^2}{2(\alpha^2 + \beta^2)}\right\}\right.$$
$$\left. -\sqrt{\frac{\pi}{2(\alpha^2 + \beta^2)}}\,\dot{\rho}(t)\,\mathrm{Erfc}\left(\frac{\dot{\rho}(t)}{\sqrt{2(\alpha^2 + \beta^2)}}\right)\right] \quad (9)$$

and $\rho(t) = A\sin(2\pi t/Q)$. (See, for example, Figure 2(c)). In [10] a similar result is proved for the upcrossing FPT density. (See Figure 2(d)).

It should be stressed that the analytic and the simulation results constitutes only a preliminary step towards the construction of neuronal models based on non-Markov processes. Nevertheless, the unveiling of properties of the asymptotic behavior of FPT may turn out to be useful also for the description of neuronal activities at small times whenever the intrinsic time scale of the microscopic events involved during the neuron's evolution is much smaller than the macroscopic observation time scale, or when the asymptotic regime is exhibited also in the case of firing thresholds not too distant from the resting potential, similarly to what was already pointed out by us in connection with the OU neuronal model [14].

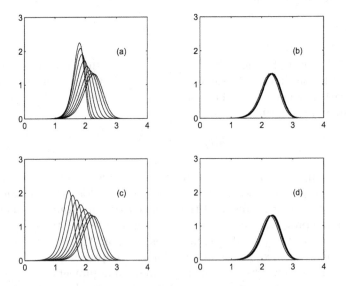

Fig. 3. For different choices of ϑ in the interval $[0.008, 1.024]$, with threshold $S(t) = -t^2/2 - t + 5$, plot of the simulated $\tilde{g}_u(t)$ is shown in (a) and plot of $\tilde{g}_u(t)$ for the OU-model in (c). Same in (b) and (d) for values of ϑ in $[2.048, 200]$

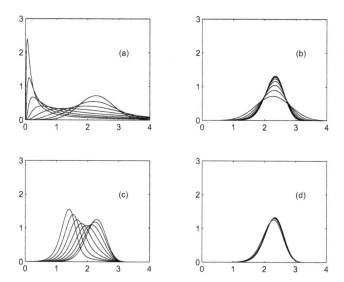

Fig. 4. For different choices of ϑ in the interval $[0.008, 1.024]$, with the same boundary as in Figure 3, plot of $\tilde{g}_u(t)$ for the Wiener model is shown in (a) and plot of $q(t)$ for the Kostyukov-model in (c). Same in (b) and (d) for values of ϑ in $[2.048, 200]$

4   An Alternative Approach

Within the context of single neuron's activity modeling a completely different, apparently not well known, approach was proposed by Kostyukov *et al.* ([16]) in which a non-Markov process of a Gaussian type is assumed to describe the time course of the neural membrane potential. The model due to Kostyukov (K-model) makes use of the notion of correlation time. Namely, let $X(t)$ be a stationary Gaussian process with zero mean, unit variance and correlation function $R(t)$. Then [21],

$$\vartheta = \int_0^{+\infty} |R(\tau)| \, d\tau < +\infty$$

is defined as the correlation time of the process $X(t)$. Under some assumptions on the threshold and by using a sort of diffusion approximation, Kostyukov works out a numerical evaluation $q(t)$ to the upcrossing FPT pdf. This approximation is obtained as solution of an integral equation that can be solved by routine methods. The relevant feature of this approach is that the unique parameter ϑ characterizes the considered class of stationary standard Gaussian processes.

In [2] and in [5] we analyzed this method pinpointing similarities and differences with respect to our models. Recently [12], we have again made use of the K-model and compared the obtained results with those worked out by numerically solving the integral equation holding for Gauss-Markov processes in the case of the OU and Wiener models, see [5] for details, and with the results obtained via the simulations of Gaussian processes. A variety of thresholds and of ϑ values has been considered. For some values of ϑ between 0.008 and 200, our results have been plotted in Figures 3 and 4.

Our investigations in this direction suggest that the validity of approximations of the firing densities in the presence of memory effects by the FPT densities of Markov type depends on the magnitude of the correlation time. Hence, an object of our present research has been the investigation of those models whose asymptotic behavior becomes increasingly similar as the correlation time ϑ grows larger.

References

1. Buonocore, A., Nobile, A.G., Ricciardi, L.M.: A new integral equation for the evaluation of first-passage-time probability densities. Adv. Appl. Prob. 19 (1987) 784–800
2. Di Crescenzo, A., Di Nardo, E., Nobile, A.G., Pirozzi, E., Ricciardi L.M.: On some computational results for single neurons' activity modeling. BioSystems, 58 (2000) 19–26
3. Di Nardo, E., Pirozzi, E., Ricciardi, L.M., Rinaldi, S.: Vectorized simulations of normal processes for first crossing-time problems. Lecture Notes in Computer Science, 1333 (1997) 177–188
4. Di Nardo, E., Nobile, A.G., Pirozzi, E., Ricciardi, L.M., Rinaldi, S.: Simulation of Gaussian processes and first passage time densities evaluation. Lecture Notes in Computer Science, 1798 (2000) 319-333
5. Di Nardo, E., Nobile, A.G., Pirozzi, E., Ricciardi, L.M.: On a non-Markov neuronal model and its approximation. BioSystems, 48 (1998) 29–35
6. Di Nardo, E., Nobile, A.G., Pirozzi, E., Ricciardi, L.M.: Evaluation of upcrossing first passage time densities for Gaussian processes via a simulation procedure. In: Atti della Conferenza Annuale della Italian Society for Computer Simulation. (1999) 95-102
7. Di Nardo, E., Nobile, A.G., Pirozzi, E., Ricciardi, L.M.: Parallel simulations in FPT problems for Gaussian processes. In: Report 2001, Science and Supercomputing at CINECA, (2001) 405–412
8. Di Nardo, E., Nobile, A.G., Pirozzi, E., Ricciardi, L.M.: A computational approach to first-passage-time problem for Gauss-Markov processes. Adv. Appl. Prob. 33 (2001) 453–482
9. Di Nardo, E., Nobile, A.G., Pirozzi, E., Ricciardi, L.M.: Computer-aided simulations of Gaussian processes and related asymptotic properties. Lecture Notes in Computer Science, 2178 (2001) 67–78
10. Di Nardo, E., Nobile, A.G., Pirozzi, E., Ricciardi, L.M.: Gaussian processes and neuronal models: an asymptotic analysis. In Trappl, R. (ed.): Cybernetics and Systems 2002, Vol. 1. Austrian Society for Cybernetics Studies, Vienna (2002) 313–318
11. Di Nardo, E., Nobile, A.G., Pirozzi, E., Ricciardi, L.M.: On the asymptotic behavior of first passage time densities for stationary Gaussian processes and varying boundaries. Methodology and Computing in Applied Probability, 5 (2003) 211-233

12. Di Nardo, E., Nobile, A.G., Pirozzi, E., Ricciardi, L.M.: Computational Methods for the evaluation of Neuron's Firing Densities. Lecture Notes in Computer Science, 2809 (2003) 394-403

13. Di Nardo, E., Nobile, A.G., Pirozzi, E., Ricciardi, L.M.: Towards the Modeling of Neuronal Firing by Gaussian Processes. Scientiae Mathematicae Japonicae 58 No. 2 (2003) 255-264 (e8, 497-506)

14. Giorno, V., Nobile, A.G., Ricciardi, L.M.: On the asymptotic behavior of first-passage-time densities for one-dimensional diffusion processes and varying boundaries. Adv. Appl. Prob. 22(1990) 883–914

15. Franklin, J.N.: Numerical simulation of stationary and non stationary gaussian random processes. SIAM Review 7 (1965) 68–80

16. Kostyukov, A. I. , Ivanov, Yu.N. , Kryzhanovsky, M.V.: Probability of Neuronal Spike Initiation as a Curve-Crossing Problem for Gaussian Stochastic Processes. Biological Cybernetics. 39 (1981) 157-163

17. Nobile, A.G., Ricciardi, L.M., Sacerdote, L.: Exponential trends of Ornstein-Uhlenbeck first passage time densities. J. Appl. Prob. 22 (1985) 360–369

18. Nobile, A.G., Ricciardi, L.M., Sacerdote, L.: Exponential trends of first-passage-time densities for a class of diffusion processes with steady-state distribution. J. Appl. Prob. 22 (1985) 611–618

19. Ricciardi, L.M., Sato, S.: On the evaluation of first-passage-time densities for Gaussian processes. Signal Processing. 11 (1986) 339-357

20. Ricciardi, L.M.: Diffusion Processes and Related Topics in Biology. Springer-Verlag New York (1977)

21. Stratonovich, R.L.: Topics in Theory of Random Noise. Vol. 1. Gordon and Breach, New York (1963)

22. Yaglom, A.M.: Correlation Theory of Stationary Related Random Functions. Vol. I. Basic Results. Springer-Verlag New York (1987)

On the Moments of Firing Numbers in Diffusion Neuronal Models with Refractoriness

Virginia Giorno[1], Amelia G. Nobile[1], and Luigi M. Ricciardi[2]

[1] Dipartimento di Matematica e Informatica, Università di Salerno,
Via Ponte don Melillo, 84084 Fisciano (SA), Italy
{giorno,nobile}@unisa.it
[2] Dipartimento di Matematica e Applicazioni, Università di Napoli Federico II,
Via Cintia, 80126 Napoli, Italy
luigi.ricciardi@unina.it

Abstract. For diffusion neuronal models, the statistical features of the random variable modeling the number of neuronal firings are analyzed by including the additional assumption of the existence of random refractoriness. For long times, the asymptotic behaviors of the mean and variance of the number of firings released by the neuron are determined. Finally, simple asymptotic expressions are obtained under the assumption of exponentially distributed firing times.

1 Introduction

Diffusion processes have long played a relevant role in biological modeling and first passage time problems for such processes have been extensively investigated in view of their usefulness to interpret neuronal firing activity (see, for instance, [12] and references therein). In particular, a quantitative description of the behavior of the neuron's membrane potential as an instantaneous return process for one dimensional diffusions has been the object of various investigations [6], [9], [10], [11], under the assumption that after each firing the membrane potential is either reset to a unique fixed value, or that the reset value is characterized by an assigned probability density function (pdf). Recently, in [1] and [2] the presence of refractoriness has been included in the mathematical characterization of the membrane potential as an instantaneous return process within diffusion models of single neuron's activity, assuming that the firing threshold acts as an elastic barrier, that is 'partially transparent', in the sense that its behavior is intermediate between total absorption and total reflection. Instead, in [3] and [13] the return process paradigm for the description of the time course of the membrane potential is analyzed by assuming that the neuronal refractoriness period is described by a random variable with a preassigned pdf.

The purpose of this paper is to implement certain theoretical results considered in [13] in order to provide a quantitative description of the input–output behavior of single neurons subject to a diffusion–like dynamics. Indeed, in Section 2 the statistical features of the random process modeling the number of firings released by the neuron up to time t will be analyzed by means of the above

J. Mira and J.R. Álvarez (Eds.): IWINAC 2005, LNCS 3561, pp. 186–194, 2005.

mentioned return process under the assumption of the existence of random refractoriness. In particular, the Laplace transforms of the first two moments of
the number of firings released by the neuron up to time t will be determined.
Furthermore, in Section 3 an asymptotic analysis of the first two moments and
of the variance of the number of neuronal firings will be provided. Finally, in
Section 4 for long times and large firing thresholds simple asymptotic expressions for mean and variance of the number of firing released by the neuron up
to time t are obtained under the assumption of exponentially distributed firing
times.

2   Mathematical Background

Let $\{X(t), t \geq 0\}$ be a regular, time-homogeneous diffusion process defined over
the interval $I = (r_1, r_2)$, characterized by drift and infinitesimal variance $A_1(x)$
and $A_2(x)$, respectively, that we assume to satisfy Feller condition [5]. Let $h(x)$
and $k(x)$ denote scale function and speed density of $X(t)$:

$$h(x) = \exp\left\{ -2 \int^x \frac{A_1(z)}{A_2(z)} \, dz \right\}, \quad k(x) = \frac{2}{A_2(x) \, h(x)}$$

and denote by

$$H(r_1, y] = \int_{r_1}^y h(z) \, dz, \qquad K(r_1, y] = \int_{r_1}^y k(z) \, dz$$

scale and the speed measures, respectively.

As is well-known, the random variable "first passage time" (FPT) of $X(t)$
through S $(S \in I)$ with $X(0) = y < S$ is defined as follows:

$$T_y = \inf_{t \geq 0} \{t : X(t) \geq S\}, \qquad X(0) = y < S. \tag{1}$$

Then,

$$g(S, t \mid y) = \frac{\partial}{\partial t} P(T_y < t), \qquad y < S \tag{2}$$

is the FPT probability density function of $X(t)$ through S conditional upon
$X(0) = y$.

In the neuronal modeling context the state S represents the neuron's firing
threshold, the FPT through S the firing time and $g(S, t \mid y)$ the firing pdf. In
the sequel we assume that one of the following cases holds:

(i) r_1 is a natural nonattracting boundary and $K(r_1, y] < +\infty$
(ii) r_1 is a reflecting boundary or it is an entrance boundary.

Under such assumptions, for $y < S$ the first passage probability $P(S \mid y)$ from y
to S is unity and the moments FPT are finite and given by (cf. [14]):

$$t_n(S \mid y) := \int_0^{+\infty} t^n \, g(S, t \mid y) \, dt = n \int_x^S h(z) \, dz \int_{r_1}^z k(u) \, t_{n-1}(S \mid u) \, du$$

$$(n = 1, 2, \ldots), \tag{3}$$

with $t_0(S \mid y) \equiv P(S \mid y) = 1$.

We assume that after each firing a period of refractoriness of random duration occurs, during which either the neuron is completely unable to respond or it only partially responds to the incoming stimulations.

We now construct the return process $\{Z(t), t \geq 0\}$, describing the time course of the membrane potential, in (r_1, S) as follows. Starting at a point $\eta \in (r_1, S)$ at time zero, a firing takes place when $X(t)$ attains the threshold S for the first time, after which a period of refractoriness of random duration occurs. At the end of the period of refractoriness, $Z(t)$ is instantaneously reset to a certain fixed state η. The subsequent evolution of the process then goes as described by $X(t)$, until the boundary S is again reached. A new firing then occurs, followed by the period of refractoriness, and so on.

The process $\{Z(t), t \geq 0\}$ thus consists of recurrent cycles $\mathcal{F}_0, \mathcal{R}_1, \mathcal{F}_1, \mathcal{R}_2, \mathcal{F}_2,$ ..., each of random duration, where the durations F_i of $\mathcal{F}_i$ $(i = 0, 1, \ldots)$ and the durations of refractory period R_i of $\mathcal{R}_i$ $(i = 1, 2, \ldots)$ are independently distributed random variables. Here, F_i $(i = 0, 1, \ldots)$ describes the length of the firing interval, i.e. of the time interval elapsing between the i-th reset at the value η and the $(i + 1)$-th FPT from η to S. Instead, R_i $(i = 1, 2, \ldots)$ describes the duration of i-th refractory period. Since the diffusion process $X(t)$ is time-homogeneous, the random variables $F_0, F_1, \ldots$ can be safely assumed to be independent and identically distributed, each with pdf $g(S, t \mid \eta)$ depending only on the length of the corresponding firing interval. Furthermore, we assume that $R_1, R_2, \ldots$ are independent and identically distributed (iid) random variables, each with pdf $\varphi(t)$ depending only on the duration of the refractory period.

Hereafter, we shall provide a description of the random process $\{M(t), t \geq 0\}$ representing the number of firings released by the neuron up to time t. To this purpose, for $\eta \in (r_1, S)$, let

$$q_k(t \mid \eta) = P\{M(t) = k \mid Z(0) = \eta\} \qquad (k = 0, 1, \ldots) \qquad (4)$$

be the probability of occurrence of k firings up to time t. Since $X(t)$ is time-homogeneous and $R_1, R_2, \ldots$ are iid random variables, the following relations can be seen to hold:

$$q_0(t \mid \eta) = 1 - \int_0^t g(S, \tau \mid \eta)\, d\tau$$

$$(5)$$

$$q_k(t \mid \eta) = \left[g(S, t \mid \eta) * \varphi(t)\right]^{(k)} * \left[1 - \int_0^t g(S, \tau \mid \eta)\, d\tau\right]$$
$$+ g(S, t \mid \eta) * \left[\varphi(t) * g(S, t \mid \eta)\right]^{(k-1)} * \left[1 - \int_0^t \varphi(\tau)\, d\tau\right]$$
$$(k = 1, 2, \ldots),$$

where $(*)$ means convolution, exponent (r) indicates (r)-fold convolution, $g(S, t \mid \eta)$ is the FPT pdf of $X(t)$ through S starting from $X(0) = \eta < S$ and $\varphi(t)$ is the pdf of the refractoriness period.

Probabilities $q_k(t \mid \eta)$ can be used to explore the statistical characteristics of the random variable that describes the number of firings. Let now

$$E\{[M(t)]^r \mid \eta\} := \sum_{k \geq 1} k^r \, q_k(t \mid \eta) \qquad (r = 1, 2, \ldots) \tag{6}$$

be the r-th order moment of the number of firings released by the neuron up to time t and let

$$\psi_r(\lambda \mid \eta) := \int_0^{+\infty} e^{-\lambda t} \, E\{[M(t)]^r \mid \eta\} \, dt \qquad (r = 1, 2, \ldots) \tag{7}$$

be its Laplace transform. As proved in [13], denoting by $g_\lambda(S \mid \eta)$ and by $\Phi(\lambda)$ the Laplace transforms of $g(S, t \mid \eta)$ and $\varphi(t)$, respectively, from (5) and (7) for $\lambda > 0$, one obtains:

$$\psi_1(\lambda \mid \eta) = \frac{g_\lambda(S \mid \eta)}{\lambda \left[1 - g_\lambda(S \mid \eta) \, \Phi(\lambda)\right]} \, ,$$

$$\psi_2(\lambda \mid \eta) = \frac{g_\lambda(S \mid \eta) \left[1 + g_\lambda(S \mid \eta) \, \Phi(\lambda)\right]}{\lambda \left[1 - g_\lambda(S \mid \eta) \, \Phi(\lambda)\right]^2} \, . \tag{8}$$

Let now $I_0, I_1, I_2, \ldots$ denote the random variables describing the interspike intervals, with I_0 representing the time of occurrence of the first firing and $I_k \ (k = 1, 2, \ldots)$ the duration of the time interval elapsing between k-th and $(k+1)$-th firing. Furthermore, let $\gamma_k(t)$ denote the pdf's of $I_k \ (k = 0, 1, \ldots)$. Therefore, I_0 identifies with the FPT through the threshold S starting at initial state $X(0) = \eta < S$, so that $\gamma_0(t) \equiv g(S, t \mid \eta)$. Due to time-homogeneity of $X(t)$ and since $R_1, R_2, \ldots$ are independent and identically distributed, the interspike intervals $I_1, I_2, \ldots$ are iid random variables having pdf

$$\gamma(t) \equiv \gamma_k(t) = \int_0^t \varphi(\vartheta) \, g(S, t - \vartheta \mid \eta) \, d\vartheta \qquad (k = 1, 2, \ldots). \tag{9}$$

Hence, by virtue of (9), the first three moments of the interspike intervals $I_1, I_2, \ldots$ are given by:

$$
\begin{aligned}
E(I) &= t_1(S \mid \eta) + E(R), \\
E(I^2) &= t_2(S \mid \eta) + E(R^2) + 2 \, E(R) \, t_1(S \mid \eta) \\
E(I^3) &= t_3(S \mid \eta) + E(R^3) + 3 \, E(R) \, t_2(S \mid \eta) + 3 \, E(R^2) \, t_1(S \mid \eta).
\end{aligned}
\tag{10}
$$

where $E(I^r)$ is the r-th order moment of the interspike intervals, $t_r(S \mid \eta)$ is the r-th order moment of the FPT of $X(t)$ through S conditional upon $X(0) = \eta$ and where $E(R^r)$ is the r-th order moment of refractory periods.

Denoting by $\Gamma(\lambda)$ the Laplace transform of $\gamma(t)$, from (9) one has

$$\Gamma(\lambda) = g_\lambda(S \mid \eta) \, \Phi(\lambda), \tag{11}$$

so that (8) leads to:

$$\psi_1(\lambda \mid \eta) = \frac{g_\lambda(S \mid \eta)}{\lambda\left[1 - \Gamma(\lambda)\right]}, \tag{12}$$

$$\psi_2(\lambda \mid \eta) = \frac{g_\lambda(S \mid \eta)\left[1 + \Gamma(\lambda)\right]}{\lambda\left[1 - \Gamma(\lambda)\right]^2}. \tag{13}$$

If $g_\lambda(S \mid \eta)$ and $\Gamma(\lambda)$ are known, the right-hand-sides of (12) and (13) can be calculated. One can then arise to $E\{M(t) \mid \eta\}$ and $E\{[M(t)]^2 \mid \eta\}$ via the inverse Laplace transforms of (12) and (13), respectively. It is worth to point out that even though the inverse Laplace transforms of the functions $\psi_1(\lambda \mid \eta)$ and $\psi_2(\lambda \mid \eta)$ cannot be calculated, they can nevertheless provide useful information on the asymptotic behavior for long time of mean and variance of number of firing released by neuron up to time t, as we shall see in Sections 3 and Sections 4.

3   Asymptotic Behavior

In this Section we analyze the asymptotic behavior for long times of the first two moments and of the variance related to the number of firing released by neurons up to time t under the assumption that the first three moments of the refractoriness period are finite.

Proposition 1. *For long times, the first two moments of the number of firing released by the neuron up to time t are approximatively:*

$$E\{M(t) \mid \eta\} \simeq \frac{1}{E(I)}\, t + \frac{1}{2}\frac{E(I^2)}{E^2(I)} - \frac{t_1(S \mid \eta)}{E(I)}, \tag{14}$$

$$E\{[M(t)]^2 \mid \eta\} \simeq \frac{1}{E^2(I)}\, t^2 + \left[\frac{2\,E(I^2)}{E^3(I)} - \frac{1}{E(I)} - \frac{2\,t_1(S \mid \eta)}{E^2(I)}\right] t$$

$$+ \frac{3}{2}\frac{[E(I^2)]^2}{E^4(I)} - \frac{2}{3}\frac{E(I^3)}{E^3(I)} - \frac{1}{2}\frac{E(I^2)}{E^2(I)} + \frac{t_1(S \mid \eta)}{E(I)}$$

$$+ \frac{t_2(S \mid \eta)}{E^2(I)} - \frac{2\,E(I^2)}{E^3(I)}\, t_1(S \mid \eta), \tag{15}$$

where the first three moments of the interspike intervals are given in (10).

Proof. The proof goes along the lines indicated in [4]. First of all, we shall prove that (14) holds. We note that

$$1 - \Gamma(\lambda) = \sum_{r \geq 0} \frac{(-1)^r \lambda^{r+1}}{(r+1)!}\, E(I^{r+1}),$$

$$\tag{16}$$

$$\frac{\lambda}{1 - \Gamma(\lambda)} = \left[\sum_{r \geq 0} \frac{(-1)^r \lambda^r}{(r+1)!}\, E(I^{r+1})\right]^{-1} = \frac{1}{E(I)} \sum_{k \geq 0} c_k\, \lambda^k,$$

where (cf. [7], p. 14, n. 0.313)

$$c_0 = 1, \qquad c_k = -\frac{1}{E(I)} \sum_{r=0}^{k} \frac{(-1)^r}{(r+1)!} E(I^{r+1}) c_{k-r} \qquad (k = 1, 2, \ldots). \qquad (17)$$

Furthermore, one has:

$$g_\lambda(S \mid \eta) = \sum_{r \geq 0} \frac{(-\lambda)^r}{r!} t_r(S \mid \eta), \qquad (18)$$

with the r-th order moment of the firing time given in (3). Hence, making use of (16) and (18), from (12) we have:

$$
\begin{aligned}
\psi_1(\lambda \mid \eta) &= \frac{1}{\lambda^2} g_\lambda(S \mid \eta) \frac{\lambda}{1 - \Gamma(\lambda)} \\
&= \frac{1}{E(I)} \frac{1}{\lambda^2} \sum_{r \geq 0} \frac{(-\lambda)^r}{r!} t_r(S \mid \eta) \sum_{k \geq 0} c_k \lambda^k, \qquad (19)
\end{aligned}
$$

so that

$$\psi_1(\lambda \mid \eta) \simeq \frac{1}{E(I)} \left\{ \frac{1}{\lambda^2} + \left[c_1 - t_1(S \mid \eta) \right] \frac{1}{\lambda} + \left[c_2 - c_1 t_1(S \mid \eta) + \frac{t_2(S \mid \eta)}{2} \right] \right\} + O(\lambda), \qquad (20)$$

with

$$c_1 = \frac{E(I^2)}{2\,E(I)}, \qquad c_2 = \left[\frac{E(I^2)}{2\,E(I)} \right]^2 - \frac{E(I^3)}{6\,E(I)} \qquad (21)$$

and where $O(\lambda)$ denotes an infinitesimal quantity of order λ or higher. Taking the inverse Laplace transform of (20) and making use of (21), for long times one obtains (14).

We now shall prove that (15) holds. Making use of (12), (16) and (19), from (13) then follows:

$$
\begin{aligned}
\psi_2(\lambda \mid \eta) &= \frac{1}{\lambda} \psi_1(\lambda \mid \eta) \frac{\lambda}{1 - \Gamma(\lambda)} \left[1 + \Gamma(\lambda) \right] \\
&= \frac{1}{E^2(I)} \frac{1}{\lambda^3} \sum_{r \geq 0} \frac{(-\lambda)^r}{r!} t_r(S \mid \eta) \left[\sum_{k \geq 0} c_k \lambda^k \right]^2 \left[2 + \sum_{j \geq 1} \frac{(-\lambda)^j}{j!} E(I^j) \right],
\end{aligned}
$$
$$(22)$$

so that

$$
\begin{aligned}
\psi_2(\lambda \mid \eta) \simeq \frac{1}{E^2(I)} \bigg\{ & \frac{2}{\lambda^3} + \left[4c_1 - E(I) - 2 t_1(S \mid \eta) \right] \frac{1}{\lambda^2} + \left[4c_2 + 2c_1^2 \right. \\
& \left. -2c_1 E(I) - 4c_1 t_1(S \mid \eta) + \frac{E(I^2)}{2} + E(I) t_1(S \mid \eta) + t_2(S \mid \eta) \right] \frac{1}{\lambda} \bigg\}
\end{aligned}
$$

$$+\left[4\,c_3 + 4\,c_1\,c_2 - (2\,c_2 + c_1^2)\,[E(I) + 2\,t_1(S\mid\eta)] - \frac{E(I^3)}{6}\right.$$

$$-\frac{E(I^2)}{2}\,t_1(S\mid\eta) + 2\,c_1\left(\frac{E(I^2)}{2} + E(I)\,t_1(S\mid\eta) + t_2(S\mid\eta)\right)$$

$$\left.\left.-\frac{E(I)}{2}\,t_2(S\mid\eta) - \frac{t_3(S\mid\eta)}{3}\right]\right\} + O(\lambda). \tag{23}$$

Taking the inverse Laplace transform of (23) and recalling (21), for long times one has:

$$E\{[M(t)]^2\mid\eta\} \simeq \frac{1}{E^2(I)}\,t^2 + \left[2\,\frac{E(I^2)}{E^3(I)} - \frac{1}{E(I)} - \frac{2}{E^2(I)}\,t_1(S\mid\eta)\right]t$$

$$+\frac{1}{E^2(I)}\left\{4\,c_2 + 2\,c_1^2 - 2\,c_1\,E(I) - 4\,c_1\,t_1(S\mid\eta)\right.$$

$$\left.+\frac{E(I^2)}{2} + E(I)\,t_1(S\mid\eta) + t_2(S\mid\eta)\right\}. \tag{24}$$

By virtue of (21) one obtains

$$4\,c_2 + 2\,c_1^2 - 2\,c_1\,E(I) - 4\,c_1\,t_1(S\mid\eta)$$

$$= \frac{3}{2}\left[\frac{E(I^2)}{E(I)}\right]^2 - \frac{2}{3}\frac{E(I^3)}{E(I)} - E(I^2) - \frac{2\,E(I^2)}{E(I)}\,t_1(S\mid\eta), \tag{25}$$

so that (15) immediately follows from (24). Proposition 1 is then proved.   □

Proposition 1 shows that for long times the mean and the second order moment of the number of firing released by the neuron up to time t are approximatively linear and quadratic functions of t, respectively.

Proposition 2. *For long times, the variance of the number of firing released by the neuron up to time t is approximatively given by:*

$$\mathrm{Var}\{M(t)\mid\eta\} \simeq \frac{\mathrm{Var}(I)}{E^3(I)}\,t + \frac{5}{4}\frac{[E(I^2)]^2}{E^4(I)} - \frac{2}{3}\frac{E(I^3)}{E^3(I)} - \frac{1}{2}\frac{E(I^2)}{E^2(I)}$$

$$+\frac{t_1(S\mid\eta)}{E(I)} + \frac{t_2(S\mid\eta)}{E^2(I)} - \frac{E(I^2)}{E^3(I)}\,t_1(S\mid\eta) - \frac{t_1^2(S\mid\eta)}{E^2(I)}. \tag{26}$$

Proof. Making use of (14) and (15), one easily obtains:

$$\mathrm{Var}\{M(t)\mid\eta\} = E\{[M(t)]^2\mid\eta\} - \left[E\{M(t)\mid\eta\}\right]^2$$

$$\simeq \frac{1}{E^3(I)}\left\{2\,E(I^2) - E^2(I) - 2\,E(I)\,t_1(S\mid\eta) - E(I^2) + 2\,E(I)\,t_1(S\mid\eta)\right\}t$$

$$+\frac{5}{4}\frac{[E(I^2)]^2}{E^4(I)} - \frac{2}{3}\frac{E(I^3)}{E^3(I)} - \frac{1}{2}\frac{E(I^2)}{E^2(I)} + \frac{t_1(S\mid\eta)}{E(I)} + \frac{t_2(S\mid\eta)}{E^2(I)}$$

$$-\frac{E(I^2)}{E^3(I)}\,t_1(S\mid\eta) - \frac{t_1^2(S\mid\eta)}{E^2(I)}. \tag{27}$$

By virtue of (10), the following identity holds

$$2\,E(I^2) - E^2(I) - 2\,E(I)\,t_1(S\mid\eta) - E(I^2) + 2\,E(I)\,t_1(S\mid\eta) \equiv \mathrm{Var}(I), \quad (28)$$

so that (26) immediately follows from (27).   □

Proposition 2 shows that for long times the variance of the number of firing released by the neuron up to time t is approximatively a linear function of t, whose coefficients depend only from the first three moments of the firing time and from the first three moments of refractory period.

4   Exponential Behavior of the Firing Time

If $\{X(t), t \geq 0\}$ possesses a steady state distribution, we expect that for large firing thresholds an exponential behavior of the firing time pdf $g(S, t \mid \eta)$ takes place (cf. [8]):

$$\lim_{S\to+\infty} t_1(S\mid\eta)\,g\{S, t\,t_1(S\mid\eta)\mid\eta\} = e^{-t}, \quad (29)$$

where $t_1(S \mid \eta)$ is the mean of the firing time. Hence, for large firing thresholds from (29) one has:

$$g(S, t\mid\eta) \simeq \frac{1}{t_1(S\mid\eta)}\,\exp\left\{-\frac{t}{t_1(S\mid\eta)}\right\}. \quad (30)$$

Proposition 3. *For long times and large firing thresholds, the mean and variance of the number of firing released by neuron up to time t are approximatively:*

$$E\{M(t)\mid\eta\} \simeq \frac{1}{E(I)}\,t + \frac{1}{2}\,\frac{E(R^2)}{E^2(I)}, \quad (31)$$

$$\mathrm{Var}\{M(t)\mid\eta\} \simeq \frac{\mathrm{Var}(R) + t_1^2(S\mid\eta)}{E^3(I)}\,t + \frac{1}{E^4(I)}\left\{\frac{5}{4}\,[E(R^2)]^2 - \frac{1}{2}\,E^2(R)\,E(R^2)\right.$$
$$\left. + \frac{3}{2}\,E(R^2)\,t_1^2(S\mid\eta) + E(R)\,E(R^2)\,t_1(S\mid\eta) - \frac{2}{3}\,E(R^3)\,E(I)\right\}. \quad (32)$$

Proof. It follows from (14) and (26) by noting that, under the assumption (30), for large firing thresholds $t_2(S \mid \eta) \simeq 2\,t_1(S \mid \eta)$ and $t_3(S \mid \eta) \simeq 6\,t_1(S \mid \eta)$.   □

Hence, for long times and large firing thresholds the mean and variance of the number of firing released by neuron up to time t are linear functions of t, whose coefficients depend only from the mean of firing time $t_1(S \mid \eta)$ and from the first three moments of refractory period.

    In conclusion, the knowledge of the first three moments of the firing time and of the first three moments of refractory period can play an essential role in suggesting some general statistical features of the mean and variance of the number of firings released by the neuron up to time t. Extension of random refractoriness to special neuronal diffusion models of interest in neurobiological context will be the object of future works.

References

1. Buonocore A., Giorno V., Nobile A.G., Ricciardi L.M.: Towards modeling refractoriness for single neuron's activity. In Trappl, R. (ed.): Cybernetics and Systems 2002, Vol. 1. Austrian Society for Cybernetics Studies, Vienna (2002) 319-324

2. Buonocore, A., Giorno, V., Nobile, A.G., Ricciardi, L.M.: A neuronal modeling paradigm in the presence of refractoriness. BioSystems **67** (2002) 35-43

3. Esposito, G., Giorno, V., Nobile, A.G., Ricciardi, L.M., Valerio, C.: Interspike analysis for a single neuron's activity in presence of refractoriness. In: Trappl, R. (ed.): Cybernetics and Systems. Vol. 1. Austrian Society for Cybernetics Studies, Vienna (2004) 199-204

4. Feller W.: On probability problems in the theory of counters. In: *Studies and Essays, A Volume for the Anniversary of Courant.* Interscience Publishers, Inc., New York (1948) 105-115

5. Feller W.: The parabolic differential equations and the associated semi–groups of transformations. Ann. Math. **55** (1952) 468-518

6. Giorno V. , Lánský P. , Nobile A.G., Ricciardi L.M.: Diffusion approximation and first–passage–time problem for a model neuron. III. A birth–and–death process approach. Biol. Cybern. **58** (1988) 387-404.

7. Gradshteyn I.S., Ryzhik I.M.: Tables of Integrals, Series and Products. (1980) Academic Press, New York

8. Giorno V., Nobile A.G., Ricciardi L.M.: On the asymptotic behavior of first–passage–time densities for one–dimensional diffusion processes and varying boundaries. Adv. Appl. Prob. **22** (1990) 883-914

9. Giorno V., Nobile A.G., Ricciardi L.M.: Instantaneous return process and neuronal firings. In Trappl, R. (ed.): Cybernetics and Systems Research 1992. World Scientific (1992) 829-836

10. Giorno V., Nobile A.G., Ricciardi L.M.: On asymptotic behaviors of stochastic models for single neuron's activity. In Trappl, R. (ed.): Cybernetics and System 1996. Austrian Society for Cybernetic Studies (1996) 524-529

11. Ricciardi L.M., Di Crescenzo A., Giorno V., Nobile A.G.: On the instantaneous return process for neuronal diffusion models. In Marinaro, M., Scarpetta, G. (eds.): Structure: from Physics to General Systems. World Scientific (1992) 78-94

12. Ricciardi, L.M., Di Crescenzo, A., Giorno, V., Nobile, A.G.: An outline of theoretical and algorithmic approaches to first passage time problems with applications to biological modeling. Math. Japonica **50** No. 2 (1999) 247-322

13. Ricciardi, L.M., Esposito, G., Giorno, V., Valerio, C.: Modeling Neuronal Firing in the Presence of Refractoriness. In: Mira, J., Alvarez, J.R. (eds): Computational Methods in Neural Modeling. IWANN 2003. Lecture Notes in Computer Sciences, Vol. 2686. Springer-Verlag (2003) 1-8

14. Siegert A.J.F.: On the first passage time probability problem. Phys. Rev. **81** (1951) 617-623

Fluctuation Dynamics in Electroencephalogram Time Series

In-Ho Song and Doo-Soo Lee

Department of Electrical and Computer Engineering, Hanyang University,
Haengdang-dong, Seongdong-ku, Seoul, 133-791, Korea
dslee@bme.hanyang.ac.kr
http://bme.hanyang.ac.kr

Abstract. We investigated the characterization of the complexity of electroencephalogram (EEG) fluctuations by monofractals and multifractals. We used EEG time series taken from normal, healthy subjects with their eyes open and their eyes closed, and from patients during epileptic seizures. Our findings showed that the fluctuation dynamics in EEGs could be adequately described by a set of scales, and characterized by multifractals, in both healthy and pathologic conditions. Multifractal formalism based on the wavelet transform modulus maxima (WTMM) appears to be a good method for characterizing EEG dynamics.

1 Introduction

Neuroscientists have reported in recent years that the brain has multiple feedback loops and that numerous components of the brain are mutually interactive [1]. Researchers have also begun to recognize that electroencephalogram (EEG) signals are generated by highly nonlinear systems [1] [2] [3]. Various nonlinear dynamic methods such as correlation dimension (D2), Kolmogorov entropy, largest Lyapunov exponents (LLE) and detrended fluctuation analysis (DFA) have been used to analyze EEG signals [1] [4] [5] [6]. Recent research has suggested that EEGs can be characterized in some situations by a single dominant time scale, indicating time scale invariance, and a $1/f$ fractal structure [5] [6]. However, many physiological time series fluctuate in a complex manner and are inhomogeneous, suggesting that different parts of the signal have different scaling properties [7] [8]. Our preliminary results suggested that EEGs are characterized by multifractal long-range correlations, without dominant time scales.

The aim of the present study is to investigate whether the EEG can be characterized by a multifractal spectrum using the wavelet transform modulus maxima (WTMM). To investigate whether pathological brain states affect the phenomenon of multifractality in the EEG dynamics, we compared the multifractal properties of EEG signals obtained during epileptic seizures and in healthy subjects in a relaxed states, both with their eyes open and with their eyes closed. In addition, we used a surrogate data method to assess whether EEGs have nonlinear deterministic structures.

J. Mira and J.R. Álvarez (Eds.): IWINAC 2005, LNCS 3561, pp. 195–202, 2005.

2   Materials and Methods

2.1   Materials

We used three sets of EEG data files [9]. Set A consists of 100 extracranial EEG segments taken from the scalps of five healthy subjects, using a standardized electrode location scheme. The subjects were relaxed, awake, and with their eyes open. Set B consists of 100 segments taken from five healthy subjects with the same methods as set A, but with the subject's eyes closed. Set E consists of 100 segments obtained from five patients during epileptic seizures. Set E was recorded intracranially. Each segment of 23.6 s duration was selected and cut out from continuous multichannel EEG recordings after visual inspection for artifacts and was chosen to satisfy a criterion of weak stationarity. All EEG signals were sampled at 173.61 Hz, 12 bits/sample. Band-pass filter settings were 0.53-40 Hz (12 dB/octave). All recording parameters were fixed except different recording electrodes were used for extracranial and intracranial EEG registration [9].

In this study, 30 segments from set A, 30 segments from set B and 30 segments form set E were randomly selected. The set of 30 segments selected from set A was denoted as AA. Similarly, the set of 30 segments selected from set B was denoted as BB, and the set of 30 segments selected from set E was denoted as EE.

2.2   WTMM-Based Multifractal Formalism

As stated above, manay physiological time series are inhomogeneous, fluctuating in an irregular manner. In order to characterize such signals, we used a WTMM-based multifractal formalism. Waveletes can remove polynomial trends masking the local singularities in the signal [7]. The wavelet transform of a signal $f(t)$ is defined as

$$T_\psi[f](x_0, a) = \frac{1}{a} \int_{-\infty}^{\infty} f(x)\psi(\frac{x - x_0}{a})dx \qquad (1)$$

where x_0 is the position parameter, a is the scale parameter, and $\psi(t)$ is the mother wavelet function [10]. We used the third derivative of the Gaussian function as the analyzing wavelet.

A partition function $Z_q(a)$ was defined as the sum of the power of order q of the local maxima of $|T_\psi[f](x_0, a)|$ at scale a [10]. For small scales, it was expressed as

$$Z_q(a) \sim a^{\tau(q)} \qquad (2)$$

where $\tau(q)$ is is the scaling exponent (multifractal spectrum). For monofractal signals, $\tau(q)$ is a linear function: $\tau(q) = qh(q) - 1$, where $h(q) = d\tau(q)/dq = constant$, and is the global Hurst exponent [7] [8]. For multifractal signals, $\tau(q)$ is a nonlinear function and $h(q) = d\tau(q)/dq = constant$ not constant [7] [8]. For positive values of q, $Z_q(a)$ characterizes the scaling of the large fluctuations and strong singularities, whereas for negative values of q, $Z_q(a)$ characterizes the scaling of the small fluctuations and weak singularities.

The singularity spectrum $D(h)$ can be expressed using a Legendre transform [7] [8] [10].

$$D(h) = \frac{q\tau(q)}{dq} - \tau(q) \qquad (3)$$

$D(h)$ can quantify the statistical properties of the different subsets characterized by different exponents, h. Nonzero $D(h)$ and $h = 0.5$ imply that the fluctuations in signal exhibit uncorrelated behavior. $h > 0.5$ corresponds to correlated behavior, while values of h in the range $0 < h < 0.5$ imply anticorrelated behavior [7] [8].

2.3   Surrogate Time Series and Statistical Analysis

To assess the presence of nonlinearity in the time series, an appropriate null hypothesis is that the original time series arise from a Gaussian linear process measured through a static monotonic, possibly nonlinear function [11]. In this study, the iterative amplitude adjusted Fourier transform (iAAFT) was used for each original time series. The surrogate time series generated by the iAAFT preserve the Fourier amplitudes of the original time series but randomize the phase [11].

To examine the differences between the mean values of the local Hurst exponents with maximum $D(h)$ for all sets, one-way analysis of variance (ANOVA) and post hoc analyses were performed. A paired-sample t-test was used to compare the mean values of each set between the original data and the surrogate data. The significant level was 0.05.

3   Results

Multifractal spectra and singularity spectra were computed using the WTMM for the three sets and the surrogate data. Ensemble averaged multifractal spectra, $\tau(q)$, for set AA and the surrogate data were computed, and each singularity spectrum $D(h)$ was computed through a Legendre transform from the ensemble-averaged $\tau(q)$. The two shapes of $\tau(q)$ for set AA and the surrogate data were nearly identical. The two shapes of $D(h)$ for set AA and for the surrogate data were broad. The range of local Hurst exponents with $D(h)$ greater than 0.8 was $0.1 < h < 0.4$ for set AA. The local Hurst exponents in the range $0.1 < h < 0.4$ corresponded to $D(h)$ greater than 0.8 for the surrogate data. The mean values of local Hurst exponents with maximum $D(h)$ for set AA and for the surrogate data were 0.232 and 0.222, respectively. Fig. 1 shows the ensemble averaged multifractal spectra and the singularity spectra for sets AA and the surrogate data.

Likewise, an ensemble averaged multifractal spectra, $\tau(q)$, for set BB and for the surrogate data were computed, and each singularity spectrum $D(h)$ was

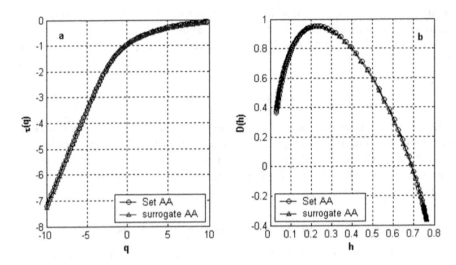

Fig. 1. (a) Ensemble averaged multifractal spectra, $\tau(q)$, for set AA and the surrogate data. Two curves have nearly identical curvature. (b) Sigularity spectra, $D(h)$, for the set AA and the surrogate data. $D(h)$ was computed through a Legendre transform from ensemble-averaged $\tau(q)$. Two curves have nearly identical curvature

computed through a Legendre transform from ensemble-averaged $\tau(q)$. They are shown in the Fig. 2. The two shapes of $\tau(q)$ for set BB and for the surrogate data were also nearly identical. The two shapes of $D(h)$ for set BB and for the surrogate data were also broad. The range of local Hurst exponents with $D(h)$ greater than 0.8 was $0.11 < h < 0.37$ for set BB. For the surrogate data, the range of local Hurst exponents with $D(h)$ greater than 0.8 was almost identical with that for set BB.The mean values of local Hurst exponents with maximum $D(h)$ for set BB and for the surrogate data were 0.217 and 0.219, respectively.

For set EE and for surrogate data, an ensemble averaged multifractal spectra, $\tau(q)$, were computed. Each singularity spectrum $D(h)$ was computed through a Legendre transform from the ensemble-averaged $\tau(q)$. Fig. 3 shows the ensemble averaged multifractal spectra and the singularity spectra for set EE and the surrogate data. For set EE, the range of local Hurst exponents with $D(h)$ greater than 0.6 was $0.2 < h < 0.96$ and that of local Hurst exponents with $D(h)$ greater than 0.8 was $0.4 < h < 0.7$. For the surrogate data of set EE, the shape of $D(h)$ was broad but was reduced in size. The range of local Hurst exponents with $D(h)$ greater than 0.6 was reduced to $0.22 < h < 0.67$ and that with $D(h)$ greater than 0.8 was reduced to $0.3 < h < 0.56$ for the surrogate data. The mean values of local Hurst exponents with maximum $D(h)$ for set EE and for the surrogate data were 0.552 and 0.414, respectively. The results of ANOVA are shown in Table 1.

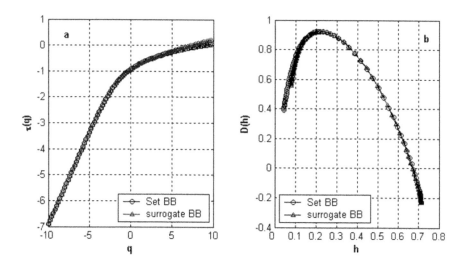

Fig. 2. (a) Ensemble averaged multifractal spectra, $\tau(q)$, for set BB and the surrogate data. Two curves have nearly identical curvature. (b) Sigularity spectra, $D(h)$, for the set BB and the surrogate data. $D(h)$ was computed through a Legendre transform from ensemble-averaged $\tau(q)$. Two curves have nearly identical curvature

Table 1. Comparison of $D(h)$ for sets AA, BB and EE. The results are presented as mean values $\pm$ SD. ANOVA was performed, followed by Sidak post hoc analysis

	a) Set AA	b) Set BB	c) Set EE	sidak post hoc analysis
$D(h)$	0.232 ± 0.691	0.217 ± 0.529	0.552 ± 0.171	$c > a, b \ (p < 0.0001)$
				$No\ significance$ a,b $p = 0.849$

4   Discussion and Conclusion

Fig. 4 shows the singularity spectra for the three sets and the surrogate data. The shapes of $D(h)$ for the three sets were broad, indicating that the EEG could not be characterized as monofractal, and was therefore multifractal. We suggest that the EEGs from the three sets have high complexity and inhomogeneous. We found that the subsets characterized by local Hurst exponents in the range $0.1 < h < 0.4$ were statistically dominant for sets AA and BB. Post hoc analysis showed no significant difference between sets AA and BB ($p = 0.98$). Therefore, the dynamics of the EEGs taken at rest with eyes open and with eyes closed exhibit anticorrelated behavior, indicating that large values are more likely to be followed by small values and vice versa [12]. The results observed after the surrogate test showed no significant change in the shape of $D(h)$ and no significant change in the values of the local Hurst exponents with maximum $D(h)$ for sets AA and BB. The paired-sample t-test showed no significant differences between set AA

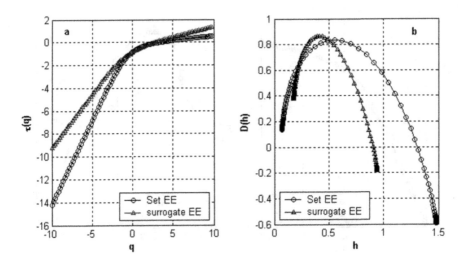

Fig. 3. (a) Ensemble averaged multifractal spectra, $\tau(q)$, for set EE and the surrogate data. Note that there are clear differences between the $\tau(q)$ curves for set EE and the surrogate data. (b) Sigularity spectra, $D(h)$, for the set BB and the surrogate data. $D(h)$ was computed through a Legendre transform from ensemble-averaged $\tau(q)$. The different shape of $D(h)$ for the surrogate data reflects different fluatuations associated with phase correlation

and the surrogate data, and between set BB and the surrogate data, respectively ($p > 0.05$). Since the iAAFT preserves the linear characteristics of the original time series but removes nonlinearities, we suggest that the EEGs taken during rest with eyes open and with eyes closed have no significant phase correlation and exhibit quasilinear structures. These results are consistent with previous studies [5] [9].

Statistically dominant subsets were characterized by local Hurst exponents of $0.3 < h < 0.85$ for set EE. We found that the dynamics of the epileptic EEGs exhibited anticorrelated, correlated, and uncorrelated behaviors. The mean values of local Hurst exponents with maximum $D(h)$ for sets AA, BB and EE were 0.232, 0.217 and 0.552, respectively. ANOVA showed a significant difference in the mean value of the local Hurst exponents with maximum $D(h)$ ($F(2, 87) = 87.21, p < 0.0001$). Post hoc analysis showed that the mean value of the local Hurst exponents with maximum $D(h)$ for set EE differed significantly ($p < 0.0001$) from that for sets AA and BB. Moreover, the range of dominant local Hurst exponents for set EE was wider than that of set AA or set BB. These findings indicate that the epileptic EEG has more inhomogeneity than EEGs taken at rest with eyes open or with eyes closed. We found that multifractality could quantify the fluctuation dynamics from different pathological brain states. From the results of the surrogate data for set EE, we found that the local Hurst exponents for the dominant subsets were significantly changed to $0.26 < h < 0.61$. In addition, we found that the mean value of local Hurst

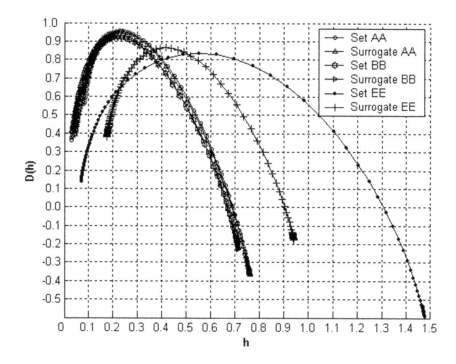

Fig. 4. Comparison of singularity spectra for the three sets and the surrogate data

exponents with maximum $D(h)$ was significantly changed from 0.552 to 0.414 ($p < 0.0001$). The surrogate data, however, remained a multifractal process. These findings indicate that epileptic EEGs have significant nonlinear determinostic structures and phase correlation. This finding is in accord with other studies [9]. As shown Fig. 1, 2, The two curves of the ensemble averaged $\tau(q)$ for the original data and the surrogate data were almost the same as for sets AA and BB. These observations suggest that there are no significant differences in the $D(h)$ of the original data and the surrogate data for sets AA and BB. On the other hand, Fig. 3 showed that the two curves of the ensemble averaged $\tau(q)$ for set EE and the surrogate data differed significantly. This observation demonstrates that there were different fluctuations and different characteristics for the singularities between set EE and the surrogate data.

In summary, we investigated the possibility that the EEGs from the three sets exhibit higher complexity than $1/f$ scaling. Our findings showed the existence of inhomogeneities in the three sets. We found that the EEGs taken during rest with the eyes open exhibited anticorrelated behavior and had linear structures. The epileptic EEGs exhibited anticorrelated, correlated, and uncorrelated behaviors, and the epileptic EEGs had nonlinear deterministic structures rather than linear structures. We suggest that the complexity of the EEG fluctuations can best described by a set of scales rather than a single dominant scale.

Acknowledgements. This study was supported by a grant of the project "Development of the Core Technology of Medical Devices for Elderly", Ministry of Science and Technology (MOST), Republic of Korea.

References

1. Fell, J., Röschke, J., Mann, K., Schäffner, C.: Discrimination of sleep stages: a comparison between spectral and nonlinear EEG measures. Electroencephalogr. Clin. Neurophysio. **98** (1996) 401–410
2. Thomasson, N., Hoeppner,T.J., Webber Jr, C.L., Zbilut, J.P.: Recurrence quantification in epileptic EEGs. Phys. Lett. A. **279**. (2001) 94–101
3. Song, I.H., Lee, D.S., Kim, Sun.I.: Recurrence quantification analysis of sleep electroencephalogram in sleep apnea syndrome in humans. Neurosci. Lett. **366** (2004) 148–153
4. Krystal, A.D., Zaidman, C., Greenside, H.S., Weimer, R.D., Coffey, C.E.: The largest Lyapunov exponent of the EEG during ECT seizures as a measure of ECT adequacy. Electroencephalogr. Clin. Neurophysiol. **103** (1997) 599–606
5. Pereda, E., Gamundi, A., Rial, R., González, J.: Non-linear behaviour of human EEG: fractal exponent versus correlation dimension in awake and sleep stages. Neurosci. Lett. **250** (1998) 91–94
6. Watters, P.A.: Time-invariant long-range correlations in electroencephalogram dynamics. Int. J. Sys. Sci **31** (2000) 819–825
7. Ivanov, P.Ch., Amaral, Luís A.N., Goldberger, A.L., Havlin, S., Rosenblum, M.G., Struzik, Z.R., Stanley, H.E.: Multifractality in human heartbeat dynamics. Nature. **399** (1999) 461–465
8. Ivanov, P.Ch., Amaral, Luís A.N., Goldberger, A.L., Havlin, S., Rosenblum, M.G., Stanley, H.E., Struzik, Z.R.: From $1/f$ noise to multifractal cascades in heartbeat dynamics. Chaos, **11** (2001) 641–652
9. Andrzejak, R.G., Lehnertz, K., Mormann, F., Reike, C., David, P., Elger, C.E.: Indications of nonlinear deterministic and finite-dimensional structures in time series of the electrical activity: Dependence on recording region and brain state. Phys. Rev. E. **64** (2001) 061907-1–061907-8
10. Muzy, J.F., Bacry, E., Arneodo, A.: The multifractal formalism revisited with wavelet. Int. J. Bifurc. chaos. **4** (1994) 245-302
11. Schreiber, T., Schmitz, A.: Surrogate time series. Physica D **142** (2000) 346-382
12. Beran, J.: Statistics for Long-Memory Processes. Mew York: Chapman & Hall. (1994)

Modelling of Dysfunctions in the Neuronal Control of the Lower Urinary Tract

Daniel Ruiz Fernández, Juan Manuel García Chamizo,
Francisco Maciá Pérez, and Antonio Soriano Payá

Dpto Tecnologia Informática y Computación, University of Alicante,
PO 99, 03080, Alicante, Spain

Abstract. In this article a model of the regulator system of the lower urinary tract is presented discussing, in particular, the dysfunctions associated with the neurogenic bladder. The design and implementation of the model has been carried out using distributed artificial intelligence, more specifically a multi-agent system. Each agent is modelled so that its behaviour is similar to that of a neuronal centre. By means of this design, the behaviour of the neuronal regulator of the lower urinary tract is simulated using a model with a similar structure to the organisation of the biological system, conferring on it emergent properties. We compare the results obtained using the model in situations with several neurological dysfunctions with experimental results obtained from patients suffering from the analysed dysfunctions. The data, obtained using the model, are consistent with the existing real clinic studies in medicine related to the same dysfunctions.

1 Introduction

The complexity of operation of the biological structures related to the control of the organic systems gives rise to the search for technological support to facilitate this study. In this association between computer technology and medicine, the support systems for the diagnosis become particularly relevant. These systems make available to the specialist complementary information whose objective is to help him in the diagnostic task that is, on occasion, extremely complicated. Urology is a medical discipline in which a system of help with the diagnosis can come in useful to the specialist and to the patient. In this case - the study of dysfunctions of the lower urinary tract due to neurological causes [1], [2], [3] - can facilitate the study and the understanding of the pathology for both specialist and patient.

As stated in the International Society of Continence [4], the lower urinary tract is common to the bladder and the urethra. These two elements form a functional unit with an evident interaction between them [5]. The voluntary nervous system (identified with the somatic nerves) and the involuntary nervous system (identified with the sympathetic and parasympathetic systems) participate in the control of the functions of the lower urinary tract (storage and voiding of the bladder).

J. Mira and J.R.Álvarez (Eds.): IWINAC 2005, LNCS 3561, pp. 203–212, 2005.

The complexity of the interaction between mechanical elements subjected to physical laws of visco-elasticity and hydro-dynamics, and neuronal control (arising from neuronal centres), is related to an important number of associate dysfunctions. These dysfunctions are distributed between dysfunctions with exclusively mechanical implications and dysfunctions with a neuronal component. The existence of imprecise areas in the different models for diagnostic of obstructions and the high casuistry related to the typical differences between individuals (different bladder capacities, sensations of urgency,...), justify the use of the system suggested in this article.

The system also presents other functional aspects besides help with diagnosis. Firstly, the possibility of experimenting with the model using values outside of range, experimentation that would not be possible in normal individuals. Secondly, it is possible to find a didactic component since the system is presented as a simulator of the lower urinary tract, showing the patient the possible reasons for their urologic dysfunctions, as well as the mechanism to cure them or alleviate them by means of medication or surgery. The proposed architecture is based on models referred to in the bibliography and they incorporate static and dynamic properties of the bladder and the urethra [6], [7], [8].

The paradigm of intelligent agents has been used in the design and development of the system, more concretely agents PDE (Perception-Deliberation-Execution)with the capacity to perceive, to deliberate and to execute tasks. This paradigm confers on the system a great adaptability and a good approximation to the biological operation of the neuronal control centres [9].

This paper will explain the paradigm of agents and the relationships between this paradigm and the neuronal regulator of the lower urinary tract. Details will follow of the neuronal control model that has been designed, showing its application in situations with dysfunctions, which are previously described. Lastly the main conclusions are presented as well as the future lines of development.

2   The Agents Paradigm and the Biological Model

Traditionally biological systems modelling or the simulation of their functions has been carried out by means of mathematical models. Techniques of classic control have been used to model the biological regulators or techniques allied to artificial intelligence, the latter necessitating a base of expert knowledge (heuristic based on fuzzy logic or artificial neuronal networks). Intelligent agents are identified with an architecture based on artificial intelligence and confirmed by the knowledge of an expert [10].

The concept of agent as used in this work is based on a structure of perception, deliberation and execution [9], [11]. Each one of the neuronal centres that is part of the neuronal regulator is represented by an agent so that it perceives the situation of the elements in the environment (urinary system and other neuronal centres); using the information obtained and the internal state of the agent, the agent takes working decisions; and, finally, the agent executes the decision that is represented, in our system by efferent signals to the mechanical system and

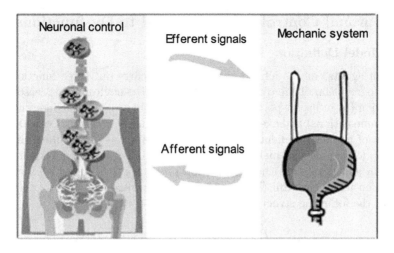

Fig. 1. Control system of the lower urinary tract

internal signals to other neuronal centres (Fig. 1). The decision at this stage is based on functional rules that define the behaviour of each one of the neuronal centres and that, in this case, they are based on the cognitive capacity that is conferred on the agents by a memorisation function that allows them to store an internal state.

The use of intelligent agents to model neuronal control confers numerous advantages on the model, such as:

1. Adaptability. The model can easily be improved by looking deeply into the workings of the neuronal centres that participate in the control of the urological system. Other neuronal centres can be included immediately, as can modification of the internal control of each one of the centres: it is only necessary to modify the system of rules that governs the mechanism of decision.
2. Distributed computing. Modelling using agents allows a distributed computing scheme of calculation, distributing the computing capacity and the decision-making power among the different agents that configure the system; the model is closer to the biological system of neuronal control than another centralized computing scheme with two components, one mechanical and another neuronal, that interact but which didn't offer a distributed idea of control.
3. Asynchronous behaviour. The existence of several agents that interact in the system points directly to the possibility of a synchronous or an asynchronous operation. Since the biological system works in an asynchronous way we have chosen this operational way to develop the system. This implies that each operation of an agent is independent in time, only coordinating its operation by means of the inherent behaviour of each neuronal centre and not by means of an external coordinating mechanism in the system. This being so, the results obtained are characteristic of a system with emergent properties.

3   Neuronal Control Model Applied to Dysfunctions

3.1   Model Definition

Intelligent agents can be defined as entities that carry out three functions continually: perception of the dynamic conditions of the environment, reasoning or deliberation (to interpret perceptions, to solve problems, to make inferences and to determine actions) and execution of the particular actions (that can affect the condition of the environment). This focus type is known as architecture PDE. In the model, capacity of memorisation has been incorporated into each centre, so that an added value in the deliberation phase is provided, which makes the deliberation more powerful (Fig. 2). A neuronal centre can thus be defined by means of the following structure:

$$\alpha = (\phi_\alpha, S_\alpha, Percept_\alpha, Mem_\alpha, Decision_\alpha, Exec_\alpha) \tag{1}$$

where ϕ_α is the perception set, S_α the internal states set and the rest are functions of perception, memorizing, decision and execution respectively.

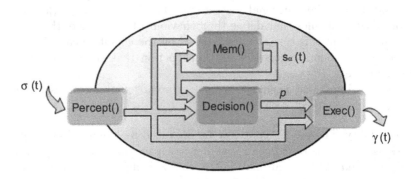

Fig. 2. Structure of a neuronal centre modelled as a PDE agent

Each agent tries to reach its own target by interacting with other agents (neuronal centres) or with the environment (mechanical subsystem). Input function (t) shows the state of the system at a given time, and the execution of a certain task (p) will be the one that determines the agent's interaction with the rest of the system. The neuronal centres receive as input the group of neuronal afferent signals (information generated by the mechanical system) and internal signals that come from other neuronal centres, and they provide as output neuronal efferent (that act on the mechanical system) and internal signals, which have other centres as their destinations [7]. The dynamics of operation of the system would be defined by the following equations:

$$\sigma(t+1) = React\left(\sigma(t), \bigcup_{i=1}^{n} Exec\left(Decision_i\left(\sigma_i(t), s_i(t)\right), \sigma(t)\right)\right) \tag{2}$$

$$s_1 (t + 1) = Mem (\sigma_i (t), s_i (t))$$

$$\cdot$$
$$\cdot$$
$$\cdot$$

$$s_n (t + 1) = Mem (\sigma_n (t), s_n (t))$$

The functions of the lower urinary tract are urine storage and later voiding, functions that are controlled by the parasympathetic, sympathetic and somatic nervous systems [7], [12]. Each one of these systems is formed by one or several neuronal groups that present afferent, efferent and internal signals [8], [13]. The neuronal model on which the system is based incorporates nine neuronal centres, classified in three different areas according to its neuroanatomical situation [1], [7]:

- In the suprapontine zone the preoptic area centre (PA), the periaqueductal grey area (PAG) and the cortic-diencefalic centre (CD) are found.
- At pontine level the pontine storage centre (PS) and the pontine micturition centre (PM) can be observed. These centres are considered as information transmitters between the suprapontine and the spinal centres.
- In the spinal area, we can observe four neuronal centres: the thoracolumbar storage centre (TS) located in the spinal segment T11-L2, and in the sacral area, the dorsal grey comisure centre (DGC) located in the sacral segment S2-S4, beside the sacral micturition centre (SM), and the sacral storage centre (SS) located in the segment S1-S3.

The relation between the neuronal centres and its afferent and efferent signals can be observed in fig. 3.

3.2   Dysfunctions Characterization

The set of dysfunctions associated with the lower urinary tract covers a wide spectrum of changes with diverse symptoms and origins. This wide spectrum is due to the diverse control and action implications that can be found in the micturition mechanism. The variety of elements and rules that operate in the working of the lower urinary tract means that the distribution in categories or subgroups of the dysfunctions that can be presented facilitates its study and its later analysis, diagnosis and treatment.

To study the dysfunctions associated with the system of the lower urinary tract, there exist diverse classifications according to the neuroanatomical localization of the lesion, the urodynamic discoveries, the clinical symptoms or the phase of the bladder function affected. This last classification is simple and clinically useful, distributing the dysfunctions depending on whether they are caused by the bladder or the sphincter and whether they are in the retention or voiding phase [14].

A failure in retention can originate in the bladder or to be consequence of some sphincter dysfunction or possibly a combination of both and can give rise

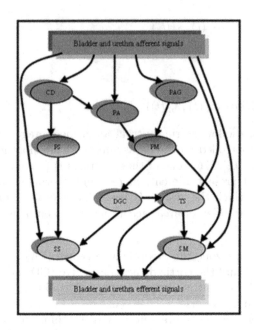

Fig. 3. Diagram of the neuronal control of the lower urinary tract

to several dysfunctions from different causes. Among the bladder causes we can find involuntary contractions of the detrusor that can cause a hyperreflexia of the detrusor or a dysinergy; a decrease in the bladder straining can cause low bladder accommodation and incontinence; a bladder hypersensitivity will give rise to an incontinence of urgency. The more common sphincter cause of dysfunction associated with the storage is a failure in the sphincter strength that frequently implies situations of stress incontinence.

Voiding failures usually appear as a consequence of a contractile decrease of the bladder (hyporeflexia) or an obstruction in the exit tract, due to a physical body or to a detrusor-sphincter dysinergy. Another classification type is the neuroanatomical that organizes the dysfunctions in groups according to the localization of the neuronal components that cause the dysfunction. Although this classification is not very much used by the urologists because from the uro-dynamical point of view it is difficult to study each level of the system separately, its use would be of interest because it provides an adjusted idea of the origin of the dysfunction, enabling the clinician to treat the cause (if this is possible) and not only its consequence.

4 Simulation of Dysfunctions Related to the Lower Urinary Tract

Several dysfunctions have been studied according to the following neuroanatomical classification: suprapontine lesions, spinal suprasacral lesions and sacral le-

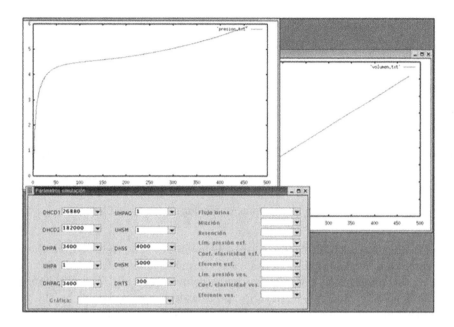

Fig. 4. Monitoring interface of the neuronal regulator simulation of the lower urinary tract

sions. This classification has been used because it is the most closely related to the space distribution of neuronal centres and with their communication with each other and with the mechanical subsystem. To obtain the urodynamic results corresponding to the dysfunctions introduced artificially in the model an application has been used (fig. 4). The software has been designed to modify the variables that participate in the operation of the lower urinary tract and to be able to simulate dysfunctions.

The simulator prototype has been carried out on the Linux platform and using Java as the programming language. This application consists of a module of control of parameters where it is possible to modify the thresholds of the different agents that act as neuronal centres and several variables related to the mechanical system: the coefficients of elasticity of the urethra and the bladder, the input flow from the ureters to the bladder, etc. [15]. The other component of the simulator is the graphic monitor that allows visualisation of the evolution during the micturitional cycle of the values of the variables related to the model (bladder pressure, urethral pressure, urine volume, efferent and afferent signals, etc).

4.1   Spinal Suprasacral Lesion

At this point we present the urodynamic situation as a consequence of a spinal suprasacral lesion in an upper level of the T11-L2 segment, so that the operation of the storage thoracolumbar centre would not be affected while the reflex communication with the pontine centres is damaged, the involuntary defense

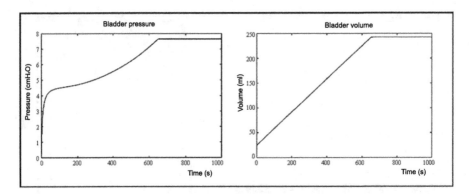

Fig. 5. Pressure and volume graphs for a supraspinal lesion

reflex (associated with the sacral centres) being intact. The result obtained with
the simulation is quite similar to that of the previous case but, on this occasion
(fig. 5), the urine losses begin with a lower volume (245 ml). The obtaining of
similar results to the case of the suprapontine lesion is explained by the fact
recognized in several models that the pontine centres are transmitter centres of
the information obtained by the higher centres.

4.2   Sacral and Supraspinal Lesion

Finally we propose the simulation of a lesion that affects some of the sacral
centres, for example to the sacral storage centre. In the pressure graph (Fig. 6)
increases in the bladder pressure as a consequence of the involuntary contractions
of the detrusor are observed, causing incontinence situations. It is observed in
figure 6 how the voluntary control disappears totally and the micturition takes
place with a very low volume of urine (140 ml). The cause of this is explained in

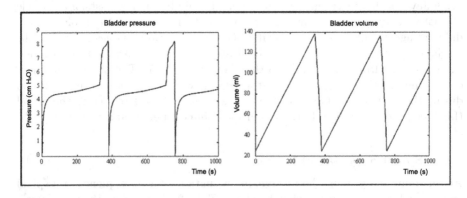

Fig. 6. Pressure and volume graphs in a supraspinal lesion joined to a dysfunction in
the sacral storage centre

the model by the inhibition in the simulation of the activity of the sacral storage centre (besides the loss of the communication with the higher centres).

5   Conclusion

In this work, we present an alternative focus for the simulation of biological processes, designing the regulator system of the operation of the lower urinary tract using a model based on intelligent agents. We have observed how the use of this methodology facilitates the modelling of a biological system essentially distributed as the neuronal regulator of the lower urinary tract. The model has been implemented and a simulation system has been created to verify the operation of the lower urinary tract when some of its components operate with values outside of the normal ranges of action; to simulate dysfunctions in the system we only act on the agent corresponding to the neuronal centre that is affected by the dysfunction. To check the validity of the model we have chosen a classification that is useful in distributing the dysfunctions anatomically and treats their cause in a more specific way. This classification is not normally used for the diagnostic activity because of the difficulty associated with the study of the different neuronal centres that affect this system, but it can be easily applied to the dysfunctions study using the designed model. We have been observed as the results obtained with the model in several dysfunctions coincide with the expected behaviour according to the data that can be found in the specialized literature. In a continuation of this work we seek to improve the deliberative capacity of the intelligent agents by means of the inclusion of heuristic, at the same time as we improve their structure with new data that appear in the bibliography about the neuronal centres implied. Future works propose the design of a hardware platform that puts together in its architecture the intrinsic characteristics of the multi-agent systems. This architecture is directed to the modelling of neuronal system regulators.

References

1. Blok, B.F.M., Holstege, G.: The central control of micturition and continence: implications for urology. British Journal of Urology, 83 (1999) 1-6
2. Yoshimura, N., de Groat, W.C.: Neural Control of the Lower Urinary Tract, International Journal of Urology, 4 (1997) 111-125
3. Inatomi, Y., Itoh, Y., Fujii, N., Nakanishi, K.: The spinal cord descending pathway for micturition: analisis in patients with spinal cord infarction, Journal of Neurological Sciences, 157 (1998) 154-157
4. Abrams, P., Blaivas, J.G., Stanton, S., Andersen, J.T.: The Standardisation of Terminology of Lower Urinary Tract Function, Neurourology and Urodynamics, 7 (1988) 403-426
5. de Groat, W.C.: Anatomy and Physiology of the Lower Urinary Tract, Urologic Clinics of North America 20 (3) (1993) 383-401

6. van Duin, F., Rosier, PF.W.M., Bemelmans, B.L.H., Wijkstra, H., Debruyne, F.M.J., van Oosterom, A.: Comparison of Different Computer Models of the Neural Control System of the Lower Urinary Tract, Neurourology and Urodynamics 19 (2000) 289-310

7. García, J.M., Romero, J., Maciá, F., Soriano, A.: Modelado y simulación del regulador neuronal del tracto urinario inferior, Urodinámica Aplicada, 15 (2) 2002

8. Heldoorn, M., van Leeuwen, J.L., Vanderschoot, J.: Modelling the biomechanics and control of sphincters, The Journal of Experimental Biology, 204 (2001) 4013-4022

9. García, J.M., Maciá, F.: A Mobile Agent-based Model for a Node Regeneration System, International Conference on Knowledge Based Computer Systems (KBCS 2000). Mumbai (India),. (2000) 82-93

10. García, J.M., Maciá, F., Soriano, A., Flórez, F.: A Multi-Agent System uses Artificial Neural Networks to Model the Biological Regulation for the Lower Urinary Tract, WSES Conference on Neural Networks and Applications. Interlaken (Switzerland). (2002), 162-167

11. García, J.M., Soriano, A., Maciá, F., Ruiz, D.: Modelling of the sacral micturition centre using a deliberative intelligent agent, Proceedings of the IV International Workshop on Biosignal Interpretation (BSI 2002). Como (Italy). (2002), 451-454

12. Kinder, M.V., Bastiaanssen, E.H.C., Janknegt, R.A., Marani, E.: The Neuronal Control of the Lower Urinary Tract: A Model of Architecture and Control Mechanisms, Archives of Physiology and Biochemistry, 107 (1999) 203-222

13. Bastiaanssen, E.H.C., van Leeuwen, J.L., Vanderschoot, J., Redert, P.A.: A Myocybernetic Model of the Lower Urinary Tract, Journal of Theoretical Biology 178 (1996) 113-133

14. Micheli, F., Nogu'es, M.A., Asconap'e, J.J., Pardal, M.M.F., Biller, J.: Tratado de Neurología Clínica, Ed. Panamericana, 222-234, 2002

15. Rowan, D., Douglas, E., Kramer, A.E.J.L., Sterling, A.M., Seul, P.F.: OUrodynamic equipment: technical aspects, Journal of Medical Engineering and Technology, 11 (2) (1987) 57-64

Coding Strategies in Early Stages of the Somatosensory System

Juan Navarro[1], Eduardo Sánchez[2], and Antonio Canedo[1]

[1] Departamento de Fisioloxía, Facultade de Medicina,
Universidade de Santiago de Compostela,
15782 Santiago de Compostela, Spain
{fsjna, fsancala}@usc.es
[2] Grupo de Sistemas Intelixentes (GSI),
Departamento de Electrónica e Computación, Facultade de Físicas,
Universidade de Santiago de Compostela,
15782 Santiago de Compostela, Spain
eduardos@usc.es
http://www-gsi.usc.es/index.html

Abstract. This paper explores the information coding performed by the local circuit of the Cuneate Nucleus (CN). Based on physiological data, we have developed a realistic computational model and studied its output after presenting different types of plausible cutaneous stimuli. Computer simulations show that (1) static stimuli are encoded in progressive spatio-temporal patterns made up of single-spike trains generated by each stimulated neuron, and (2) moving stimuli are encoded with a bursting discharge of those units responding to the leading edge of the stimulus. These results suggest that the role of CN could be the transformation of the stimulus representation in order to facilitate both discrimination and classification in later processing stages.

1 Introduction

Our interest is focused on the role of the Cuneate Nucleus (CN), a prethalamic structure. The cuneate and gracile constitute the Dorsal Column Nuclei (DCN), which receive input from primary afferent fibres (PAF) carrying somesthetic information from cutaneous receptors located in the limbs and trunk. The middle region of the DCN is the first stage of the Dorsal Column-Medial Lemniscus system, which processes fine tactile information. After being processed by the sensory receptors and the DCN, the cutaneous ascending information follows the Medial Lemniscus and make a relay in the Ventro-Postero-Lateral thalamic nucleus (VPL) [5] before reaching the primary somatosensory cortex.

The middle cuneate posses two main types of neurons [3]: cuneolemniscal (CL) or relay cells, and local neurons or interneurons. The former receive cutaneous input and project to the VPL nucleus through the Medial Lemniscus, hence the name cuneolemniscal; the later receive input from various sources

J. Mira and J.R. Álvarez (Eds.): IWINAC 2005, LNCS 3561, pp. 213–222, 2005.
© Springer-Verlag Berlin Heidelberg 2005

(mainly cutaneous and proprioceptive) and exert their influence over the cu-neothalamic neurons as well as other interneurons.

A key point to better understand the cuneate function relies on the precise knowledge of the receptive field (RF) structure of the different cell types im-plicated in the CN circuitry. The current knowledge is limited mainly to CL cells. It has been recently shown that their RF has an excitatory center and an inhibitory surround [4]. This spatial arrangement is generated through somato-topically organized afferent organization consisting of direct excitatory input on CL cells and GABaergic mediated inhibition from surrounding areas [4]. It has been demonstrated that when this circuit is constructed with realistic models of cuneate cells [11], it can perform edge detection as well as motion discrimination [10]. Moreover, the relay cells with matched receptive fields monosynaptically activate each other through recurrent collaterals re-entering the nucleus, while inhibiting other projection neurons with different RFs [1]. Their function could be to reinforce and synchronize the activity of CL cells in order to increase the robustness of the transmission of information to the VPL [1, 9]. Finally, the pres-ence of a second interneuronal type, the GLYcinergic (GLY) interneurons, which constitute about 30% of total neuronal population in the rat [6], introduces fur-ther complexity into the circuit. It is known that GLY neurons evoke facilitation of relay cells by inhibiting GABaergic interneurons [1, 2].

The cuneate circuitry made up with the aforementioned neurons can perform a spatio-temporal coding of the cutaneous stimulus [12]. Computational simula-tions show that the information associated to the stimulus could be transmitted to the thalamus in a progressive way, starting at regions with higher contrast, probably the more informative, and finishing at those with lower contrast. Fur-thermore, it has been shown that a clear dependency exists between the time required for the progressive coding and the spatial features of the stimulus, such as contrast and texture [7, 8].

The goal of this paper is to explore the coding possibilities of the CN by considering the complete circuit studied in Sanchez et al. (2004), and Navarro et al. (2004a,2004b), with the realistic models of CL cells derived from Sanchez et al. (2003). Specifically, we wonder: (1) whether the progressive fill-in coding performed by the circuit remains when realistic models of CL cells, which intro-duce a complex firing behaviour, are considered in the circuit, and (2) whether bursting activity would be a significant code for moving stimuli.

2   Computational Methods

The model consists of 40,000 units distributed over three main layers represent-ing: (1) projection or cuneolemniscal (CL) neurons, (2) GABaergic recurrent interneurons, and (3) GLYcinergic interneurons (figure 1). CL units show an excitatory centre - inhibitory surround afferent 3x3 RF, as well as recurrent inhibition mediated through GAB interneurons. These units have a 7x7 ring-shaped RF derived from CL cells that are second-order neighbours. Finally, the GLY interneurons present a fully excitatory 9x9 RF deriving from CL cells with

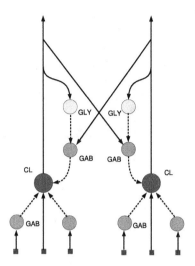

Fig. 1. The CN circuit

overlapped RFs. The RFs sizes were selected such that their combination gives the more stable results. In addition, the interneurons produce shunting inhibition on CL neurons thus achieving robust edge detection against stimulus intensity.

As the rich dynamics of cuneolemniscal neurons is key to understand the coding capabilities of the CN, we have modeled each neuron n_{ij} in a realistic way with the following firing condition:

$$y_{ij}(t) = \begin{cases} 1 \text{ if } v_{ij}(t) > \Theta_{spike} \\ 0 \text{ otherwise} \end{cases} \qquad (1)$$

where $v_{ij}(t)$ is the membrane potential at time t and Θ_{spike} a positive threshold value (see figure 2). The membrane potential is updated by the following equation:

$$v_{ij}(t) = v_{ij}(t-1) + I_{ij}^{total}(t) - I_{ij}^{ionic}(t) \qquad (2)$$

The total afferent input $I_{ij}^{total}(t)$ of the CL neuron n_{ij} is computed by multiplying every input I_{ij} inside its receptive field by its corresponding synaptic weight w_{ij}, thus $I^{total} = [W][I]$.

The ionic current term I_{ij}^{ionic} is a linear combination of the contributions of a sodium current, I_{Na}, which is activated under depolarization and generates sodium spikes; a potassium current, I_K, which repolarizes the membrane potential; a hyperpolarization-activated cationic current I_h, which impedes a greater hyperpolarization; and a low-threshold calcium current I_T, which is de-inactivated after hiperpolarization and generates after further depolarization an slow rebound potential. Each current was computed according to the generalized Ohm's law $I = g * (v - v_{ion})$, where g denotes the current conductance and v_{ion}

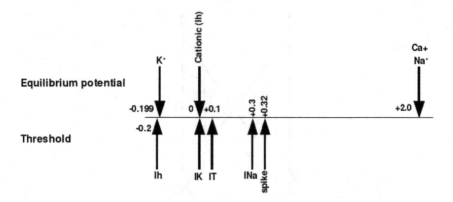

Fig. 2. Equilibrium potentials and thresholds for the ionic currents

the equilibrium potential for such ion. Figure 2 shows the values of the equilibrium potential and threshold for each current. These equilibrium potential parameters do not represent the real reversal potentials, but the values to which the membrane potential is attracted when the corresponding ionic current is activated. In a realistic model, g is usually described by means of Hodgkin-Huxley formalism but here we have used simplified descriptions of the conductance dynamics:

$$g_{ij}^{Na}(t) = \begin{cases} 0 \text{ if } y_{ij}(t - k \triangle t) = 1 \; \forall k; 0 < k \leq a \\ 1 \text{ otherwise} \end{cases} \tag{3}$$

$$g_{ij}^{K}(t) = \begin{cases} \frac{g_{ij}^{K}(t-1)}{\alpha} \text{ if } y_{ij}(t - k \triangle t) = 1 \text{ and } v_{ij}(t) > 0 \; \forall k; 0 < k \leq a \\ 0 \qquad \text{otherwise} \end{cases} \tag{4}$$

$$g_{ij}^{h}(t) = \begin{cases} 1 \text{ if } v_{ij}(t) < v_{ij}(t-1) \text{ and } v_{ij}(t) < \Theta_h \\ 0 \text{ otherwise} \end{cases} \tag{5}$$

$$g_{ij}^{T}(t) = \begin{cases} 1 \text{ if previous hyperpolarization and } v_{ij}(t) > v_{ij}(t-1) \\ 0 \text{ otherwise} \end{cases} \tag{6}$$

The interneurons have been described as McCulloch-Pitts units, in which the output of the neuron j is $y_j = \Psi(\sum w_{ji}x_i)$, with the activation function Ψ being of the threshold type. Contribution of synapse between neuron i and neuron j is modelled through weight w_{ji}.

The computational simulations update the network units in cycles of three stages. In the first stage, the output of the afferent GAB interneurons as well as the total input for the CL cells is computed. In the second stage, the membrane potential $v_{ij}(t)$ is updated based on the contribution of the CL ionic currents during a fixed time interval. Finally, in the third stage, the output of recurrent

GAB and GLY interneurons is calculated. Each GAB and GLY unit possesses a synaptic buffer that temporally stores the output of CL cells during the second stage. Two values are considered: the number of spikes generated during the time interval of the second stage, and the number of consecutive spikes, which are considered as bursts. In the simulations, the occurrence of each burst is weighted by 10, while each single spike by 1, in order to consider the different impact of single spikes and bursts as inputs of the GAB and GLY interneurons. In the results section, each stage in the update process is called iteration.

Experiments were performed with stimuli of different forms, sizes and textures over a white background. Both the stimulus and the output intensity are represented in grey scale, thus taking values from 0 to 255. Stimulus textures are obtained from .bmp files produced with GIMP, a Linux image-processing application. Quasi-random frames with a repetitive pattern of hexagonal tiles were generated, and then combined to build a mosaic. The GIMP function gaussian blur was used to modify the degree of stimulus contrast. Network responses were characterized based on 3 features: robustness of the edge detection process, processing time required to reach a stationary state, and bursting discharge in response to a moving stimulus.

3 Results

In the first experiment we want to know (1) whether the progressive coding performed by the CN is independent of the form of the stimulus, or, in other words, whether it depends on the shape and orientation of the receptive fields in our model, and (2) how a moving stimulus would be encoded by the circuit. Figure 3 shows 12 outputs of the CL cells during a simulation with an stimulus that contains three geometric objects. The label on each inset indicates the iteration number during the simulation process, and those labels starting at letter m represent a moving stimulus. Black points represent firing units generating a single spike after reaching threshold, while white points represent silent units. The upper row (1 to 52) illustrates the output after the presentation of a static stimulus. Afther that, the stimulus is moving to the right with a speed of 1 pixel per iteration from iteration 72 until iteration 180. The middle row (m79 to m179) shows 4 iterations corresponding to two different parts of the moving sequence. The firing units in insets m80 and m179 (colored in red in the electronic version of the paper) show a bursting activity made up of two consecutive spikes. The last row illustrates 3 iterations when the stimulus has stopped and the network recovers the output before the onset of the motion. Intermediate iterations have been omitted in order to show the relevant activity. In this simulation different coding strategies can be observed: (1) a progressive coding of static stimulus, initiated with the representation of the complete surface of the stimulus (iteration 1), followed up by the processing of regions with highest contrast (edges in iteration 7) and a fill-in effect that converges to the initial state; and (2) bursting activity encoding a dynamic stimulus in those units in the leading front of the stimulus (iterations m80 to m179). It is important to notice that there is no memory effect of the bursting activity when the motion finishes, as the output of iteration 187 is the same as the one of iteration 1.

Figure 4A plots a histogram of the total number of CL units ("global output") activated for each iteration of the sequence shown in figure 3. The plot shows an oscillatory pattern with the different states of the output: (1) the input state, in which the output reproduces the presented stimulus, (2) the silent state, in which no activity is observed in CL cells, and (3) the fill-in or progressive state, where new CL units are recruited over time. This effect is clearly observed from iteration 7 to iteration 60. Figure 4B shows an histogram of the total number of units in which bursting activity has been detected during the sequence shown in figure 3. This type of discharge only appears when the stimulus starts a displacement to the right (see figure 3). It completely disappears when the stimulus stops after iteration 180.

The bursting discharge associated to a moving stimulus does not depend on the direction of motion or on the object form. Figure 5 shows a stimulus with an

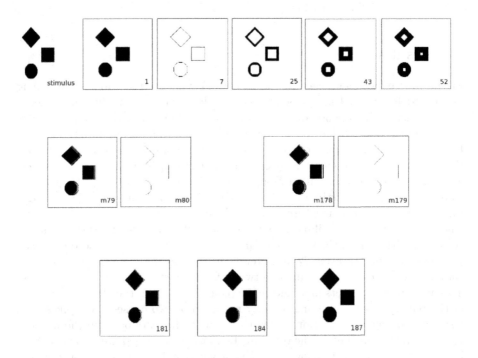

Fig. 3. Progressive coding and bursting discharge. Progressive fill-in coding appears when static stimuli are presented (top row). The coding changes to bursting activity when the stimuli starts moving (middle row). Bursts only appears in those neurons responding to the leading edges of the moving stimuli. Progressive coding is recovered when the stimuli cease their motion (bottom row)

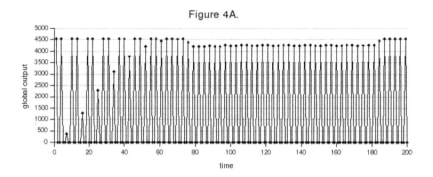

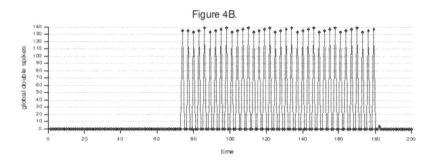

Fig. 4. Histograms of the global activity of CL units for the experiment of figure 3. The top plot illustrates the total number of CL units that generate a single spike on each iteration. Progressive coding evolves from 400 activated units to 4500 ones between iteration 1 to 52, and it stops when the stimuli start moving. The bottom plot shows the total number of CL units that generate a burst. This type of activity only appears when moving stimuli is presented

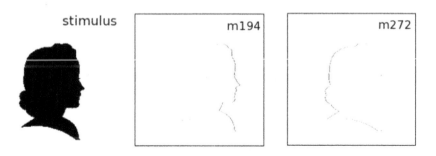

Fig. 5. Bursting discharge is not affected by the form nor the direction of the moving stimulus. An stimulus with an irregular shape moving to the right (m194), and then to the left (m272) induces bursting behaviour in both leading edges of the stimulus

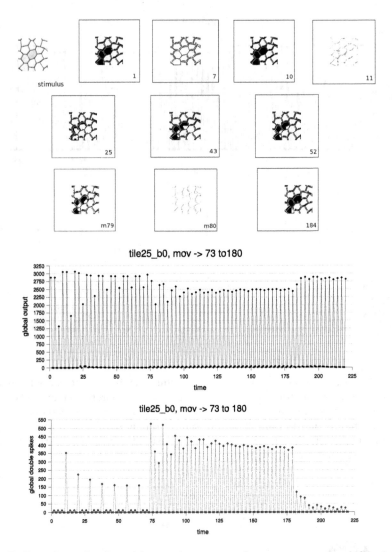

Fig. 6. Coding of an stimulus with complex texture. Longer progressive coding is observed when an stimulus with complex texture is presented (top histogram). In this example, some spontaneous bursting units are also observed when the stimulus is static (bottom histogram)

irregular shape moving initially to the right (iteration m194), and then to the left (m272). The output units clearly highlight the direction of motion in both cases. Moreover, the activity of such units was of the bursting type, as it was expected.

Figure 6 shows the network coding of an stimulus made up of hexagonal tiles, whose size is randomly generated. The grey scale represents the intensity of each point. The upper rows represent the output of the network when the stimulus is static (from the beginning until iteration 72) and when it starts moving at a speed of 1 pixel per iteration (from iteration 73 to 180). The middle plot represents the total number of single-spike firing units over time, while the bottom plot illustrates the total number of bursting discharge firing units over time. The coding is similar to that shown in figures 3 and 4. The main differences are: (1) some bursting activity appears when there is no movement at the input, and (2) the bursting histogram shows a higher activity both at the start and end of the moving episode. It is important to notice that the bursting activity during the static episode affects a reduced number of units when compared with the moving episode.

4   Discussion

Nowadays, it is assumed that the Dorsal Column Nuclei play a minor role in the processing of somesthetic information along the Somatosensory System. Our computational models show, on the contrary, that these nuclei can perform a complex coding of the sensory information that would be a key element for later perceptual processes. Based on the computational simulations presented in this paper, we believe that the role of the CN is to transform the representation of incoming cutaneous stimuli by generating a new spatio-temporal neural code, which is complex enough to represent different features of the stimuli, such as their size, texture, contrast and motion.

There exist numerous evidences supporting the idea of spatio-temporal codes in sensory systems. In the auditory system it has been reported that the inner hair cells of the cochlea generate a spatio-temporal pattern encoding the tones of the incoming sound signals [13]. These patterns are processed later by the central auditory neurons in order to categorize the characteristic frequency. In the visual system, a rank-based coding has been proposed based on the order of the firing of the first spike for each neuron [14].

As future work, we are interested in analyzing the characteristics of the progressive coding in a formal way by using concepts from information theory. The calculation of the entropy and mutual information associated to each iteration will provide an indication about how compact and informative is this code over time. Another research line regards with the cathegorization possibilities of the generated code, or, in other words, how this code facilitates the further classification of sizes, textures and other somesthetic features.

Acknowledgments

We would like to thank to **Laboratorios de Neurociencia y Computación neuronal (LANCON)**, the environment in which this work has been developed, and the support from the Ministry of Science and Technology through grant BFI 2003-01940.

References

1. Aguilar J, Soto C, Rivadulla C, Canedo A: The lemniscal-cuneate recurrent excitation is suppressed by strychnine and enhanced by $GABA_A$ antagonists in the anaesthetized cat. Eur J Neurosci. Vol. 16 (2002) 1697-1704.
2. Aguilar J, Rivadulla C, Soto C, Canedo A: New corticocuneate cellular mechanisms underlying the modulation of cutaneous ascending transmission in anesthetized cats. J Neurophysiol. Vol. 89 (2003) 3328-39.
3. Berkley KJ, Badell RJ, Blomqvist A, Bull M: Output Systems of the Dorsal Column Nuclei in the cat. Brain Res. Rev. Vol. 11 (1986) 199-225.
4. Canedo A, Aguilar J, Mariño J: Lemniscal recurrent and transcortical influences on cuneate neurons. Neuroscience. Vol. 97 **2** (2000) 317-334
5. Kuypers HGJM, Tuerk JD: The distribution of the cortical fibers within the nuclei cuneatus and gracilis in the cat. J. Anat. Vol. 98 (1964) 143-162
6. Lue JH, Luai SM, Wang TJ, Shieh JY, Wen CY: Synaptic relationships between corticocuneate terminals and glycine-immunoreactive neurons in the rat cuneate nucleus. Brain Res. Vol. 771 (1997) 167-171
7. Navarro J, Canedo A, Sánchez E: Progressive Coding in the dorsal column nuclei. Proc. FENS. Abstr. Vol. 2 (2004a) A.190.12
8. Navarro J, Sánchez E, Canedo A: Spatio-temporal information coding in the cuneate nucleus. Proc. Brain Inspired Computational Systems (BICS). Vol. 2 (2004b) 1-7.
9. Nuñez A, Panetsos F, Avendaño C: Rhythmic neuronal interactions and synchronization in the rat dorsal column nuclei. Neuroscience Vol. 100 **3** (2000) 599-609
10. Sánchez E, Barro S, Canedo A: Edge Detection and Motion Discrimination in the Cuneate Nucleus. Lect. Notes in Comp. Sci. Vol. 2415 (2002) 198-203
11. Sánchez E, Barro S, Mariño J, Canedo A: A computational model of cuneothalamic projection neurones. Network: Comput. neural syst. Vol. 14. (2003) 211-231
12. Sánchez E, Aguilar J, Rivadulla C, Canedo A: The Role of Glycinergic Interneurons in the Dorsal Column Nuclei. Neurocomputing. Vol. 58-60 (2004) 1049-1055.
13. Shamma S: Spatial and temporal processing in central auditory networks. In Methods in Neuronal Modeling. Eds: C. Koch e I. Segev. (1998) 411-460.
14. Thorpe SJ, Gautrais J: Rank order coding: A new coding scheme for rapid processing in neural networks. In Computational Neuroscience: Trends in Research. Ed. J. Bower. (1998) 113-119.

Auditory Nerve Encoding of High-Frequency Spectral Information

Ana Alves-Pinto[1], Enrique A. Lopez-Poveda[1], and Alan R. Palmer[2]

[1] Instituto de Neurociencias de Castilla y León, Universidad de Salamanca,
Avenida Alfonso X El Sabio s/n, 37007 Salamanca, Spain
{aalvespinto, ealopezpoveda}@usal.es
http://web.usal.es/~ealopezpoveda/
[2] MRC Institute of Hearing Research, University Park,
Nottingham NG7 2RD, United Kingdom
alan@ihr.mrc.ac.uk

Abstract. We have recently shown [1] that our ability to discriminate between a flat-spectrum noise and a similar noise with a high-frequency spectral notch deteriorates for levels around 70-80 dB SPL. The present paper explores an underlying physiological mechanism for this result. The hypothesis is that discriminability relies on the sensitivity of the auditory nerve to changes in the stimulus spectrum. A good correlation was found between behavioural results and sensitivity functions for two auditory nerve fibers with different dynamic ranges and with characteristic frequencies within the notch band. Although preliminary, these results suggest that the sensitivity of the auditory nerve to spectral notches is a non-monotonic function of stimulus level.

1 Introduction

Research on the functional properties of the peripheral auditory system is motivated, at least in part, by the need to develop better speech processors or robots with hearing capabilities. The underlying assumption is that any new knowledge on the functioning of the peripheral auditory system may contribute to developing better sound processors for artificial systems.

A physiological stage that is crucial for auditory perception is the auditory nerve (AN), since it is the only transmission path of auditory information from the peripheral to the central auditory system. Indeed, our ability to detect and employ the auditory information contained in the spectrum of an acoustic stimulus requires that the spectrum be adequately represented in the response of the AN.

In the AN, spectral features may be represented either in the temporal pattern of the fibers' activity or in terms of rate profiles, that is, as the distribution of activity over a population of AN fibers with different characteristic frequencies (CFs) [15]. Spectral features with frequencies above the cutoff of phase locking (>4000 Hz; [10]) must be represented in terms of AN discharge rate alone [13].

J. Mira and J.R. Álvarez (Eds.): IWINAC 2005, LNCS 3561, pp. 223–232, 2005.

The majority of AN fibers are of the high spontaneous rate (HSR) type. These have low thresholds (<10 dB SPL) and dynamic ranges of approximately 30-40 dB ([6], [14]). Fibers with low spontaneous rates (LSR) have higher thresholds (>15 dB SPL) and wider dynamic ranges (~ 50-60 dB; [6], [14]). The existence of these two fiber types with distinct thresholds and dynamic ranges have led several investigators ([5],[11],[13],[15]) to suggest that high-frequency spectral features could be encoded in the rate profile of HSR fibers at low levels and in that of LSR fibers at high levels. This would provide the auditory system with the ability to encode spectral features over a wide range of levels.

The present work investigates whether this applies to the encoding of high-frequency spectral notches. The study focuses on spectral notches because they are important cues for sound localization ([3], [4], [9]). Summarised is a psychophysical experiment [1] that shows that the threshold notch depth for discriminating between a flat-spectrum noise and a similar noise with a spectral notch at 8 kHz is a *non*-monotonic function of the stimulus level. Specifically, spectral discrimination deteriorates for levels around 70-80 dB SPL.

To elucidate the physiological mechanisms underlying this result, estimates of threshold notch depths based on guinea pig auditory nerve sensitivity to changes in the noise spectra are compared with the behavioral data. A strong correlation will be shown to exist between behavioral and physiological notch-depth thresholds. It is suggested that LSR fibers are responsible for the improvement in discrimination at very high levels, possibly as a result of suppression. The implications of these results for the design of bio-inspired speech processors are discussed.

2   Psychophysical Experiment

The ability to detect high-frequency spectral notches was studied by measuring the threshold notch depth necessary for listeners to discriminate between a broadband (20-16000 Hz) noise with a notch centered at 8000 Hz (target stimulus) and a similar noise with a flat spectrum (standard stimulus). The notch had a rectangular shape and a bandwidth of 4000 Hz. Threshold notch depths were measured for overall levels of the noise ranging from 30 to 100 dB SPL.

The presence of the notch in the target stimulus makes its overall level lower than that of the standard stimulus. To minimize for the possibility that listeners use this level difference as a cue for discrimination, the standard and the target noise bursts were presented with equal overall level. This was done by reducing the spectrum level of the flat spectrum noise as appropriate (e.g., a reduction of 0.58 dB would be required for a 2000-Hz wide, 27-dB deep notch). The details of the experimental procedure are provided elsewhere [1].

The main result is shown in Fig. 1(b). Despite the large variability of the data across listeners, the most significant feature is the non-monotonic variation of threshold notch-depth with stimulus level. As the stimulus level increases up to 70-80 dB SPL threshold notch depth also increases, meaning that spectral

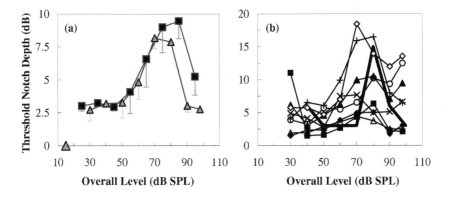

Fig. 1. Threshold notch depths for discriminating between a flat-spectrum noise and a notch noise as a function of stimulus overall level. **(a)** Results for a single listener for two conditions (notch bandwidth = 2000 Hz): with equal levels for the standard and target stimuli (triangles); with levels being roved (squares). The triangle on the abscissa indicates the absolute threshold for the flat spectrum noise. **(b)** Threshold notch depths for eight listeners (notch bandwidth = 4000 Hz). Each symbol illustrates the results for a different listener. The continuous bold curve represents threshold notch depths computed from d' values based on the response of AN fibers to the flat-spectrum noise and the notch noise stimuli (Sec. 3.3)

discrimination becomes harder. However, for levels between 80-100 dB SPL, notch depth thresholds decrease with increasing level; that is, discrimination becomes easier.

A similar pattern was observed when the overall level of the noise was roved across trials (Fig. 1(a)). This suggests that discrimination must be based on comparing the AN rate profiles rather than by monitoring the discharge rate of specific AN fibers with CFs within the notch band. Furthermore, the non-monotonic variation of the threshold notch depth *vs.* level function suggests that the AN conveys clearer information for spectral discrimination at levels lower and higher than 70-80 dB SPL. The next section investigates the validity of this interpretation.

3   Auditory Nerve Response to High-Frequency Spectral Notches

The physiological mechanisms underlying the non-monotonic character of the threshold notch depth *vs.* level function (Fig. 1) were investigated by measuring the response of guinea pig AN fibers to the same stimuli used in the psychophysical experiment. Though the obvious separation between the two species (i.e., human and guinea pig) imposes a special care when establishing any relations between psychophysical and physiological results, the following data provide information that is fundamental to establish a consistent and reliable theory of how high-frequency spectral notches are encoded in the human AN.

3.1   Experimental Procedure

Stimuli consisted of wideband noise bursts (20-16000 Hz). Their spectra were
either flat or contained a rectangular notch centered at a high frequency (7000
or 9000 Hz), with a bandwidth of 4000 Hz and with varying depth. The duration
of the noise bursts was 220 ms. The notch center frequency was 7000 and 9000
Hz during the first and second halves of the stimulus duration respectively. The
change between the two notch center frequencies was abrupt. The notch depths
tested ranged from 3 to 9 dB in 3-dB steps and from 9 to 27 dB in 6-dB steps.
Noise levels ranged from 40 to 100 dB SPL in steps of 10 dB. Stimuli were
presented every 880 ms. Notch depth and noise level varied randomly between
presentations. Unlike in the behavioral experiments, the level difference between
the flat-spectrum and the notch noise bursts was not made equal. The surgical
procedure and the recording methods are similar to those employed by [6].

AN responses were collected for at least four stimulus presentations. The
results presented below are average discharge rates over the last 110 ms of the
response, hence correspond to the portion of the stimulus with the spectral notch
centered at 9000 Hz. The onset response is, therefore, excluded from the analysis.

3.2   Rate-Level Functions for HSR and LSR Fibers

Figure 2 illustrates rate *vs.* levels functions (RLFs) for two AN fibers, one HSR
and one LSR. The CFs of both fibers are well within the notch band (7000-
11000 Hz).

Let us first consider the RLFs for the flat-spectrum noise (filled diamonds
in Fig. 2). Their shapes are clearly different for the two fibers. For the HSR
fiber (Fig. 2(a)), it is representative of AN fibers of the saturating type [16].
The maximum discharge rate is reached at a level of approximately 20 dB above
threshold. The RLF for the LSR fiber (Fig. 2(b)), however, clearly shows a wider

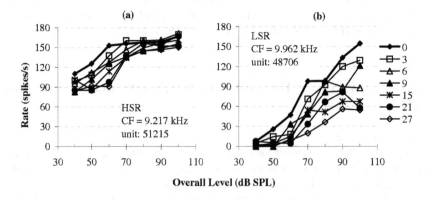

Fig. 2. Rate-level functions for the notch noise stimuli for two AN fibers with CFs well
within the notch band. **(a)** HSR fiber; **(b)** LSR fiber. Symbols denote the notch depths
in dB *re.* reference spectrum level

dynamic range and does not saturate over the range of levels tested. It is typical of the non-saturating type of AN fibers [16].

Since both fibers have CFs near the notch center frequency, they both effectively receive less energy from the notch noise stimulus than from the flat-spectrum noise stimulus. For any given notch depth, the energy difference between the two stimuli is the same regardless of the noise level. Therefore, one would expect this energy difference to shift the level at which their rate threshold occurs, but *not* the slope of the RLFs. In other words, one would expect that the effect of the notch be to shift the RLF in parallel to higher levels. This is clearly *not* the case. For the LSR fiber, the deeper the notch the shallower the associated RLF. As a result, the vertical distance between the RLF for any given notch depth and that for the flat-spectrum noise increases gradually as the overall level of the noise increases. As for the HSR fiber, its discharge rate also increases more slowly with increasing the noise level, mainly for deeper notches. Furthermore, saturation seems to occur at a lower rate and for a level higher than when stimulated with the flat-spectrum noise.

As the level of the wideband noise increases, fibers not only receive a stronger stimulation as a result of their frequency response areas becoming wider with increasing level, but they may also become gradually more affected by suppression effects [2]. Hence, it is possible that the differences between the RLFs for the flat-spectrum and the notch noise stimuli reflect the combined effect of suppression and energy reduction associated to the notch.

The rate reduction produced by the notch introduces a cue for discriminating between the flat-spectrum noise and the notch noise stimuli. The largest differences between the rates for the two stimuli occur between 40 and 60 dB SPL for the HSR fiber, and between 90-100 dB SPL for the LSR fiber (Fig. 2). This may be why spectral discrimination is easier over these ranges of level (Fig. 1). This idea is further explored in the following section.

3.3   Neural Sensitivity to the Spectral Notch

Although the spectral notch produces a rate reduction in fibers with CFs within the notch band, it is difficult to evaluate whether the decrease is enough for discriminating between the flat-spectrum and the notch noise stimuli. A common numerical measure of the discriminability between two stimuli is the d' parameter (see, for example, [7] and [13]). d' values are computed as follows:

$$d' = \frac{R_0 - R}{\sqrt{0.5 \times (\sigma_0^2 + \sigma^2)}}$$

where (R_0, σ_0) and (R, σ) are the mean discharge rates and the associated standard deviations in response to the flat-spectrum and the notch noise stimuli, respectively, over a number of stimulus presentations. d' takes into account the variability in the response inherent to each fiber and, therefore, provides a more accurate measure of the fibers sensitivity to stimulus variations. Discrimination, whether behavioral or physiological, is considered to occur when $|d'| \geq 1$. There-

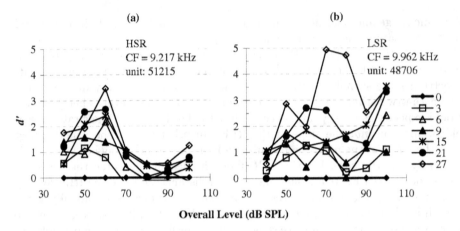

Fig. 3. d' values as a function of noise overall level, calculated from the discharge rates presented in Fig. 2, for the same **(a)** HSR and **(b)** LSR fibers. Symbols denote the notch depths in dB *re.* reference spectrum level

fore, the analysis of d' values makes it possible to compare psychophysical and physiological discrimination thresholds.

Figure 3 illustrates d' values as a function of noise overall level for the two fibers whose RLFs are illustrated in Fig.2. For both HSR and LSR fibers, d' values change non-monotonically with noise overall level. The non-monotonic character of the function gets more obvious the deeper the notch.

The non-monotonic shape of d' *vs.* level functions suggests that the ability of the fiber to discriminate spectral differences should correspondingly vary non-monotonically with stimulus level. Discrimination based on the information provided by HSR fibers would be optimal at levels between 50-70 dB SPL, where $|d'| \geq 1$, and would be almost impossible from 70-90 dB SPL. LSR fibers show a less consistent trend. Their associated d' values are generally higher than corresponding values for HSR fibers and exceed unity for most notch depths at almost all levels except 40 dB SPL. Discrimination based on the information provided by LSR fibers seems to be optimal at around 100 dB SPL.

Imagine, now, a system comprised of these two AN fibers only. Its ability to discriminate between the flat-spectrum and the notch noise stimuli would be the result of combining the discrimination capabilities of each of the two fibers. This can be quantified by calculating a global d' as follows [8]:

$$d' = \sqrt{d'_{HSR}{}^2 + d'_{LSR}{}^2}$$

Figure 4 illustrates global d' values as a function of level for different notch depths. Also plotted are the minimum notch depths for which this global $|d'|$ exceeds unity at each of the overall levels considered.

The global d' also varies non-monotonically with stimulus level, as would be expected, since it is calculated from two functions that already are non-monotonic (Fig. 3). A comparison of the global d' values (Fig. 4) with the

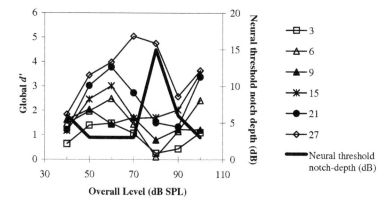

Fig. 4. Global d' values as a function of noise overall level for different notch depths (inset). The bold continuous line (right ordinate axis) illustrates the minimum notch depth, at each noise level, for which the global $|d'|$ exceeds 1. Symbols denote the notch depths in dB *re.* reference spectrum level

individual d' values of each fiber type (Fig. 3) reveals that the global d' is essentially determined by d'_{HSR} and d'_{LSR} at low and high levels, respectively. This suggests, therefore, that discrimination based on the combined sensitivity of these two fibers, essentially relies on HSR fibers at low stimulus levels and on LSR fibers at high levels.

Approximate threshold notch depths for each stimulation level can be computed from the global d' values of Fig. 4 by calculating the minimum notch depth that produces a d' value greater than one. The results are shown in Fig. 1(b), where they can be directly compared with the behavioral notch depth thresholds. Notice the excellent agreement between the physiological and behavioral results. Not only the peak in the two functions occur around 80 dB SPL, but also the threshold notch depth estimates and the behavioural data are similar in magnitude over the range of noise levels tested.

4   Discussion

Psychophysical discrimination between a flat-spectrum noise and a similar noise with a notch centered at 8000 Hz was shown to vary non-monotonically with stimulus level. The activity of guinea pig AN fibers in response to the stimuli employed in the psychophysical experiment was measured with the aim to provide a physiological explanation for the behavioral data. Physiological data have been shown for two representative fibers only rather for a large population of fibers. Hence, the present study must be considered as preliminary and the conclusions should be taken cautiously. However, the data illustrate real properties of HSR and LSR AN fibers; properties that are likely to apply to human AN fibers.

A close agreement was found between the physiological and the psychophysical results (Fig. 1(b)). This supports the hypothesis that the main limitations for behavioral spectral discrimination come from the sensitivity of the AN to changes in the stimulus spectrum.

For both HSR and LSR fibers, differences were observed in the RLFs in response to the flat-spectrum noise and the notch noise. These differences are likely to reflect a combined effect of suppression and a reduction of energy due to the spectral notch. However, understanding the way these two factors, especially suppression, affects the activity of these fibers in response to the notch noise stimulus needs further investigation. The decrease in the slope of the RLFs observed for both HSR and LSR fibers increases the dynamic range of the two fiber types in response to the notch noise stimuli. This is likely to be responsible for the non-monotonic variation of the global d' values with stimulus level, and by extension, for the non-monotonic variation of the physiologically-estimated threshold notch depths with stimulus level.

Suppression may affect not only the rate of fibers with CFs within the notch-band, but also the activity of fibers with CFs at the notch boundaries. Indeed, activity of the latter may be less suppressed when the notch is present than when it is absent, hence their rate may be higher than that of other fibers with CFs well away (in frequency) from the notch. As has been previously suggested (e.g. [12]), this serves to enhance the notch boundaries. The influence of these fibers on threshold notch depths has not been considered in the present study.

Suppression plays an important role for encoding not only high-frequency spectral features but also low-frequency features. Specifically, suppression linked to LSR fibers is essential for encoding the spectrum of vowels (of both their formants and troughs). As noted by [11], "...under conditions where HSR rate representation is degraded by saturation effects, LSR fibers are able to sustain a differential response to formants and troughs because their trough-driven discharge rates are suppressed by off-BF energy in the surrounding formants."

The fact that the interpretation given to psychophysical results agrees with the results derived from physiological responses of only two AN fibers does not contradict the conclusion of Alves-Pinto and Lopez-Poveda ([1]) that high-frequency spectral notches are probably represented in the AN rate profile, that is in the activity of a larger group of fibers. A global d' based on a population of fibers would probably yield physiological threshold notch depths that are closer to the behavioral data. More data are currently being collected to test this hypothesis.

Finally, it should be noted that the d' values for the LSR fiber are generally larger than those for the HSR fiber. This suggests that LSR fibers are more sensitive than HSR fibers to changes in the spectrum, which may compensate for the scarcity of LSR fibers. Having said that, it remains unclear (to us at least) whether discriminability benefits more from having few fibers producing large d' values than from having many fibers producing smaller d' values or vice versa.

4.1   Implications for the Design of Bio-inspired Speech Processors

It has been shown that HSR fibers by themselves can not explain the improvement in spectral discrimination performance observed at levels higher than 70-80 dB SPL. Furthermore, it has been discussed that such improvement probably occurs as the result of suppression acting upon LSR fibers, which have higher thresholds and wider dynamic ranges. This imposes minimum requirements for the speech processors of artificial intelligence systems that aim to mimic human performance at spectral discrimination tasks, namely they must reproduce the effects of both HSR and LSR fibers as well as AN suppression effects.

There already exist computational models that simulate these properties of the peripheral auditory system (e.g., [17-19]). They, therefore, constitute a good starting point for building bio-inspired speech processors for artificial intelligence systems with human-like hearing capabilities.

5   Conclusions

The main conclusion is that the combined sensitivity of HSR and LSR AN fibers to the presence of a high-frequency spectral notch is a non-monotonic function of stimulus level. The good correlation between the psychophysical data and the present analysis of auditory nerve responses to flat-spectrum and notch noise stimuli suggests that spectral discrimination depends directly on information conveyed by the *difference* rate profile rather than on the quality of the individual rate profiles.

Acknowledgments

Work supported by the Spanish *Fondo de Investigaciones Sanitarias* (grant refs. PI02/0343 and G03/203) and by European Regional Development Funds.

References

1. Alves-Pinto A. and Lopez-Poveda, E. A.: Detection of high-frequency spectral notches as a function of level. J. Acoust. Soc. Am. (submitted)
2. Arthur R.M., Pfeiffer R.R., and Suga N.: Properties of 'two-tone inhibition' in primary auditory neurones. J. Physiol. **212** (1971) 593–609.
3. Butler, R. A., and Belendiuk, K.: Spectral cues utilized in the localization of sound in the median sagittal plane. J. Acoust. Soc. Am. **61** (1977) 1264–1269.
4. Butler, R. A., and Humanski, R. A.: Localization of sound in the vertical plane with and without high-frequency spectral cues. Percept. Psychophys. **51** (1992) 182–186.
5. Delgutte, B., and Kiang, N. Y. S.: Speech coding in the auditory nerve: III. Voiceless fricative consonants. J. Acoust. Soc. Am. **75** (1984) 887–896.
6. Evans, E. F., and Palmer, A. R.: Relationship between the dynamic range of cochlear nerve fibers and their spontaneous activity. Exp. Brain Res. **40** (1980) 115–118.
7. Green, D. M. and Swets, J.A. Signal Detection Theory and Psychophysics. New York; Wiley, 1966.

8. Hancock, K. E. and Delgutte, B.: A physiologically based model of Interaural Time difference discrimination. The Journal of Neuroscience **24** (2004) 7110–7117.

9. Hebrank, J., and Wright, D.: Spectral cues used in the localization of sound sources on the median plane. J. Acoust. Soc. Am. **56** (1974) 1829–1834.

10. Johnson, D. H.: The relationship between spike rate and synchrony in responses of auditory-nerve fibers to single tones. J. Acoust. Soc. Am. **68** (1980) 1115–1122.

11. May, B. J.: Physiological and psychophysical assessments of the dynamic range of vowel representations in the auditory periphery. Speech Communication **41** (2003) 49–57.

12. Poon, P. W. F., and Brugge, J. F.: Sensitivity of auditory nerve fibers to spectral notches. J.Neurophysiol. **70** (1993) 655–666.

13. Rice, J. J., Young, E. D., and Spirou, G. A.: Auditory-nerve encoding of pinna-based spectral cues: Rate representation of high-frequency stimuli. J.Acoust.Soc.Am. **97** (1995) 1764–1776.

14. Sachs, M. B., and Abbas, P. J.: Rate versus level functions for auditory-nerve fibers in cats: tone-burst stimuli. J. Acoust. Soc. Am. **56** (1974) 1835–1847.

15. Sachs, M. B., and Young, E. D.: Encoding of steady-state vowels in the auditory nerve: Representation in terms of discharge rate. J. Acoust. Soc. Am. **66** (1979) 470–479.

16. Winter, I. M., Robertson, D., and Yates, G. K.: Diversity of characteristic frequency rate-intensity functions in guinea pig auditory nerve fibres. Hear. Res. **45** (1990) 191–202.

17. Sumner, C., Lopez-Poveda, E. A., O'Mard, L. O. P., and Meddis, R.: A revised model of the inner hair cell and auditory nerve complex. J. Acoust. Soc. Am. **111** (2002) 2178–2188.

18. Zhang, X., Heinz, M. G., Bruce, I. C., and Carney, L. H.: A phenomenological model for the responses of auditory-nerve fibers: I. Nonlinear tuning with compression and suppression. J. Acoust. Soc. Am. **109** (2001) 648–670.

19. Robert, A., and Eriksson, J. L.: A composite model of the auditory periphery for simulating responses to complex sounds. J. Acoust. Soc. Am. **106** (1999) 1852–1864.

Multielectrode Analysis of Information Flow Through Cat Primary Visual Cortex

Luis M. Martinez[1] and Jose-Manuel Alonso[2]

[1] Departamento de Medicina, Universidade da Coruña,
Campus de Oza, 15006 A Corua, Spain
lmo@neuralcorrelate.com
[2] Department of Biological Sciences,
State University of New York, College of Optometry,
33 West 42nd Street, New York, NY 10036, USA
http://www.neuralcorrelate.com

Abstract. Cells in cortical layer 4 contact layer 2+3 neurons that, in turn, send their main intracolumnar output to the infragranular layers. This simple outline of feedforward connections summarizes the main route followed by visual information through cat primary visual cortex. We studied the patterns of connectivity among physiologically identified neurons recorded at different depths within the same, or nearby, cortical columns using multielectrode recordings and cross-correlation analysis. Our preliminarily results suggest that the projection from layer 4 to layers 2+3 is actually split in parallel pathways of information flow that are defined by differences in the dynamic properties of the connections and that may be used to optimally decode the temporal information embedded on every spike train.

1 Introduction

To a first approximation, information flow through the cortex follows a simple, feedforward route from layer 4 to the infraganular layers via a relay in the superficial layers (figure 1a). Interestingly, new functional response properties emerge at each successive cortical stage, suggesting that different cortical layers, with their specific circuitry, are specialized for different tasks (see [5, 6] for review). For example, in rodent somatosensory cortex, receptive field structure, orientation selectivity and plasticity change according to laminar location [2, 4, 10]. Likewise, in the auditory cortex, response properties such as inhibitory side-band structure and bandwidth also seem to vary with position in the cortical microcircuit [11]. Finally, receptive field structure, orientation tuning and dynamics and the relative orientation preference of excitatory and inhibitory inputs vary substantially according to the laminar location of the recorded cells in primary visual cortex [7, 14, 15, 17, 19]. In spite of the extensive experimental and theoretical effort, we still discuss the precise synaptic circuitry and mechanisms that build the different receptive field structures and functional response properties found

J. Mira and J.R. Álvarez (Eds.): IWINAC 2005, LNCS 3561, pp. 233–241, 2005.

at successive stages of cortical integration. Here, we use multielectrode recording and cross-correlation analysis to investigate the connectivity and synaptic
dynamics of physiologically identified neurons in different layers of the cat primary visual cortex (V1). Our goal is to explore the different strategies of neural
processing used by V1 to analyze specific features of the visual scene. Our preliminary results show evidence of at least two parallel circuits that run from cells
in layer 4 with simple receptive fields to complex cells in the superficial layers
that show distinct functional response properties. These circuits are defined by
differences in the dynamic properties of the monosynaptic connections involved.
Thus, in each case, the synaptic efficacy of the layer 4 simple cell in driving the
complex cell, in layers 2+3, is modulated by the timing history of the presynaptic spikes. Specifically, for some pairs, the synaptic efficacy is highest when
the presynaptic spikes are separated by short interspike intervals and, in other
instances, the synaptic efficacy grows with increasingly large interspike intervals. These differences in synaptic physiology may reflect a circuit split in the
connection from layer 4 to layers 2+3 that could be used to separately process
different temporal aspects of the input (e.g. transient and sustained phases of
visual responses).

2   Methods

2.1   General

Cats (2.5-3 kg) were anesthetized with ketamine and thiopental sodium and
paralyzed with Norcuron. Temperature, EKG, EEG and expired CO_2 were continuously monitored throughout the experiment. A more general description of
the experimental preparation has been previously reported [1, 13]. All procedures conformed to the University of A Coruña (Spain) ethical board and are in
accordance with The Declaration of Helsinki, the guidelines of the National Insti-
tutes of Health (USA) and the animal care and use committee of the Rockefeller
University and the State University of New York (New York, USA).

2.2   Cortical Recordings and Histology

Cortical cells were recorded with a multielectrode matrix; seven thin electrodes
were introduced into the medial bank of the cortex, parallel to each other and
parallel to the cortical surface. The location of the recording sites was assessed
by combining histology and electrode depth readings. The tips of the electrodes
were coated with fluorescent dyes (DiI and DiO) and the tracks reconstructed at
the end of the experiments (contiguous sections were processed for cresyl violet
to precisely establish laminar borders) (figure 1b). Each electrode in the array
was moved independently with a vertical accuracy of 1 micron, allowing isolation
of single units.

Visual Stimuli. Cortical receptive fields were mapped with white-noise or
sparse-noise protocols or with moving bars at different orientations. White-noise

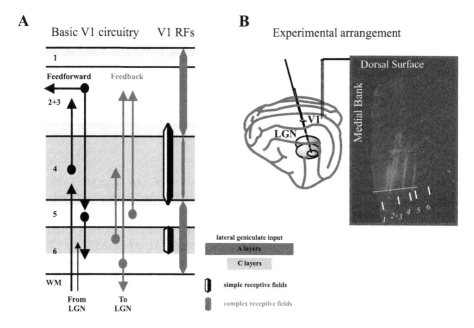

Fig. 1. (A) Schematic diagram illustrating the main feedforward and feedback connections through cat primary visual cortex (left) and the laminar distribution of the main cortical receptive field types. (B) Right, situation of the recording electrodes. Left, coronal section (thickness, 40 microns) of cat visual cortex showing five electrode tracks in the medial bank. Because each of these electrodes could be moved independently, we could record from cells located at the position represented by the dotted line at different cortical layers (track thickness, 100 microns; electrode thickness, 80 microns; vertical misalignment for this case, < 100 microns). Adapted from [1]

consisted of a series of 16 x 16-pixel pseudo-random 'checkerboards' (0.4 degrees/pixel) presented each during 20 ms. Sparse noise consisted of light and dark squares flashed singly for 40 ms in pseudo-random order, 16 times each, on a 16 x 16 grid (square size, 0.8-1.2 degrees). All correlograms were obtained by stimulating the cortical cells with a slowly moving bar (usually 0.25-0.5 Hz).

Data Analysis. Connectivity was estimated by cross-correlating the spike trains of simultaneously recorded cells. The presence of a fast, narrow, asymmetric peak superimposed on a broader, stimulus dependent correlation, was indicative of a monosynaptic connection. The synaptic efficacy was calculated as: (peak - baseline) / Total number of spikes produced by the postsynaptic cell (see [1] for further details). To check for potential changes in synaptic efficacy depending on the cells firing patterns, "log-ISIHs" (logarithmic interspike interval histograms) were constructed from the spike trains of the presynaptic cells in layer 4. All cells showed log-ISIH with two or three peaks. The temporal courses of each ISI-peak were independent of the different temporal profiles of the stimulus used to drive cortical cells (white-noise, sparse-noise or moving bars). For each pair, we computed the cross-correlogram from the full response and from response subsets. Spikes in each subset were all preceded by ISIs from a single ISI group (denoted short [S], medium [M] or long [L] and selected by visual inspection). Since each subset contained a different number of spikes, their synaptic strengths were normalized by comparing them to their expected values if all spikes contributed the same to the raw cross-correlogram (noted real and expected, respectively, in the figures).

3   Results

Cross-correlation analysis can be used to characterize the temporal relations between pairs of neurons and, ultimately, to explore the synaptic structure of neural circuits. Here, connectivity was estimated by crosscorrelating the spike trains of simultaneously recorded cells at different depths (in layer 4 and layers 2+3) within the same or nearby orientation columns. We recorded from 92 pairs of cortical neurons with overlapping receptive fields (overlap > 50%) and similar orientation preferences. In 17 cases, we were able to collect a large number of spikes to measure correlations using different interspike intervals (see Methods). Because our cells had overlapping receptive fields, all our correlograms showed slow stimulus-dependent correlations. These correlations were very broad and reproduced the temporal frequency of the slowly moving bars that we used as stimulus (usually 0.25-0.5 Hz). A much faster, narrow and asymmetric peak was superimposed on the broad correlations. In fact, these asymmetric peaks were much narrower than any of our stimulus-dependent correlations or any correlation produced by changes in excitability [3]. The slow stimulus-dependent correlation was calculated by fragmenting the spike train in stimulus cycles and shuffling these fragments [18]. The significance level for a positive correlation was set to 2.5 times the standard deviation of the shuffle correlogram that served as a baseline (for details, see [1]). The correlogram asymmetry was calculated, after subtracting the baseline, as R-L/R+L, where R is the number of paired spikes at the right side of the correlogram and L is those at the left side, within 5ms on each side. The baseline was defined as the average value in the correlogram at 10 ms on each side. We identified monosynaptic connections by using criteria similar to our previous studies [1, 13]: (1) Presence of a positive narrow peak displaced from the zero of in the correlogram (correlogram asymmetry > 0.4). The strongest peaks had a dip on the left side of the correlogram that matched the autocorrelogram of the presynaptic cell and represents the refractory period of the connection (see figure 2c). (2) The narrow peak had a short latency and a fast rise time that matched the peak time and rise time of synaptic potentials recorded intracellularly in vitro (latency + rise time = 2.36-4.96 ms [see [16, 9, 20, 12]]. Cross-correlation analysis assumes that the two spike trains are stationary. Therefore, it follows that the synaptic properties of the connection between the two recorded cells do not change over time. This assumption is hard to justify given that cortical synapses are subject to clear spike-time dependent plasticity rules [8]. Thus, to check for potential changes in synaptic efficacy depending on the cells firing patterns, we constructed "log-ISIHs" (logarithmic

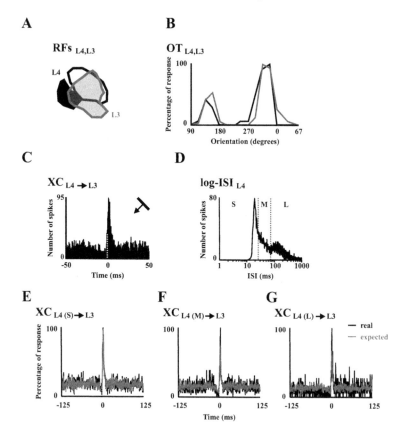

Fig. 2. A layer 4 simple cell simultaneously recorded with a layer 3 complex cell. (A) Superimposed receptive fields of the simple cell (on-subregions in white and off-subregions in black) and the complex cell (on-off receptive field in gray) mapped with a sparse-noise protocol. (B) Orientation tuning curves of the simple cell and the complex cell obtained with a moving bar. Black line, simple cell; gray line, complex cell. (C) Cross-correlation between the firing patterns of the simple cell and the complex cell. The ordinate gives the number of simple-cell spikes that occurred before (right) or after (left) a complex-cell spike. This correlogram shows a peak displaced to the right of zero, indicating that the simple cell tended to fire before the complex cell. The characteristics of the positive peak are consistent with a direct excitatory connection from the simple cell to the complex cell (see results). Both cells were simultaneously stimulated by a bar moved at their preferred orientation. Strength, 6.1%. (D) Log-ISI histogram showing the distribution of interspike intervals in the response of the simple cell to the moving bar. The full spike train can be split in three subsets of spikes corresponding to short, S, medium, M, and long, L, ISIs. (E), (F) and (G), show real (in black) vs. expected (in gray) cross-correlograms computed separately for each ISI subset (see methods). Modulation in synaptic strengths, −7.7% in (E), +35.3% in (F) and +44.9% in (G)

interspike interval histograms) from the spike trains of the presynaptic cells in layer 4 (see Methods). All our layer 4 simple cells showed log-ISIH with usually three peaks (figures 2d and 3d) in response to the visual stimuli used in this study (white-noise, sparse-noise or moving bars). For each pair, we computed synaptic efficacy for the full response and compared it to that obtained for the different ISI subsets (denoted short [S], medium [M] or long [L]). Since each subset contained a different number of spikes, their synaptic strengths were normalized by comparing them to their expected values if all spikes contributed the same to the raw cross-correlogram (noted real and expected, respectively, in the figures). Figures 2 and 3 show two characteristic examples of our results. Results obtained from a simple cell located in mid layer 4 and a complex cell recorded at the bottom half of layers 2+3 are shown in figure 2. Both cells had similar orientation preferences (figure 2b) and overlapped receptive fields (RF). Both receptive fields could be mapped with static stimuli, like the sparse-noise, and the simple cell had a RF comprising an on-subregion, in white, and an off-subregion, in black, and the complex cell produced on- and off-responses throughout the entire RF (gray) (figure 2a). Cross-correlation analysis indicated that the simple cell was monosynaptically connected to the complex cell (figure 2c) and that the connection facilitated when presynaptic spikes arrived at long interspike intervals (compare the correlograms shown in figure 2e and g). Other 7 pairs, of the 17 considered in this study, showed a similar time-dependent modulation of synaptic efficacy.

Figure 3 shows results obtained from a simple cell recorded in layer 4 and a complex cell located in the upper tier of layers 2+3. The two cells have similar orientation preferences and well overlapped RFs but in this case, only the simple cell could be mapped with static stimuli (figure 3a, black and white represent off and on-subregions, respectively) and the complex RF had to be studied with moving bars instead (gray rectangle in figure 3a). Again, cross-correlation analysis indicated that the simple cell was monosynaptically connected to the complex cell (figure 3c), but in this case the connection was more efficient when presynaptic spikes arrived at short interspike intervals (compare the correlograms shown in figure 3e and g). The remaining 10 pairs considered in this study showed a similar preference for short interspike intervals.

4   Discusion

Here we show how cross-correlation analysis can be used to study the dynamics of synaptic connections in cat primary visual cortex. In particular, our results suggest that the influence of spike history, represented by the different ISI subsets, on synaptic efficacy changes according to sub-laminar location and functional response properties. Essentially, when layer 4 simple cells fire at low frequencies (for instance below 20 or 10 Hz) their synapses onto complex cells at the top of layer 2+3 potentiate while their synapses onto complex cells at the bottom of the layer seem to depress; and the converse is true when they fire at higher frequencies. Interestingly, the more superficial complex cells are different from

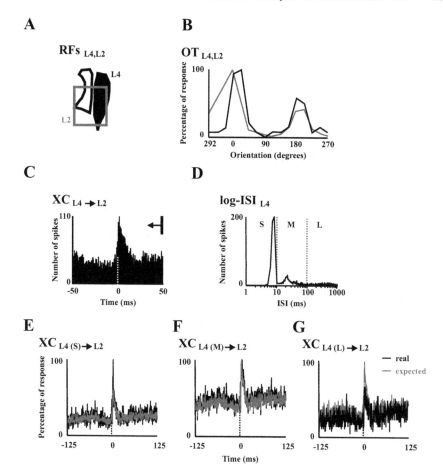

Fig. 3. A layer 4 simple cell simultaneously recorded with a layer 2 complex cell. (A) Superimposed receptive fields of the simple cell (on-subregions in white and off-subregions in black) mapped with a sparse-noise protocol and the complex cell (on-off responses represented by the gray rectangle) mapped with a moving bar. (B) Orientation tuning curves of the simple cell and the complex cell obtained with a moving bar. Black line, simple cell; gray line, complex cell. (C) Cross-correlation between the firing patterns of the simple cell and the complex cell. The characteristics of the positive peak are consistent with a direct excitatory connection from the simple cell to the complex cell (see results). Both cells were simultaneously stimulated by a bar moved at their preferred orientation. Strength, 5.8%. (D) Log-ISI histogram showing the distribution of interspike intervals in the response of the simple cell to the moving bar. The full spike train can be split in three subsets of spikes corresponding to short, S, medium, M, and long, L, ISIs. (E), (F) and (G), show real (in black) vs. expected (in gray) cross-correlograms computed separately for each ISI subset (see methods). Modulation in synaptic strengths, +70.6% in (E), −0.1% in (F) and −35.5% in (G)

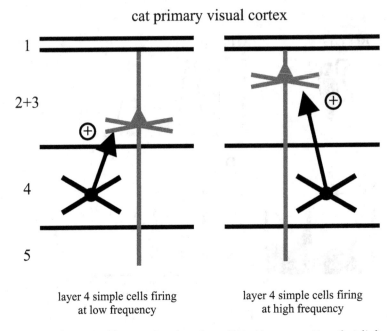

Fig. 4. Proposed circuit diagram showing the split in the connection that links simple cells in layer 4 to complex cells in the superficial layers

those located near the border between layers 2+3 and 4; while the former cells do not usually respond to static stimuli, most of the later cells do [15]. Thus, the flow of information through primary visual cortex appears to follow two parallel pathways originating in layer 4 that are selected for on the basis of the input pattern determined by the visual stimulus (see figure 4).

Acknowledgments

This study was supported by the NIH (JMA) and Human Frontier Science Program Organization (LMM).

References

1. Alonso JM and Martinez LM (1998). Functional connectivity between simple cells and complex cells in cat striate cortex. Nature Neurosci. 1:395-403.
2. Brecht M, Roth A and Sakmann B (2003). Dynamic receptive fields of reconstructed pyramidal cells in layers 3 and 2 of rat somatosensory barrel cortex. Journal of Physiology 553:243-265.
3. Brody CD (1998). Slow covariations in neuronal resting potentials can lead to artefactually fast cross-correlations in their spike trains. J Neurophysiol. 80:3345-3351.
4. Brumberg JC, Pinto DJ and Simons DJ (1999). Cortical columnar processing in the rat whisker-to-barrel system. Journal of Neurophysiology 82:1808-1817.

5. Callaway EM (1998). Local circuits in primary visual cortex of the macaque monkey. Annual Review of Neuroscience 21:47-74.
6. Callaway EM (2004). Feedforward, feedback and inhibitory connections in primate visual cortex. Neural Netw. 17:625-32.
7. Chisum HJ, Mooser F and Fitzpatrick, D (2003). Emergent properties of layer 2/3 neurons reflect the collinear arrangement of horizontal connections in tree shrew visual cortex. Journal of Neuroscience 23:2947-2960.
8. Dan Y, Poo MM (2004) Spike timing-dependent plasticity of neural circuits. Neuron. 44:23-30.
9. Deuchars J, West DC and Thomson AM (1994). Relationships between morphology and physiology of pyramid-pyramid single axon connections in rat neocortex in vitro. J. Physiol. (Lond.) 478:423-435.
10. Diamond ME, Huang W and Ebner FF (1994). Laminar comparison of somatosensory cortical plasticity. Science 265:1885-1888.
11. Linden JF and Schreiner CE (2003). Columnar transformations in auditory cortex? A comparison to visual and somatosensory cortices. Cerebral Cortex 13:83-89.
12. Markram H, Lubke J, Frotscher M, Roth A and Sakmann B (1997). Physiology and anatomy of synaptic connections between thick tufted pyramidal neurones in the developing rat neocortex. J. Physiol. (Lond.) 500:409-440.
13. Martinez LM and Alonso JM (2001). Construction of complex receptive fields in cat primary visual cortex. Neuron. 32:515-525.
14. Martinez LM, Alonso JM, Reid RC and Hirsch JA (2002). Laminar processing of stimulus orientation in cat visual cortex. J Physiol. 540:321-33.
15. Martinez LM, Wang Q, Reid RC, Pillai C, Alonso JM, Sommer FT and Hirsch JA (2005). Receptive field structure varies with layer in the primary visual cortex. Nature Neurosci. 8:372-379.
16. Mason A, Nicoll A and Stratford K (1991). Synaptic transmission between individual pyramidal neurons of the rat visual cortex in vitro. J. Neurosci. 11:72-84.
17. Mooser F, Bosking WH and Fitzpatrick D (2004). A morphological basis for orientation tuning in primary visual cortex. Nature Neurosci. 8:872-879.
18. Perkel DH, Gerstein GL and Moore GP (1967). Neuronal spike trains and stochastic point processes. I. The single spike train. Biophys. J. 7:391-418.
19. Ringach DL, Shapley R and Hawken MJ (2002). Orientation selectivity in macaque V1: diversity and laminar dependence. Journal of Neuroscience 22, 5639-5651.
20. Stratford KJ, Tarczy-Hornoch K, Martin KAC, Bannister NJ and Jack JJ (1996). Excitatory synaptic inputs to spiny stellate cells in cat visual cortex. Nature 382:258-261.

Towards Evolutionary DNA Computing*

Christiaan V. Henkel[1,2] and Joost N. Kok[1]

[1] Leiden Institute of Advanced Computer Science, Leiden University,
Niels Bohrweg 1, 2333 CA Leiden, The Netherlands
[2] Institute of Biology, Leiden University,
Wassenaarseweg 64, 2333 AL Leiden, The Netherlands
henkel@rulbim.leidenuniv.nl, joost@liacs.nl

Abstract. In the last decade, the implementation of computations using DNA has attracted a lot of attention from both computer scientists and biologists. Molecular computation has many advantages, including small size, a biological interface and massive parallelism. This potential offers interesting possibilities for the implementation of evolutionary computation using DNA. We discuss the potential of evolutionary DNA computing and experimental progress.

1   Introduction

DNA is a natural choice for the construction of molecular computers. It is amongst the most intensively studied molecules, and very well characterized in comparison to other complex macromolecules. More importantly, it clearly shows support for information technology through its role in genetics, and a large assortment of molecular biological methods is available for its modification and analysis (a more detailed discussion of these in the context of molecular computing can be found in [1]).

The basic computational operation for biological macromolecules can viewed as pattern recognition, with subsequent conformational change and often catalysis. Suitably complex molecules have many more states than just the binary 'on' and 'off', and the exploration of complementary shape is actually a highly parallel optimization procedure on an energy landscape. For nucleic acids, the basic pattern recognition operation is basepairing. The molecules are polymers of four different nucleotides, abbreviated A, T, C and G. Pair wise interactions of these bases, according to the rules A–T and C–G, can cause two complementary DNA molecules to anneal or hybridize and form a double helical assembly.

Molecular computers that operate through pattern recognition are necessarily three-dimensional in nature, implying the free movement and interaction of elements in all directions. This paradigm of free diffusion is central to molecular computing. The interactions of many molecules in solution in theory allow

---

* Invited paper.

J. Mira and J.R. Álvarez (Eds.): IWINAC 2005, LNCS 3561, pp. 242–257, 2005.

for massive parallelism: for example, everyday molecular biological procedures use picomolar quantities of molecules (10^{12}–10^{14} molecules), which could all act as simple computer processors. In addition, because of the thermal noise these molecules experience, the search for correct interactions is thermodynamically free. Only the final act of (irreversible) information transformation requires the dissipation of energy, in biochemistry often coupled to ATP hydrolysis. This way, biological computers could be remarkably efficient with energy, performing up to 10^{19} operations per Joule [2,3]. These features have prompted many suggestions for applications of DNA computing, for example the processing of biomedical data, in parallel computing, in molecular databases, in diagnostic systems and in nanotechnology.

This paper is organized as follows. Section 2 provides a brief overview of experimental research in the area of solving (NP-hard) optimization problems with DNA computing. Although progress has been made, it has become clear that DNA computers will have their difficulties. From a theoretical point of view, it is obvious that even the massive parallelism offered by molecules is not enough. But there are also many practical challenges, mainly on the difficulty of dealing with the intrinsic noisiness of (bulk) biochemical operations. These will be discussed in Section 3. In the case of optimization problems, evolutionary computation has demonstrated that noisy and non-deterministic operators like mutation and recombination hold considerable computational potential. A combination of DNA computation, with its potential for huge population sizes and intrinsic fuzziness, and evolutionary optimization may be very promising, as will be discussed in Section 4. In Section 5, the potential for experimental implementation of evolutionary DNA computations is examined in more detail. Several designs for evolutionary DNA computing are presented in Section 6. Section 7 provides conclusions.

2   Experimental DNA Computing

The first example of an artificial DNA based computing system was presented in 1994 [2]. This special purpose DNA computer solved a small instance of a hard computational problem, the Hamiltonian Path Problem. Given a graph with vertices and edges, this problem asks for a path with fixed start and end vertices, that visits every vertex exactly once (Fig. 1a). To solve this problem, every vertex was encoded as a 20 nucleotide oligomer. Edges were encoded as 20-mers, with the first 10 nucleotides complementary to the last 10 nucleotides of the entry vertex, and the last 10 complementary to the first 10 of the exit vertex. This way, an edge oligonucleotide can bring the two vertices it connects together by acting as a splint (Fig. 1b). By mixing and ligating oligomers corresponding to every edge and every vertex, concatenates are formed that represent paths through the network (Fig. 1c). Not just any path through the graph is a solution to the problem. Random ligation will form many paths that do not meet the conditions set. Therefore, several selection steps are required (Fig. 1d, e): first, use selective amplification (PCR) to retain only those paths that start and end at the right vertex; then, keep only paths of correct length (seven vertices times 20 nucleotides); and finally, confirm the presence of every vertex sequence (using affinity separation). If any DNA remains after the last separation, this must correspond to a Hamiltonian path. Experimental implementation of this protocol indeed recovered a single species of oligonucleotide, which was shown to encode the only possible Hamiltonian path through the graph (Fig. 1f). From a computer science point of view, the path-construction phase of the algorithm is the most impressive. Because of the huge number of oligonucleotides used (50 picomol per species), all potential solutions are formed in parallel, in a single step, and through random molecular encounters.

Since 1994, many designs for DNA based computers have been presented, and many actual computations have been performed in biochemical laboratories. All these experiments are of a 'proof of principle' scale, and serve to evaluate algorithms and techniques. A number of representative benchmark experiments are summarized in Tables 1 and 2 (see [17] for a more comprehensive version). Fig. 2 illustrates how information is stored in linear molecules and operated on in several 'computer architectures'.

Table 1. Selected DNA computations

Refs.	Problem	Dimension	Solved for	Data pool generation		Selection	
				Initial species	Steps	Species	Steps
[2]	Directed Hamiltonian Path	n vertices, m edges	$n = 7$, $m = 14$	$n + m$	1	n	n
[4]	Maximal Clique	n vertices, m edges	$n = 6$, $m = 11$	$2n$	1	n enzymes	$\frac{1}{2}(n^2 - n) - m$
[5]	3-Satisfiability	n variables, m clauses	$n = 4$, $m = 10$	$4 + 4n$	$n - 2$		m
[6]	Satisfiability	n variables, m clauses	$n = 9$, $m = 5$		n	$2n$	m
[7]	Maximum Independent Set	n vertices, m edges	$n = 6$, $m = 4$	1	m	n enzymes	1
[8]	Satisfiability	n variables, m clauses	$n = 4$, $m = 4$		2^n	2^n	m
[9]	3-Satisfiability	n variables, m clauses	$n = 6$, $m = 10$	$3m$	1		1
[10]	Satisfiability	n variables, m clauses	$n = 4$, $m = 5$		2^n	2^n	m
[11]	3-Satisfiability	n variables, m clauses	$n = 20$, $m = 24$		n	$2n$	m
[12]	3-Satisfiability	n variables, m clauses	$n = 4$, $m = 4$		2^n	$m2^{n-3}$	m

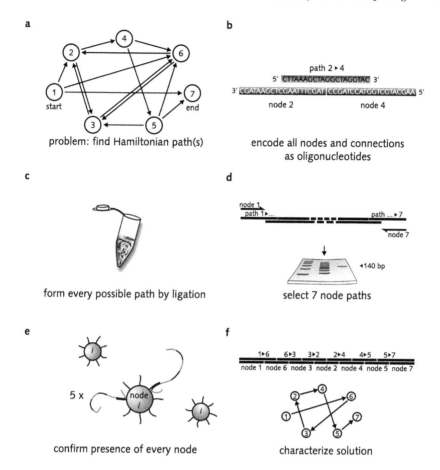

a

problem: find Hamiltonian path(s)

b

path 2 ▸ 4

5' CTTAAAGCTAGGCTAGGTAC 3'
3' CGATAAGCTCGAATTTCGAT | CCGATCCATGGTCGTACGAA 5'

node 2   node 4

encode all nodes and connections
as oligonucleotides

c

form every possible path by ligation

d

node 1
path 1 ▸...   path ...▸ 7
                              node 7

◂140 bp

select 7 node paths

e

5 x   node *i*

confirm presence of every node

f

1▸6   6▸3   3▸2   2▸4   4▸5   5▸7
node 1 node 6 node 3 node 2 node 4 node 5 node 7

characterize solution

Fig. 1. DNA solution to a Hamiltonian Path Problem instance (from [2]). (a) The graph used. (b) Encoding strategy. The oligonucleotides encoding two vertices (5'→3') and the edge connecting (3'→5') them are shown. (c) Mixing all seven vertices and 14 edges results in the ligation of all possible paths through the graph. Vertices 1 and 7 are not strictly necessary, as their presence in the final paths can be encoded by the splint oligos alone. Incorporation promotes the formation of paths both entering and leaving these vertices. (d) Selection of paths of correct length. The ligation product was amplified by PCR, using the oligonucleotide encoding vertex 7 and the complement of vertex 1 as primers. After gel electrophoresis, products of 140 bp (seven vertices) long were selected. (e) This product was denatured, and only those strands containing vertex 2 were selected using beads coated with the complement of vertex 2. This procedure was repeated for vertices 3 to 6. (f) Output of the computation. Presence and position in the path of every vertex was verified by PCR using complements of this vertex and the oligo for vertex 1 as primers. Only the correct Hamiltonian path was retrieved from the ligation mixture

Table 2. Technical aspects of DNA computations

Refs.	Library generation	Molecule	Selection criteria	Selection technology	Output technology	Architecture
[2]	splint ligation	dsDNA	length, subsequence	electrophoresis, PCR, subsequence selection	PCR, electrophoresis	
[4]	overlap assembly	dsDNA	subsequence, length	restriction enzymes, electrophoresis	cloning, sequencing	
[5]	splint ligation	ssDNA	subsequence	subsequence selection	PCR, electrophoresis	[13]
[6]	combinatorial chemical synthesis	ssDNA	subsequence	duplex RNase	cloning, PCR, electrophoresis	Lipton[14]
[7]	combinatorial subsequence removal	plasmid DNA	length	electrophoresis	cloning, sequencing	aqueous
[8]	chemical synthesis	immobilized ssDNA	entire sequence	hybridization, ssDNA nuclease	PCR, array hybridization	surface
[9]	ligation	ssDNA	subsequence	hairpin formation, PCR, length selection	PCR, cloning, sequencing	
[10]	chemical synthesis	immobilized ssDNA	entire sequence	hibridization, ssDNA nuclease	enzymatic cleavage, FRET, array	surface
[11]	combinatorial chemical synthesis	ssDNA	subsequence	subsequence selection	PCR, electrophoresis	Lipton[14], sticker[15]
[12]	chemical synthesis	ssDNA	subsequence	hybridization	FCS	blocking[16]

3   Implementation Issues

Application of DNA computing to large combinatorial optimization problems is promising in theory, but quite a challenge to implement in practice. A number of practical difficulties arise in the implementation of computations in DNA. Many computations suffer errors because of undesired molecular interactions.

There are several reasons for this, the most important being intrinsic molecular behaviours and suboptimal protocols.

The first source of errors is nearly impossible to eradicate completely. On the protocol side, however, a lot can be improved. Most of molecular biology is not concerned at all with 100% reliability, as most computing purposes require: DNA handling protocols are designed for acceptable results in reasonable time, where acceptable may even signify a 5% success rate for some operations. DNA computing is therefore a catalyst in the optimization of protocols for reliability, reproducibility and accuracy (see also Table 2). However, handling molecules is still largely a stochastic activity.

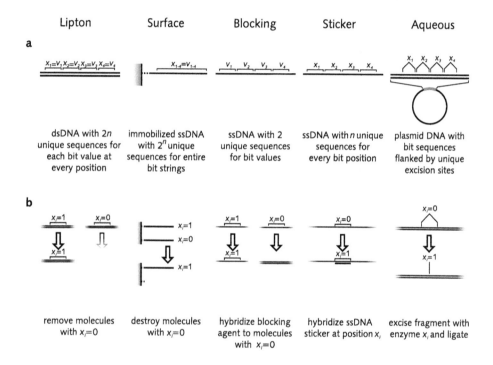

Fig. 2. Architectures for DNA based parallel search algorithms. Shown are (a) representations of bit strings (values v_i for bits x_i, $1 < i < n$, here for $n = 4$) and (b) operation on those bits (setting bit x_i to 1) in five major models: the method of Lipton [14], the surface based approach [8], the blocking algorithm [16], the sticker model [15] and aqueous (plasmid) computing [7]. All except the sticker model have been implemented experimentally. The five models fall into two categories. Lipton, surface and blocking start with a mixture of all possible bit strings (basically a read-only memory) and set bit values by discarding those molecules that have another value for that bit. In the sticker and aqueous algorithms it is possible to reversibly alter the value of a bit in every molecule. These models start with a single species of molecule, typically with every bit set to an identical value. Bit operations on subsets of this molecular random access memory generate a library, which is then searched using other techniques

An important issue in the implementation of DNA computers is careful sequence design. The coding of potential solutions is often done in the nucleotide sequence, but other considerations also affect sequence choice. Depending on the design of the DNA computer, different types of molecular interaction are required which put constraints on sequence design [18]. Examples include melting behaviour, subsequence distinctiveness, enzyme recognition sites and three-dimensional folding. Several software packages have been developed to aid in sequence design (see for example [19]).

Computations such as those listed in the previous section are still performed by manual pipetting. This hinders scaling of computations to larger instances, but it also introduces human error. Several strategies for the automation of computations have been presented, e.g. liquid handling robotics [20] and microfluidic lab-on-a-chip procedures [21,22]. Application of such technology will certainly enhance the power and reliability of DNA computing. Another strategy to avoid human error is to let the computations proceed autonomously, without the need for intervention between input and output [23].

Another important drawback of current DNA computation concerns the output mechanisms. Output molecules are generally analysed using 'crude' detection methods (predominantly gel electrophoresis, see Table 2). Sensitive high-throughput screening technologies need to be developed in order to overcome the limitations imposed by existing readout methods.

Finally, DNA computers are relatively slow. Not only do laboratory operations take anywhere from seconds to hours to execute, there are lower bounds on the switching times of macromolecules. Molecular processes such as electron transfer and optical switching can be very fast, but the timescales for pattern recognition and conformational changes are typically on the microsecond scale.

4   Evolutionary DNA Computing

It has been proposed (early on in [24,25], later in more detail in [26,27,28]) that the limitations as sketched in the previous section may be circumvented by the implementation of evolutionary algorithms in DNA computers. Using careful encoding, such systems could take advantage of biochemical noise by using it as a source of variation. Iteration of a selection cycle could yield a directed search through sequence space. Evolutionary DNA computing could be modelled after *in vitro* directed evolution of proteins, but with different selection criteria. Fig. 3 illustrates the phases of a hypothetical evolutionary computation using DNA.

Evolutionary DNA computers are potentially much more powerful than *in silico* evolutionary problem solving approaches. Advantages may include vastly larger population sizes (10^{12} *in vitro*, typically 10^3 *in silico*), better evolutionary performance (*in silico* evolutionary algorithms are abstractions of biochemical realities) and true non-determinism. With sufficiently superior evolutionary performance, the lack of speed of biochemical operations would become irrelevant.

The relationship between huge population sizes and superior performance, however, is not guaranteed. Simulations of a DNA based evolutionary computa-

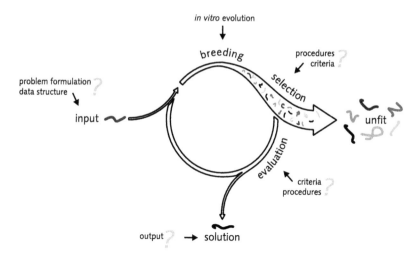

Fig. 3. Evolutionary DNA computing. Candidate solutions are generated by amplification and diversification through mutation and recombination (molecular breeding). A selection procedure filters out the 'fittest' candidates. If these are satisfactory, the computation ends; if not, they are used as input for another iteration of the cycle. Implementation issues are indicated by arrows

tion to Maximal Clique instances have been performed [28] to assess the effect of increased population size. In the most cases, a correlation between large populations and better optimization results was indeed observed. Other simulations also suggest that evolutionary DNA computation is feasible [29].

5   Experimental Issues in Evolutionary DNA Computing

We will discuss three experimental issues in evolutionary DNA computing in subsections: on suitable encodings, on diversification procedures and on fitness selection protocols.

5.1   Problems, Encodings and Algorithms

Most DNA computations work on instances of Satisfiability problems (see Table 1) and require iterated selection procedures for local properties of the solution molecules, with the number of iterations dependent on the problem size. These complicated selections are equivalent to the combined selection and evaluation in a single evolutionary cycle (Fig. 3). As the computations listed represent the state of the art in selection, it is clear that at present repeated cycles are not feasible. Selection procedures for evolutionary DNA computations should consist of a limited number of steps (ideally only one), but they may allow for larger errors than those used in deterministic computation. The evaluation procedure should be equally limited in time, but needs to be more precise.

We can also look at the data structures and test problems used. The majority relies on either a surface-based [8] architecture or Lipton [14] architecture (Fig. 2), both of which are currently inadequate as candidates for evolutionary DNA computing. The surface-based methods suffer from an intrinsic bound on evolutionary search space: the number of iterations of the evolutionary loop is determined by the chosen surface area instead of by the appearance of satisfactory solutions. Still, if methods are developed to generate and recombine strands on a surface, immobilized DNA might support evolutionary searches. The Lipton encoding would be more difficult to adapt, as it is dependent on local properties (subsequences) of the solution strands. Populations would have to be subjected to serial subsequence inspection for every selection iteration, which is an unfeasible scenario.

5.2   Diversification

Evolutionary computation employs mutation or recombination operators, or a combination of these, to achieve diversification of a population of candidate solutions. They are originally modelled after biological processes, but their *in silico* version can not be directly translated back to evolutionary DNA computing. For example, a standard mutation operator flips the value of a bit. In biochemistry, mutation is a name for many processes, including deletion and insertion of nucleotides, and nucleotide substitution. Changing the value of a bit is nearly impossible using *in vitro* mutation. Even if bit values are encoded in single nucleotides (which is very uncommon in experimental DNA computing), a random nucleotide substitution will produce a nonsense encoding in 2 out of 3 cases. The chances for a typical 20 nucleotide bit sequence switching to its inverse value are very low. Recombination has roughly the same meaning *in vivo* and *in silico*: the exchange of genetic material between two molecules. Still, the mechanisms are very different. These major differences make it unlikely that evolutionary algorithms can be easily ported between software and 'wetware'. *In vitro* mutation and recombination parameters will be useless, and molecular diversity will have to be achieved by different means.

A promising direction is directed molecular evolution. Such *in vitro* evolution has proved to be a valuable strategy in searching for macromolecules with certain desired properties. The approach uses recursive cycles of recombination, mutation and selection on a population of candidate nucleic acids to enrich this pool in 'successful' species. Specific protocols have been developed for the generation of molecular diversity, for instance artificial recombination by 'DNA shuffling' [30]. Searches can be conducted for novel or improved protein functions [31], as well as for catalytic nucleic acid species (ribozymes) and nucleic acid ligands [32].

5.3   Fitness Selection

Several selectable properties of DNA molecules can be used as a fitness indicator for evolutionary computation, for instance molecule length and nucleotide sequence. Not every analytical technology is well suited, for example DNA sequencing yields very precise data on the molecules involved, but is an essentially sequential procedure on single molecular species. What is needed is a fast and highly parallel procedure.

An interesting candidate is gel electrophoresis. In this procedure, an electric field is applied to a gel matrix containing a sample of macromolecules. These molecules move through the matrix at a certain rate, dependent on their charge, shape and size, and on gel conditions. For double-stranded DNA molecules, the movement is predictable and only related to the length of the molecule, i.e. the number but not the identity of the constituent nucleotides matters. The procedure can be applied to large and complex populations of DNA species, and molecules of the same length end up in the same 'band' in the gel. These molecules can then be rescued from the gel by simple excision of the band.

For nucleic acids of identical length, a modified gel electrophoresis technique can be used to separate molecules based on sequence. During electrophoresis, perfect double-stranded (homoduplex) DNA migrates through a gel at a predictable rate. However, DNA containing nucleotide mismatches (heteroduplex) and single-stranded DNA migrate at anomalous rates, caused by secondary structure formation (ssDNA) or helix distortion (dsDNA). Such structures experience specific, but unpredictable, resistances when migrating through the gel matrix. Heteroduplex mobility is lower than that of homoduplexes of equal length and as a result bands end up higher on the gel; single strands migrate faster. Several well-established and sensitive mutation detection techniques exploit this effect, such as single strand conformational polymorphism (SSCP), temperature or denaturing gradient gel electrophoresis (TGGE, DGGE) and heteroduplex analysis [33].

The use of this phenomenon as a separation procedure for evolutionary DNA computing was first proposed by Wood and Chen [34]. In these designs, a ssDNA molecule represents a candidate solution, and it is evaluated by adding another strand which acts as a reference strand. The combination of these (if they combine at all) is analysed using gel electrophoresis. Experimental evidence has been presented for molecules representing potential solutions to Max 1s [34] and Royal Road [35] problems. However, the computational power of the design is limited: optimal solutions are those that are perfectly complementary to the reference strand. Since the reference strand is known, the functions to be optimized are trivial.

Differential gel migration can also be used to implement the so-called blocking algorithm [16,36]. In this design, the reference strands ('blockers') correspond to falsifying conditions instead of perfect solutions. The optimization function is then inverted: better candidate solutions are those that are able to escape hybridization to an ensemble of blockers. Experimental evidence has been given for a 4 variable 3SAT instance [36], however the error rates are quite high ($\pm 15\%$), implying that the procedure should be followed by a more precise output technique. Also, in the current technological incarnation, the number of blockers needed can become problematic —it scales exponentially with respect to input size. The application of artificial 'universal' nucleotide analogues [37] in theory circumvents this difficulty.

6   Proposals

We discuss a number of designs for evolutionary DNA computing.

6.1   Satisfiability, Based on Hybridization Detection

The gel migration experiments discussed above provide the basis for the design outlined in Fig. 4. Satisfiability problems are in principle open to evolutionary optimization, but care must be taken in the design of the algorithms [38]. *In silico* approaches often use a bit string representation, which is equivalent to the molecular bit string of the blocking algorithm. Variation is best achieved using the mutation operator, which is biochemically unfeasible. However, it is quite probable that the behaviour of DNA based evolutionary algorithms is very different from *in silico* counterparts.

DNA hybridization is a promising candidate for selection [34,35]. The duplex migration assay [36] may make a good selection procedure. Key requirements for this stage in the computation are that it takes only a few steps, is highly parallel, and does not destroy molecules in the process. Somewhat problematically, the procedure is not entirely accurate, because of the lack of an absolute correlation between electrophoretic mobility and DNA complementarity (i.e. relatively highly mismatched ensembles may migrate closer to perfect duplexes than those with only a single mismatch). This can complicate the definition of a fitness function. However, for mismatch ratios of 5% and up, the migration behaviour does become proportional [39]. Also, the duplex migration assay is generally recommended for molecules of over 100 bp in length, as mismatch deformation is probably accentuated by longer duplex arms. The 4 variable SAT instance reported used 36 bp candidate solutions [36]. Therefore, the precision of the selection step may even become higher for larger problem instances. While electrophoretic behaviour may be used as an enrichment procedure, or a first relatively crude fitness selection on huge numbers of candidate solutions, a more

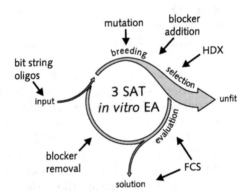

Fig. 4. An evolutionary DNA algorithm for Satisfiability problems. Abbreviations: HDX, heteroduplex gel migration assay; FCS, fluorescence correlation spectroscopy

precise method is required for subsequent inspection of a limited number of candidate solutions. In [12], the application of spectroscopy method (fluorescence correlation spectroscopy, FCS), is described which appears to be well suited for this purpose. Although this is a so-called single-molecule detection technique, and therefore essentially dependent on sequential instead of parallel examination of sample molecules, it is highly sensitive and precise: using the same 4 variable SAT problem used in [36] as a benchmark, a 0% error rate has been achieved.

Further research for the implementation of this design should focus on parameter choice (mutation rates, population sizes, selection pressure), molecular breeding methods to achieve bit flipping, and finally integration of components.

6.2 Knapsack, Based on Length Detection

Besides sequence, nucleic acid length can be used as a molecular phenotype. Several experimental computations have been implemented in which the absence or presence of a DNA sequence represents bit value [4,7]. In these cases, solution molecules are those with a minimal or maximal number of bits set to one, and therefore longer or shorter than all other molecules. Simple gel electrophoresis can be used as a highly parallel selection procedure.

A class of hard optimization problems that appears to suit this type of encoding particularly well is that of knapsack problems. Given a volume of fixed capacity and a set of items of fixed size, knapsack problems ask for the subset of items that exactly fills the knapsack, or else approach its capacity as closely as possible. The presence of items of a particular size in the knapsack can be identified with the presence of certain DNA sequences of particular length in a DNA molecule, for example a plasmid. In [17], a computation to a knapsack instance with seven items is solved experimentally using the aqueous computing architecture (see Fig. 2).

This design can be modified to yield the *in vitro* (evolutionary) algorithm shown in Fig. 5. Here, items are added to an initially empty knapsack (represented on the molecular level by linear dsDNA and plasmids, respectively). After addition, the length of the knapsacks can be checked in parallel by electrophoresis. If the knapsacks are still packed far below their capacity, new items can be added. If knapsacks start to approach their capacity, items can be exchanged between knapsacks by (artificial) recombination, until satisfactory knapsack contents are evolved. The contents of these knapsack plasmids can then be determined by nucleotide sequencing.

6.3 Special Purpose Evolutionary Computations

In analogy to 'classical' DNA computing (Tables 1 and 2), the abstract computations outlined above may serve multiple purposes: not only to produce output, but also as benchmarks that produce valuable information on the techniques and mechanisms involved. The real applications of evolutionary DNA computing may be other than formal computation.

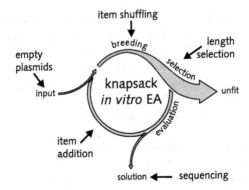

Fig. 5. An evolutionary DNA algorithm for knapsack problems

The clearest example of this is to examine one of the inspirations for evolutionary DNA computing: *in vitro* evolution of protein function. This set of technologies has been developed because the search spaces involved are of astronomical proportions, making *de novo* design of functional molecules all but impossible. In fact, models of protein design are known to be NP-complete problems [40,41]. Recursive selection and recombination can 'compute' answers to such molecular design problems [24]. This analogy not only suggests that other hard problems may well be open to molecular evolutionary optimization, but also that directed molecular evolution actually already is a successful molecular implementation of an evolutionary algorithm.

Another example is on the design of DNA computations. As is discussed in Section 3, a major area of investigation in DNA computing is the search for good encodings. This is also a difficult problem: the choice of any nucleotide sequence of length n for use in computing or otherwise may require the evaluation of all $4n$ possible sequences. [42] presents the design and experimental verification of a molecular algorithm that generates non-crosshybridizing oligonucleotides, i.e. sets of DNA molecules that are as different from each other as possible (see also 26,43]. These molecules can be employed in computations to minimize unwanted interactions. The design uses selective amplification by polymerase chain reaction (PCR) as a selection procedure. Multiple cycles of PCR may be used to execute an evolutionary search. In this case, like in more biologically inspired instances of *in vitro* evolution, there is no need to find a fitness function that matches a formal problem. Furthermore, the molecular format of the computation allows the elimination of approximations and models used in *in silico* searches for such sets of molecules.

7   Conclusions and Perspectives

Evolutionary computation seems promising to circumvent many implementation difficulties in DNA computing, and with considerable investments of time and effort the designs outlined in Section 5 could work. However, if implemented, these molecular evolutionary computers would still be single purpose: programming would be difficult. For example letting the knapsack algorithm run on a different input could require a long time for the synthesis, cloning, and integrity checks of new DNA items.

Then again, these molecular algorithms would be marvels of molecular control. The best recent illustration is the experimental computation by Braich et al. [11], where a single species of DNA could be isolated from a pool of over a million other, but very similar ones. This is currently the pinnacle of precise detection in an ocean of molecular noise. Such developments are themselves already illustrations of the value of DNA computing, and may be useful for diagnostic or analytical purposes outside of computation.

Acknowledgements

We would like to thank our colleagues Thomas Bäck, Grzegorz Rozenberg, Herman Spaink and Kristiane Schmidt for many interesting discussions on (evolutionary) DNA computing. This work was supported by the Netherlands Organization for Scientific Research (NWO), Council for Exact Sciences.

References

1. Păun, G., Rozenberg, G., Salomaa, A.: DNA Computing: New Computing Paradigms. Springer-Verlag, Berlin Heidelberg (1998)
2. Adleman, L.M.: Molecular Computation of Solutions to Combinatorial Problems. Science 266 (1994) 1021–1024
3. Schneider, T.D.: Theory of Molecular Machines. II. Energy Dissipation from Molecular Machines. J. Theor. Biol. 148 (1991) 125–137
4. Ouyang, Q., Kaplan, P.D., Liu, S.M., Libchaber, A.: DNA Solution of the Maximal Clique Problem. Science 278 (1997) 446–449
5. Yoshida, H., Suyama, A.: Solution to 3-SAT by Breadth First Search. In: WInfree, E., Gifford, D.K. (eds.): DNA Based Computers V. American Mathematical Society, Providence, RI (2000) 9–22
6. Faulhammer, D., Cukras, A.R., Lipton, R.J., Landweber, L.F.: Molecular Computation: RNA Solutions to Chess Problems. Proc. Natl. Acad. Sci. USA 91 (2000), 1385–1389
7. Head, T., Rozenberg, G., Bladergroen, R.S., Breek, C.K.D., Lommerse, P.H.M., Spaink, H.P.: Computing with DNA by Operating on Plasmids. Biosystems 57 (2000) 87–93
8. Liu, Q., Wang, L., Frutos, A.G., Condon, A.E. Corn, R.M., Smith, L.M.: DNA Computing on Surfaces. Nature 403 (2000) 175–179
9. Sakamoto, K., Gouzu, H., Komiya, K., Kiga, D., Yokoyama, S., Yokomore, T., Hagiya, M.: Molecular Computation by DNA Hairpin Formation. Science 288 (2000) 1223–1226
10. Wang, L., Hall, J.G., Lu, M., Liu, Q., Smith, L.M.: A DNA Computing Readout Operation Based on Structure-Specific Cleavage. Nat. Biotechnol. 19 (2001) 1053–1059

11. Braich, R.S., Chelyapov, N., Johnson, C., Rothemund, P.W.K., Adleman, L.: Solution to a 20-Variable 3-SAT Problem on a DNA Computer. Science 296 (2002) 499–502

12. Schmidt, K.A., Henkel, C.V. Rozenberg, G., Spaink, H.P.: DNA Computing Using Single-Molecule Hybridization Detection. Nucleic Acids Res. 32 (2004) 4962–4968

13. Ogihara, M., Ray, A.: DNA-Based Parallel Computing by 'Counting'. In: Rubin, H., Wood, D.H. (eds.): DNA Based Computers III: DIMACS Workshop, June 23–25, 1997. American Mathematical Society, Providence, RI (1997) 265–274

14. Lipton, R.J.: DNA Solution of Hard Computational Problems. Science 268 (1995) 542–545

15. Roweis, S. Winfree, E., Burgoyne, R., Chelyapov, N.V., Goodman, M.F., Adleman, L.M., Rothemund, P.W.K.: A Sticker-Based Model for DNA Computation. J. Comput. BIol. 5 (1998) 615–629

16. Rozenberg, G., Spaink, H.: DNA Computing by Blocking. Theor. Comput. Sci. 292 (2003) 653–665

17. Henkel, C.V.: Experimental DNA Computing. PhD thesis, Leiden University (2005)

18. Mauri, G., Ferretti, C.: Word Design for Molecular Computing: a Survey. In: Chen, J., Reif, J. (eds.): DNA Computing, 9th International Workshop on DNA Based Computers. Springer-Verlag, Berlin Heidelberg (2004) 37–47

19. Heldkamp, U., Rauhe, H., Banzhaf, W.: Software Tools for DNA Sequence Design. Genetic Programming and Evolvable Machines 4 (2003) 153–171

20. Hagiya, M.: From Molecular Computing to Molecular Programming. In: Condon, A., Rozenberg, G. (eds.): DNA Computing, 6th International Workshop on DNA-Based Computers. Springer-Verlag, Berlin Heidelberg (2001) 89–102

21. Gehani, A., Reif, J.: Micro Flow Bio-Molecular Computation. Biosystems 52 (1999) 197–216

22. McCaskill, J.S.: Optically Programming DNA Computing in Microflow Reactors. Biosystems 59 (2001) 125–138

23. Benenson, Y., Paz-Elizur, T., Adar, R., Keinan, E., Livneh, Z., Shapiro, E.: Programmable and Autonomous Computing Machine Made of Biomolecules. Nature 414 (2001) 430–434

24. Stemmer, W.P.C.: The Evolution of Molecular Computation. Science 270 (1995) 1510

25. Adleman, L.M.: On Constructing a Molecular Computer. In: Lipton, R.J., Baum, E. (eds.): DNA Based Computers, DIMACS 27. American Mathematical Society, Providence, RI (1996), 1–21

26. Deaton, R., Murphy, R.C., Rose, J.A., Garzon, M., Franschetti, D.R., Stevens Jr., S.E.: A DNA Based Implementation of an Evolutionary Search for Good Encodings for DNA Computation. In: Proceedings of the Fourth IEEE Conference on Evolutionary Computation, Indianapolis, IN. IEEE Press, Piscataway, NJ (1997) 267–271

27. Chen, J., Wood, D.H.: Computation with Biomolecules. Proc. Natl. Acad. Sci. USA 97 (2000) 1328–1330

28. Bäck, T., Kok, J.N., Rozenberg, G.: Evolutionary Computation as a Paradigm for DNA-Based Computation. In: Landweber, L.F., Winfree, E. (eds.): Evolution as Computation. DIMACS Workshop, Princeton, January 1999. Springer-Verlag, Berlin Heidelberg (2003) 15–40

29. Nuser, M., Deaton, R.: Simulations of DNA Computing with In Vitro Selection. Genetic Programming and Evolvable Machines 4 (2003) 173–183

30. Stemmer, W.P.C.: Rapid Evolution of a Protein In Vitro by DNA Shuffling. Nature 370 (1994) 389–391
31. Minshull, J., Stemmer, W.P.C.: Protein Evolution by Molecular Breeding. Curr. Opin. Chem. Biol. 3 (1999) 284–290
32. Joyce, G.F.: Directed Evolution of Nucleic Acid Enzymes. Annu. Rev. Biochem. 73 (2004) 791–836
33. Nataraj, A.J., Olivos-Glander, I., Kusukawa, N., Highsmith, W.E.: Single-Strand Conformation Polymorphism an Heteroduplex Analysis for Gel-Based Mutation Detection. Electrophoresis 20 (1999) 1177–1185
34. Wood, D., Chen, J., Antipov, E., Lemieux, B., Cedeno, W.: A DNA Implementation of the Max 1s Problem. In: Banzhaf, W., Eiben, A.E., Garzon, M.H., Honavar, V., Jakiela, M., Smith, R.E. (eds.): Proceedings of the Genetic and Evolutionary Computation Conference 1999. Morgan Kaufman, San Francisco (1999) 1835–1841
35. Goode, E., Wood, D.H., Chen, J.: DNA Implementation of a Royal Road Fitness Evaluation. In: Condon, A., Rozenberg, G. (eds.): DNA Computing, Proceedings 6th International Workshop on DNA-Based Computers. Springer-Verlag, Berlin Heidelberg (2001) 247–262
36. Henkel, C.V., Rozenberg, G., Spaink, H.P.: Application of Mismatch Detection Methods in DNA Computing. In: Ferretti, C., Mauri, G., Zandron, C. (eds.): DNA10, Tenth International Meeting on DNA Computing, Preliminary Proceedings. Università di Milano-Bicocca (2004) 183–192
37. Loakes, D.: The Application of Universal DNA Base Analogues. Nucleic Acids Res. 29 (2001) 2437–2447
38. Gottlieb, J., Marchiori, E., Rossi, C.: Evolutionary Algorithms for the Satisfiability Problem. Evol. Comput. 10 (2002) 35–50
39. Upchurch, D.A., Shankarappa, R., Mullins, J.I.: Position and Degree of Mismatches and the Mobility of DNA Heteroduplexes. Nucleic Acids Res. 28 (2000) e69
40. Fraenkel, A.S.: Protein Folding, Spin Glass and Computational Complexity. In: Rubin, H., Wood, D.H. (eds.): DNA Based Computers III, Proceedings DIMACS Workshop. American Mathematical Society, Providence, RI (1999) 101–121
41. Pierce, N.A., Winfree, E.: Protein Design is NP-Hard. Protein Eng. 15 (2002) 779–782
42. Deaton, R., Chen, J., Bi, H., Rubin, H., Wood, D.H.: A PCR-Based Protocol for In Vitro Selection of Non-Crosshybridizing Oligonucleotides. In: Hagiya, M., Ohuchi, A. (eds.): DNA Computing, 8th International Workshop on DNA-Based Computers. Springer-Verlag, Berlin Heidelberg (2003) 196–204
43. Chen, J., Deaton, R., Garzon, M., Kim, J.W., Wood, D., Bi, H., Carpenter, D., Wang, Y.Z.: Characterization of Non-Crosshybridizing DNA Oligonucleotides Manufactured In Vitro. In: Ferretti, C., Mauri, G., Zandron, C. (eds.) DNA10, Tenth International Meeting on DNA Computing, Preliminary Proceedings. Università di Milano-Bicocca (2004) 132–141

A Linear Solution of Subset Sum Problem by Using Membrane Creation

M.A. Gutiérrez-Naranjo, M.J. Pérez-Jiménez, and F.J. Romero-Campero

Research Group on Natural Computing,
Department of Computer Science and Artificial Intelligence,
University of Sevilla,
Avda. Reina Mercedes s/n, 41012, Sevilla, Spain
{magutier, marper, fran}@us.es

Abstract. Membrane Computing is a branch of Natural Computing which starts from the assumption that the processes taking place in the compartmental structure of a living cell can be interpreted as computations. In this framework, the solution of NP problems is obtained by generating an exponential amount on workspace in polynomial time and using parallelism to check simultaneously all the candidates to solution. We present a solution to the Subset Sum problem for P systems where new membranes are generated from objects.

Keywords: Natural Computing, Membrane Computing, Cellular Complexity Classes, Subset Sum Problem.

1 Introduction

Membrane Computing is a cross-disciplinary field with contributions by computer scientists, biologists, formal linguists and complexity theoreticians, enriching each others with results, open problems and promising new research lines.

This emergent branch of Natural Computing was introduced by Gh. Păun in [11]. Since then it has received important attention from the scientific community. In fact, Membrane Computing has been selected by the Institute for Scientific Information, USA, as a fast *Emerging Research Front* in Computer Science, and [10] was mentioned in [14] as a highly cited paper in October 2003.

This new non-deterministic model of computation starts from the assumption that the processes taking place in the compartmental structure of a living cell can be interpreted as computations. The devices of this model are called *P systems*.

Roughly speaking, a P system consists of a cell-like membrane structure, in the compartments of which one places multisets of objects which evolve according to given rules in a synchronous non-deterministic maximally parallel manner[1].

In this paper we present a contribution to this new discipline from the computational eficiency point of view. We introduce a family of P systems, constructed

---

[1] A layman-oriented introduction can be found in [12] and further bibliography at [15].

J. Mira and J.R. Álvarez (Eds.): IWINAC 2005, LNCS 3561, pp. 258–267, 2005.
© Springer-Verlag Berlin Heidelberg 2005

in an *uniform way*, that solves the Subset Sum problem. The search of polynomial solution to NP-complete problems is done by trading time by space: An
exponential amount of membranes (workspace) is built in polynomial time.

Inspired in living cells, P systems abstract the way of obtaining new membranes. These processes are basically two: *mitosis* (membrane division) and *autopoiesis* (membrane creation). Both ways of generating new membranes have
given rise to different variants of P systems: *P systems with active membranes*,
where the new workspace is generated by membrane division, and *P systems
with membrane creation*, where the new membranes are created from objects.

Both models are universal from a computational point of view, but tech-
nically, they are pretty different. In fact, nowadays there does not exist any
theoretical result which proves that these models can simulate each other in
polynomial time.

P systems with active membranes have been successfully used to design solutions to well-known **NP**-complete problems, as SAT [9], Subset Sum [6], Knapsack [7], Bin Packing [8] and Partition [2], but as Gh. Păun pointed in [13]
*"membrane division was much more carefully investigated than membrane cre-
ation as a way to obtain tractable solutions to hard problems"*. Recently, the first
results related to the power and design of algorithms to solve NP problems in
these model have arisen (see [3, 4]).

The paper is organized as follows: first P systems with membrane creation
are recalled in the next section. In section 3 recognizer P systems (devices that
capture the intuitive idea underlying the concept of algorithm) are presented.
The solution in the framework of *membrane creation* to the Subset Sum problem
is given in section 4. Finally, some formal details and conclusions are given in
the last sections.

2   P Systems with Membrane Creation

Since Gh. Păun presented the cellular computation with membranes, many different variants have been proposed. If the membrane structure is considered to
set a classification among these different variants, two big groups are obtained:
P systems where the initial structure does not change along computations and
P systems where the tree structure of the membranes vary (or can do it) along
computation. The decrease of the number of membranes is made by applying a
so-called *dissolution rule* $[a]_e \rightarrow b$ in which the object a inside a membrane with
label e produces the dissolution of the rule, a disappears and a new element b and
the rest of the multiset in the membrane go to its father (more precisely, they
go to the closest non-dissolved ancestor in the membrane hierarchy, since several
membranes can dissolve in the same step). Increasing the number of membranes
are usually made via division of existing ones or creating new ones from objects[2]

---

[2] Recently, new operations to change the membrane structure have been explored as
merging membranes or the operations of *endocytosis, exocytosis* or *gemmation*.

Membranes are created in living cells, for instance, in the process of vesicle mediated transport and in order to keep molecules close to each other to facilitate their reactions. Membranes can also be created in a laboratory - see [5]. Here we abstract the operation of creation of new membranes under the influence of existing chemical substances to define P systems with membrane creation. Recall that a *P system with membrane creation* is a construct of the form $\Pi = (O, H, \mu, w_1, \ldots, w_m, R)$ where:

1. $m \geq 1$ is the initial degree of the system;
2. O is the alphabet of *objects*;
3. H is a finite set of *labels* for membranes;
4. μ is a *membrane structure* consisting of m membranes labelled (not necessarily in a one-to-one manner) with elements of H;
5. $w_1, \ldots, w_m$ are strings over O, describing the *multisets of objects* placed in the m regions of μ;
6. R is a finite set of *rules*, of the following forms:
   (a) $[a \rightarrow v]_h$ where $h \in H$, $a \in O$ and v is a string over O describing a multiset of objects. These are *object evolution rules* associated with membranes and depending only on the label of the membrane.
   (b) $a[\,]_h \rightarrow [b]_h$ where $h \in H$, $a, b \in O$. These are *send-in communication rules*. An object is introduced in the membrane possibly modified.
   (c) $[a]_h \rightarrow [\,]_h\, b$ where $h \in H$, $a, b \in O$. These are *send-out communication rules*. An object is sent out of the membrane possibly modified.
   (d) $[a]_h \rightarrow b$ where $h \in H$, $a, b \in O$. These are *dissolution rules*. In reaction with an object, a membrane is dissolved, while the object specified in the rule can be modified.
   (e) $[a \rightarrow [v]_{h_2}]_{h_1}$ where $h_1, h_2 \in H$, $a \in O$ and v is a string over O describing a multiset of objects. These are *creation rules*. In reaction with an object, a new membrane is created. This new membrane is placed inside of the membrane of the object which triggers the rule and has associated an initial multiset and a label.

Rules are applied according to the following principles:

- Rules from (a) to (e) are used as usual in the framework of membrane computing, that is, in a maximal parallel way. In one step, each object in a membrane can only be used for one rule (non deterministically chosen when there are several possibilities), but any object which can evolve by a rule of any form must do it (with the restrictions below indicated).
- If a membrane is dissolved, its content (multiset and interior membranes) becomes part of the immediately external one. The skin membrane is never dissolved.
- All the elements which are not involved in any of the operations to be applied remain unchanged.
- The rules associated with the label h are used for all membranes with this label, irrespective of whether or not the membrane is an initial one or it was obtained by creation.

- Several rules can be applied to different objects in the same membrane simultaneously. The exception are the rules of type (d) since a membrane can be dissolved only once.

3   Recognizer P Systems with Membrane Creation

Recognizer P systems were introduced in [7] and are the natural framework to study and solve decision problems, since deciding whether an instance has an affirmative or negative answer is equivalent to deciding if a string belongs or not to the language associated with the problem.

In the literature, recognizer P systems are associated in a natural way with P systems with *input*. The data related to an instance of the decision problem has to be provided to the P system in order to compute the appropriate answer. This is done by codifying each instance as a *multiset*[3] placed in an *input membrane*. The output of the computation (*yes* or *no*) is sent to the environment. In this way, P systems with input and external output are devices which can be seen as black boxes, in which the user provides the data before the computation starts and the P system sends to the environment the output in the last step of the computation. Another important feature of P systems is the non-determinism. The design of a family of recognizer P system has to consider it, because all possibilities in the non-deterministic computations have to output the same answer. This can be summarized in the following definitions (taken from [1]).

Definition 1. *A P system with input is a tuple (Π, Σ, i_Π), where: (a) Π is a P system, with working alphabet Γ, with p membranes labelled by $1, \ldots, p$, and initial multisets $w_1, \ldots, w_p$ associated with them; (b) Σ is an (input) alphabet strictly contained in Γ; the initial multisets are over $\Gamma - \Sigma$; and (c) i_Π is the label of a distinguished (input) membrane.*

Let m be a multiset over Σ. The *initial configuration of (Π, Σ, i_Π) with input m* is $(\mu, w_1, \ldots, w_{i_\Pi} \cup m, \ldots w_p)$.

Definition 2. *A recognizer P system is a P system with input, (Π, Σ, i_Π), and with external output such that:*

1. *The working alphabet contains two distinguished elements yes, no.*
2. *All its computations halt.*
3. *If C is a computation of Π, then either some object yes or some object no (but not both) must have been released into the environment, and only in the last step of the computation. We say that C is an accepting computation (respectively, rejecting computation) if the object yes (respectively, no) appears in the external environment associated to the corresponding halting configuration of C.*

---

[3] Representing the data via multiset is inspired in the multiset of chemical compounds inside living cells.

We denote by $\mathcal{MC}$ the class of recognizer P systems with membrane creation.

In the next section we present a solution to the Subset Sum problem in linear time in the sense of the following definition.

Definition 3. *Let $\mathcal{F}$ be a class of recognizer P systems. We say that a decision problem $X = (I_X, \theta_X)$ is solvable in polynomial time by a family $\mathbf{\Pi} = (\Pi(n))_{n \in \mathbb{N}}$, of $\mathcal{F}$, and we denote this by $X \in \mathbf{PMC}_{\mathcal{F}}$, if the following is true:*

- *The family $\mathbf{\Pi}$ is polynomially uniform by Turing machines; that is, there exists a deterministic Turing machine constructing $\Pi(n)$ from $n \in \mathbb{N}$ in polynomial time.*
- *There exists a pair (cod, s) of polynomial-time computable functions over the set of instances I_X such that:*
  - *for each instance $u \in I_X$, $s(u)$ is a natural number and $cod(u)$ is an input multiset of the system $\Pi(s(u))$.*
  - *the family $\mathbf{\Pi}$ is polynomially bounded with regard to (X, cod, s); that is, there exists a polynomial function p, such that for each $u \in I_X$ every computation of $\Pi(s(u))$ with input $cod(u)$ is halting and, moreover, it performs at most, $p(|u|)$ steps.*
  - *the family $\mathbf{\Pi}$ is sound with regard to (X, cod, s); that is, for each $u \in I_X$, if there exists an accepting computation of $\Pi(s(u))$ with input $cod(u)$, then $\theta_X(u) = 1$.*
  - *the family $\mathbf{\Pi}$ is complete with regard to (X, cod, s); that is, for each $u \in I_X$, if $\theta_X(u) = 1$, then every computation of $\Pi(s(u))$ with input $cod(u)$ is an accepting one.*

In the above definition we have imposed to every P system $\Pi(n)$ to be *confluent*, in the following sense: every computation of a system with the *same* input must always give the *same* answer.

It can be proved that $\mathbf{PMC}_{\mathcal{F}}$ is closed under polynomial–time reduction and complement, see [9]. In this paper we will deal with the class $\mathcal{MC}$ of recognizer P systems with membrane creation.

4   Solving Subset Sum in Linear Time with Membrane Creation

In this section we present a family $\prod$ of recognizer P systems that solves the Subset Sum problem in linear time.

The Subset-Sum problem can be settled as follows: *Given a finite set, A, a weight function, $w : A \rightarrow N$, and a constant $k \in \mathbb{N}$, determine whether or not there exists a subset $B \subseteq A$ such that $w(B) = k$. If A has n elements with weights $w_1, \ldots, w_n$, one instance of the problem can be encoded as $(n, (w_1, \ldots, w_n), k)$.*

As usual in the framework of P systems, the solution of the problem is based on an algorithm of brute force where an exponential amount of workspace is built in linear time. The algorithm is split in the following stages:

- *Generation stage and weight calculation stage*: for every subset of A, a membrane is created. In each working membrane the weight of the associated subset is calculated.
- *Checking stage*: in each membrane it is checked whether or not the weight of its associated subset is exactly k.
- *Output stage*: when the previous stage has been completed in all membranes, the system sends out the answer (*yes* or *not*) to the environment.

- Working alphabet:

$$\Gamma = \left\{ \begin{array}{l} e_2, \ldots e_n, e_1^+, \ldots, e_n^+, e_1^-, \ldots, e_n^-, c_0, c_1, k_1, k_2, k_3, t_0, \ldots, t_{2n+2k+7}, \\ s, s^+, s^-, s_1, \ldots, s_n, s_2^+, \ldots, s_n^+, s_2^-, \ldots, s_n^-, d_1, \ldots, d_{k+1}, \\ a, p_0, p_1, p_2, p_3, q, q_0, q_1, yes_0, yes_1, yes, no_0, no_1, no_2, no \end{array} \right\}$$

- Initial membrane structure: $\mu = [\ ]_0$
- Set of labels: $H = \{0, p, n, t, f, r, s, c\}$
- Initial Multiset: $w_0 = t_0 e_1^+ e_1^-$
- Input: $s_1^{w_1} \ldots s_n^{w_n}$
- The set of evolution rules, $R(\langle n, k \rangle)$, consists of the following rules (recall that λ denotes the empty string):

1. $[e_j^+ \rightarrow [e_{j+1} k_0]_p]_l$ for $l = 0, p, n$ $j = 1, \ldots, n-1$
    $[e_j^- \rightarrow [e_{j+1} k_0]_n]_l$ for $l = 0, p, n$ $j = 1, \ldots, n-1$
    $[e_j \rightarrow e_j^+ e_j^-]_l$ for $l = p, n$ $j = 2, \ldots, n$
    $[e_n^+ \rightarrow [c_0]_t]_l$ for $l = 0, p, n$
    $[e_n^- \rightarrow [c_0]_f]_l$ for $l = 0, p, n$

The goal of these rules is to create one membrane for each possible subset of A. The new membrane with label p, represents the partial subset in which we place the the sum of the object a_1; on the other hand the new membrane with label n, represents the partial subset in which a_1 is not considered.

2. $[k_i \rightarrow k_{i+1}]_l$ for $l = p, n$ $i = 0, 1, 2$
    $[k_3]_l \rightarrow \lambda$ for $l = p, n$
    $[t_i \rightarrow t_{i+1}]_0$ for $i = 0, \ldots, 2n + 2k + 6$
    $[t_{2n+2k+7} \rightarrow [no_0]_c]_0$

These rules manage the counters k and t. When a new membrane labelled with p or n is created, an object k_0 is placed inside it. When this counter reaches k_3, the membrane is dissolved in the next stage. The counter t creates a new membrane with the object no_0 after an appropriate number of steps. If this membrane is not dissolved by an element yes_1, the answer no will be sent to the environment.

3. $s_j^+ [\ ]_p \rightarrow [s_{j-1}]_p$ for $j = 2, \ldots, n$
    $s_j^- [\ ]_n \rightarrow [s_{j-1}]_n$ for $j = 2, \ldots, n$
    $[s_j \rightarrow s_j^+ s_j^-]_l$ for $l = 0, p, n$ $j = 2, \ldots, n$
    $[s \rightarrow s^+ s^-]_l$ for $l = 0, p, n$
    $s^+ [\ ]_l \rightarrow [s]_l$ for $l = p, t$
    $s^- [\ ]_l \rightarrow [s]_l$ for $l = n, f$
    $[s_1 \rightarrow s^+]_l$ for $l = 0, p, n$

These rules manage the weights of the elements and are applied simultaneously with the rules of the set **1** which create the new workspace. By using the duplication of symbols and controlling the labels we can place the weight of each different subset into a membrane. For that we need an exponential amount of membranes.

4. $[c_0 \rightarrow c_1]_l$ for $l = t, f$
    $[c_1 \rightarrow d_1]_l$ for $l = t, f$
    $[s \rightarrow [\,]_s]_l$ for $l = t, f$
    $d_i[\,]_s \rightarrow [d_i]_s$ for $i = 1, \ldots, k$
    $[d_i]_s \rightarrow d_{i+1}$ for $j = 1, \ldots, k$

For each possible subset of A we have one membrane in which the weight of the subset is represented in unary form via the object s. With help of these rules we create as many new membranes labelled by s as objects s there are in the membrane. If k is greater or equal to the number of objects s it the membrane, an object d_{k+1} appears. If not, the computation inside this membrane halts.

5. $[d_{k+1} \rightarrow a\, q]_l$ for $l = t, f$
    $[a \rightarrow [p_0]_r]_l$ for $l = t, f$
    $q[\,]_s \rightarrow [q_0]_s$
    $[p_i \rightarrow p_{i+1}]_r$ for $i = 0, 1, 2$
    $[q_0]_s \rightarrow q_1$
    $q_1[\,]_r \rightarrow [q_1]_r$
    $[q_1]_r \rightarrow \lambda$
    $[p_3]_r \rightarrow yes_0$

As we saw above, if an object d_{k+1} appears in a membrane, then the weight of the associated subset is less or equal to k. This set of rules check that both amounts are the same. If so, an object yes_0 is produced in the membrane.

6. $[yes_0]_l \rightarrow yes_1$ for $l = t, f$
    $yes_1[\,]_c \rightarrow [yes_1]_c$
    $[yes_1]_c \rightarrow yes$
    $[yes]_0 \rightarrow yes[\,]_0$
    $[no_i \rightarrow no_{i+1}]_c$ for $l = 0, 1$
    $[no_2]_c \rightarrow no$
    $[no]_0 \rightarrow no[\,]_0$

There is a counter no_i in the membrane labelled with c. If an object yes_1 is obtained from one (or more) of the exponential amount of membranes which check the subsets of A, this object will stop the counter no_i and send the object yes to the environment. If not, i.e., if in the checking stage none of the membranes output yes_1, the counter no_i will not be stopped and an object no will be sent to the environment.

4.1 An Overview of the Computation

First of all we prepare an input for the instance $u = (n, (w_1, \ldots, w_n), k)$ of the Subset Sum problem. This instance will be processed by $\Pi(s(u))$, being

$s(u) = < n, k > = \frac{(n+k)(n+k+1)}{2} + n$. The input is the multiset $cod(u) = s^{w_1} \ldots s^{w_n}$ and we place it in the unique membrane of the initial configuration of $\Pi(s(u))$ and the computation starts.

At the beginning, the counter t_i is started from t_0 and when it reaches the object $t_{2n+2k+7}$ it will create a new membrane with the object no_0 inside. If the process is not stopped, the evolution of that object no_0 will send the answer no to the environment. This process only can be stopped if an object yes_0 is produced in the checking stage, i.e., we will create one membrane for each subset of the initial set A and check if the whole sum in that subset is equal to k. If this happens in one membrane, this membrane produces yes_1. This object dissolves the membrane where the counter no_i is placed and send the object yes to the environment.

In the first stage, from each object e_j we obtain two copies: e_j^+ and e_j^-. These object create new membranes labelled, respectively, with p and n. Since the index j vary along the computation, we obtain an exponential amount of membranes in linear time (it only depends on n).

Simultaneously, the multiset $s^{w_1} \ldots s^{w_n}$ which codifies the *input* trigger the appropriate rules to copy the weights in the new membranes. The use of the labels allow us to handle the flow of elements from one membrane to others and after $2n$ steps we have 2^n membranes which the appropriate number of objects s inside. At this point, the checking stage starts.

In this stage, for each membrane we have an object d_1 and as many membranes with label s as the weight of the subset associated to the membrane. The element d_1 goes inside one of this membranes, dissolves it and is changed to d_2. This new object d_2 does the same: it dissolves a membrane with label s and changes to d_3. If the number of membranes labelled with s is greater or equal to k, a new object d_{k+1} is created after $2k$ steps.

Then we have to check if k is greater of equal to number of membranes labelled with s. In other words, if there remains any membrane labelled with s after the apparition of d_{k+1}. This is checked by the rules of the set 5. If there does not remain any membrane; i.e., if the weight of the subset associated to this membrane is equal to k an object yes_0 is produced. If not, the computation in this membrane halts.

When the checking stage finish, for each of the 2^n checking membranes we have one of the following cases: The membrane produces yes_1 in the skin or the computation has halted and nothing has been sent to the skin. At this point the output stage starts.

In the skin we have a membrane labelled by c, produced by the counter t_i when it reached $t_{2n+2k+7}$, with a counter no_i inside. If an element yes_1 has been produced, this means that at least in one of the 2^n checking membrane the weight of the associated subset is equal to k. In this case, the object yes_1 goes inside the membrane with label c, it dissolves it (it stops the counter no_i) and finally the object yes is sent to the environment. If not, the counter no_i is not stopped and in the last step, the object no is sent to the environment.

5   Some Formal Details

In the previous section we have presented a uniform family Π of recognizer P systems which solves the Subset Sum problem. For each n and k a P system $\Pi(\langle n, k \rangle)$ is constructed, where n is the number of elements of the initial set and k is the constant to be reached. First of all, observe that the evolution rules of $\Pi(\langle n, k \rangle)$ are defined in a recursive manner from n and k. The necessary resources to construct the P system are polynomially bounded by n and k, therefore a Turing machine can build the P system in polynomial time with respect to n and k. It can also be proved that the family Π solves the Subset Sum problem in the sense of definition 3 in section 3. Recall that the input of $\Pi(<n, k>)$ is given in a unary representation.

Finally, a formal description of the computation let prove that the P system always halts and sends to the environment the object *yes* or *no* in the last step. The number of steps of the P system is $2n + 2k + 11$ if the output is *yes* and $2n + 2k + 12$ if the output is *no*, therefore there exists a linear bound for the number of steps of the computation.

From the above discussion we deduce that Subset Sum belongs to $\mathbf{PMC}_{\mathcal{MC}}$, therefore since this class is closed under polynomial-time reduction and complement we have $\mathbf{NP} \cup \mathbf{co\text{-}NP} \subseteq \mathbf{PMC}_{\mathcal{MC}}$.

Recently, we have proved that this variant of P systems with membrane creation is $\mathbf{PSPACE}$ powerful; this result will be published in a forthcoming paper.

6   Conclusions and Future Work

Membrane Computing is a young branch of Natural Computing which has reached an important success in its short life. In these years many results have been presented related to the computational power of membrane devices, but up to now no implementation in electronic or biochemical media has been carried out. This paper deals with the study of *algorithms* to solve well-known problems and in this sense it is placed between the theoretical results, mainly related to computational completeness and computational efficiency, and the real implementation of the devices. Moreover this paper represents a new step in the study of algorithms in the framework of P systems because it exploits membrane creation (a variant poorly studied) to solve $\mathbf{NP}$-complete problems. The next steps are, on the one hand, a deeper study of the processes inside living cells in order to improve the models and make them closer to Biology and, on the other hand, to go on with theoretical and computational aspects which allow us to improve the designs and algorithms.

Acknowledgement

This work is supported by Ministerio de Ciencia y Tecnología of Spain, by *Plan Nacional de I+D+I (2000–2003)* (TIC2002-04220-C03-01), cofinanced by

FEDER funds, and by a FPI fellowship (of the third author) from the University of Seville.

References

1. Gutiérrez-Naranjo,M.A; Pérez-Jiménez, M.J. ; Riscos-Núñez, A.: Towards a programming language in cellular computing. *Proceedings of the 11th Workshop on Logic, Language, Information and Computation* (WoLLIC'2004), July 19-22, 2004, 1-16 Campus de Univ. Paris 12, Paris, France.
2. Gutiérrez-Naranjo, M.A.; Pérez-Jiménez, M.J.; Riscos-Núñez, A.: A fast P system for finding a balanced 2-partition, *Soft Computing*, in press.
3. Gutiérrez-Naranjo, M.A.; Pérez-Jiménez, M.J.; Romero-Campero, F.J.: Solving SAT with Membrane Creation. Accepted paper for CiE 2005.
4. Gutiérrez-Naranjo, M.A.; Pérez-Jiménez, M.J.; Romero-Campero, F.J.: A linear solution for QSAT with Membrane Creation. Submitted, 2005.
5. Luisi, P.L.: The Chemical Implementation of Autopoiesis, *Self-Production of Supramolecular Structures* (G.R. Fleishaker et al., eds.), Kluwer, Dordrecht, 1994
6. Pérez-Jiménez, M.J.; Riscos-Núñez, A.: Solving the Subset-Sum problem by active membranes, *New Generation Computing*, in press.
7. Pérez-Jiménez, M.J.; Riscos-Núñez, A.: A linear solution for the Knapsack problem using active membranes, in C. Martín-Vide, G. Mauri, Gh. Păun, G. Rozenberg and A. Salomaa (eds.), *Membrane Computing*, Lecture Notes in Computer Science, **2933**, 2004, 250–268.
8. Pérez-Jiménez, M.J.; Romero-Campero, F.J.: Solving the BIN PACKING problem by recognizer P systems with active membranes, *Proceedings of the Second Brainstorming Week on Membrane Computing*, Gh. Păun, A. Riscos, A. Romero and F. Sancho (eds.), Report RGNC 01/04, University of Seville, 2004, 414–430.
9. Pérez-Jiménez, M.J.; Romero-Jiménez, A.; Sancho-Caparrini, F.: A polynomial complexity class in P systems using membrane division, *Proceedings of the 5th Workshop on Descriptional Complexity of Formal Systems, DCFS 2003*, E. Csuhaj-Varjú, C. Kintala, D. Wotschke and Gy. Vaszyl (eds.), 2003, 284-294.
10. Păun, A.; Păun, Gh.: The power of communication: P systems with symport/antiport, *New Generation Computing*, **20**, 3 (2002), 295–305
11. Păun, Gh.: Computing with membranes, *Journal of Computer and System Sciences*, **61**, 1 (2000), 108–143.
12. Păun, Gh.; Pérez-Jiménez, M.J.: Recent computing models inspired from biology: DNA and membrane computing, *Theoria*, **18**, 46 (2003), 72–84.
13. Păun, Gh.: Further Open Problems in Membrane Computing, *Proceedings of the Second Brainstorming Week on Membrane Computing*, Gh. Păun, A. Riscos, A. Romero and F. Sancho (eds.), Report RGNC 01/04, University of Seville, 2004, 354–365.
14. ISI web page `http://esi-topics.com/erf/october2003.html`
15. P systems web page `http://psystems.disco.unimib.it/`

A Study of the Robustness of the EGFR Signalling Cascade Using Continuous Membrane Systems

M.J. Pérez-Jiménez and F.J. Romero-Campero

Research Group on Natural Computing,
Department of Computer Science and Artificial Intelligence,
University of Sevilla,
Avda. Reina Mercedes s/n, 41012, Sevilla, Spain
{marper, fran}@us.es

Abstract. Many approaches to anticancer treatment have had a limited success. A fundamental hurdle to cancer therapy is the robustness of the signalling networks involved in tumourgenesis. The complexity of networks of biological signalling pathways is such that the development of simplifying models is essential in trying to understand the wide-ranging cellular responses they can generate. In this paper a model of the epidermal growth factor receptor signalling cascade is developed using continuous membrane systems. This model is used to study the robustness of this signalling cascade which is known to play a key role in tumour cell proliferation, angiogenesis and metastasis.

Keywords: membrane computing, EGFR signalling network, signal transduction, robustness.

1 Introduction

Membrane Computing is an emergent branch of Natural Computing introduced by Gh. Păun in [9]. Since then it has received important attention from the scientific community. In fact in 2003 the Institute for Scientific Information (ISI) has considered the seminal paper [9] as *fast breaking* and Membrane Computing has been selected as a *fast emerging area* in computer science.

This new non-deterministic model of computation starts from the assumption that the processes taking place in the compartmental structure of a living cell can be interpreted as computations. The devices of this model are called *P systems*. Roughly speaking, a P system consists of a cell-like membrane structure, in the compartments of which one places multisets of objects which evolve according to given rules in a synchronous non-deterministic maximally parallel manner.

Most variants of membrane systems have been proved to be computationally complete, that is equivalent in power to Turing machines, and computationally efficient, that is being able to solve computationally hard problems in polynomial time trading time for space. P systems as a discrete model of computation have

J. Mira and J.R. Álvarez (Eds.): IWINAC 2005, LNCS 3561, pp. 268–278, 2005.

also been used to model biological phenomena (see the volume [1]), and as a continuous model in [7]. A first formalization of non-discrete P system and a way to approximate them was introduced in [2].

In this paper we use a continuous variant of P systems, different from that in [7], to model the epidermal growth factor receptor (EGFR) signalling cascade. Up to now the usual mathematical formalization of biochemical signalling networks has been done using differential equations which are focused on the description of the change in concentration of chemical compounds. Here we use a new formalization of these phenomena in a computational framework which focuses on the compartmental structure (membrane structure) of the cell and on the chemical reactions (rules) that take place in different regions of the cell. Thus this new approach makes possible a topological and modular modelling of intracellular signalling networks. In this framework expansion of an existing model is done by adding new rules (reactions) and, if it is necessary, new membranes to represent new regions (organella) of the cell; therefore the previous model does not need to be changed. Besides modularity and easy extensibility, in favour of our approach we also mention the easy understandability and programmability, features which are not easily achieved in models which use differential equations.

The epidermal growth factor receptor (EGFR) belongs to the tyrosine kinase family of receptors. Binding of the epidermal growth factor (EGF) to the extracellular domain of EGFR induces receptor dimerization and autophosphorylation of intracellular domains. Then a multitude of proteins are recruited starting a complex signalling cascade and the receptor follows a process of internalization and degradation in endosomals. Two principal pathways lead to activation of Ras-GTP by hydrolization of Ras-GDP. One of these pathways depends on the concentration of the Src homology and collagen domain protein (Shc) and the other one is Shc-independent. Ras-GTP acts like a *switch* that stimulates the Mitogen Activated Protein (MAP) kinase cascade by phosphorylating the proteins Raf, MEK and ERK. Subsequently phosphorylated ERK regulates several cellular proteins and nuclear transcription factors. Disregulated EGFR expression, ligand production and signalling have been proved to have a strong association with tumourgenesis. As a result of this, EGFR has been identified as a key biological target for the development of novel anticancer therapies.

The paper is organised as follows. Continuous P systems are introduced in the next section. In section 3 the EGFR signalling cascade is briefly described. We present our model using continuous P systems in section 4. Results and discussion are exposed in section 5. Finally, conclusions and future work are given in the last section.

2   Continuous P Systems

Usual variants of P systems are discrete models of computation where in every step the rules are applied in a maximal way an integer number of times, we refer to [10] for details. Here we use a variant whose systems can evolve in every instant applying a maximal set of rules a positive *real number of times* determined by a certain function $\mathcal{K}$. This variant is inspired by the fact that in vivo chemical reactions evolve in a continuous way following a *rate* that depends on the concentration of the reactants.

Roughly speaking, a continuous P system consists of a membrane structure, a hierarchically arranged set of membranes, where one places multisets of objects that represent the concentration of chemical substances. Usual P systems deal with discrete multisets over an alphabet Σ but here we work with continuous multisets (mappings from Σ to $\mathbf{R}^+$, the set of non-negative real numbers). These multisets evolve according to a finite set of rules that represent chemical reactions.

Next we give a formal definition of continuous P systems. A *continuous P system* is a construct, $\mathbf{\Pi} = (\Sigma, \mu, w_1, \ldots, w_n, \mathcal{R}, \mathcal{K})$, where:

1. $n \geq 1$ is the degree of the system (number of membranes).
2. $\Sigma = \{c_1, \ldots, c_m\}$ is the alphabet of *objects*.
3. μ is a *membrane structure* (a rooted tree) consisting of n membranes (nodes of the tree) labelled with $1, \ldots, n$ (often, we identify the membranes with labels from a finite set H).
4. $w_1, \ldots, w_n$ are continuous multisets associated with each membrane of the membrane structure μ.
5. $\mathcal{R}$ is a finite set of *rules* of the form:

$$u\,[\,v\,]_i \;\rightarrow\; u'[\,v'\,]_i,$$

where $u, v \in \Sigma^*$ represent the reactants, $u', v' \in \Sigma^*$ represent the products, and $i \in H$ is the label of the relevant membrane of the reaction that is modelled.
6. $\mathcal{K}$ is the *rate of application function* which associates with each rule and multiplicity of the objects in μ, a non-negative real number considered as the rate of application of the rule:

$$\mathcal{K} : \mathcal{R} \times \mathcal{M}_{n \times m}(\mathbf{R}^+) \rightarrow \mathbf{R}^+,$$

where $\mathcal{M}_{n \times m}(\mathbf{R}^+)$ is the set of matrices of order $n \times m$ over $\mathbf{R}^+$.

A *configuration* of a continuous P system $\mathbf{\Pi}$ is a matrix of $\mathcal{M}_{n \times m}(\mathbf{R}^+)$ where the object in row i and column j, $a_{i,j}$, represents the multiplicity of the object c_j in the membrane i. We interpret the configurations as assignments of continuous multisets to the membranes of the system, that is, the association of each region with the concentration of chemical substances present in it.

For usual P systems we talk about *computations* but for continuous P systems we prefer to think of *evolutions*. An *evolution* of a continuous P system is a mapping from $\mathbf{R}^+$ to $\mathcal{M}_{n \times m}(\mathbf{R}^+)$. That is, an evolution E associates with each instant $t \in \mathbf{R}^+$ an instantaneous configuration $E(t)$ of the system:

$$E(t) = (a_{i,j}(t))_{\substack{1 \leq i \leq n \\ 1 \leq j \leq m}}$$

For each $t \in \mathbf{R}^+$ and i, $1 \leq i \leq n$, we denote by $v_i(t)$ the continuous multisets over $\Sigma = \{c_1, \ldots, c_m\}$ defined as follows: $(v_i(t))(c_j) = a_{ij}(t)$ for $1 \leq j \leq m$. That is, we can describe $E(t)$ by a tuple $(v_1(t), \ldots, v_n(t))$.

The way a continuous P system, $\mathbf{\Pi} = (\Sigma, \mu, w_1, \ldots, w_n, \mathcal{R}, \mathcal{K})$, evolves is determined by the initial multisets $w_1, \ldots, w_n$ and the rate of application function $\mathcal{K}$. We define the *initial configuration* of $\mathbf{\Pi}$ as the tuple $(w_1, \ldots, w_n)$.

The rules are applied during the evolution of the system in a continuous way according to the *rate of application function* $\mathcal{K}$. At an instant $t \in \mathbf{R}^+$, a rule $r \in \mathcal{R}$ is applied exactly $\mathcal{K}(r, E(t))$ *times* (in this sense, we can say that the rules are applied in a $\mathcal{K}$-maximal way); that is, $\mathcal{K}(r, E(t))$ units of the reactants are consumed and $\mathcal{K}(r, E(t))$ units of the products are produced. Observe that the effect of the rule r decreases the multiplicity (concentration) of its reactants and increases the multiplicity (concentration) of its products. More precisely, we define the effect of a rule r during an interval of time $[t, T]$ as follows:

$$Ef(r, t, T) = \int_t^T \mathcal{K}(r, E(s)) \, ds.$$

More formally, given an object $c_j \in \Sigma$, $1 \leq j \leq m$, and a membrane i, $1 \leq i \leq n$, we denote by $production_i(c_j)$ (resp. $consumption_i(c_j)$) the set of rules where c_j is a product in membrane i (resp. a reactant). Therefore the real number $(v_i(t))(c_j)$, denoted by $|c_j|_i(t)$, is determined by the next formula:

$$|c_j|_i(t) = |c_j|_i(0) + \sum_{r \in production_i(c_j)} \int_0^t \mathcal{K}(r, E(s)) \, ds -$$
$$- \sum_{r \in consumption_i(c_j)} \int_0^t \mathcal{K}(r, E(s)) \, ds,$$

with $v_i(0) = w_i$, that is, $|c_j|_i(0) = w_i(c_j))$.

In computers, real numbers are represented by a finite set of rational numbers. Therefore, like in most continuous models we need to develop approximations in order to simulate evolutions of continuous P systems in computers.

As shown above, in order to determine the effect of a rule on the evolution of a system during an interval of time $[t, T]$ we only need to compute an integral of the rate of application function $\mathcal{K}$. Hence, in order to approximate the evolution of a continuous P systems in a finite set of instants $t_0, \cdots, t_q$ we can use any suitable known numerical method to approximate integrals. Here for simplicity we use the rectangle rule; that is, we suppose $t_{l+1} - t_l = p$ is *small enough* to assume that $\mathcal{K}$ remains constant and equal to $\mathcal{K}(r, E(t_l))$ in the interval $[t_l, t_{l+1}]$ for $l = 0, \ldots, q - 1$. With this assumption we can approximate the effect of a rule during an interval of time of length p by $Ef(r, t_l, t_{l+1}) \approx p\mathcal{K}(r, E(t_l))$.

By doing this approximation we reach an usual P systems that performs q steps $(t_0, \ldots, t_q)$ and in each steps the rules are applied $p\mathcal{K}(r, E(t_l))$ times. Therefore, we have approximated the evolution of a continuous P system by the computation of an usual discrete P system working in a $p\mathcal{K}$ bounded parallel manner.

3   EGFR Signalling Cascade

In this section we will describe briefly the EGFR signalling cascade following the network depicted below. During the signal transduction which takes place in this cascade, the information about the concentration of the EGF in the outside of the cell is translated into kinetic information inside the cell by EGFR phosphorylation.

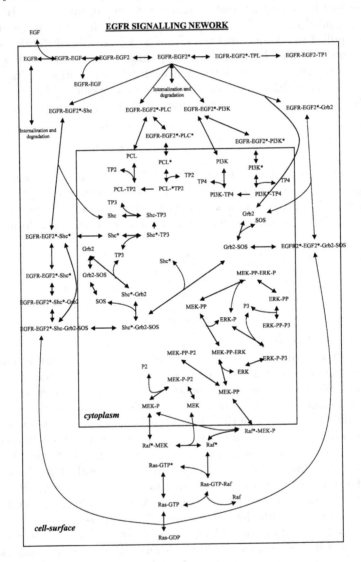

The epidermal growth factor receptor (EGFR) belongs to the tyrosine kinase family of receptors. The binding of the epidermal growth factor (EGF) to the extracellular domain of EGFR induces receptor dimerization and autophosphorylation of intracellular domains. Then, on the one hand, a multitude of proteins are recruited starting a complex signalling cascade and, on the other hand, the receptor follows a process of internalization, ubiquitination and degradation.

In our model we consider two marginal pathways and two principal pathways starting from the phosphorylated receptor.

In the first marginal pathway phospholipase C-γ (PLC$_\gamma$) binds to the phospholyrated receptor, then it is phosphorylated (PLC$_\gamma^*$) and released into the cytoplasm where it can be translocated to the cell membrane or desphosphorylated. In the second marginal pathway the protein PI3K binds to the phospholyrated receptor, then it is phosphorylated (PI3K*) and released into the cytoplasm where it regulates several proteins that we do not include in our model.

Both principal pathways lead to activation of Ras-GTP. The first pathway does not depend on the concentration of the Src homology and collagen domain protein (Shc). This pathway consist of a cycle where the proteins growth factor receptor-binding protein 2 (Grb2) and Son of Sevenless homolog protein (SOS) bind to the phosphorylated receptor. Later the complex Grb2-SOS is released in the cytoplasm where it dissociates into Grb2 and SOS.

In the other main pathway Shc plays a key role, it binds to the receptor and it is phosphorylated. Then either Shc* is released in the cytoplasm or the proteins Grb2 and SOS binds to the receptor yielding a four protein complex (EGFR-EGF2*-Shc*-Grb2-SOS). Subsequently this complex dissociates into the complexes Shc*-Grb2-SOS, Shc*-Grb2 and Grb2-SOS which in turn can also dissociate to produce the proteins Shc*, Grb2 and SOS.

Finally, Ras-GTP is activated by these two pathways and in turn it stimulates the Mitogen Activated Protein (MAP) kinase cascade by phosphorylating the proteins Raf, MEK and ERK. Subsequently phosphorylated ERK regulates several cellular proteins and nuclear transcription factors that we do not include in our model.

There exist *cross-talks* between different parts and cycles of the signalling cascade which suggest a strong robustness of the system.

For a more detailed description of the cascade see the literature listed in the bibliography.

4   Modelling EGFR Signalling Cascade by Continuous P Systems

We have developed a model of the signalling cascade described in the previous section using a continuous P system, $\Pi_{EGF} = (\Sigma, \mu, w_e, w_s, w_c, \mathcal{R}, \mathcal{K})$. Our model consists of more that 60 proteins and complexes of proteins and 160 chemical reactions. *Supplementary information and details* about the model are available on the web page www.gcn.us.es/egfr.pdf.

• **Alphabet:** In the alphabet Σ we collect all the proteins and complexes of proteins that take part in the signalling cascade. In table 1 of the *Supplementary* *information* all the objects of the alphabet and the chemical compounds that they represent are listed.

- **Membrane Structure:** In the EGFR signalling cascade described in the previous section, there are three relevant regions, namely the *environment*, the *cell surface* and the *cytoplasm*. We represent them in the membrane structure as the membranes labelled with: e for the environment, s for the cell surface and c for the cytoplasm.

- **Initial Multisets:** In the initial multisets we represent the initial concentrations of the chemical substances in the environment, the cell surface and the cytoplasm. These concentration has been obtained from the references listed in the bibliography. A detailed presentation of initial multisets is shown in table 2 of the *Supplementary information*.

- **Rules and Rate of application function:** In the rules we model the chemical reactions described which form the signalling cascade. To model the reactions we use the *Law of Mass Action* which states that the rate of a reaction is proportional to the product of the concentrations of the reactants. That is, if we have a reaction of the form:

$$r_1 + \cdots + r_k \rightarrow p_1 + \cdots + p_{k'},$$

then the rate of this reaction is $k|r_1| \cdots |r_n|$, where k is called *kinetic constant*.

In tables 3-11 of the *Supplementary information* (`www.gcn.us.es/egfr.pdf`) all the rules are listed as well as the kinetic constants and the references from where they were taken. As an example of the procedure we have followed to develop our model, we next present the derivation of one of the 160 rules.

Let us consider the binding of EGF to EGFR:

$$\text{EGF EGFR} \rightarrow \text{EGF-EGFR}$$

We know from biological experiments that EGF, which is present in the environment, binds to EGFR, which is present in the cell surface at a rate of $0.003\,nM^{-1}s^{-1}$. According to this, the relevant membrane in this reaction is the cell-surface because it separates the two regions involved in this reaction. Besides following the Mass Action Law the reaction takes place at a velocity of $0.003|EGF||EGFR|$. Therefore, in our model we represent this chemical reaction by the following rule and rate of application:

$$\text{EGF } [\text{EGFR}]_s \rightarrow [\text{EGF-EGFR}]_s \quad \mathcal{K}(r, E(t)) = 0.003|EGF(t)|_e|EGFR(t)|_s$$

5   Results and Discussion

The model presented in the previous section has been implementing using CLIPS, a productive development and delivery expert system tool which provides a complete environment for the construction of rule and/or object based expert systems.

To implement our model we have approximated the evolutions of the continuous P system Π_{EGF} by computations of an usual P system working in a bounded parallel manner. The parameter p, chosen for the approximation, was fixed to 10^{-3} after testing different values until the results obtained did not change.

We study the effect of different EGF concentrations on the signalling cascade. To illustrate this effect we depict the evolution of the concentration of the most relevant proteins in the signalling cascade over time.

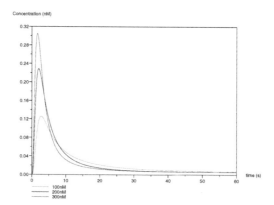

Above it can be seen that the receptor activation by autophophorylation is clearly concentration dependent showing a high peak in the first 5 seconds to decay rapidly afterwards to very low levels of concentration. According to the variance in the receptor activation it is intuitive to expect different cell responses to different EGF concentrations. Here we will show that this is not the case. Next we show the evolution of the phosphorylation of Shc after binding to the receptor over 60 seconds.

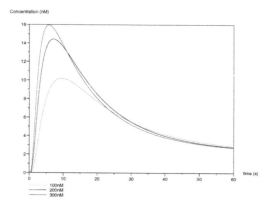

It can be observed that the responses to EGF stimulation get more sustained as we get deeper in the cascade and that the network has almost managed to attenuate the overstimulation with EGF. Observe that the red and blue line are almost identical.

The activation of the Ras protein is a key node in the EGFR signalling cascade. We depict its evolution over 180 seconds in the next graphic.

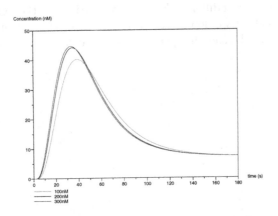

Note that at this point the response to the overstimulation with EGF has been completely attenuated and that an amplification of the response to the low concentration has been performed.

Finally, the goal of the cascade is the activation of the mitogenic kinases MEK and ERK. These proteins regulates the transcription of several proto-oncogenes like c-fos and c-Jun. Next the evolution of phosphorylated MEK over 180 second is presented.

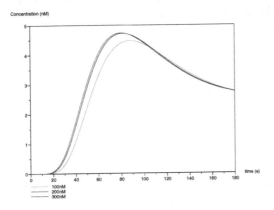

In this picture it is shown the surprising robustness of the signalling cascade. The signals from outside due to EGF concentration have been either attenuated or amplified to get the same concentration of the most relevant kinases. Observe that after 100 seconds, when the response gets sustained, the three line representing the response to different external EGF concentrations are identical. For more graphics of the evolution of the different proteins in the dynamics of the cascade see the *Supplemtary information* on the web page `www.gcn.us.es/egfr.pdf`.

Currently, the robustness of the EGFR signalling cascade is proposed to be a product of receptor internalization and cross-talk between different pathways in the cascade. According to the literature listed in the bibliography receptor internalization produces a signal attenuation by protection from high external EGF concentration, meanwhile an amplification of the signal due to low EGF concentrations is performed in several nodes of the cascade where there exists a cross-talk between different pathways of the cascade. Our model is in accordance with this hypothesis. This outcome shows the reliability of our model to make post-diction and supports the possibility of using our model to produce new hypotheses and predictions about the behaviour of this relevant network in the cell cycle and tumourgenesis.

6   Conclusions and Future Work

In this paper we have developed a topological and modular model of the EGFR signalling cascade which consists in more than 60 proteins and complexes of proteins and 160 chemical reactions. Our model can provide insight into the dynamics of the MAP kinase cascade which is activated by EGFR autophosphorylation. It can be also useful to formulate hypotheses that can be tested experimentally. Actually, our model suggests that the the cascade is robust to variation in the EGF concentration; therefore in the next future we intend to study the influence of kinase inhibition at different cytoplasmic nodes of the signalling cascade.

The results obtained using this model are in well agreement with experimental data. This shows that continuous membrane systems are a reliable framework for modelling networks of biochemical signalling cascades.

Currently our model is being translated into *SBML* (System Biology Markup Language) a computer-readable format like XML for representing models of biochemical reaction networks. Moreover an user friendly interface for the CLIPS implementation is being designed using JAVA. We hope that these two current works will help to spread our model in the scientific community and so the authors, who are not biologists, can get some feedback from specialists in networks of biochemical signalling cascades.

Finally, this model takes a first step (PI3K phosphorylation) towards the MDM2-p53 feedback loop; we intend to expand our model to comprise this interaction between proteins which it is known to play a key role in the regulation of cell cycle and tumourgenesis.

Acknowledgement

This work is supported by Ministerio de Ciencia y Tecnología of Spain, by *Plan Nacional de I+D+I (2000–2003)* (TIC2002-04220-C03-01), cofinanced by FEDER funds, and by a FPI fellowship from the University of Seville.

References

1. Ciobanu, G.; Păun, Gh.; Pérez-Jiménez, M.J., eds: *Applications of Membrane Computing*, Springer-Verlag (2005) in press.
2. Cordón-Franco, A.; Sancho-Caparrini, F.: Approximating Non-Discrete P Systems. In G. Mauri, Gh. Păun, M.J. Pérez-Jiménez, Gr. Rozenberg, A. Salomaa (eds.) *Membrane Computing, International Workshop, WMC5, Milan, Italy, June 2005, Revised Papers*. Lecture Notes in Computer Science, Springer, **3365**, in press.
3. El-Masri, H.A.; Portier, C.J.: Replication Potential of Cells via the Protein Kinase C-MAPK pathway: Application of a Mathematical Model, *Bulletin of Mathematical Biology* **61** (1999) 379–398.
4. Kholodenko, B.G. et al.: Quantification of Short Term Signalling by the Epidermal Growth Factor Receptor. *J. Biol. Chem.*, **274** (1999) 30169–30181.
5. Miller, J.H.; Zheng, F.: Large-scale Simulations of Cellular Signaling Processes, *Parallel Computing* **30** (2004) 1137–1149.
6. Moehren G. et al: Temperature Dependence of the Epidermal Growth Factor Receptor Signaling Network Can Be Accounted for by a Kinetic Model, *Biochemistry* **41** (2002) 306–320.
7. Nishida, T.Y.: Simulations of Photosynthesis by a K-Subset Transforming System with Membranes, *Fundamenta Informaticae*, **49**, 1 (2002), 101–113.
8. Păun, A.; Păun, Gh.: The Power of Communication: P Systems with Symport/Antiport, *New Generation Computing*, **20**, 3 (2002), 295–305.
9. Păun, Gh.: Computing with Membranes, *Journal of Computer and System Sciences*, **61**(1) (2000) 108 – 143.
10. Păun, Gh.: *Membrane Computing. An Introduction*, Springer-Verlag Berlin, 2002.
11. Pérez-Jiménez, M.J.; Romero-Campero, F.J.: Modelling EGFR Signalling Network using Continuous Membrane Systems, Accepted paper in the Workshop on Computational Methods in Systems Biology 2005
12. Plank, M.J.; Sleeman, B.D.: Tumour-Induced Angiogenesis: A Review, *Journal of Theoretical Medicine*, in press.
13. Schoeberl, B. et al.: Computatinal Modeling of the Dynamics of the MAP Kinase Cascade Activated by Surface and Internalized EGF Receptors, *Nature Biotech.* **20** (2002) 370–375.
14. Sermon, B.A. et al: The Importance of Two Conserved Arginine Residues for Catalysis by the Ras GTPas-activating Protein, Neurofibromin, *The Journal of Biological Chemistry* **273** (1998) 9480–9485.
15. Starbuck, C.; Lauffenburger, D.A.: Mathematical Model for the Effects of Epidermal Growth Factor Receptor Trafficking Dynamics on Fibroblast Proliferation Responses, *Biotechnol. Prog.* **8** (1992) 132–143.
16. Sydor, J.R. et al: Transient Kinetic Studies on the Interaction of Ras and the Ras-Binding Domain of c-Raf-1 Reveal Rapid Equilibration of the Complex, *Biochemistry* **37** (1998) 14292–14299.
17. Wiley, H.S.: Anomalous Binding of Epidermal Growth Factor to A431 Cells is Due to the Effect of High Receptor Densities and a Saturable Endocytic System, *The Journal of Cell Biology* **107** (1988) 801–810.
18. ISI web page `http://esi-topics.com/erf/october2003.html`
19. P systems web page `http://psystems.disco.unimib.it/`

A Tool for Implementing and Exploring SBM Models: Universal 1D Invertible Cellular Automata

Joaquín Cerdá, Rafael Gadea, Jorge Daniel Martínez, and Angel Sebastiá

Group of Digital Systems Design, Dept. Of Electronic Engineering,
Polithecnic University of Valencia, 46022 Valencia, Spain
{joacerbo, rgadea, jormarp1, asebasti}@eln.upv.es
http://dsd.upv.es

Abstract. The easiest form of designing Cellular Automata rules with features such as invertibility or particle conserving is to rely on a partitioning scheme, the most important of which is the 2D Margolus neighborhood. In this paper we introduce a 1D Margolus-like neighborhood that gives support to a complete set of Cellular Automata models. We present a set of models called Sliding Ball Models based on this neighborhood and capable of universal computation. We show the way of designing logic gates with these models, propose a digital structure to implement them and finally we present SBMTool, a software development system capable of working with the new models.

1 Introduction

Cellular Automata (CA) model massively parallel computation and physical phenomena [1]. They consist of a lattice of discrete identical sites called *cells*, each one taking a value from a finite set, usually a binary set. The values of the cells evolve in discrete time steps according to deterministic rules that specify the value of each site in terms of the values of the neighboring sites. This is a parallel, synchronous, local and uniform process [1, 2].

CA are used as computing and modeling tools in biological organization, self-reproduction, image processing and chemical reactions. They have proved themselves to be useful tools for modeling physical systems such as gases, fluid dynamics, excitable media, magnetic domains and diffusion limited aggregation [3, 4, 5]. CA have been also applied in VLSI design in areas such as generation of pseudorandom sequences and their use in built-in self test (BIST), error-correcting codes, private-key cryptosystem, design of associative memory and testing finite state machines [6, 7, 8].

1.1 Invertible Cellular Automata

A CA is invertible when its global function is a bijection, i.e., if every configuration – which, by definition, has exactly one successor – also has exactly one

J. Mira and J.R. Álvarez (Eds.): IWINAC 2005, LNCS 3561, pp. 279–289, 2005.

predecessor [9]. In the context of dynamical systems, invertibility coincides with microscopic reversibility. One of the most common ways of constructing invertible cellular automata is by using a partitioning schema. Partitioning Cellular Automata (PCA) are based on a different kind of local map that takes as input the contents of a region and produces as output the new state of the whole region (rather than of a single cell). So, the state space is completely divided into non-overlapping regions. Clearly, information cannot cross a partition boundary in a single time step. In order to exchange information between regions, partitions must change at the next step.

The partitioning format is specially good for many applications because it makes very easy to construct invertible rules. If the local map is invertible then the corresponding global map is also invertible. Moreover, it is possible to construct rules which conserve "particles", "momentum" and other desired quantities by only impose some constraints to the local map.

1.2   Margolus Neighborhood

The most important partitioning scheme is the Margolus Neighborhood, introduced in [10]. In this neighborhood each partition is 2x2 cells as shown in figure 1. We alternate between even partitions (solid lines) and odd partitions (dotted line) in order to couple them all. Periodic boundary conditions are assumed.

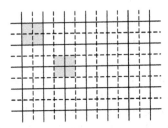

Fig. 1. Margolus Neighborhood: even (*solid lines*) and odd (*dotted lines*) partitions of a two-dimensional array into 2x2 blocks. One block in each partition is shaded

In previous works we extensively studied Margolus neighborhood from a theoretical an a practical point of view, paying attention to the possibilities of designing rules with definite properties [11, 12].

1.3   Billiard Ball Model

There are a lot of evolution rules based on Margolus neighborhood. Among them we must mention the BBMCA rule (*Billiard-Ball Model Cellular Automata*), introduced by Margolus itself in [10]. This rule traduces a computational model from E. Fredkin called *Billiard-Ball Model* [13]. This Cellular Automata has the characteristic of computational universality. Recently, Margolus has presented a modified version of this model, called *Soft Sphere Model*[14].

2 Sliding Ball Models

It is easy to generalize Margolus neighborhood to more than 2D and arbitrary block sizes. It is even possible to associate several layers to make it more complex. In our study we present a 1D version of Margolus neighborhood. We dispose the cells in a linear array. Partitions are 4-cell blocks. We alternate partitions between even and odd steps, as we show in figure 2. Alternating partitions we allow information to move between them. Periodic boundary conditions are assumed.

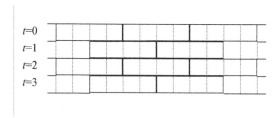

Fig. 2. 1D version of the Margolus neighborhood. The partitions alternate between even and odd steps. Solid lines delimit a partition

To completely specify a rule using this neighborhood it is necessary to map all possible combinations of four cells. This allows a total number of rules $n_R = 2^{64} \approx 1,8*10^{19}$. From this incredibly large number of possibilities, only $n_{RI} = 16! \approx 2,1*10^{13}$ are invertible, and less ($n_{RCP} = 4!*6!*4! = 414720$) are particle conserving.

1D Margolus neighborhood gives support to a family of Cellular Automata Models based on considerations similar to BBMAC models. These models also have the important property of computational universality. We call them *Sliding Ball Models*, SBM.

These models are based on the idea of reflecting the behaviour of a system of moving particles in a 1D region. Particles move in uniform, constant unit speed in one of its two possible senses (left and right). The situation is similar to that shown on figure 3.

To translate this model in terms of Cellular Automata we allow in each partition the simultaneous existence of up to two particles travelling in each sense. Each partition has four cells, so two cells (A and B) indicate right sense

Fig. 3. Sliding Ball Models

and the remaining two cells (C and D) indicate left sense. With this convention, a particle placed on cell A will move to cell C in the next sep, and a particle on cell C will move to cell A. The same states for cells B and D, as shown in figure 4.

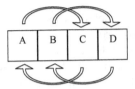

Fig. 4. Senses on movement in SBM: In the next step a particle placed on one cell will move to the cell shown by the arrow

Among all the possible evolution rules that can be designed in 1D Margolus neighborhood we must impose three constraints to the SBM models: first of all, rule must be invertible, which implies designing an invertible local map. Secondly, rule must preserve the number of particles. Finally, when there is just a particle in the partition, the rule must conserve its liner momentum, leading to a primitive Principle of Inertia. The way in which the various particles in a partition interact will define the specific properties of the model that we consider. SBM models were first introduced in [15].

2.1   No Interaction Model

The simplest model that we can design is called "No Interaction Model", and it is shown in figure 5. In this model we establish that the evolution when two or more particles collide into a partition takes place just as a superposition of the movement of every single particle. This leads to no possible interaction and no possibility of giving support to logic. This model is interesting from a theoretical point of view but it lacks of practical application.

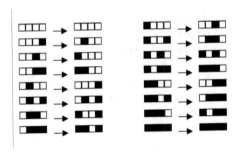

Fig. 5. No Interaction Model

2.2   Static Structures Model

If we want to obtain models capable of perform computation it is necessary to rely on the process of collision between particles, to give support to logic functions. This can be achieved by modifying the evolution rule. In this case we establish a mechanical analogy in which particles are represented by spheres of a given radius. When two particles traveling in opposite directions collide they interchange its velocities. If the radii of the spheres are not zero, the trajectory followed by the particles is different from that corresponding to a non collision situation, as it is shown in figure 6 (dotted line indicates the trajectory followed if there is not a collision).

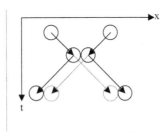

Fig. 6. Elastic collision of rigid spheres (dotted line indicates the trajectory followed if there is not a collision)

To express all this in terms of Cellular Automata we just have to slightly modify the evolution rule, as it is shown in figure 7, thus modifying the cases in which there is a collision.

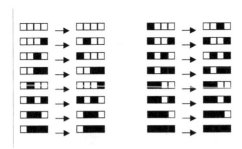

Fig. 7. Static Structures Model

With these interactions it is possible for us to design logic gates. For instance, an implementation of the Interaction Gate is given in figure 8.

The Interaction Gate, introduced in [13], shows itself to be computational universal. Information is placed as follows: we have two logia variables p and

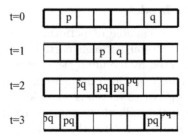

Fig. 8. Implementation of the Interaction Gate in the Static Structures Model

q, each one into a different but adjacent partition. p is codified as the B cell of the left partition, while q corresponds to the C cell of the right partition. All other cells must be at '0'. Performing the evolution of the automata leads to a distribution of the information like that shown in figure 8. As a demonstration we show the evolution for all tour combinations of p and q in figure 9.

Fig. 9. Every possible combination of p and q variables in the Interaction Gate in the Static Structures Model

This rule includes the possibility of defining static structures, which gives its name to the model. When in a determined partition there are two particles moving in the same sense, we have decided that they form structures that do not change with time. The possibility of implementing logic with these static structures depends on what happens when a free particle collides with the structure. In this case we choose that the particle inverts its velocity, just as if it bounces. The result can be seen in figure 10. These static structures are capable of retaining information over a definite time, playing the role of a memory. A possible implementation of a memory can be seen in figure 11.

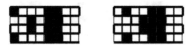

Fig. 10. A particle bouncing on a static structure

2.3   Impulse Conservation Model

The existence of static structures in which free particles bounce implies the loose of momentum conservation. When a particle of mass m and speed v bounds,

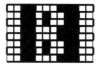

Fig. 11. Implementation of a memory cell in the Static Structures Model

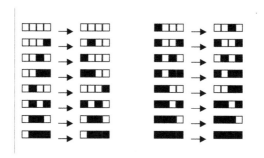

Fig. 12. Impulse Conservation Model

there is an increment of momentum of $2mv$. The rule presented in the previous paragraph presents this effect. It is possible for us to design a new rule with the condition of momentum conservation, but this constraint lead us to forget the possibility of having static structures. A new rule to do that is shown on figure 12. The implementation of Interaction Gate shown in the previous paragraph is still correct with this model. The Impulse Conservation Model is especially valid to perform physical simulations.

3   Implementation

As we show in [15]there is a very simple digital structure that can implement the SBM models. Each cell of the array can be represented by a set of two registers and two logic functions like the ones showed in figure 13.

For each cell in the matrix we need to define two different functions: the even one and the odd one. A multiplexer selects between results on even or odd branches depending on the present cycle. Also, to easily supply initial data to the circuit, we have included a second multiplexer for data synchronous load. Finally, an Enable terminal has been added to hold and start the evolution.

If we fix the proposed architecture for all the cells in the matrix, the only difference between classes is the way the neighbor cells are connected to the inputs of the even/odd functions. This digital structure has been implemented over several programmable devices such as FPGAs, and its performances have shown to be very good. Results are given in [15].

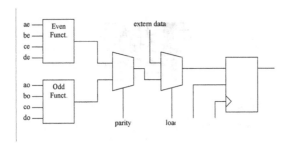

Fig. 13. Digital structure for implementing a cell. Each one needs two different functions: the even one and the odd one

4   SBMTool

Working with SBM models is sometimes a tedious process because it implies making a lot of calculations. In this way, it is useful to develop a software tool that simplifies our work. To accomplish this we have developed SBMTool, a program that simplifies the process of designing new rules and applying them to a definite structure to obtain new logic gates and logic resources.

A general view of this program can be seen in figure 14.

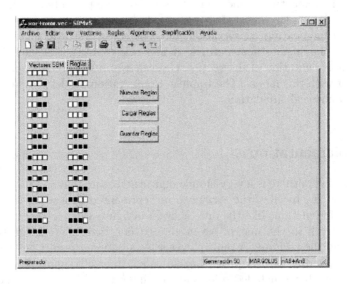

Fig. 14. The program SBMTool: definition of the rule

First of all, the program requires the definition of the rule to work with. This definition is performed graphically, in the way that can be seen in figure 13. The program displays all possible combinations in the 1D Margolus neighborhood, and the user must map this input combination over an output combination. This way, the user can easily design rules that are invertible or particle conserving. To simplify even more this process, the program also permits to store and read data from previous designed rules in a proper format called .rul.

Once the rule has been defined it is possible to perform the evolution from any initial condition. This condition consists on a integer number of 4-cell blocks that are assumed to be connected in a cyclic way, leading to periodic boundary conditions, the most common in SBM models. First of all, the user must select how many 4-cell blocks must be included in the initial condition array, and when this has been done it is possible to define the particular state of every cell in the array. By defect every cell is al '0' state, but can be set to '1' state or include a logic variable (named sequentially as 'A', 'B', 'C' and so on). When the initial condition array has been completely defined, it can be stored in a proper format called .vec that can be stored on disk to be read in the future.

After that the user can run the rule over this initial condition to see how the information is modified as the system evolves. A typical evolution can be seen in figure 15, in which we show an initial condition composed of two 4-cell blocks: the first one (left) contains two '0' cells, one '1' cell and a logic variable (called A). The second block (right) contains two '1' cells, one '0' cell and the second variable (called B). As the system evolves information moves between the cells interacting with itself to form several logic functions, as shown in figure.

Fig. 15. The program SBMTool: Evolving a rule from a definite initial condition

The way the logic function is evaluated is very simple: as soon as a new variable is introduced on the array, the program duplicates internally the number of arrays created, each one for a definite combination of the inputs. So, if 2 variables are set on the array the program creates 4 internal arrays (2^2). In general, for an array of n variables 2^n internal copies are needed. Each evolution of the general array implies $2n$ internal evolutions, one for each combination of the inputs. The logic function implemented on a cell in a concrete step is evaluated by logic simplification of the cases in which this cell is set to '1' for any possibility. Logic simplification is performed using very well-known procedures as Karnaugh Tables or Quine-McCluskey Method.

The program allows defining the appearance of the results to make this information visible, as well as showing the logic function that appears in every cell of the array for any future step.

5   Conclusions

We have introduced the SBM CA models of computation. To do this we have needed to use a 1D partitioning neighborhood that is a variation of the 2D Margolus neighborhood. We have studied the foundations of the models and the way of designing local maps with some definite properties. After that we have shown three different models interesting from a theoretical and a practical point of view: the No Interaction Model, the Static Structures Model, and the Impulse conservation Model. We have show how to implement a logic gate, like the Interaction Gate, in some of these models. After that, we have presented a digital structure that allows us to implement any SBM model. Finally, we have introduced SBMTool, a software tool that allows the user to design any SBM rule and explore its properties in a simple and powerful way.

References

1. Wolfram, S.: A New Kind of Science. Wolfram media (2002)
2. Wolfram, S.: Statistical mechanics of cellular automata. Reviews of Modern Physics, 55 (1983) 601–644
3. Chopard, B., Droz, M..: Cellular Automata Modelling of Physical Systems. Cambridge University Press (1998)
4. Popovici, A. , Popovici, D.: Cellular Automata in Image Processing. Proceedings of MTNS 2002 (2002)
5. Toffoli, T.: Cellular Automata as an alternative to (rather than an approximation of) Differential Equations in Modeling Physics. Physica 10D (1984) 117-127
6. Wolfram, S.: Cryptography with Cellular Automata. Advances in Cryptology: Crypto '85 Proceedings, Lecture Notes in Computer Science, 218 (Springer-Verlag, 1986) 429–432
7. Sarkar, P.: A brief history of cellular automata. ACM Computing Surveys, Vol. 32, Issue 1 (2000) 80–107
8. Shackleford, B., Tanaka, M., Carter, R.J., Snider, G.: FPGA Implementation of Neighborhood-of-Four Cellular Automata Random Number Generators. Proceedings of FPGA 2002 (2002) 106–112
9. Toffoli, T., Margolus, N.: Invertible cellular automata: a review. Physica D, Nonlinear Phenomena, 45 (1990) 1–3
10. Margolus, N.: Physics-Like models of computation. Physica 10D (1984) 81-95
11. Cerdá, J., Gadea, R. Payá, G.: Implementing a Margolus neighborhood Cellular Automata on a FPGA. Artificial Nets Problem solving Methods. Lecture Notes in Computer Science 2687 (2003) 121-128

12. Cerdá, J., Gadea, R. Herrero, V. Sebastiá, A.: On the Implementation of a Margolus neighborhood Cellular Automata on FPGA. Field-Programmable Logic and Applications. Lecture Notes in Computer Science 2778 (2003) 776-785
13. Fredkin, E., Toffoli, T.: Conservative Logic. International journal of Theoretical Physics 21 (1982) 219-253
14. Margolus, N.: Universal CA's based on the collisions of soft spheres. In Adamatzky ed. Collision Based Computation. Springer (2002)
15. Cerdá, J.: Arquitecturas VLSI de Autómatas Celulares para modelado físico. PhD Thesis. Universidad Politécnica de Valencia (2004)

Network of Evolutionary Processors with Splicing Rules

Ashish Choudhary and Kamala Krithivasan

Dept of Computer Science and Engineering,
Indian Institute of Technology Madras,
Chennai, India 600036
ashish@meenakshi.cs.iitm.ernet.in
kamala@iitm.ernet.in

Abstract. In this paper we consider networks of evolutionary processors with splicing rules (NEPS) as language generating and computational devices. Such a network consists of several processors placed on the nodes of a virtual graph and are able to perform splicing (which is a biologically motivated operation) on the words present in that node, according to the splicing rules present there. Each node is associated with an input and output filter. When the filters are regular languages one gets the computational power of Turing machines with networks of size two. We also show how these networks can be used to solve $NP-$complete problems in linear time.

1 Introduction

The current trend in computer science is to have the computation done in a distributed and/or parallel manner. Formal language and automata theory has always tried to give suitable models which will represent the real life situations. Grammar systems[2], distributed automata[9], membrane systems[10], splicing systems[11] are some of the models defined to capture distribution and parallelism.

To have a model which is more similar to Connection Machine[8] and Logic Flow paradigm[5], in[1] Networks of Evolutionary processors were defined. Networks of language processors[4] are closely related to parallel communicating grammar systems[2], but here the main idea is to place a language generating device in any node of an underlying graph. In each node we have some strings and some rewriting rules, according to which the strings are rewritten. Then the strings are communicated to the neighboring nodes using filters for which generally regular sets are used.

The above idea also has its motivation from cell biology. Each processor placed in a node is a very simple processor performing simple operations. In[1], as motivation from cell biology, it is mentioned that each node may be viewed as a cell having genetic information encoded in DNA sequence which may evolve by local evolutionary events, that is point mutations. But in[1], they are mainly

J. Mira and J.R. Álvarez (Eds.): IWINAC 2005, LNCS 3561, pp. 290–299, 2005.

looked as language processors, making use of Chomsky type grammars. Several NP-complete problems could be solved in polynomial time by such network of processors. The generative power is equal to that of type 0 grammars[1].

With the reference to motivation from cell biology, it looks more appropriate to consider splicing rules in each processor rather than rewriting rules. In this paper, we consider network of evolutionary processors as defined in[1] but where the rules applied in a processor are splicing rules. We have a processor at each node of an underlying graph. Each node has a set of strings to begin with and a set of splicing rules. Evolution takes place using splicing rules and regular sets are used as filters to transfer strings from one node to the neighbor. Each node has an input filter and an output filter. A string can enter a node only if it belongs to the input filter and can leave the node only if it belongs to the output filter. One of the nodes is designated as the output node where we collect the output strings.

We can look at such a network of evolutionary processors as a language generating device, if we look at the strings collected in the output node. We can also look at them as doing some computation. It should be noted that this model is similar to test tube distributed systems based on splicing [3] with slight difference. In test tube systems, we keep on using the splicing rules on the strings till no further splicing rules can be applied, after that only the strings are communicated. But here we have alternate evolutionary and communication step.

We show how these networks can solve NP-complete problems in linear time. We also show that all recursively enumerable languages can be generated using these networks having just two processors. The rest of the paper is organized as follows: In section two, we give some preliminary definitions and the main definition of our paper. In section three, we show the computational completeness of these networks by showing that any recursively enumerable language can be generated by using network of evolutionary processors with splicing rules. In section four, we show how to solve some NP-complete problems related to graphs using these networks. In section five, we informally explain how to solve Exact cover by 3 sets problem which is a NP-complete problem using this model. The paper ends with a brief conclusion.

2 Preliminaries

The set of all strings over V is denoted by V^* while $alph(x)$ denotes the minimal alphabet W such that $x \in W^*$. Consider an alphabet V and two special symbols, $\#, \$$, not in V. A *splicing rule* (over V) is a string of the form

$$r = u_1 \# u_2 \$ u_3 \# u_4,$$

where $u_1, u_2, u_3, u_4 \in V^*$. For a splicing rule $r = u_1 \# u_2 \$ u_3 \# u_4$ and strings $x, y, z \in V^*$ we write

$$(x, y) \vdash_r z \text{ iff } x = x_1 u_1 u_2 x_2,$$

$$y = y_1 u_3 u_4 y_2,$$

$$z = x_1 u_1 u_4 y_2,$$

$$\text{for some } x_1, x_2, y_1, y_2 \in V^*.$$

A *network of evolutionary processors with splicing rules* (NEPS for short) of size n is a construct

$$\Gamma = (V, N_1, N_2, , \ldots, N_n, G),$$

where V is an alphabet and for each $1 \leq i \leq n$, $N_i = (M_i, A_i, PI_i, PO_i)$ is the i−th evolutionary node processor of the network. The parameters of every processor are:

- M_i is a set of splicing rules which may be finite or infinite. Each rule in M_i is a splicing rule of the form $V^* \# V^* \$ V^* \# V^*$ where $\{\#, \$\} \notin V$.
- A_i is a finite set of strings over V. The set A_i is the set of initial strings in the i−th node. We consider that each string appearing in any node at any step has an arbitrarily large number of copies in that node.
- PI_i and PO_i are subsets of V^* representing the input and output filter respectively. These filters are defined by membership condition, namely a string $w \in V^*$ can pass the input filter (the output filter) if $w \in PI_i$ ($w \in PO_i$).

In the theory of networks some types of underlying graphs are common, for example, rings, stars, grids, etc. Thus, a NEPFS is said to be a *star, ring or complete* NEPFS if its underlying graph is a star, ring or complete graph respectively.

By a configuration (state) of a NEPFS as above we mean an n−tuple $C = (L_1, L_2, \ldots L_n)$, with $L_i \subseteq V^*$ for all $1 \leq i \leq n$. A configuration represents the sets of strings (each string appears in an arbitrarily large number of copies) which are present in any node at a given moment. The initial configuration of the network is $C_0 = (A_1, A_2, \ldots, A_n)$. A configuration can either change by an evolutionary step or by a communicating step. When changing by an evolutionary step, each component L_i of the configuration is changed in accordance with the splicing rules associated with the node i.

Formally we say that the configuration $C_1 = (L_1, L_2, \ldots, L_n)$ directly changes into the configuration $C_2 = (L'_1, L'_2, \ldots, L'_n)$ by a evolutionary step, written as $C_1 \Rightarrow C_2$ if L'_i is the set of strings obtained by applying the splicing rules of N_i to the strings in N_i. Since, we are assuming that each string has got multiple copies in a node, so after an evolutionary step in each node one gets an arbitrarily large number of copies of any string which can be obtained by using all possible splicing rules associated with that node.

When changing by communication step, each node processor N_i sends all copies of the strings it has which are able to pass its output filter to all the node processors connected to N_i and receives all copies of the strings sent by any node processor connected with N_i provided they can pass its input filter. Formally, we say that the configuration C' is obtained in *one communication step* from configuration C, written as $C \vdash C'$, iff

$$C'(x) = (C(x) - \tau_x(C(x))) \cup \bigcup_{\{x,y\} \in E} (\tau_y(C(y)) \cap \rho_x(C(y))) \text{ for all } x \in G.$$

Here E is the set of edges of the underlying graph of the network. Let Γ be a NEPFS, a *computation* in Γ is a sequence of configurations $C_0, C_1, \ldots$, where C_0 is the initial configuration of Γ, $C_{2i} \Longrightarrow C_{2i+1}$ and $C_{2i+1} \vdash C_{2i+2}$, for all $i \geq 0$. If the sequence is finite, we have a finite computation. The result of any finite or infinite computation is a language which is collected in a designated node called the output (master) node of the network. If one considers the output node of the network as being the node k, and if $C_0, C_1, \ldots$ is a computation, then all strings existing in the node k at some step t - the k–th component of C_t belongs to the language generated by the network. We denote this language by $L_k(\Gamma)$.

The time complexity of computing a finite set of strings Z is the minimal number of steps t in a computation $C_0, C_1, \ldots, C_t \ldots$ such that Z is a subset of the k–th component of C_t.

3   Computational Completeness

We shall consider NEPS's having infinite regular languages as filters. The following lemma is taken from[11].

Lemma 1. (The Basic University Lemma) *Every recursively enumerable language $L \subseteq T^*$ can be written in the form $L = L' \cap T^*$ for some $L' \in H_1(FIN, REG)$.*

Here, H_1 denotes iterated splicing system and FIN and REG denote the family of finite and regular languages respectively. Using the proof of the above lemma, we can state the following theorem:

Theorem 1. *Each recursively enumerable language can be generated by a complete NEPS of size 2 where the splicing rules are of type regular and filters are of type infinite regular languages.*

Proof. Let $G = (N, T, S, P)$ be an arbitrary phrase-structure grammar. Denote $U = N \cup T \cup \{B\}$, where B is a new symbol. We construct the following NEPS of size 2 having a complete underlying graph:

$$\Gamma = (V, N_0, N_1, K_2),$$

where

$$V = N \cup T \cup \{B, X, Y, Z, X'\} \cup \{Y_\alpha | \alpha \in U\}$$

The parameters for different nodes are defined as follows:

- $N_0 = (\emptyset, \emptyset, T^*, N^*)$
- $N_1 = (M_1, \{XBSY, ZY, XZ,\} \cup \{ZvY | u \to v \in P\} \cup \{ZY_\alpha, X'\alpha Z | \alpha \in U\}, \emptyset, T^*)$ and M_1 contains rules of the following forms :

Simulate: 1.	$Xw\#uY\$Z\#vY,$	for $u \to v \in P, w \in U^*,$
Rotate : 2.	$Xw\#\alpha Y\$Z\#Y_\alpha,$	for $\alpha \in V, w \in U^*,$
Rotate : 3.	$X'\alpha\#Z\$X\#wY_\alpha,$	for $\alpha \in V, w \in U^*,$

$$\begin{array}{lll} Rotate:4. & X'\alpha w\#Y_\alpha\$Z\#Y, & \text{for } \alpha \in V, w \in U^*, \\ Rotate:5. & X\#Z\$X'\#wY, & \text{for } w \in U^*, \\ Terminate:6. & \#ZY\$XB\#wY, & \text{for } w \in T^*, \\ Terminate:7. & \#Y\$XZ\#. & \end{array}$$

The working of the above system is exactly similar to **Lemma 1 (The Basic University Lemma)** as given in[11]. Once a terminal string is generated in the node N_1, it is passed on to the node N_0 using filters. No other strings are allowed to leave N_1. Thus, we can easily see that $L(G) = L(\Gamma)$.   □

4   Solving NP-Problems Using NEPS

In this section we describe some algorithms to solve some NP-complete problems related to graphs using NEPS. For further details regarding these problems, the reader is referred to [7] and [6].

4.1   3-Vertex Colorability

Here we discuss about the vertex colorability of a graph. The problem is as follows: We have to assign colors to the vertices of the graph in such a way that no two adjacent vertices get the same color. The problem of finding whether three colors are sufficient to achieve such a coloring for an arbitrary graph is NP-complete, see [7].

Let $G = (\{1, 2, \ldots, n\}, E)$ be an arbitrary graph. Let c_1, c_2 and c_3 be the three colors. We will first generate all possible strings of the form $p_1 x_1 p_2 x_2 \ldots p_n x_n X_1$ where for all $i, 1 \leq i \leq n, p_i$ uniquely encodes "vertex i" and each x_i is any of the "colors" c_1, c_2 or c_3. Each such string represents one possible assignment of colors to the vertices of the graph in which, for each $i, 1 \leq i \leq n$, color x_i is assigned to the vertex i. We construct the following NEPS of size $n + 1$ where the underlying graph is fully connected:

$$\Gamma = (V, N_0, N_1, \ldots, N_n, K_{n+1}),$$

where

$$V = P \cup T \cup A \cup \{X, X', B, Y, Z\} \cup \{Y_\alpha | \alpha \in U\} \cup N,$$

where

$$U = P \cup T \cup \{B\}, \quad P = \{p_1, p_2, \ldots, p_n\}, \quad T = \{c_1, c_2, c_3\},$$

$$A = \{a_1, a_2, \ldots, a_n\} \text{ and } N = \{X_1, X_2, X_3, \ldots, X_n, X_{n+1}\}.$$

Here B is a new symbol. The parameters for each node are as follows :

- $N_0 = (M_0, \{XBp_1a_1p_2a_2 \ldots p_na_nX_1, ZX_1, XZ, ZY\} \cup \{ZvX_1 | v \in T\} \cup \{ZY_\alpha, X'\alpha Z | \alpha \in U\}, (PA)^n X_1, (PT)^n X_1)$, where M_0 contains the following rules:

Simulate: 1.	$Xw\#uX_1\$Z\#vX_1,$	$w \in U^*, u \in A, v \in T,$
Rotate : 2.	$Xw\#aX_1\$Z\#Y_\alpha,$	for $\alpha \in U, w \in U^*,$
Rotate : 3.	$X'\alpha\#Z\$X\#wY_\alpha,$	for $\alpha \in U, w \in U^*,$
Rotate : 4.	$X'\alpha w\#Y_\alpha\$Z\#X_1,$	for $\alpha \in U, w \in U^*,$
Rotate : 5.	$X\#Z\$X'\#\alpha wX_1,$	for $w \in U^*,$
Terminate : 6.	$\#ZY\$XB\#p_1x_1p_2x_2\ldots p_nx_nX_1,$ for $x_1, x_2, \ldots, x_n \in T.$	

The parameters for the rest of the nodes are defined as follows:

- $N_i = (M_i, ZX_{i+1}, (PT)^nX_i, PO_i)$, $1 \le i \le n$, where PO_i is defined as follows for $1 \le i \le n - 1$:

$PO_i = \{(PT)^nX_{i+1} - (PT)^{i-1}(p_ic_1)(PT)^k(p_{k+i+1}c_1)(PT)^{n-k-i-1}X_{i+1}|$
$(i, k+i+1) \in E_1\}\cup\{(PT)^nX_{i+1} - (PT)^{i-1}(p_ic_2)(PT)^l(p_{l+i+1}c_2)(PT)^{n-l-i-1}$
$X_{i+1}|(i, l + i + 1) \in E_1\} \cup \{(PT)^nX_{i+1} - (PT)^{i-1}(p_ic_3)(PT)^m(p_{m+i+1}c_3)$
$(PT)^{n-m-i-1}X_{i+1}|(i, m + i + 1) \in E_1\}$ where $0 \le k \le n - i - 1, 0 \le l \le$
$n - i - 1, 0 \le m \le n - i - 1$ and

$PO_n = \emptyset.$

M_i contains the following rule, for $1 \le i \le n - 1$:

$$p_1x_1p_2x_2\ldots p_nx_n\#X_i\$Z\#X_{i+1}$$, where $x_1, x_2, \ldots, x_n \in T.$

M_n contains the following rule:

$$p_1x_1p_2x_2\ldots p_nx_n\#X_n\$ZX_{n+1}\#$$, where $x_1, x_2, \ldots, x_n \in T.$

We now give the explanation about the working of the system. Initially, only N_0 can start the computation. The node N_0 generates all possible strings of the form $p_1x_1p_2x_2\ldots p_nx_nX_1$ where each $x_i \in \{c_1, c_2, c_3\}$; i.e node N_0 generates all possible ways of coloring the vertices of the graph. The generation of all such possible strings takes place in parallel due to multiple copies of the strings that are present in node N_0. It should be noted that before the generation of the strings of the form $p_1x_1p_2x_2\ldots p_nx_nX_1$ in N_0, no strings are allowed to leave N_0 or enter N_0. It is only after that all such possible strings are generated, they are allowed to leave N_0. The total number of steps required to generate all such possible strings is $18n+10$ which comes as follows: Suppose that $n \ge 2$. First, we will apply rule 1 to the symbol a_n which takes 2 steps (one evolutionary step and one communication step, where nothing is communicated). Suppose that $x_n \in T$ is substituted in place of a_n. After that we have to rotate x_n which takes 8 steps (4 evolutionary steps and 4 communication steps, where nothing is communicated). After this we have to rotate p_n which also takes 8 steps (4 communication and 4 evolutionary steps). Now we have a_{n-1} at the rightmost position. We will susbtitute $x_{n-1} \in T$ for a_{n-1} which takes 2 steps (one communication and one evolutionary step). So, total number of steps spent between substituting x_n for a_n and x_{n-1} for a_{n-1} is 18. This process has to be repeated $n - 1$ times for the pairs $(p_n, a_n), (p_{n-1}, a_{n-1}), \ldots, (p_2, a_2)$ respectively. After this, we have to spend 8 steps (4 communication and 4 evolutionary steps) for rotating each of the symbols B, p_1 and x_1 where $x_1 \in T$ is the symbol substituted for a_1. Finally, rule 6 is applied which takes 2 steps (one communication and one evolutionary step). So the total number of steps required is $18(n-1) + 3.8 + 2.2$ which is $18n + 10$. When $n = 1$, the total number of steps spent in N_0 is 4.

After all possible combinations are generated we have to check that whether any of such combinations represents a 3-vertex coloring or not. This checking is done by the nodes N_1 to N_{n-1} one by one. In general, node N_j will receive strings of the form $p_1 x_1 p_2 x_2 \ldots p_n x_n X_j$, where each $x_i \in T$ and $1 \le j \le n-1$. Node N_j will check that if in a candidate solution, vertex j is colored with $x_j \in \{c_1, c_2, c_3\}$, then any other vertex k, with $k > j$ which is adjacent to vertex j should not be colored with the same color c_j. This checking is done parallely on all the candidate solutions present in N_j by defining the output filter in such a way that only the strings which satisfy the above property are allowed to leave N_j and enter N_{j+1}.

Finally in N_n we get strings of the form $p_1 x_1 p_2 x_2 \ldots p_n x_n X_n$ from N_{n-1}. If we don't get any string in N_n from N_{n-1} then G does not satisfy 3-vertex coloring. If we get any string in N_n then G does satisfy 3-vertex coloring. In N_n, we erase X_n. So finally we get strings of the form $p_1 x_1 p_2 x_2 \ldots p_n x_n$ which satisfy the 3-vertex coloring, where vertex i is assigned the color $x_i \in \{c_1, c_2, c_3\}$.

The total number of steps required by the strings to go from N_0 to N_n is $2n$ and hence the total number of steps required by the entire NEPS is $20n + 10$. So the overall time required is $O(n)$ which is linear. It should be noted that instead of using a fully connected underlying graph, we could have used a linearly connected underlying graph, where N_i is connected to N_{i+1} for $0 \le i \le n-1$, thus reducing the complexity of the underlying graph.

4.2 Hamiltonian Path Problem

A Hamiltonian path in a graph is a path which traverses each vertex of the graph exactly once. Given an arbitrary graph $G = (\{1, 2, \ldots, n\}, E)$, the Hamiltonian path problem is to decide whether there exists any Hamiltonian path in the graph. The Hamiltonian path problem is a $NP-$complete problem [6]. Using NEPS we can solve this problem in linear time. It should be noted that here we are not fixing the initial and final node.

We will construct a NEPS of size $2n+1$ consisting of the nodes $N_0, N_1, \ldots, N_{2n}$, where the underlying graph is linearly connected, with N_i connected to N_{i+1}. The alphabet symbol V is same as that in 3-coloring problem, except that here $T = \{1, 2, \ldots, n\}$. The parameter for the node N_0 is same as that of the node N_0 in 3-coloring problem. Thus, in N_0, we generate all possible combinations of numbers 1 to n. From these combinations, we will generate all possible permutations of numbers 1 to n, with the help of the processor in nodes N_1 to N_{n-1}. In general, in node N_i, we check that if in a combination, integer $j \in T$ is occurring at position i, then it should not occur at any other position k, where $i+1 \le k \le n$. This checking is done by defining the output filter of node N_i in such a way that only the strings which satisfy the above property are allowed to leave N_i and enter N_{i+1}. This process is repeated one by one in the nodes N_1 to N_{n-1}.

Each string in the node N_n represents a permutation of 1 to n, which in turn is a candidate solution. To check whether these candidate solutions represent a Hamiltonian path, we use the processors in nodes N_{n+1} to N_{2n-1}. In general, at node N_{n+i} we check if in a candidate solution, vertex $x_i \in T$ is traversed in the ith step and vertex $x_{i+1} \in T$ is traversed in $(i+1)^{th}$ step, then $(x_i, x_{i+1}) \in E$. Again, to perform this checking, we define the output filter of the node N_{n+i} in such a way that only the strings which satisfy the above property are allowed to leave N_{n+i} and enter N_{n+i+1}. This checking is done one by one in the nodes N_{n+1} to N_{2n-1}. Finally, the strings from the node N_{2n-1} will leave N_{2n-1} and will be received by N_{2n} where the output filter is defined in such a way, that the strings remains there forever.

Thus, if we take N_{2n} as the output node, then all the strings encoding a Hamiltonian path will be collected in N_{2n}. The total number of steps required by the above network is $22n+10$ which comes as follows: Similar to the 3-coloring problem, the total number of steps performed in the node N_0 is $18n + 10$. After this $4n$ steps are performed for the strings to come from N_0 to N_{2n}. So, the overall time required is $22n + 10$ when $n \geq 2$ which is linear.

4.3   Subgraph Isomorphism Problem

The subgraph isomorphism problem is as follows: Given two graphs $G_1 = (\{v_1, v_2, \ldots, v_n\}, E_1)$ and $G_2 = (\{u_1, u_2, \ldots, u_t\}, E_2)$, is G_2 a sub-graph of G_1 ? Without loss of generality, we assume that $t \leq n$. G_2 will be a sub-graph of G_1 written as $G_2 \subseteq G_1$ iff $V_2 \subseteq V_1$ and $E_2 \subseteq E_1$, where V_2 and V_1 are the set of vertices of G_2 and G_1 respectively.

We will construct a NEPS of size $n+t+1$ consisting of the nodes $N_0, N_1, \ldots, N_{n+t}$ where the underlying graph is linearly connected as in Hamiltonian path problem. The alphabet V is similar to that in Hamiltonian path problem except that here

$$P = \{u_1, u_2, \ldots, u_n\} \quad \text{and} \quad T = \{v_1, v_2, \ldots, v_n\}$$

Similar (except that here P is different) to the Hamiltonian path problem, the first n nodes (N_0 to N_{n-1}) generate all possible strings of the form $u_1 x_1 u_2 x_2 \ldots u_n x_n X_n$ in the node N_n where each $x_i \in T$ is unique. For $1 \leq j \leq t$, a string $u_1 x_1 u_2 x_2 \ldots u_n x_n X_n$ is interpreted as a mapping between the vertices of G_1 and G_2, associating vertex u_j with vertex $x_j, x_j \in \{v_1, v_2, \ldots, v_n\}$. We have to check whether such a mapping satisfies the edge subset property; i.e., if $(u_i, u_j) \in E_2$ then $(x_i, x_j) \in E_1$, $x_i, x_j \in \{v_1, v_2, \ldots, v_n\}, 1 \leq i, j \leq t$. This checking is done step by step in the nodes $N_{n+1}, N_{n+2}, \ldots, N_{n+t}$, for each vertex $u_t \in V_2$, by defining the output filter as in Hamiltonian path problem. In general, the node N_{n+i} performs the checking for the edge subset property for the edges $(u_i, u_j), u_i, u_j \in V_2$ where $1 \leq i \leq t$ and $i \leq j$.

Finally all the strings which represent a valid mapping between G_1 and G_2 get collected in N_{n+t}, which we consider as the output node. The total number of steps performed to bring the strings from the N_{n+1} to N_{n+t} is $2t$ and hence the total time required by the overall network is $20n + 2t + 10$ which is $O(n+t)$ and hence linear.

5   Solving Exact Cover by 3-Sets Problem

Here we will see how to solve exact cover by 3-sets problem using NEPS in linear time. The problem is as follows: Given a finite set X with $|X| = 3q$ and a collection C of 3-elements subsets of X, does C contain an exact cover for X, that is, is there a sub collection $C' \subseteq C$ such that every element of X occurs in exactly one member of C'? We assume that $X = \{x_1, x_2, \ldots, x_n\}$. We construct a NEPS of size $n + q + 3$ consisting of the nodes $N_0, N_1, \ldots, N_{n+q+2}$, where $n = 3q$ and the underlying graph is linearly connected as in the Hamiltonian path problem.

The alphabet V of the NEPS is same as that of the Hamiltonian path problem except that here $T = \{x_1, x_2, \ldots, x_n\}$. Also, V contains a set W, where

$$W = \{w | w = e_{[ijk]} \text{ where } \{x_i, x_j, x_k\} \in C\} \text{ and } q \leq |W| \leq \binom{n}{3}.$$

We informally give the description of the network: We begin the computation in the node N_0 and similar to the Hamiltonian path problem, we generate all possible strings of the form $XBp_1x_1p_2x_2 \ldots p_nx_nY$ in the node N_{n-1}, where each $x_i \in T$ occurring at position $p_i \in P$ is distinct. These strings leave N_{n-1} and are received by N_n. In the node N_n, we erase all the elements of the set P from all the strings present in it. All these strings are then sent to the node N_{n+1}.

In the node N_{n+1}, we insert elements from the set W in the strings present in N_{n+1} after each third element starting from the second symbol. After the insertion of the symbols from the set W, we erase the symbols X and B from all the strings. All the strings will now leave N_{n+1} and enter N_{n+2}.

In N_{n+2}, we first replace Y by X_2 in all the strings present in the node. We then check that there exists a string of the form $x_{a_1}x_{a_2}x_{a_3}e_{[ijk]}((T)^3W)^{q-1}X_2$ present in the node, such that $\{x_{a_1}, x_{a_2}, x_{a_3}\} = \{x_i, x_j, x_k\}$; i.e., we are checking that whether the collection of the first 3 elements occurring in a candidate solution is really a member of the set C. If it is a member of C, then it can be part of the set-cover for X. If there exists such a string then it is allowed to leave the node, which will be received only by N_{n+3}. All the remaining strings do not satisfy the set-cover property and hence remain in N_{n+2} forever.

In N_{n+3}, we perform the same checking, as done in N_{n+2} with $x_{a_4}, x_{a_5}, x_{a_6}$ and w_2, where $x_{a_4}, x_{a_5}, x_{a_6} \in T$ represent the next group of three elements from the set X, present in the strings in the node and $w_2 \in W$ is the symbol present after x_{a_6}.

The above process is repeated in the nodes $N_{n+4}, N_{n+5}, \ldots, N_{n+q+1}$ step by step. Finally in the node N_{n+q+2}, we get strings of the form $x_{a_1}x_{a_2}x_{a_3}e_{[a_1a_2a_3]}$ $x_{a_4}x_{a_5}x_{a_6}e_{[a_4a_5a_6]} \ldots x_{a_{n-2}}x_{a_{n-1}}x_{a_n}e_{[a_{n-2}a_{n-1}a_n]}Y$ from the node N_{n+q+1} where each $x_{a_i} \in X$ will occur exactly at one position. The set cover is the collection of sets $\{\{x_{a_1}, x_{a_2}, x_{a_3}\}, \{x_{a_4}, x_{a_5}, x_{a_6}\}, \ldots, \{x_{a_{n-2}}, x_{a_{n-1}}, x_{a_n}\}\}$ where each $x_{a_i} \in X$ and each x_{a_i} occurs in exactly one of the sets in the above collection of sets.

It can be shown that similar to the subgraph isomorphism problem, the overall time required by the above network is $O(n + q)$ which is linear.

6   Conclusions

In this paper, we have considered a mechanism inspired from cell biology, namely networks of evolutionary processors with splicing rules (NEPS). These networks consist of nodes which are very simple processors and are able to perform splicing operation on the strings present in the node. The nodes are endowed with input and output filter which are defined by a regular language. Networks with two nodes are able to generate all recursively enumerable languages.

Using this mechanism we are able to solve hard problems in polynomial time. We presented linear time solution for NP-complete problems namely "3-coloring problem", "Hamiltonian path problem", "Subgraph isomorphism problem" and "Exact cover by 3 sets problem". What other complicated problems could be solved in this framework is another point of interest which might be further investigated.

The remarks made about "evolutionary processors" in[1] holds good here also. Definitely, the computational process described here is not exactly an evolutionary process in Darwinian sense. But the rewriting operations we have considered might be interpreted as mutations and the filtering process might be viewed as a selection process. Furthermore, we are not concerned here with a possible biological implementation, though a matter of great importance.

References

1. J. Castellanos, C. Martin-Vide, V.Mitrana, and J. Sempere. Networks of evolutionary processors. *Acta Informatica*, 39:517–529, 2003.
2. E. Csuhaj-Varju, J.Dassow, J.Keleman, and Gh. Păun. *Grammar Systems: A Grammatical Approach to Distribution and Cooperation.* Gordon and Breach, London, 1994.
3. E. Csuhaj-Varju, L. Kari, and Gh. Păun. Test tube distributed systems based on splicing. *Computers and AI*, 15(2-3):211–232, 1996.
4. E. Csuhaj-Varju and A. Salomaa. Networks of parallel language processors. In *New Trends in Formal Languages*, pages 299–318, 1997.
5. L. Errico and C. Jesshope. Towards a new architecture for symbolic processing. In *Artificial Intelligence and Information Control Systems of Robots'94*, pages 31–40, 1994.
6. M. R. Garey and D. S. Johnson. *Computers and Intractability, A Guide to the Theory of NP-Completeness.* W. H. Freeman and Company, 1979.
7. A. M. Gibbons. *Algorithmic Graph Theory.* Cambridge University Press, 1985.
8. W. D. Hillis. *The Connection Machine.* MIT Press, Cambridge, 1985.
9. K. Krithivasan, M. Sakthi Balan, and P. Harsha. Distributed processing in automata. *International Journal of Foundation of Computer Science*, 10(4):443–464, 1999.
10. Gh. Păun. Computing with membranes. *Journal of Computer and System Sciencea*, 61(1):108–143, 2000.
11. Gh. Păun, G. Rozenberg, and A. Salomaa. *DNA Computing.* Springer-Verlag, 1998.

Network of Evolutionary Processors with Splicing Rules and Forbidding Context

Ashish Choudhary and Kamala Krithivasan

Dept of Computer Science and Engineering,
Indian Institute of Technology, Madras,
Chennai, India 600036
ashish@meenakshi.cs.iitm.ernet.in, kamala@iitm.ernet.in

Abstract. In this paper we consider networks of evolutionary processors with splicing rules and forbidding context (NEPFS) as language generating and computational devices. Such a network consists of several processors placed on the nodes of a virtual graph and are able to perform splicing (which is a biologically motivated operation) on the words present in that node, according to the splicing rules present there. Before applying the splicing operation on words, we check for the absence of certain symbols (forbidding context) in the strings on which the rule is applied. Each node is associated with an input and output filter. When the filters are based on random context conditions, one gets the computational power of Turing machines with networks of size two. We also show how these networks can be used to solve $NP-$complete problems in linear time.

1 Introduction

The current trend in computer science is to have the computation done in a distributed and/or parallel manner. Formal language and automata theory has always tried to give suitable models which will represent the real life situations. Grammar systems[2], distributed automata[8], membrane systems[10], splicing systems[11] are some of the models defined to capture distribution and parallelism.

To have a model which is more similar to Connection Machine[7] and Logic Flow paradigm[4], in[1] Networks of Evolutionary processors were defined. Networks of language processors[3] are closely related to parallel communicating grammar systems[2], but here the main idea is to place a language generating device in any node of an underlying graph. In each node we have some strings and some rewriting rules, according to which the strings are rewritten. Then the strings are communicated to the neighboring nodes using filters for which generally regular sets are used.

The above idea also has its motivation from cell biology. Each processor placed in a node is a very simple processor performing simple operations. In[1], as motivation from cell biology, it is mentioned that each node may be viewed as a cell having genetic information encoded in DNA sequence which may evolve

J. Mira and J.R. Álvarez (Eds.): IWINAC 2005, LNCS 3561, pp. 300–309, 2005.

by local evolutionary events, that is point mutations. But in[1], they are mainly looked as language processors, making use of rewriting rules. Several NP-complete problems could be solved in polynomial time by such network of processors. The generative power is equal to that of type 0 grammars[1].

With the reference to motivation from cell biology, it looks more appropriate to consider splicing rules in each processor rather than rewriting rules. In this paper, we consider network of evolutionary processors as defined in[1] but where the rules applied in a processor are splicing rules. We have a processor at each node of an underlying graph. Each node has a set of strings to begin with and a set of splicing rules. Evolution takes place using splicing rules. Moreover, before applying a splicing rule, we check for the absence of some symbols (forbidding context) in the strings on which, we are applying the splicing rule. Each node has an input filter and an output filter, which are based on random context conditions, as defined in[9]. A string can enter a node only if it can pass the input filter and can leave the node only if it can pass the output filter. One of the nodes is designated as the output node where we collect the output strings.

We can look at such a network of evolutionary processors as a language generating device, if we look at the strings collected in the output node. We can also look at them as doing some computation. We show how these networks can solve NP-complete problems in linear time. We also show that all recursively enumerable languages can be generated using these networks having just two processors.

The rest of the paper is organized as follows: In section two, we give some preliminary definitions and the main definition of our paper. In section three, we show the computational completeness of these networks by showing that any recursively enumerable language can be generated by using network of evolutionary processors with splicing rules and forbidding context. In section four, we show how to solve some NP-complete problems related to graphs using these networks. The paper ends with a brief conclusion.

2   Preliminaries

The set of all strings over V is denoted by V^* and the empty string is denoted by ϵ. The length of a word x is denoted by $|x|$, while $alph(x)$ denotes the minimal alphabet W such that $x \in W^*$. Consider an alphabet V and two special symbols, $\#, \$$, not in V. A *splicing rule* (over V) is a string of the form $r = u_1 \# u_2 \$ u_3 \# u_4$, where $u_1, u_2, u_3, u_4 \in V^*$. For a splicing rule $r = u_1 \# u_2 \$ u_3 \# u_4$ and strings $x, y, z \in V^*$ we write $(x, y) \vdash_r z$ iff $x = x_1 u_1 u_2 x_2, y = y_1 u_3 u_4 y_2, z = x_1 u_1 u_4 y_2$, for some $x_1, x_2, y_1, y_2 \in V^*$. For a splicing rule $r = u_1 \# u_2 \$ u_3 \# u_4$ and strings $x, y, z, w \in V^*$ we write $(x, y) \models_r (z, w)$ iff $x = x_1 u_1 u_2 x_2, y = y_1 u_3 u_4 y_2, z = x_1 u_1 u_4 y_2, w = y_1 u_3 u_2 x_2$ for some $x_1, x_2, y_1, y_2 \in V^*$. For a set of splicing rules R, we define the *radius* of R as $rad(R) = \max \{|x|, x = u_i, 1 \leq i \leq 4,$ for some $u_1 \# u_2 \$ u_3 \# u_4 \in R\}$. A splicing rule with forbidding context (over V) is a triple of the form $p = (r; D_1, D_2)$ with $r = u_1 \# u_2 \$ u_3 \# u_4$ being a splicing rule over V and D_1, D_2 being finite subsets of V^*. Note that here we only consider systems with finite components. For $x, y, z, w \in V^*$ and $p = (r; D_1, D_2)$, we define $(x, y) \models_p (z, w)$ iff $(x, y) \models_r (z, w)$, with no element of D_1 appearing as a sub-string in x and no element of D_2 appearing as a sub-string in y; when $D_1 = \emptyset$ or $D_2 = \emptyset$, then no condition on x, respectively y, is imposed.

For two disjoint subsets P and F of an alphabet V and a word over V, we define the predicates $\varphi^{(1)}$ and $\varphi^{(2)}$ as follows:

$$\varphi^{(1)}(w; P, F) \equiv P \subseteq alph(w) \wedge F \cap alph(w) = \emptyset$$

and

$$\varphi^{(2)}(w; P, F) \equiv alph(w) \cap P \neq \emptyset \wedge F \cap alph(w) = \emptyset.$$

The construction of these predicates is based on *random-context conditions* defined by the two sets P (*permitting contexts*) and F (*forbidding contexts*). It should be noted that the forbidding contexts defined here are different from the forbidding contexts which are used in each splicing rule. For every language $L \subseteq V^*$ and $\beta \in \{(1), (2)\}$, we define:

$$\varphi^{\beta}(L, P, F) = \{w \in L | \varphi^{\beta}(w; P, F)\}.$$

A *network of evolutionary processors with splicing rules and forbidding context* (NEPFS for short) of size n is a construct

$$\Gamma = (V, N_1, N_2, , \ldots, N_n, G),$$

where V is an alphabet and for each $1 \leq i \leq n$, $N_i = (M_i, A_i, PI_i, FI_i, PO_i, FO_i, \beta_i)$ is the i-th evolutionary node processor of the network and G is called the *underlying graph* of the network. The parameters of every processor are:

- M_i is the set of splicing rules with forbidding context of the ith processor.
- A_i is the finite set (multiset) of words over V present initially in the ith processor.
- $PI_i, FI_i \subseteq V$ are the *input* permitting/forbidding contexts of the ith processor.
- $PO_i, FO_i \subseteq V$ are the *output* permitting/forbidding contexts of the ith processor.
- $\beta_i \in \{(1), (2)\}$ defines the type of the *input/output filters* of a node. More precisely, for every node, $x \in G$, we define the following filters: the input filter is given as

$$\rho_x(.) = \varphi^{\beta(x)}(.; PI_x, FI_x),$$

and the output filter is defined as

$$\tau_x(.) = \varphi^{\beta(x)}(.; PO_x, FO_x).$$

That is $\rho_x(w)$ (resp. τ_x) indicates whether or not the word w can pass the input (resp. output) filter of x. More generally, $\rho_x(L)$ (resp. $\tau_x(L)$) is the set of words of L that can pass the input (resp. output) filter of x.

By a configuration (state) of a NEPFS as above we mean an $n-$tuple $C = (L_1, L_2, \ldots L_n)$, with $L_i \subseteq V^*$ for all $1 \leq i \leq n$. A configuration represents the sets of strings (each string appears in an arbitrarily large number of copies) which are present in any node at a given moment. The initial configuration of the network is $C_0 = (A_1, A_2, \ldots, A_n)$. A configuration can either change by an evolutionary step or by a communicating step. When changing by an evolutionary step, each component L_i of the configuration is changed in accordance with the splicing rules associated with the node i.

Formally we say that the configuration $C_1 = (L_1, L_2, \ldots, L_n)$ directly changes into the configuration $C_2 = (L'_1, L'_2, \ldots, L'_n)$ by a evolutionary step, written as $C_1 \Rightarrow C_2$ if L'_i is the set of strings obtained by applying the splicing rules of N_i to the strings in N_i. Since an arbitrarily large number of copies of each string is available in every node, after an evolutionary step in each node one gets an arbitrarily large number of copies of any string which can be obtained by using all possible splicing rules associated with that node. By definition, if L_i is empty for some $1 \leq i \leq n$, then L'_i is empty as well.

When changing by communication step, each node processor N_i sends all copies of the strings it has which are able to pass its output filter to all the node processors connected to N_i and receives all copies of the strings sent by any node processor connected with N_i providing they can pass its input filter. Formally, we say that the configuration C' is obtained in *one communication step* from configuration C, written as $C \vdash C'$, iff

$$C'(x) = (C(x) - \tau_x(C(x))) \cup \bigcup_{\{x,y\}\in E} (\tau_y(C(y)) \cap \rho_x(C(y))) \text{ for all } x \in G.$$

Here E is the set of edges of the underlying graph of the network. Let Γ be a NEPFS, a *computation* in Γ is a sequence of configurations $C_0, C_1, \ldots$, where C_0 is the initial configuration of Γ, $C_{2i} \Longrightarrow C_{2i+1}$ and $C_{2i+1} \vdash C_{2i+2}$, for all $i \geq 0$. If the sequence is finite, we have a finite computation. The result of any finite or infinite computation is a language which is collected in a designated node called the output (master) node of the network. If one considers the output node of the network as being the node k, and if $C_0, C_1, \ldots$ is a computation, then all strings existing in the node k at some step t - the $k-$th component of C_t belongs to the language generated by the network. We denote this language by $L_k(\Gamma)$.

The time complexity of computing a finite set of strings Z is the minimal number of steps t in a computation $C_0, C_1, \ldots, C_t \ldots$ such that Z is a subset of the $k-$th component of C_t.

3   Computational Completeness

In this section, we show that NEPFS with only two nodes are computationally complete, irrespective of the underlying graph structure. The following lemma is taken from[11].

Lemma 1. $RE \subseteq EH_2(FIN, f[2])$.

where $EH_2([n], f[m]), n, m \geq 1$, denotes the family of languages $L(\gamma)$ generated by extended H systems with forbidding contexts, $\gamma = (V, T, A, R)$ with $card(A) \leq n$ and $rad(R) \leq m$, where $rad(R)$ is the maximum radius of the splicing rules r in triples $(r; D_1, D_2)$. When no restriction on the number of axioms or on the maximal radius is imposed (except that these numbers are still finite), we replace $[n]$ or $[m]$ by FIN, see[11].

Using the proof of the above lemma, we can state the following theorem:

Theorem 1. *Each recursively enumerable language can be generated by a complete NEPFS of size 2 where the splicing rules are of type regular.*

Proof. Consider a type-0 grammar $G = (N, T, S, P)$ in the Kuroda normal form, namely P only contains rules of the following forms:

$$A \rightarrow a \ , \ A \rightarrow BC \ , \ AB \rightarrow CD \ , \ A \rightarrow \epsilon,$$

where A, B, C and D are nonterminals and a is a terminal. We denote by P_1 the set of context free rules in P and the set of non context-free rules in P by P_2. Let us denote $U = N \cup T \cup \{B\}$, where B is a new symbol. We construct the following NEPFS of size two having the underlying graph as fully connected:

$$\Gamma = (V, N_1, N_2, K_2),$$

where

$$V = N \cup T \cup \{B, X, X', Y, Z, Z', Z''\} \cup \{Y_r | r \in P_2\} \cup \{Z_r | r \in P\} \cup \{Y_\alpha | \alpha \in U\}.$$

Here K_2 denotes a fully connected graph of size two. The parameters for different nodes are defined as follows:

- $N_1 = (M_1, A_1, N, T, T, \{N, B, X, X', Z, Z', Z'', Y\}, (2)\}$, where

$$A_1 = \{XBSY, ZY, XZ, Z', Z''\} \cup \{ZY_\alpha, X'\alpha Z | \alpha \in U\} \cup \{Z_r xY | r : C \rightarrow x \in P_1\}$$

$$\cup \{ZY_r, Z_r EFY | r : CD \rightarrow EF \in P_2\}$$

and M_1 contains rules of the following forms:

Simulate: 1.	$(\#CY\$Z_r\#x; \{X'\}, \emptyset)$,	for $r : C \rightarrow x \in P_1$,
Simulate: 2.	$(C\#DY\$Z\#Y_r; \{X'\}, \emptyset)$,	
Simulate: 3.	$(\#CY_r\$Z_r\#EF; \{X'\}, \emptyset)$, for $r : CD \rightarrow EF \in P_2$,	
Rotate: 4.	$(\#\alpha Y\$Z\#Y_\alpha; \{X'\}, \emptyset)$,	
Rotate: 5.	$(X\#\$X'\alpha\#Z; \{Y\} \cup \{Y_\beta | \beta \in U, \beta \neq \alpha\} \cup \{Y_r | r \in P_2\}, \emptyset)$,	
Rotate: 6.	$(\#Y_\alpha\$Z\#Y; \{X\}, \emptyset)$, for $\alpha \in U$,	
Rotate: 7.	$(X'\#\$X\#Z; \{Y_\beta | \beta \in U\} \cup \{Y_r | r \in P_2\}, \emptyset)$,	
Terminate : 8.	$(XB\#\$\#Z'; \{Y_\beta | \beta \in U\} \cup \{Y_r | r \in P_2\}, \emptyset)$,	
Terminate : 9.	$(\#Y\$Z''\#; \{X, B\}, \emptyset)$.	

- $N_2 = (\emptyset, \emptyset, T, N \cup \{B, X, X', Z, Z', Z'', Y\}, N, \emptyset, (2))$.

The working of the above system is exactly similar to **Lemma 1** as given in[11]. Once a terminal string is generated in the node N_1, it is passed on to the node N_2 using filters. No other strings are allowed to leave N_1. Thus we have $L(G) = L(\Gamma)$.   □

It should be noted that we are using a complete graph containing two nodes. Since any complete graph with two nodes is equivalent to a star or ring graph with two nodes, the above proof holds even when the underlying graph is star or ring with two nodes.

4   Solving NP-Complete Problems Using NEPFS

In this section we describe some algorithms to solve some NP-complete problems related to graphs using NEPFS. For further details regarding these problems, the reader is referred to [6, 5].

4.1   Subgraph Isomorphism Problem

The subgraph isomorphism problem is as follows: Given two graphs $G_1 = (\{v_1, v_2, \ldots, v_n\}, E_1)$ and $G_2 = (\{u_1, u_2, \ldots, u_m\}, E_2)$, is G_2 a sub-graph of G_1 ? Without loss of generality, we assume that $m \leq n$. G_2 will be a sub-graph of G_1 written as $G_2 \subseteq G_1$ iff $V_2 \subseteq V_1$ and $E_2 \subseteq E_1$, where V_2 and V_1 are the set of vertices of G_2 and G_1 respectively.

Theorem 2. *Given graphs $G_1 = (\{v_1, v_2, \ldots, v_n\}, E_1)$ and $G_2 = (\{u_1, u_2, \ldots, u_m\}, E_2)$ where $m \leq n$ and $|E_2| = r$, where each edge e_t is of the form (u_{i_t}, u_{j_t}), $1 \leq t \leq r$, $1 \leq i_t, j_t \leq m$, we can find whether G_2 is isomorphic to G_1 by using a complete NEPFS of size $(n^2 + r + 2)$ in $O(n + r)$ time.*

Proof. We define the alphabet

$$U = \{u_1, u_2, \ldots, u_n\} \cup \{1, 2, \ldots, n\} \cup \{X_1, X_2, \ldots, X_n\}$$

$$\cup \{X_{e_1}, X_{e_2}, \ldots, X_{e_r}, X_{e_{r+1}}\} \cup \{a_1, a_2, \ldots, a_n\} \cup \{p\}$$

We construct a complete NEPFS of size $(n^2 + r + 2)$, where the nodes are labelled as $In, N_{ij}, N_{e_t}, 1 \leq t \leq r$ and Out, where $1 \leq i, j \leq n$. The parameters for different nodes are defined as follows:

- $In = (\emptyset, \{u_1 a_1 u_2 a_2 \ldots u_n a_n X_1\}, \{a_1, a_2, \ldots, a_n\}, \emptyset, \emptyset, \emptyset, (1))$
- $N_{1i} = (M_{1i}, \{pi, X_2\}, \{X_1\}, \{i\}, \{i, X_2\}, \{a_1, p, X_1\}, (1)), 1 \leq i \leq n$, where M_{1i} contains the following rules:

$$(\#a_1\$p\#i; \emptyset, \emptyset), (i\#\$pa_1\#; \{p\}, \emptyset), (\#X_1\$\#X_2; \{a_1, p\}, \emptyset).$$

- $N_{2i} = (M_{2i}, \{pi, X_3\}, \{X_2\}, \{X_1, i\}, \{i, X_3\}, \{a_2, p, X_2\}, (1)), 1 \leq i \leq n$, where M_{2i} contains the following rules:

$$(\#a_2\$p\#i; \emptyset, \emptyset), (i\#\$pa_2\#; \{p\}, \emptyset), (\#X_2\$\#X_3; \{a_2, p\}, \emptyset).$$

$-\ \dots\dots\dots\dots$

- $N_{ni} = (M_{ni}, \{pi, X_{e_1}\}, \{X_n\}, \{X_1, X_2, \dots, X_{n-1}\} \cup \{i\}, \{i, X_{e_1}\}, \{a_n, p, X_n\}, (1))$ where M_{n1} contains the following rules:

$$(\#a_n\$p\#i; \emptyset, \emptyset), (i\#\$pa_n\#; \{p\}, \emptyset), (\#X_n\$\#X_{e_1}; \{a_n, p\}, \emptyset).$$

- $N_{e_t} = (M_{e_t}, \{X_{e_{t+1}}\}, \{X_{e_t}\}, \emptyset, \{X_{e_{t+1}}\}, FO_{e_t}, (1)), 1 \le t \le r$, where M_{e_t} contains the following rule:

$$(\#X_{e_t}\$\#X_{e_{t+1}}; \emptyset, \emptyset)$$

and

$$FO_{e_t} = \{x, y | x, y \in \{1, 2, \dots, n\} \text{ and } (v_x, v_y) \notin E_1 \text{ and } (u_{i_t}, u_{j_t}) \in E_2\}.$$

where x, y are the symbols occurring after the symbols u_{i_t}, u_{j_t} respectively in the strings present in N_{e_t}.
- $Out = (\emptyset, \emptyset, \{X_{e_{r+1}}\}, \emptyset, \{a_1, a_2, \dots, a_n\}, \emptyset, (1))$

We now give the explanation about the working of the above network: We begin the computation with the axiom $w = u_1a_1u_2a_2 \dots u_na_nX_1$ in the node In. The first phase of the computation involves generating the strings of the form $w' = u_1x_1u_2x_2 \dots u_nx_nX_{e_1}$ in the node N_{e_1} through the processing of the nodes $N_{11}, N_{12}, \dots, N_{1n}, N_{21}, N_{22}, \dots, N_{2n}, \dots, N_{n1}, N_{n2}, \dots, N_{nn}$ where each $x_i \in \{1, 2, \dots, n\}$ is distinct for $1 \le i \le n$. In other words, we will be generating all possible permutations of the numbers from 1 to n in the node N_{e_1}. In general, the node N_{ij}, replace a_i with j, provided j is not substituted in place of any of the symbols $a_1, a_2, \dots, a_{j-1}$, $1 \le i \le n$, $1 \le j \le n$. The sequence of splicing operations that take place in the node N_{ij} is as follows:

1. $(u_1x_1u_2x_2 \dots u_ia_iu_{i+1}a_{i+1} \dots u_na_nX_i, pj) \models (u_1x_1u_2x_2 \dots u_ij, pa_iu_{i+1}a_{i+1}$

   $\dots u_na_nX_i)$ where each $x_i \in \{1, 2, \dots, n\}$ is distinct .

2. $(u_1x_1u_2x_2 \dots u_ij, pa_iu_{i+1}a_{i+1} \dots u_na_nX_i) \models (u_1x_1u_2x_2 \dots u_iju_{i+1}a_{i+1} \dots$

   $u_na_nX_i, pa_i).$

3. $(u_1x_1u_2x_2 \dots u_iju_{i+1}a_{i+1}u_na_nX_i, X_{i+1}) \models (u_1x_1u_2x_2 \dots u_iju_{i+1}a_{i+1}u_na_n$

   $X_{i+1}, X_i).$

The generation of such permutations takes $6n + 2$ steps in total. Any permutation $w = u_1x_1u_2x_2 \dots u_nx_nX_{e_1}$ is interpreted as a mapping between the vertices of G_1 and G_2 where vertex u_1 is mapped to vertex v_{x_1}, vertex u_2 is mapped to the vertex v_{x_2} and so on. We have to check whether such a mapping satisfies the edge subset property or not; i.e, we have to check that if in a mapping, vertex u_i is mapped to vertex v_{x_i} and vertex u_j mapped to vertex v_{x_j} and if $(u_i, u_j) \in E_2$, then (v_{x_i}, v_{x_j}) should also belong to E_1. If this property is satisfied for all $1 \le i < j \le m$, then such a mapping between the vertices of

G_1 and G_2 satisfy the sub-graph isomorphism property. This checking is done by the nodes $N_{e_1}, N_{e_2}, \ldots, N_{e_r}$ one by one. In general, the node $N_{e_i}, 1 \leq i \leq r$ checks the edge subset property for the edge e_i of G_2. This checking is done once for each edge. The total number of steps required for this checking is $2r$. Finally, if we get any string in the node Out, then the graph G_2 is isomorphic to a subgraph of G_1, where the mapping between the vertices is represented by the strings present in the node Out. The total number of steps required by the above network is $6n + 2r + 4$ which is $O(n + r)$. $\square$

4.2   3-Vertex Colorability

Here we discuss about the vertex colorability of a graph. The problem is as follows: We have to assign colors to the vertices of the graph in such a way that no two adjacent vertices get the same color. The problem of finding whether three colors are sufficient to achieve such a coloring for an arbitrary graph is NP-complete, see[6].

Theorem 3. *The "3-colorability problem" can be solved in $O(m + n)$ time by a complete NEPFS (network of evolutionary processors with splicing rules and forbidding contexts) of size $7m + 2$, where n is the number of vertices and m is the number of edges of the input graph.*

Proof. The network is exactly similar to the network used for solving the 3-colorability problem in[1]. The only difference here is that here we use splicing rules in place of normal rules. Also, the filters are different here.

   Let $G = (\{1, 2, \ldots, n\}, \{e_1, e_2, \ldots, e_m\})$ be a graph and assume that $e_t = \{i_t, j_t\}, 1 \leq i_t < j_t \leq n, 1 \leq t \leq m$. We consider the alphabet

$$U = V \cup V' \cup T \cup \{X_1, X_2, \ldots, X_{m+1}\} \cup \{p_1, p_2, \ldots, p_n\},$$

where $V = \{b_1, r_1, g_1, \ldots, b_n, r_n, g_n\}, T = \{a_1, a_2, \ldots, a_n\}$. Here V' is the primed copy of V, that is the set formed by the primed copies of all letters in V. We construct the following complete NEPFS of size $7m + 2$ having nodes:

- $N_0 = (M_0, A_0, T \cup X_1, \emptyset, \{X_1\}, T, (1))$, where

$$A_0 = \{a_1 a_2 \ldots a_n X_1\} \cup \{c_n X_1, c_{n-1} c_n X_1, \ldots, c_1 c_2 \ldots c_n X_1\} \text{ where}$$

$$c_i \in \{b_i, r_i, g_i\},$$

   M_0 contains rules of the following forms:

$$(\#a_n X_1 \$ \# c_n X_1; \emptyset, \{T\}), (\#a_{n-1} c_n X_1 \$ \# c_{n-1} c_n X_1; \emptyset, \{T\}),$$

$$\cdots\cdots\cdots\cdots$$

$$(\#a_i c_{i+1} c_{i+2} \ldots c_n X_1 \$ \# c_i c_{i+1} c_{i+2} \ldots c_n X_1; \emptyset, \{T\}) , 1 \leq i \leq n - 1.$$

- $N_{e_t}^{(Z)} = (M_{e_t}^{(Z)}, p_t Z'_{i_t}, \{X_t\}, U - V - \{X_t\}, \{X_t, Z'_{i_t}\}, \{p_t, Z_{i_t}\}, (1)), Z \in \{b, r, g\},$
  $1 \le t \le m$, where $M_{e_t}^{(Z)}$ contains the following rules:
  $$(\#Z_{i_t}\$p_t\#Z'_{i_t}; \emptyset, \emptyset), (Z'_{i_t}\#\$p_t Z_{i_t}\#; \{p_t\}, \emptyset).$$

- $N_{e_t}^{(b')} = (M_{e_t}^{(b')}, \{p_{j_t}r'_{j_t}, p_{j_t}g'_{j_t}\}, \{X_t, b'_{i_t}\}, \{r'_{j_t}, g'_{j_t}, b_{j_t}\}, \{X_t\}, \{b_{j_t}, r_{j_t}, g_{j_t}, p_{j_t}\},$
  $(1))$, where $M_{e_t}^{(b')}$ contains the following rules:
  $$(\#r_{j_t}\$p_{j_t}\#r'_{j_t}; \emptyset, \emptyset), (r'_{j_t}\#\$p_{j_t}r_{j_t}\#; \{p_{j_t}\}, \emptyset),$$
  $$(\#g_{j_t}\$p_{j_t}\#g'_{j_t}; \emptyset, \emptyset), (g'_{j_t}\#\$p_{j_t}g_{j_t}\#; \{p_{j_t}\}, \emptyset).$$

- $N_{e_t}^{(r')} = (M_{e_t}^{(r')}, \{p_{j_t}b'_{j_t}, p_{j_t}g'_{j_t}\}, \{X_t, r'_{i_t}\}, \{b'_{j_t}, g'_{j_t}, r_{j_t}\}, \{X_t\}, \{r_{j_t}, b_{j_t}, g_{j_t}, p_{j_t}\},$
  $(1))$, where $M_{e_t}^{(r')}$ contains the following rules:
  $$(\#b_{j_t}\$p_{j_t}\#b'_{j_t}; \emptyset, \emptyset), (b'_{j_t}\#\$p_{j_t}b_{j_t}\#; \{p_{j_t}\}, \emptyset),$$
  $$(\#g_{j_t}\$p_{j_t}\#g'_{j_t}; \emptyset, \emptyset), (g'_{j_t}\#\$p_{j_t}g_{j_t}\#; \{p_{j_t}\}, \emptyset).$$

- $N_{e_t}^{(g')} = (M_{e_t}^{(g')}, \{p_{j_t}r'_{j_t}, p_{j_t}b'_{j_t}\}, \{X_t, g'_{i_t}\}, \{r'_{j_t}, b'_{j_t}, g_{j_t}\}, \{X_t\}, \{g_{j_t}, r_{j_t}, b_{j_t}, p_{j_t}\},$
  $(1))$, where $M_{e_t}^{(g')}$ contains the following rules:
  $$(\#r_{j_t}\$p_{j_t}\#r'_{j_t}; \emptyset, \emptyset), (r'_{j_t}\#\$p_{j_t}r_{j_t}\#; \{p_{j_t}\}, \emptyset),$$
  $$(\#b_{j_t}\$p_{j_t}\#b'_{j_t}; \emptyset, \emptyset), (b'_{j_t}\#\$p_{j_t}b_{j_t}\#; \{p_{j_t}\}, \emptyset).$$

- $N_{e_t} = (M_{e_t}, \{p_{i_t}r_{i_t}, p_{j_t}r_{j_t}, p_{i_t}b_{i_t}, p_{j_t}b_{j_t}, p_{i_t}g_{i_t}, p_{j_t}g_{j_t}, X_{t+1}\}, \{X_t\},$
  $\{r_{i_t}, b_{i_t}, g_{i_t}, r_{j_t}, b_{j_t}, g_{j_t}\}, \{X_{t+1}\}, U - V - \{X_{t+1}\}, (1))$, where M_{e_t} contains
  the following rules:
  $$(\#r'_{i_t}\$p_{i_t}\#r_{i_t}; \emptyset, \emptyset), (r_{i_t}\#\$p_{i_t}r'_{i_t}\#; \{p_{i_t}\}, \emptyset),$$
  $$(\#r'_{j_t}\$p_{j_t}\#r_{j_t}; \emptyset, \emptyset), (r_{j_t}\#\$p_{j_t}r'_{j_t}\#; \{p_{j_t}\}, \emptyset),$$
  $$(\#b'_{i_t}\$p_{i_t}\#b_{i_t}; \emptyset, \emptyset), (b_{i_t}\#\$p_{i_t}b'_{i_t}\#; \{p_{i_t}\}, \emptyset),$$
  $$(\#b'_{j_t}\$p_{j_t}\#b_{j_t}; \emptyset, \emptyset), (b_{j_t}\#\$p_{j_t}b'_{j_t}\#; \{p_{j_t}\}, \emptyset),$$
  $$(\#g'_{i_t}\$p_{i_t}\#g_{i_t}; \emptyset, \emptyset), (g_{i_t}\#\$p_{i_t}g'_{i_t}\#; \{p_{i_t}\}, \emptyset),$$
  $$(\#g'_{j_t}\$p_{j_t}\#g_{j_t}; \emptyset, \emptyset), (g_{j_t}\#\$p_{j_t}g'_{j_t}\#; \{p_{j_t}\}, \emptyset),$$
  $$(\#X_t\$\#X_{t+1}; V' \cup \{p_{i_t}, p_{j_t}\}, V).$$

- $N_1 = (M_1, Z, \{X_{m+1}\}, U - V - \{X_{m+1}\}, \emptyset, V, (1))$, where M_1 contains the
  following rule:
  $$(\#X_{m+1}\$Z\#; \{Z\}, \emptyset).$$

The working of the above NEPFS is exactly similar to the NEP constructed for solving the same problem in[1]. Node N_0 will generate all possible strings of the form $c_1 c_2 \dots c_n X_1$, where each $c_i \in \{b_i, r_i, g_i\}$, $1 \le i \le n$. This takes $2n$ steps. Once, the candidate solutions are generated, the network checks whether the colorability property is satisfied for each of the edges. This is done by means of the nodes $N_{e_t}^{(b)}, N_{e_t}^{(r)}, N_{e_t}^{(g)}, N_{e_t}^{(b')}, N_{e_t}^{(r')}, N_{e_t}^{(g')}$ and finally by N_{e_t} for each edge e_t in 18 steps. As one can see, the overall time of a computation is $18m + 2n$. Thus, the above network finds all possible colorability for the graph G (if there exists any) in $O(m + n)$ time.   □

5   Conclusions

In this paper, we have considered a mechanism inspired from cell biology, namely networks of evolutionary processors with splicing rules and forbidding context (NEPFS). These networks consist of nodes which are very simple processors and are able to perform splicing operation on the strings present in the node. Moreover, we check for the absence of certain symbols (forbidding context) in the strings before applying splicing rule on them. The nodes are endowed with input and output filter which are defined by random context conditions. Networks with two nodes are able to generate all recursively enumerable languages. Using this model we have solved two well known NP-complete problems.

Similar to the NEPFS, we can have NEPPS (Network of Evolutionary Processors with Splicing Rules and Permitting Context), where instead of checking for the absence of symbols while applying the rules, we check for the presence of certain symbols (permitting context) while applying a rule. We can also use simple splicing rules[11] in each processor. Both these models can be shown to be computationally complete. Also, we can simulate Boolean circuits using NEPFS and NEPPS.

References

1. J. Castellanos, C. Martin-Vide, V.Mitrana, and J. Sempere. Networks of evolutionary processors. *Acta Informatica*, 39:517–529, 2003.
2. E. Csuhaj-Varju, J.Dassow, J.Keleman, and Gh. Păun. *Grammar Systems: A Grammatical Approach to Distribution and Cooperation.* Gordon and Breach, London, 1994.
3. E. Csuhaj-Varju and A. Salomaa. Networks of parallel language processors. In *New Trends in Formal Languages*, pages 299–318, 1997.
4. L. Errico and C. Jesshope. Towards a new architecture for symbolic processing. In *Artificial Intelligence and Information Control Systems of Robots'94*, pages 31–40, 1994.
5. M. R. Garey and D. S. Johnson. *Computers and Intractability, A Guide to the Theory of NP-Completeness.* W. H. Freeman and Company, 1979.
6. A. M. Gibbons. *Algorithmic Graph Theory.* Cambridge University Press, 1985.
7. W. D. Hillis. *The Connection Machine.* MIT Press, Cambridge, 1985.
8. K. Krithivasan, M. Sakthi Balan, and P. Harsha. Distributed processing in automata. *International Journal of Foundation of Computer Science*, 10(4):443–464, 1999.
9. C. Martin-Vide, V. Mitrana, M. Perez-Jimenez, and F. Sancho-Caparrini. Hybrid networks of evolutionary processors. In *Proc. of GECCO 2003*, pages 401–412, 2003.
10. Gh. Păun. Computing with membranes. *Journal of Computer and System Sciencea*, 61(1):108–143, 2000.
11. Gh. Păun, G. Rozenberg, and A. Salomaa. *DNA Computing.* Springer-Verlag, 1998.

A Multiplexed Current Source Portable Stimulator Architecture for a Visual Cortical Neuroprosthesis

J.M. Ferrández[1,2], M.P. Bonomini[1], and E. Fernández[1]

[1] Instituto de Bioingeniería, Universidad Miguel Hernández, Alicante
[2] Dpto. Electrónica, Tecnología de Computadoras,
Univ. Politécnica de Cartagena, Cartagena, Spain
jm.ferrandez@upct.es

Abstract. A multiplexed current source architecture for neuroprosthetic purposes is presented. This device uses a penetrating multielectrode array which will be implanted in visual cortex, offering different signal amplitude sets with the programable current source module. This electrode array has been proved for injecting current (charge) in a safety, secure and precise way. This specifity will provide the possibility to adapt the current level to different electrodes impedances and to a particular person characteristics. The control submodule consists in a microprocessor based circuit with programable waveforms which can be configured and an adaptable power stage. With the proposed system, a wide stimuli set can be used for obtaining the optimal parameters to use in a visual neuroprosthesis using as input a retinomorphic system.

1 Introduction

Implantable stimulators such as pacemakers [1], cochlear implants [2], deep brain [3], and spinal cord devices [4] improve the quality of life of cardiac patients, deaf people, Parkinson disease affected, and decrease the pain in some diseases, so nowadays they are implanted regularly in most hospitals. However visual prosthesis are in a promising initial stage. Electrical stimulation of the visual cortex produces localized visual perceptions called phosphenes [5] [6]. The retinotopic organization of primary visual cortex would produce an ordered arrangement of phosphenes by stimulating this area through the stimulation of spatially distributed electrodes.

There exists a vast number of scientific areas which need to establish a synergic cooperation in order to give sight inside some of the tasks to perform: they include neural coding analysis, system modeling, hardware implementation, biocompability of materials, etc. The stimulator must drive the array of electrodes implanted in occipital cortex in a safe way, so balanced pulses are required, using a minimal size and low power consumption prosthesis.

In a stimulator circuit, the objective is to transfer energy from the device to the tissue in a precisely controlled process. The stimulator produces energy,

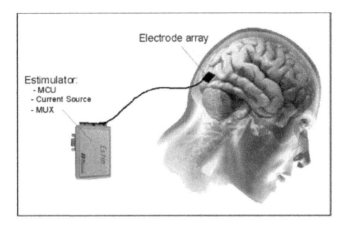

Fig. 1. The proposed visual prosthesis

which origins a voltage across the electrodes, passing current through the tissue, and the rest of the energy dissipates as heat in the system and should be minimize in order to improve the stimulation efficiency. So it is important to study the relation stimulator/electrodes defining the appropriate stimulation and configuration patterns.

The duration, frequency and either the voltage or the current amplitude must be controlled very accurately by the stimulator, however the tissue impedances are highly variable. Current injection provides more control over the injected charge and behaves similar to the natural process which elicits an action potential in excitable tissues. So, this approach has been widely used in cochlear or retinal implants.

In this paper, a multiplexed current source architecture is presented. It drives a penetrating multielectrode array which will be implanted in visual cortex offering different signal amplitude sets with the programable current source device. This specifity provides the possibility to adapt the current level to different electrodes impedances and to a particular person characteristics. This is a crucial process due to the different phosphenes threshold (variability) required by the potential users, and because of an adaptation process to the current levels provided by the brain.

2   The Electrode Array

Present design and manufacturing of intracortical electrodes may be considered as a real choice for neuroprosthetics devices intended to the recovery of motor and sensory abilities. They also provide an ideal set-up for stimulating and recording experiments from large neuron populations. Electrical stimulation of single or small populations of neurons in the central nervous system using penetrating microelectrodes arrays requires current pulsing for eliciting action potentials.

However, this approach must fulfil several conditions:

1. Alive tissue must accept the device *(biocompatibility)*
2. Electrical stimulation of nerve tissue must be induced by means of a charge displacement *(polarization)*, which is performed efficiently through current stimulation.
3. Current injection in neural tissue must be effective and safe inside a working zone delimited by electrolysis reactions nature.

The intracortical multielectrode array uses a three-dimensional architecture based on a silicon Pt/Ir ended needle array (Utah Electrode Array-UEA) [7]. This electrode array has been used extensively in acute and chronic recording experiments. It has been implanted for years and it has the FDA autorization for its use in humans.

The UEA is composed by an array of silicon p+ needles on a 0.12 mm thick monocrystaline substrate. Single electrodes are tipped by a platinum deposition and the whole, but the tips, is covered by a polyamide coating. By means of a thermomigration process,the silicon of each needle is contacted to pads laying on the other side of the substrate. These pads are ultrasonically soldered to the leading wires. The Pt tips are 50 μm height and 2 μm of radius, this expose a overall conducting area of $3,1 \cdot 10^{-4}$ cm^2 . Also, the space between adjacent electrodes are separated by 400 μm, minimizing the capacitive coupling (Fig. 2).

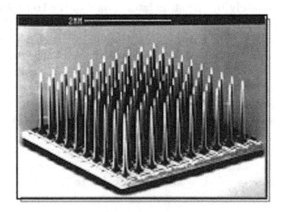

Fig. 2. Scanning electron micrograph of the Utah Electrode Array

The multielectrode array is nearly the most relevant part in a neuroprosthesis because it must must provide enough charge yo evoke action potentials,the tissue must accept the device with few reaction, so biocompatible coverings are required, it must communicate with the stimulator, and it has to be durable in order to be implanted for many years. The Utah multielectrode array verify this requirements.

3   The Current Source Scheme

The current source circuit delivers stimulation to the neurons via the electrode array. It has been used current source because a precise charge delivery is a crucial parameter for eliciting an electrical stimulation of neuron populations. A voltage source cannot maintain constant current or constant charge delivery while tissue impedance is changing over time, and hence it will produce changes in the current, charge, delivered.

Charge injection involve electron transfering across the electrode-tissue interface and therefore need that some chemical species be either oxidized or reduced. Metal electrodes must inject charge predominantly by faradic processes because the charge required to elicite a physiological response far exceeds that available from a capacitive mechanism. Faradic processes may be reversible ot not. In reversible faradic processes the chemical reactions produce no new chemical species in the electrode boundaries, then the system remains equal with a charge-balanced current stimulus. However, irreversible faradic processes involve production of chemical species that do not remain bounded to the electrode surface, these reactions lead to electrode corrosion by electrolysis reactions. A precise current source control prevent an irreversible electrode dissolution. As a conclusion, for a durable implantation of such a prosthesis, UEA material must be biocompatible and electrical-charge displacements must be in margins preferably of capacitive work, never in a irreversible-faradic working zone.

A 30-channel constant current stimulator was designed using op-amp based. Figure 3 shows the current source schematic, which allocates the analog circuitry (stimulation current source, multiplexing stage and current shunt differential amplifier). Two exhausting channels (output channels connected to ground through a low resistance) are included if it's desired to prevent charge accumulation on electrode before stimulating.

The multiplexed current source module includes the stimulation stage formed by the Digital to Analog Converter sub-system, the voltage controlled current source and the multiplexing circuitry. As is noted in the Figure below, an instrumentation stage is included to drive the current delivered in form of voltage signal to one of the control board analog inputs. Furthermore, the stimulation voltage is also driven to the other analog input, giving the designer the possibility of make measurements and take decisions depending on these parameters.

For this implementation an operational amplifier-based current source referenced to ground was used. The advantage of this topology is its high linearity and output impedance. Also, if $R4 \cdot R_{i2} = R_{i1} \cdot R5$ then IOUT $=$VOUT $/ R_{i2}$ (i.e. the transconductance gain depens only on R_{i2} , minimizing gain drift). The voltage follower is strictly recommended to ensure the integrity of impedance in positive terminal of source operational amplifier.

A drawback of this philosophy is the neccesity of power continuously the device in stimulation working mode, even in zero-current states , this generates an average consumption of 20mA, powered with +5V and a clock frecuency of 4MHz. Of course this can be minimized to over 10mA using +3.3V power. In this case is interesting to maintain this power voltage in order to get from

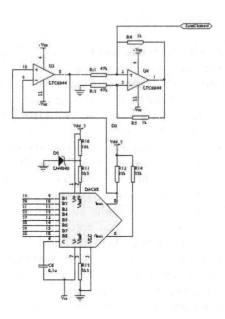

Fig. 3. The voltage controlled current source circuit

the DC/DC pump up to 10V, to determine the maximum stimulation voltage requirements, although lower voltages (over 3V) are expected for valid sensorial stimulation.

4   The Control Stage Submodule

The digital stage comprises the microprocessor, an Arizona Microchip PIC16F877, the power stage and an I2C read only memory for logging purposes. Figure 4 shows the lay out of this module. PIC16F877 is one of the most commonly used microcontroller. It has an operating speed up to 20 MHz,it includes 8 kBytes Flash Program Memory, 368 Byte RAM Data Memory, 256 Byte EEPROM Data Memory, In-Circuit Serial Programming. (ICSP) via two pins (it is importan to reprogram the chip on the board wen using SMT technology), and a wide operating voltage range: 2.0V to 5.5V. It is oriented to a stand-alone operation, powered with batteries but includes communication circuitry to link with a host PC on RS-232 port, in order to download the configuration to the device and analize the data from the stimulation and the electrodes.

The power stage has regulation circuitry to fit +5V from an external 9V battery, and DC/DC charge pump to get 10V (enough voltage for delimitation of maximum compliances). Also has an inverter to draw -5V to the analog stage if desired in future applications. The eeprom used is Microchip 24LC64, an I2C 64K CMOS memory for storing the relevants parameters of the stimulation,

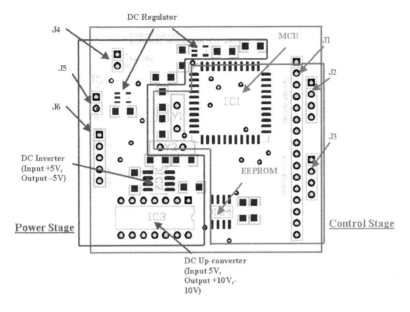

Fig. 4. The digital module layout

impedances, time, battery level. The device includes also a jumper configuration. The switch gives the possibility of power the stimulation stage and the control stage from different sources. With the upper contact the stimulation power comes from the main battery through the DC/DC up-converter, which is feeding the rest of the circuitry (J4). If it is short-cut the down-contact (by means of a 0Ohm 1206 SMD bridge or by means of a soldering drop) the battery connected to J3 powers only the stimulation stage, and the battery connected to J4 feeds the MCU. This feature provides flexibility in powering the device. The rest of the jumpers are used for connectors, as Control Bus, In Circuit Serial Programming, PC interface...

For each channel, the stimulator is capable of delivering 100 μA, 75 μA, 50 μA and 25 μA biphasic pulses at programmable duration and latencies. Each channel was capacitively-coupled to ensure the delivery of charge-balanced pulses. Channel pulses were secuentially delivered to the electrodes in an ordered process.

This control submodule is customizable from a Windows-based graphical user interface (GUI) where the amplitude, frequency and phase width for each electrode can be defined. Pulse width values from the system are 50, 100, 150, 200 ms, the frecuencias are 25 Hz, 50 Hz, 75 Hz and 100 Hz, and as mentioned, the amplitudes cover 100 μA, 75 μA, 50 μA and 25 μA. Once defined the pulse waveform for each electrode and if it is active or not and sent to the device, the program can start or stop the stimulation. It also includes built-in safety features that consist in the ability to sense the failure of any of the output drivers or other modes of operation that could result in charge imbalance and tissue damage.

 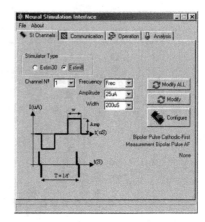

Fig. 5. The graphical user interface software, active channels (right) and pulses configuration (left)

5   Results

In order to characterize the stimulation parameters, the multielectrode array was inmersed in saline solution, the array was dipped into a ringer bath, with an Ag/AgCl reference electrode closing the circuit. A shunt resistance (Rshunt) of 150 Ω was added to give information about the current trespassing the interface.

The current waveform used for stimulation is an Anodic-First (AF) bipolar current pulse delivered at 25 Hz. This signal allows the re-polarization of the neural media in the second phase, therefore, it is easy to monitor the evoked spikes, minimizing stimulus artifacts, and as it was described earlier, it is more adequate to stimulate by means of constant current stimulus because it is more reliable to control the charge injection through any electrode-tissue interface. Because charge injection becomes independent of tissue impedances, an increase in safety is achieved.

In the following Figure an AF current pulse is shown with its associated delivered voltage waveform observed in the medium. A predominant capacitive behavior is observed since voltage linearly increases with the onset of the current waveform and a consequent negative voltage slope appears while the cathodic pulse. For larger amounts of charge, diffusion components become significant, therefore, a more resistive behaviour is observed which compromises the electrode integrity by means of no-reversible REDOX processes.

Three stimulation signal points are selected to get voltage information about Access potential and Capacitive cycles at interface. They are referred in the Figure as V Access, V anodic and V cathodic. The voltage measurement at the end of current injection gives idea about the charge storage at interface, and the measurement at Access point gives a reading of ESA (Electrochemical surface area) at electrode tip. The Figure shows the tipical voltage response of a working electrode, and denotes that for usual stimulation pulses polarization

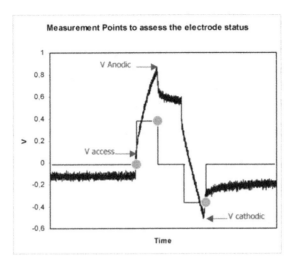

Fig. 6. Anodic First current source pulse and corresponding voltage waveform observed in the medium

access voltages work on the 1 volt region, achieving safety values. Total electrode polarization towards Ag/AgCl Reference keeps on the zone of $+/-3V$ for a large population of electrodes (large charge pulses). This voltage compliance ensures the working inside of safety margins delimited by water electrolysis levels.

6   Conclusions

A multiplexed current sourced device has been developed for neuroprosthetic purposes. In order to achieve the requirements of constant current through variable impedances, a current source circuit has been used which addresses a penetrating multielectrode array using a multiplexation stage. This electrode array has been proved for injecting current (charge) in a safety, secure and precise way. The control submodule consists in a microprocessor based circuit with programmable waveforms which can be configured using a PC connection (RS-232) and a power stage which can be used with different configurations through hardware conectors. With the proposed system, a wide stimuli set can be used for testing and obtaining the optimal parameters to used in a visual neuroprosthesis using as input a retinomorphic response of camera images.

The results concluded that the main advantage using an operational-amplifier current source is the high linearity at both high currents (some mA) and low currents (a few mA). Also, a high output impedance is easily reached only caring components matching (feedback resistance values). The main disadvantage is the dependence of operational amplifier slew-rate. With high slew rate, the common OA power consumption rises considerably and gain bandwith product decreases. Also the noise voltage is important in little current deliverings, so, also it is a relevant parameter to consider. The ideal solution in this case should be to get a design working in conmutation between stimulation cycles, and in linear zone in current delivering working zones. This will be accomplished in future designs with a BJT based circuit.

Nowadays, a new module with the multiplexing stage in the electrode is being designed which includes RF link. The BJT based circuit will obtain a higher slew rate and a better compliance voltage than the operational circuit, and will discard expensive power operational amplifiers. Better compliance towards power ratio, than in operationan amplifier topologies is expected.

This device is currently being used currently in animal experimentation using acute and cronic stimulation.

Acknowledgements

This research is being funded by E.C. QLF-CT-2001-00279 and Ministerio de Ciencia y Tecnología TIC 2003-09557-C02-02.

References

1. Woollons, D.J. "To beat or not to beat: the history and development of heart pace-makers", IEEE Engineering Science and Education Journal, 4(6), 1995, pp. 259–269.
2. G. Loeb "Cochlear prosthetics", Annual Review in Neuroscience, vol. 13, 1990, pp. 357–371.
3. Medtronic "Deep Brain Stimulation", http://www.medtronic.com/
4. North RB, Kidd DH, Olin JC, Sieracki JM. "Spinal cord stimulation electrode design: prospective, randomized, controlled trial comparing percutaneous and laminectomy electrodes-part I: technical outcomes", Neurosurgery. 2002 Aug, 51(2), pp. 381–9.
5. Schmidt E.M., Bak M.J. et al "Feasibility of a Visual Prosthesis for the Blind Based on Intracortical Stimulation of the Visual Cortex", Brain 119, 1996, pp. 507–522.
6. Normann R., Maynard E., Guillory K., Warren D "Cortical Implants for the Blind", IEEE Spectrum, May, 1996, pp.54–59.
7. Campbell P., Jones K., Huber R., Horch K. Normann R "A Silicon-Based, Three-Dimensional Neural Interface:Manufacturing Processes for an Intracortical Electrode Array", IEEE Transactions on Biomedical Engineering, Vol.38 N 8 , pp.758–767.

An Augmented Reality Visual Prothesis for People Affected by Tunneling Vision

F. Javier Toledo, J. Javier Martínez, F. Javier Garrigós, and J. Manuel Ferrández

Dpto. Electrónica, Tecnología de Computadoras y Proyectos, Universidad Politécnica de Cartagena, 30202 Cartagena, Spain
javier.toledo@upct.es

Abstract. An augmented reality based visual prothesis for people affected by tunnel vision is described in this paper. Using a head mounted display, useful information of the environment is superimposed on the own patient's view, to enhance the user's knowledge of the environment. That information is extracted by processing the images obtained from a camera. In order to provide versatility to the system and to accomplish the requirement of high performance, the image processing algorithm is performed with a cellular neural network, implemented on an FPGA device. The discrete CNN model proposed is derived from the continuous model, and details of its hardware implementation and the overall system architecture are reported.

1 Introduction

Since Chua and Yang proposed the Cellular Neural Network (CNN) in 1988 [1], a wide field of research have spread on its applications and implementation. Nowadays, the advantages of the combination of the massive parallelism and the local interaction among cells are already well known, specially for image processing.

Regarding the hardware, VLSI implementation has been the most suitable choice to exploit the CNNs' structure benefits up to now. But the emerging of the Field Programmable Gate Array (FPGA) devices offers a very interesting alternative to ASICs [2], [3], [4], because of their suitability for executing in parallel elementary digital processing tasks and their reconfigurability capabilities. Moreover, their flexibility allows to implement not only specific processing cores, but also interfaces, glue logic and even microprocessor, so making feasible the hardware/sotfware codesign in a System On a Chip.

Augmented Reality (AR) consist of a combination of the real scene viewed by a user and an artificially generated image. So, AR allows the user to see the real world supplemented with additional useful information, enhancing the user's perception and knowledge of the environment (Fig. 1). It is a highly interdisciplinary field, involving areas such as signal and image processing, wearable computing, computer vision and information visualization. The applications of

J. Mira and J.R. Álvarez (Eds.): IWINAC 2005, LNCS 3561, pp. 319–326, 2005.

AR spread on a wide range, from entertainment to military purposes, including medical visualization and training, engineering design, manufacturing and, as proposed in this work, impairment aids.

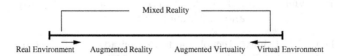

Fig. 1. Milgram's reality-virtuality continuum. It shows how real and virtual worlds can merge. Augmented reality is nearer the real world than the virtual world because the real world perception predominates over the generated data

AR usually implies the use of a Head Mounted Display (HMD), in order to increase the sense of presence in the augmented scene. Different types of HMD can be found [5]. In our application, we have adopted a solution based on a see-through HMD (Fig. 2). It merges the real world and the synthetic image optically, and so it allows the user wearing the HMD to see directly the world in front of his eyes.

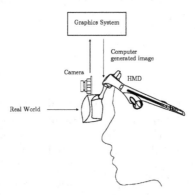

Fig. 2. Solution adopted in this work. The transmissive properties of the optical combiners inside the HMD allow the user to see simultaneously the virtual image, generated in the graphics system from the camera image, and the real world in front of his eyes

An AR-based solution for people affected by tunnel vision is proposed in this work. As it will be described next in section 2, patients can be aided by enhancing their knowledge of the environment with information obtained from a camera by image processing. We have accomplished this with a CNN, whose discrete model is deduced in section 3. In section 4 is presented the overall system architecture. The results of the implementation and some examples are shown in section 5. Finally, conclusions are presented in section 6.

2   Target Application

Tunnel vision is a visual disorder associated to several eyes diseases such as retinitis pigmentosa and glaucoma. As shown in Figure 3, it consists of the loss of the peripheral field of view, while the central field of view maitains high resolution. Consequently, patient's abilities to localize objects and to navigate are severely reduced.

Fig. 3. Simulation of patient's view affected by tunnel vision (right) and normal vision (left)

Traditional aids for affected people are based on the optical reduction of the size of the objects with a minification process. However, this reduces the usefulness of the high resolution central vision, because it just makes possible to see more objects at the expense of seeing them smaller.

A different approach has been proposed by Peli et al. [6] to overcome this disadvantage. It is an augmented reality-based method, which increases the patient's field of view by superimposing in his central field the contour information obtained from a video camera using an HMD. Contour information is extracted by an edge detection algorithm, based on a four-pixel neighbour gradient filter and a treshold function, running on a laptop PC. They report that patients consider the system useful, but a specifically designed system to perform image processing and increase frame rate is necessary.

With this aim, a system based on the FPGA implementation of the Canny edge detection algorithm to extract contour information has been recently developed [7]. Now we aim to increase the versatility of the system, making feasible to tune the output to the desired and to perform different processing algorithms just by changing a template set, while achieving high performance. We get it through a CNN, and its implementation on reconfigurable hardware.

3   The CNN Discrete Model

The dynamics defining the behaviour of Chua-Yang CNN is given by the state equation 1 and by the activation function $f(x_{ij})$ in eq. 2.

$$C\frac{dx_{ij}}{dt} = -\frac{1}{R}x_{ij} + \sum_{k,l\in Nr(ij)} A_{kl}y_{kl} + \sum_{k,l\in Nr(ij)} B_{kl}u_{kl} + I_{ij} , \tag{1}$$

$$y_{ij} = f(x_{ij}) = \frac{1}{2}(|x_{ij}+1| - |x_{ij}-1|) , \tag{2}$$

where I, u, y and x denote input bias, input, output and state variable of each cell, respectively. Neighbourhood distance r for cell (i,j) is given by $Nr(ij)$ function, where i and j denote the position of the cell in the network and k and l the position of the neighbour cell relative to the cell in consideration. B is the constant weights template for inputs feedback and A is the corresponding template for the outputs of neighbour cells. Finally, non-linear activation function of output corresponds to piecewise linear operator (PWL).

In order to obtain a discrete CNN model to be implemented on FPGA, four different approaches have been analyzed: approximation with the Euler method, with the differentia algorithm (TDA-backward), with the numeric integration algorithm (TIA-Tustin) and with the response-invariant algorithm (RIT-first order impulse). The first one presents a predicting behaviour, the second and third ones a delay in their responses, and the fourth one has its response on the ideal continuous output. The objective is to find which one best emulates the continuous model with minimun computational cost when implemented on a FPGA.

Since simulations and temporal analysis carried out show that the Euler model presents the best behaviour, this approximation process is described next.

3.1   Approximation by Euler Methods (Euler-Forward)

The equivalent Laplace transfer function of the cell, (eq. 3) is approximated using the first term of the Taylor factorizing function.

$$H(S) = \frac{\frac{1}{C}}{S + \frac{1}{RC}} . \tag{3}$$

It establishes that a time function $x(t)$ can be approximated at a subsequent time instant $x(t+h)$ by the expresion 4:

$$x(t+h) \approx x(t) + \frac{dx(t)}{dt}h , \tag{4}$$

where h represents sampling constant. This reveals the model's predictive characteristic. From this expression it can also be shown that

$$S \approx \frac{Z-1}{h} . \tag{5}$$

Substituting in the equation 3 and making $R = C = 1$ we obtain, without lost of generality, the subsequent Euler discrete CNN model:

$$X_{ij}[k] = (1-h)X_{ij}[k-1] + h\left(\sum_{k,l\in Nr(ij)} A_{kl}Y_{kl}[k-1] + \sum_{k,l\in Nr(ij)} B_{kl}U_{kl} + I_{ij} \right) . \tag{6}$$

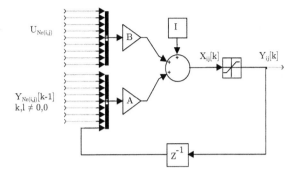

Fig. 4. The proposed discrete CNN model

Figure 4 show the structure of the model derived from equation 6 with sampling constant $h = 1$.

4   System Description

The proposed system consists basically of a camera to acquire images of the environment, a head mounting display to visualize the information that enhances the user's vision, a Xilinx Virtex-II FPGA as processor and controller unit and SRAM memory to store information.

The head mounted display used is the Sony Glasstron PLM-S700E. This is a binocular and high resolution device with SVGA/VGA input and easily adjustable see-through capability. Its design and light weight, just 120 gr. on the head, make possible the use with correction glasses and makes it easy the movement of the head and the mobility of the person. These are key characteristics for our application.

The camera is the M4088, a monochrome camera module with 8 bits digital output that provides a very low cost solution for higher quality video application. It uses OmniVision's CMOS image sensor OV5017 and outputs a 384×288 pixels image at 50 frames per second. A digital interface facilitates the configuration and inicialization of the camera, executed from the FPGA. The camera pixel clock frequency is 14.31 MHz.

The data from the camera are written on SRAM memory at the frequency determined by the camera frame rate, and simultaneously they are read from the memory to be processed by the FPGA. Due to the large size, data between the different stages of the processing can not be stored in the internal BlockRAM memory, and must be stored in external SRAM memories. The output of the last stage is also stored in SRAM memory, where data are read from at the VGA frame rate to be sent to the HMD. The memory interface for controlling the write and read operations is implemented on the same FPGA.

The CNN builds the image processing core to extract contour information. We have adopted a pixel pipeline approach for data input and output, where

5 identical stages as the one shown in Figure 4 are connected in cascade. The connectivity among cells is restricted to a 3 × 3 neighbourhood.

Once processed, data are sent to the HMD togheter with the necessary VGA synchronization signals, which are generated by the FPGA to show a 640 × 480 pixels image at 60 frames per second. Obviously, the image pixels values must be converted into analog signals. The ADV7123, a triple channel digital to analog converter for video applications from Analog Devices, has been used with this purpose.

5   Results

The design has been developed in VHDL, synthesized with Xilinx XST and implemented on a Xilinx XC2V4000 FPGA, using the Xilinx ISE 6.3i. A summary of FPGA resources used is included in Table 1. The percentage corresponds to the overall resources available on Xilinx XC2V4000 FPGA.

The results obtained in different situations, both indoor and outdoor environments, are shown in Figure 5 and 6. A residual 10° field of view has been considered to simulate the tunnel vision effect. In these examples, a 57.9° (H)$\times 45.7^{\circ}$ (V) lens has been used. A wider field of view can be acquired by the camera with the appropiate lens.

Table 1. Summary of used resources

FPGA Resources:		
Slices	1210	5.25%
Flip flops	157	0.34%
4 input LUTs	2125	4.61%
Multipliers 18×18	90	75%
Global clocks	5	31.25%

Fig. 5. Simulation of patient's view through the HMD in an indoor environment. It can be observed that the clothes stand is situated on a corner and beside a shelf

Fig. 6. Simulation of patient's view through the HMD in an outdoor environment, showing a car in a parking

6   Conclusions

An augmented reality system has been developed to aid visually impaired people. In order to achieve the requirements of performance and flexibility, a CNN has been the adopted solution to perform image processing, and an FPGA device has been chosen as the hardware platform. For it, four different approximations to the continuous CNN model have been evaluated, reaching the conclusion that the Euler method yields to the best results both in behaviour and in hardware cost. With the proposed system, the patient's limited view of the environment is enhanced superimposing on it the contour information extracted from a video camera image by means of a head mounted display with see-through capability. So, the person's abilities to localize objects, orientate and navigate are improved. The results make clear the viability and utility of the system.

However, the CNN architecture must be optimized, in order to reduce the used resources and make feasible a fully parallel implementation. Future works will mainly focus on this topic.

Acknowledgements

This research is being funded by Ministerio de Ciencia y Tecnología TIC 2003-09557-C02-02 and Fundación Séneca PI-26/00852/FS/01.

References

1. L. O. Chua and L. Yang: "Cellular neural networks: theory", IEEE Trans. Circuits and Systems, CAS-35, 1988
2. Nagy, Z., Szolgay, P.: "Configurable multi-layer CNN-UM emulator on FPGA", Proc. IEEE Int. Workshop on Cellular Neural Networks and Their Applications, Frankfurt, Germany, 2002, 164–171.

3. Malki, S., Spaanenburg, L.: "CNN image processing on a Xilinx Virtex–II 6000", Proc. European Conference on Circuit Theory and Design, 2003, Krakow, Poland, 261–264.
4. Martínez, J.J., Toledo, F.J., Ferrández, J.M: "New emulated discrete model of CNN architecture for FPGA and DSP applications", 7th Int. Work-Conference on Artificial and Natural Neural Networks, LNCS 2687, Menorca, Spain, 2003, 33–40.
5. Azuma, R.T.: "A survey of augmented reality", Presence: Teleoperators and virtual environments, vol. 6, no. 4, 1997, 355–385.
6. Vargas Martín, F., Peli, E.: "Augmented view for tunnel vision: device testing by patients in real environments", Digest of Technical Papers, Society for Information Display International Symposium, San Jose, USA, 2001, 602–605.
7. Toledo, F.J., Martínez, J.J., Garrigós, F.J, Ferrández, J.M.: "Augmented reality system for visually impaired people based on FPGA", Proc. 4th IASTED Int. Conf. on Visualization, Imaging and Image Processing, Marbella, Spain, 2004, 715-723.

Eye Tracking in Coloured Image Scenes Represented by Ambisonic Fields of Musical Instrument Sounds

Guido Bologna and Michel Vinckenbosch

University of Applied Science HES-SO,
Laboratoire d'Informatique Industrielle,
Rue de la Prairie 4, 1202 Geneva, Switzerland
`Bologna@Eig.unige.ch`

Abstract. We present our recent project on visual substitution by Ambisonic 3D-sound fields. Ideally, our system should be used by blind or visually impaired subjects having already seen. The original idea behind our targeted prototype is the use of an eye tracker and musical instrument sounds encoding coloured pixels. The role of the eye tracker is to activate the process of attention inherent in the vision and to restore by simulation the mechanisms of central and peripheral vision. Moreover, we advocate the view that cerebral areas devoted to the integration of information will play a role by rebuilding a global image of the environment. Finally, the role of colour itself is to help subjects distinguishing coloured objects or perceiving textures, such as sky, walls, grass and trees, etc ...

1 Introduction

In 2000 the World Organisation of Health estimated that 45 millions people were blind or visually impaired. For a blind person, the quality of life is appreciably improved with the use of special devices, which facilitate precise tasks of the every-day life, such as reading, moving and manipulating objects in a familiar environment.

Echolocation is a mode of perception used spontaneously by many blind people. It consists in perceiving the environment by generating sounds and then listening to the corresponding echoes. Reverberations of various types of sound, such as slapping of the fingers, murmured words, whistles, noise of the steps, or sounds from a cane are commonly used. It is thus possible to enhance spatial information of the environment by means of an auditory image.

Even if data fusion is a scientific and technical domain with increasing interest, engineers generally transpose in the devices they design, perception as an acquisition step followed by a linear chaining of treatments. Perception is much more complex; it is a mental construction organizing the various sensory signals in a whole, ensuring a maximum of coherence between the central representations of the movement and the other sensory channels [2]. This integration makes us perceive our environment like a stable and coherent whole, whereas the

J. Mira and J.R. Álvarez (Eds.): IWINAC 2005, LNCS 3561, pp. 327–337, 2005.

nervous system is organised to treat fugitive events, to produce and use changing flows of information. Thus, the coordination of the movements of the eye and the head are closely related to the retinal image processing [7], to ensure the orientation and the stabilisation of the visual glance on the one hand, the construction of stable space representations on a higher level of integration, on the other hand.

This work presents our recent project on visual substitution by representations of the spatial environment with a 3D Ambisonic sound field [10]. The basic idea is to represent every pixel of an image by a sound source located at a particular azimuth and elevation angle. Moreover, each emitted sound is assigned to a musical instrument, depending on the colour of the pixels. Our long term goal is to produce and validate an autonomous non-invasive prototype which combines an eye tracker with an auditory encoding of the space environment restoring the mechanisms of central and peripheral vision. In the following sections, section two gives a brief introduction on spatial hearing, section three presents several vision substitution systems by auditory representations, section four presents the Ambisonic spatialisation technique, section five describes our system, followed by several concluding remarks.

2   Spatial Hearing

The physical effects of the diffraction of sound waves by the human torso, shoulders, head and outer ears (pinnae) modify the spectrum of the sound that reaches the ear drums. Interaural time difference (ITD) and interaural intensity difference (IID) are important cues for the localisation of lateral sources. As shown in figure 1 at position A, the path lengths are equal, causing the wave front to arrive at the eardrums at the same time and with equal intensity. At position B, the sound source is on the right side of the listener, thus a diffracted wave front arrives to the left ear with some delay with respect to the right ear. The relation of a wave front to an obstacle of fixed size is such that the shadow effect increases with increasing frequency. For instance, a 3kHz sine wave at 90 degrees azimuth will be attenuated by about 10 dB.

Although lateralisation is able to be simulated with appropriate delays and differences of intensity between the two ears, reproducing the perception of elevation is much more difficult. The ability to disambiguate sources from above and below in cases where ITD and IID would not supply this information has brought about hypotheses regarding the role of spectral cues. In fact, the folds of the pinna cause echoes with minute time delays within a range of 0-300 μsec [5] that cause the spectral content of the eardrum to differ significantly from that of the sound source. These changes are captured by the head-related transfer functions (HRTFs), which not only varies in a complex way with azimuth, elevation, range, and frequency, but also varies significantly from person to person [1], [5]. Strong spatial location effects are produced by convolving a monaural signal with head-related impulse response (HRIR) which represents the HRTFs in the time domain.

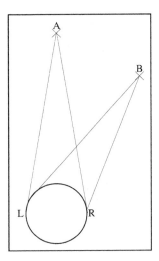

Fig. 1. An example of two oriented sound sources in the horizontal plane. A is directly ahead with respect to the listener, while B is on the right side

Several researchers have proposed replacing measured HRTFs with computational models [5]. For instance introducing Notch filters into the monaural spectrum creates definite elevation effects. However, the variability between different persons is very important and difficult to control. Externalisation is also an important problem to solve. Specifically, sounds in front appear to be too close and front/back reversals are very common. HRTFs measurements are available on the public domain. One of the most important databases is the CIPIC database[1] [1]. Release 1.0 was produced for 45 different persons; specifically, impulses responses were measured in 1250 distinct directions for each ear and for each subject.

3   Visual Substitution Devices Using Audition

The "K Sonar-Cane" combines a cane and a torch with ultrasounds [12]. With such a device, it is possible to perceive the environment by listening to a sound coding the distance and to some extent the texture of the objects which return an echo. The sound image is always centred on the axis pointed by the sonar.

"The Voice" is a system where an image is represented by 64 columns of 64 pixels [14]. Every image is processed from left to right and each column is listened for about 10 ms. Specifically, every pixel in a column is represented by

---

[1] Retrievable on the internet site http://interface.cipic.ucdavis.edu.

a sinusoidal wave with a distinct frequency. High frequencies are at the top of the column and low frequencies are at the bottom. Overall, a column is represented by a superposition of sinusoidal waves with their respective amplitudes depending on the luminance of the pixels. The signal of a column σ_j ($j = 1..64$) is given by

$$\sigma_j = \sum_{i=1}^{64} p_{ij} \cdot \sin(2\pi f_i); \tag{1}$$

where p_{ij} represents the luminance matrix and f_i the frequency at position i on column j.

More recently, Wong *et al.* modified the audio encoding of "The Voice" by adding a stereophonic component [15]. Thus, rather than feeding the columns sequentially, column j is on the left channel of the headphone, while column $max - j$ is on the right channel, with max being the number of columns of the image, splitting right and left perception fields.

Capelle *et al.* proposed the implementation of a crude model of the primary visual system [6]. The implemented device provides two resolution levels corresponding to an artificial central retina and an artificial peripheral retina, as in the real visual system. Acquisition of a visual scene results in a digitised image stored as a matrix of 124 pixels. The 64 foveal pixels have a finer spatial resolution than the 60 peripheral ones. The auditory representation of an image is similar to that used in "The Voice" with distinct sinusoidal waves for each pixel in a column and each column being presented sequentially to the listener. Experiments carried out with 24 blindfolded sighted subjects revealed that after a period of time not exceeding one hour, subjects were able to identify simple patterns such as horizontal lines, squares and letters.

A more musical model was introduced by Cronly-Dillon *et al.* [8]. First, the complexity of an image is reduced by applying several algorithms (segmentation, edge detection, ...). After processing, the image contains only black pixels. Pixels in a column define a chord, while horizontal lines are played sequentially, as a melody. When a processed image presents too complex objects, the system is able to apply segmentation algorithms to these complex objects and to obtain basic patterns such as squares, circles and polygons. Experiments carried out with normal and (elderly) blind persons showed that in many cases a satisfactory mental image was obtained. Nevertheless, this sonification model requires a very strong concentration from the subjects and thus is a source of intellectual fatigue.

Gonzalez-Mora *et al.* have been working on a prototype for the blind in the Virtual Acoustic Space project [11]. They have developed a device which captures the form and the volume of the space in front of the blind person and sends this information, in the form of a sounds map through headphones in real time. Their original contribution was to apply the spatialisation of sound in the three dimensional space with the use of HRTFs. As a result, the sound is perceived as coming from somewhere in front of the user. The first device they achieved was capable of producing a virtual acoustic space of 17x9x8 pixels covering a distance of up to 4.5 meters.

4   Synthesis of Plane Waves

The aim of a 3D sound system is the reconstruction of the acoustic sensations that a particular listener would perceive in a real environment. Multi-channel sound systems, well established in the cinema industry such as Dolby surround sound system, try to create these types of acoustic sensations. Nevertheless, these systems, although they are suitable for cinema do not provide a good localisation of sound sources in the 3D space [3].

Sounds emitted by loudspeakers at a reasonable distance from the listener can be approximated by plane waves. Our purpose is to reproduce a 3D sound field, in order to recreate as closely as possible the perception of localised sound sources. Ambisonic is a method for 3D sound production [9], [10], [13], based on the construction of the desired wave field by using several loudspeakers. Specifically, the key idea behind Ambisonic is the reconstruction of plane waves with the use of a limited number of spherical harmonics. Note that Ambisonic is able to provide a higher level of localisation due to its ability to include more information about the sound field than stereo or Dolby surround can include.

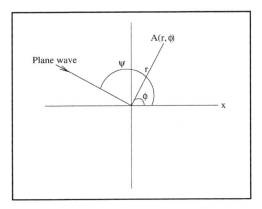

Fig. 2. An example of plane wave with an incident angle ψ. A is the listening point at a radial distance r and at an angle ϕ

For the seek of simplicity let us describe a two-dimensional case of a plane wave. Suppose that the plane wave is arriving at an angle ψ with respect to the x-axis and that the listening point is at a distance r with an angle ϕ with respect to the x-axis. The plane wave S_ψ perceived at point A of figure 2 is

$$S_\psi = P_\psi e^{ikr\cos(\phi-\psi)};\qquad(2)$$

where P_ψ is the pressure of the plane wave and k is the wave number or $2\pi/\lambda$ (with λ the wavelength).

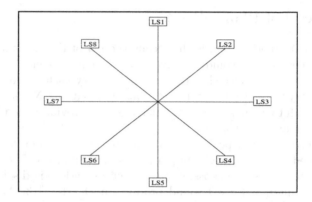

Fig. 3. An example of loudspeaker arrangement without elevation encoding. The listener is surrounded by 8 loudspeakers in the horizontal plane

With the use of Cylindrical Bessel functions $J_m(.)$, (2) becomes [9]

$$S_\psi = P_\psi \left(J_0(kr) + \sum_{m=1}^{\infty} 2i^m J_m(kr)(\cos(m\psi)\cos(m\phi) + \sin(m\psi)\cos(m\phi)) \right) (3)$$

In practice, the plane wave cannot be reproduced exactly, as the number of terms goes to infinity.

Let us assume that we would like to use an array of loudspeakers arranged at the same distance from the centre point in a regular polygon (see figure 3). With this array of loudspeaker, the purpose is to create a directional plane wave in the horizontal plane. The plane wave produced by loudspeaker S_n is

$$S_n = P_n \left(J_0(kr) + \sum_{m=1}^{\infty} 2i^m J_m(kr)(\cos(m\psi)\cos(m\phi_n) + \sin(m\psi)\cos(m\phi_n)) \right) (4)$$

with ϕ_n the angle of loudspeaker n, and P_n its associated pressure.

The constraints by which the plane wave resulting from an array of N loudspeakers will match the plane wave S_ψ are determined by imposing

$$P_\psi = \sum_{n=1}^{N} S_n. \tag{5}$$

After simplification we obtain

$$P_\psi = \sum_{n=1}^{N} P_n \tag{6}$$

Table 1. The 9 encoding channels of a 3D Ambisonic system of the second order. Symbol ψ is the azimuth angle and ϵ represents the elevation angle

Azimuth/Elevation representation	Cartesian representation
0,707107	0.707107
$\cos(\psi)\cos(\epsilon)$	x
$\sin(\psi)\cos(\epsilon)$	y
$\sin(\epsilon)$	z
$1.5\sin(\epsilon)\sin(\epsilon) - 0.5$	$1.5z^2 - 0.5$
$\cos(\psi)\sin(2\epsilon)$	$2xz$
$\sin(\psi)\sin(2\epsilon)$	$2yz$
$\cos(2\psi)\cos(\epsilon)\cos(\epsilon)$	$x^2 - y^2$
$\sin(2\psi)\cos(\epsilon)\cos(\epsilon)$	$2xy$

Table 2. A decoding matrix for an Ambisonic system of the second order with loudspeakers located at the vertex of a cube

Loudspeak. Coordinates	W	X	Y	Z	R	S	T	U	V
(0.5774; 0.5774; -0.5774)	0.1768	0.2165	0.2165	-0.2165	0.0000	-0.1875	-0.1875	0.0000	0.1875
(0.5774; -0.5774; -0.5774)	0.1768	0.2165	-0.2165	-0.2165	0.0000	-0.1875	0.1875	0.0000	-0.1875
(-0.5774; -0.5774; -0.5774)	0.1768	-0.2165	-0.2165	-0.2165	0.0000	0.1875	0.1875	0.0000	0.1875
(-0.5774; 0.5774; -0.5774)	0.1768	-0.2165	0.2165	-0.2165	0.0000	0.1875	-0.1875	0.0000	-0.1875
(0.5774; 0.5774; 0.5774)	0.1768	0.2165	0.2165	0.2165	0.0000	0.1875	0.1875	0.0000	0.1875
(0.5774; -0.5774; 0.5774)	0.1768	0.2165	-0.2165	0.2165	0.0000	0.1875	-0.1875	0.0000	-0.1875
(-0.5774; -0.5774; 0.5774)	0.1768	-0.2165	-0.2165	0.2165	0.0000	-0.1875	-0.1875	0.0000	0.1875
(-0.5774; 0.5774; 0.5774)	0.1768	-0.2165	0.2165	0.2165	0.0000	-0.1875	0.1875	0.0000	-0.1875

$$P_\psi \cos(m\psi) = \sum_{n=1}^{N} P_n \cos(m\phi_n) \tag{7}$$

$$P_\psi \sin(m\psi) = \sum_{n=1}^{N} P_n \sin(m\phi_n). \tag{8}$$

A perfect synthesis of a plane wave requires an infinity number of loudspeakers. In practice, m defines the order of the approximation. For instance, for a first order system with a homogeneous architecture of loudspeakers, it can be demonstrated that

$$P_n = \frac{1}{N}(W + 2X\cos(\phi_n) + 2Y\sin(\phi_n)); \tag{9}$$

with $W = P_\psi$, $X = P_\psi \cos(m\psi)$ and $Y = P_\psi \sin(m\psi)$.

Generalizing equations (6), (7) and (8) for a 3D Ambisonic system of the second order, we obtain the 9 encoding channels illustrated in table 1. Generally, the number of channels required to reproduce a 3D signal is $(m+1)^2$ [9].

An Ambisonic decoder depends on the placement of the loudspeakers. For instance, with eight loudspeakers located on the vertex of a cube, the decoding matrix is represented by table 2.

5   Overview of Our System

Our targeted prototype will present several components:

1. Two cameras for stereoscopic view.
2. An eye tracker.
3. A headphone.
4. A virtual Ambisonic audio rendering system.

The stereoscopic view of the cameras will estimate the distances from the objects and the eye tracker will select a sub-image (see figure 4) that will be transformed into a 3D sound field by the Ambisonic component. In our current implementation the sub-image pointed by the eye tracker has 17x9 pixels and each pixel represents a directional sound source.

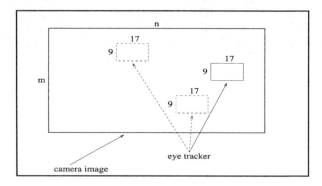

Fig. 4. Representation of an image with mxn pixels and several sub-images resulting from the movement of the eye tracker

After a sub-image has been captured, the flow of processing is illustrated by figure 5. Typically, the sound emitted by a pixel results from a monaural sound that is encoded into 9 Ambisonic channels; with parameters depending on azimuth and elevation angles. Then, the encoded Ambisonic signals are decoded for loudspeakers placed in a virtual cube layout. Finally, the physical sound is generated for headphones with the use of HRIR functions related to the directions of virtual loudspeakers.

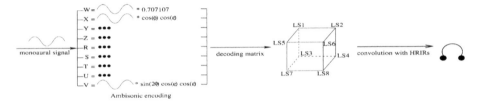

Fig. 5. The distinct steps of processing. From left to right: a monaural sound is encoded into 9 distinct channels, then decoded for loudspeakers placed in a virtual cube layout; finally, the sound is generated for headphones with the use of HRIR functions related to the directions of virtual loudspeakers

The HRIR functions we use, are those included in the CIPIC database [1]. Typically, we choose the HRIR functions giving the best impression of externalisation. In the future we will measure personalised HRIR functions, instead of using CIPIC HRIRs. One of the advantage of the virtual Ambisonic approach is that we will only have to measure those HRIRs for the positions corresponding to virtual loudspeakers, instead of measuring points every five degrees both in azimuth and elevation.

The sub-image selected by the eye tracker will have two resolution levels corresponding to an artificial high resolution macula and an artificial low resolution peripheral retina, as in the real visual system [6]. The artificial periphery will be crucial to determine new focuses of attention. Moreover, we will encode several colours by musical instrument sounds. In our current implementation we use the eight colours defined on the vertex of the RGB cube (red, green, blue, yellow, cyan, magenta, black and white). In practice a pixel in the RGB cube is approximated with the colour corresponding to the nearest vertex. Our eight colours are played on two octaves: (Do, Sol, Si, Re, Mi, Fa, La, Do). Textures of colours such as grass and trees or the sky, are represented by several instruments playing distinct notes. In some cases a particular texture will sound plentiful (major chord: Do, Sol, Mi), in other cases it will be very dissonant (for instance Si, Re, Mi, Fa). The luminance of a colour is encoded with the intensity.

With the use of Matlab, we have implemented the virtual Ambisonic component and the generation of binaural 3D sound fields for headphones. The 17x9 sub-image is actually selected by the mouse pointing on images on the screen, which emulates direct aiming by gaze position. All possible 17x9 pixels have been pre-calculated for each colour and for each azimuth and elevation angle[2]. When the mouse moves and generates a 17x9 sub-image, our program simply retrieves from memory the sound sources corresponding to the new sub-image and mixes all sounds together. Using a Pentium 4 Pc with a clock frequency of 3.0 GHz and 1 GB of RAM, the time latency between sound generation and the new sub-image is 400 ms; with musical instrument sounds sampled at 22050

---

[2] In our current implementation the sub-image is flat; in other words all sounds are emitted from the same distance.

Hz. Ideally, we would like to generate 2-3 images per second, in order to match the mechanism of fixation that typically lasts 300 ms. We are not very far from that goal; a streaming mixer implemented in C or C++ will largely reduce time latency.

6   Conclusion

Our system for visual substitution by the auditory channel presents two novel aspects. The first is the use of an eye tracker that would make it possible to activate the process of attention inherent in the vision and the cerebral areas devoted to the integration of spatial information. The second original aspect of our approach is the encoding of colours by musical instruments, in order to emphasize coloured objects and particular textures. Future experiments with blind subjects will be crucial to determine the correctness of our approach.

References

1. Algazi, V.R., Duda, R.O., Thompson, D.P., Avendano (2001). The CIPIC HRTF Database. *IEEE Proc. WASPAA01*, New Paltz, NY.
2. Andersen, R.A., Snyder, L.H., Bradley, D.C., Xing, J (1997). Multimodal Representation of Space in the Posterior Parietal Cortex and its Use in Planning. *Annu. Rev. Neurosci.*, *20*, 303–330.
3. Bamford, J.S. (1995). *An Analysis of Ambisonic Sound Systems of First and Second Order*. Master Thesis, Waterloo, Ontario, Canada.
4. Begault, R. (1994). *3-D Sound for Virtual Reality and Multimedia*. Boston A.P. Professional, ISBN: 0120847353.
5. Brown, C.P., Duda, R.O (1998). A Structural Model for Binaural Sound Synthesis. *IEEE Trans. Speech and Audio Processing, 6 (5)*.
6. Capelle, C., Trullemans, C., Arno, P., Veraart, C. (1998). A Real Time Experimental Prototype for Enhancement of Vision Rehabilitation Using Auditory Substitution. *IEEE T. Bio-Med Eng.*, *45*, 1279–1293.
7. Colby, C.L., Goldberg, M.E. (1999). Space and Attention in Parietal Cortex. *Annu. Rev. Neurosci.*, *22*, 319–349.
8. Cronly-Dillon, J., Persaud, K., Gregory, R.P.F. (1999). The Perception of Visual Images Encoded in Musical Form: a Study in Cross-Modality Information. *Proc. Biological Sciences*, *266*, 2427–2433.
9. Daniel, J. (2000). *Acoustic Field Representation, Application to the Transmission and the Reproduction of Complex Sound Environments in a Multimedia Context*. PhD thesis, University of Paris 6, (in French, English abstract available).
10. Gerzon, M.A. (1977). Design of Ambisonic Decoders for Multispeaker Surround Sound. *Journal of the Audio Engineering Society (Abstracts)*, *25*, 1064.
11. Gonzalez-Mora, J.L., Rodriguez-Hernandez, A., Rodriguez-Ramos, L.F., Dfaz-Saco, L., Sosa, N. (1999). Development of a New Space Perception System for Blind People, Based on the Creation of a Virtual Acoustic Space. *Proc. IWANN*, 321–330.
12. Kay, L. (1974). A Sonar Aid to Enhance Spatial Perception of the Blind: Engineering Design and Evaluation. *The Radio and Electronic Engineer 44*, 605–627.

13. Malham, D.G., Myatt A. (1995). 3-D Sound Spatialisation using Ambisonic Techniques. *Computer Music Journal, 19 (4)*, 58–70.
14. Meijer, P.B.L. (1992). An Experimental System for Auditory Image Representations. *IEEE Transactions on Biomedical Engineering, 39 (2)*, 112–121.
15. Wong, F.H.T., Nagarajan, R., Yaacob, S., Chekima, A., Belkhamza, N.E. (2000). A Stereo Auditory Display for Visually Impaired. *Proc. TENCON*, 377–383.

Avoidance Behavior Controlled by a Model of Vertebrate Midbrain Mechanisms

David P.M. Northmore and Brett J. Graham

Department of Psychology,
University of Delaware,
Newark, DE 19716, USA
northmor@udel.edu

Abstract. A mobile animat is simulated with a steering controller modeled on vertebrate midbrain mechanisms. As in the brain, the model optic tectum receives direct input from the eyes and is reciprocally connected to the two nucleus isthmi (NI) which respond selectively to looming objects. NI activity feeds back to tectum which controls turning movements. The animat is tested in a 3-D field of obstacles. It discriminates objects at different distances and it avoids "predators" and collision with stationary obstacles.

1   Introduction

In the vast majority of vertebrates, the optic tectum of the midbrain is the pre-eminent sensory processing center. It maintains maps of the sensory modalities that represent the spatial arrangement of objects in the animal's environment and plays a vital role in directing behavioral responses, whether to orient the animal toward objects of interest or to avoid threat. In nonmammalian vertebrates, such as fishes, the tectum is essential for normal sensory guided behavior. Because tectum processes visual, tactile and lateral-line cues, its ablation renders a fish incapable of moving through an environment of obstacles without bumping into them [7] [11]. Tectum is also important in the fine-grained analysis of visual pattern as it forms a detailed spatial representation of the environment. The role of tectum in controlling eye and body movements is also well established. Electrical stimulation of tectum elicits movements of eyes and body that resemble the orienting movements an animal makes to an object of interest such as prey; if the stimulation is strong enough, it elicits behaviors that resemble responses to threat such as predators [1] [5] [6].

It is likely that the connection between the sensory representation over the tectal surface and the premotor output of tectum involves some selective processes, allowing the animal to respond only to the most salient events. A candidate structure for mediating such selection in nonmammalian vertebrates is the nucleus isthmi (NI) of the midbrain tegmentum [8] (See Fig. 1). Each NI has reciprocal interconnections with the ipsilateral tectal lobe. In most animals, NI also projects to the contralateral tectal lobe, but in fishes, the prototype for

J. Mira and J.R. Álvarez (Eds.): IWINAC 2005, LNCS 3561, pp. 338–345, 2005.
© Springer-Verlag Berlin Heidelberg 2005

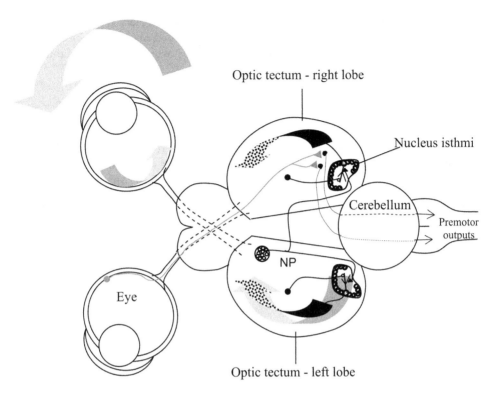

Fig. 1. Schematic of the retino-tectal system of a fish, and the connections of the optic tectum with nucleus isthmi (NI), nucleus pretectalis (NP), and the premotor pathways efferent from tectum. The gray arrows show the visuotopic mapping onto the opposite tectal lobe. Cells in tectum project excitatorily to the ipsilateral NI. NP, which receives tectal input, projects inhibitorily to NI. The stippled arrows show the projection of NI onto the ipsilateral tectal lobes

the model considered here, NI's contralateral projection is minor and will not be considered. Our electrophysiological experiments in several species of fish [2][3][4] have shown that NI (a) responds to novel visual stimulation throughout the field of the contralateral eye; (b) is excited most by object motion, particularly as an object approaches the animal; (c) discharges as a unit, probably mediated by electrical coupling between its neurons [9], and (d) broadcasts its signal across the entire tectal lobe on the same side, mainly in the deeper layers.

Recent experiments on fish with moving ball stimuli demonstrated that the average spike firing density of NI increased as an object approached the eye starting from 15-20 cm distant [2]. Moreover, the firing density on average, gauged the distance of the object, to a great extent independently of its size and speed of approach. Extended, textured surfaces also elicited increased activity as the distance to the eye decreased, suggesting that NI may also play a role in responding during the animal's approach to stationary objects. We present a model based on the fish midbrain (Fig. 2) to see whether these mechanisms could be useful in guiding an animal or animat through a course of obstacles, as well as enabling it to escape from "predators".

2   The Model

We first constructed a network that emulated the visual responses of NI in fishes to approaching and receding single objects. The network was then incorporated into a simulated animat to control its movements within a 3-D space containing visible obstacles. The architecture of this model is shown in Fig. 2. Its main features are as follows:

1. The visual field ($360°$ azimuth, $\pm\ 45°$ elevation) is sensed by two laterally-placed eyes. There is no binocular overlap of the two eyes' visual fields. The

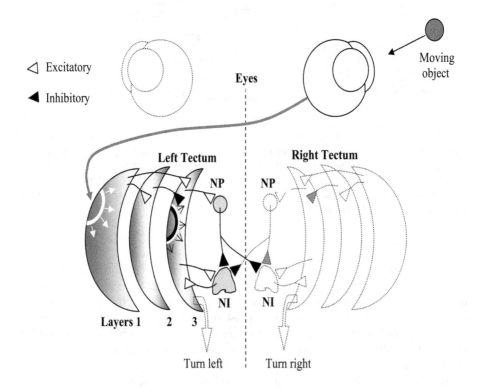

Fig. 2. Architecture of the model showing the connectivity between the eyes, the 3 tectal layers, NP and NI. An object moving toward the right eye generates on Layer 1 of left tectum an expanding boundary of excitation (white), and on Layer 3 an expanding boundary of excitation (white) enclosing a boundary of inhibition (black). Left- and right-turning behaviors are activated by Layer 3 of left and right tectal lobes

receptor spacing in the retinas is about 1°. Each retina maps onto Layer 1 of the contralateral lobe of the optic tectum. Layer 1 neurons, which are linear threshold units, respond to changes in image contrast over receptive fields of 2.4^o diameter.

2. Layer 1 projects topographically and excitatorily to tectal Layers 2 and 3. Layer 2 projects topographically and inhibitorily to Layer 3. All layers are composed of linear threshold units. Thus, object motion generates within Layer 3 an excitatory image superimposed on a delayed image of inhibition.

3. Layer 3 of each tectal lobe feeds the summation of its excitation (> 0) into NI on the same side. NI is modeled by a single linear threshold unit with noise added to its net input.

4. A second pathway from Layer 3 via NP, also a single linear threshold unit, provides inhibitory inputs to both NI. The effect is to make each NI respond more linearly to the proximity of an approaching object.

5. NI feeds its activity back to Layer 3 on the same side, adding an excitatory bias to its units.

6. The difference in summated activity of Layers 3 in the left and right tectal lobes determines the amount and direction of turning of the animat. For example, a predominance of activity in the left tectum turns the animat leftward, away from an approaching object.

3   Results

3.1   NI Responses

The model NI responded briefly to any changing stimulation in the visual field of the opposite eye, including steps in luminance level and object motion. As Fig. 3 B shows, a stimulus ball approaching the eye generated an initial burst, followed by an approximately linear ramp-up in activity as the object neared the eye. The ramp-up in the activity of the model NI increased in slope with increasing approach velocity such that the activity level represented the instantaneous distance of the ball to the eye. The NI in sunfish exhibits very similar behavior, as shown in Fig. 3 A.

Except for very small balls, which gave delayed, sharp-onset rises in model NI activity, balls of different diameters gave very similar ramp-up responses. Size independence was also seen in the responses of NI in fishes [2].

3.2   Avoidance Behavior

Tests were made on the animat with its steering controlled by tectal output as described above, together with a steady forward driving influence. The animat moved at a constant speed over a plane in simulated 3-D space. Obstacles were black or white spheres of various sizes distributed in space, on, above or below the animat's horizontal plane of movement. Fig. 4 shows paths taken by the

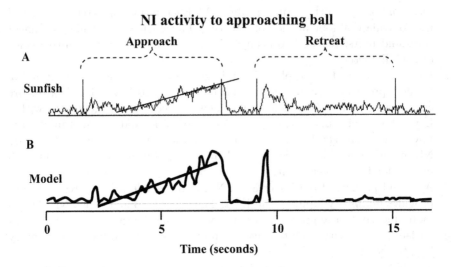

Fig. 3. A. Responses of NI in a sunfish to the approach of a 3.8 cm checkered ball. The ball traveled at a constant 5 cm/s on a horizontal path perpendicular to the body axis through the right eye, first approaching from 39 cm to 9 cm, pausing 3 s, and then retreating along the same path. The record is an average of the integrated spiking activity over 7 trials. A linear regression line is shown fitted to the activity ramp-up during approach. Data from [2]. **B.** Response of the model NI to a similar moving stimulus ball

animat from the center of a clump of spherical obstacles on different trials. With this small number of obstacles the animat rarely collided but did pass over or under obstacles when they were out of the horizontal plane.

To test the contribution of NI to the behavior, trials were run in 11 different courses with 10-42 randomly placed obstacle spheres before and after "lesioning" NI bilaterally. The effect of removing NI was a significant lengthening of the time taken to escape from the clumps of obstacle (t-test, $P = 0.0016$) and a reduced percentage of successful escapes without collision. See Fig. 5.

Tests of the animat were also conducted with a moving spherical "predator" that was released into the obstacle courses. It was targeted upon the animat and moved at the same speed. Because the animat tended to meander, the predator quickly caught up with the animat. The animat with an active NI responded defensively to the approach of the predator and realistic looking chases ensued.

Distant objects that contrasted with the background generally had a weak influence over the motion of the animat, although it steered away from clumps of them. In complex scenes, made up of objects both near and far, large distant objects, although of comparable visual angle to small near objects had less effect on behavior.

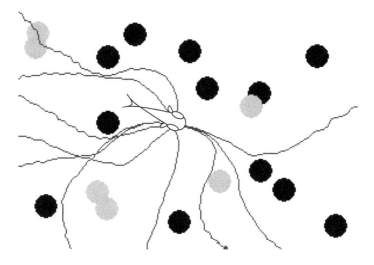

Fig. 4. The animat was released in the center of a cluster of black and white spherical obstacles in 3-D space and allowed to escape. On each trial, the animat was started at different orientations. Ten escape paths are shown. In the top left corner, the animat moved over objects of lower elevation

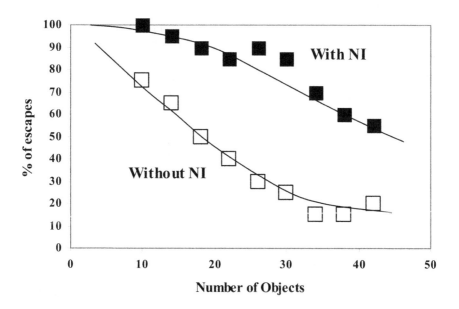

Fig. 5. Effect of lesioning NI on escape from obstacles. The animat was repeatedly placed at the center of a cluster of random obstacles consisting of 10 - 42 black and white objects (spheres). The number of successful escapes was counted both with and without functional NIs

4   Discussion

The model NI reproduces the responses of NI in fishes in all essentials. That is to say, it responded to visual change and motion, and particularly motion toward the eye. This form of motion sensitivity, looming detection, which is important for an animal's survival, is implemented by summing an expanding boundary of excitation and an expanding, but delayed boundary of inhibition. The model not only shows that looming detection can be achieved in this way, but through an additional inhibitory input from NP to NI, the NI response is linearized so as to gauge object proximity, to a great extent independently of object size and speed of approach. The neural mechanisms proposed to occur in tectum are consistent with what is known of tectal physiology.

Less understood is the action of the NI signal fed back to tectum. As NI has no other output, it must exert its effects via tectum. Electrophysiological recordings [4] show that the NI signal is broadcast over the ipsilateral tectum, appearing mainly in the deeper layers. The present model makes the simple assumption that NI activity biases neurons in tectum, increasing the strength of its premotor output, particularly of neurons that cause turning away from visual stimulation. Another possibility is that a comparison between the retinotopic image and the NI feedback results in a selection of the most salient tectal image which then controls behavior. This comparison process might involve synchronization between the firing of NI and tectal cells, which has been observed [4]. A selective synchronization of NI with parts of tectum could be important in selecting which of several objects or events to respond to. NI might then be involved in selecting which of several stimulus objects to avoid or approach.

Avoidance behavior, it can be argued, need not be accurately directed if its function is to get the animal out of harm's way. It must, however, not be evoked unnecessarily e.g. by objects that are non-threatening by virtue of their distance or their lack of movement. The activity of the model NI, and the behavior it facilitates via tectum, is appropriately selective. The mechanism ensures that little heed is paid to distant objects. Moreover the system is not misled by image size; it warns of approaching objects using monocular motion cues. The NI-tectum circuitry, apart from being extremely ancient and successful, is appealing in its simplicity, lending itself to application in mobile robots.

References

1. Al-Akel, A.S., Guthrie, D.M., Banks, J.R. Motor responses to localized electrical stimulation of the tectum in the freshwater perch (*Perca fluviatilis*).Neuroscience 13 (1986) 1381-1391.
2. Gallagher, S.P., Northmore, D.P.M. Responses of the teleostean nucleus isthmi to looming objects and other moving stimuli. (submitted).
3. Northmore, D.P.M. Visual responses of nucleus isthmi in a teleost fish (*Lepomis macrochirus*). Vision Research 31 (1991) 525-535.
4. Northmore, D.P.M., Gallagher, S.P. Functional relationship between nucleus isthmi and tectum in teleosts: Synchrony but no topography. Visual Neuroscience 20 (2003) 335-348.

5. Northmore, D.P.M., Levine E.S., Schneider G.E. Behavior evoked by electrical stimulation of the hamster superior colliculus. Experimental Brain Research 73 (1988) 595-605.

6. Pérez-Pérez, M.P., Luque, M.A., Herrero, L., Núñez-Abades, P.A., Torres, B. Afferent connectivity to different functional zones of the optic tectum of goldfish. Visual Neuroscience 20 (2003) 397-410.

7. Springer, A.D., Easter, S.S., Agranoff, B.W. The role of the optic tectum in various visually mediated behaviors in goldfish. Brain Research 128 (1977) 393-404.

8. Wang, S-R. The nucleus isthmi and dual modulation of the receptive field of tectal neurons in non-mammals. Brain Research Reviews 41 (2003) 13-25.

9. Williams B., Hernandez, N., Vanegas, H. Electrophysiological analysis of the teleostean nucleus isthmi and its relationship with the optic tectum. Journal of Comparative Physiology A 152 (1983) 545-554.

10. Xue, H.-G., Yamamoto, N., Yoshimoto, M., Yang, C.-Y., Ito, H. Fiber connections of the nucleus isthmi in the carp (*Cyprinis carpio*) and Tilapia (*Oreochromis niloticus*). Brain Behavior and Evolution 58 (2001)185-204.

11. Yager, D., Sharma, S.C., Grover, B.G. Visual function in goldfish with unilateral and bilateral tectal ablation. Brain Research 128 (1977) 267-275.

Transition Cells and Neural Fields for Navigation and Planning

Nicolas Cuperlier, Mathias Quoy, Philippe Laroque, and Philippe Gaussier

ETIS-UMR 8051, Université de Cergy-Pontoise - ENSEA,
6, Avenue du Ponceau,
95014 Cergy-Pontoise France
cuperlier@ensea.fr

Abstract. We have developped a mobile robot control system based on hippocampus and prefrontal models. We propose an alternative to models that rely on cognitive maps linking place cells. Our experiments show that using *transition cells* is more efficient than using place cells. The transition cell links two locations with the integrated direction used. Furthermore, it is possible to fuse the different directions proposed by nearby transitions and obstacles into an effective direction by using a Neural Field. The direction to follow is the stable fixed point of the Neural Field dynamics, and its derivative gives the angular rotation speed. Simulations and robotics experiments are carried out.

1 Introduction

Experiments carried out on rats have led to the definition of cognitive maps used for path planning [1]. Most of cognitive maps models are based on graphs showing how to go from one place to an other [2, 3, 4, 5, 6, 7]. They mainly differ in the way they use the map in order to find the shortest path, in the way they react to dynamical environment changes, and in the way they achieve contradictory goal satisfactions. Other works use ruled-based algorithms, classical functional approach, that can exhibit the desired behaviors, we will not discuss them in this paper, but one can refer to [8]. Instead, we will focus here on models inspired by the possible use of particular neurons of the rat's hippocampus, called *place cells* [9, 10].

We will first present the simulated environment and the animat possible behaviors (section 2), then we will describe our model based on transition cells (section 3). Finally, we will use a neural field for merging transitions and obstacle directions and select the final movement to perform (section 4). Simulations are carried out on an *animat*, and real world experiments on a Labo3 *robot*.

2 Environment and Animat Behaviors

The models studied in this paper have all been experimented using an approach inspired from the concepts of situated agents and animats [13]. We suppose

J. Mira and J.R. Álvarez (Eds.): IWINAC 2005, LNCS 3561, pp. 346–355, 2005.

our animats live in an unknown environment with several sources (like "food", "water" and "nest") and some obstacles. We use three contradictory motivations (eating, drinking, and resting) each one associated with a satisfaction level that decreases over time and increases when the animat is on the proper source. We do not provide any ad hoc description of the environment. Indeed, the animat gets only two types of information: the presence of landmarks from its visual input and the azimuth of these landmarks relatively to the north given by a compass. Animats have four possible behaviors for deciding which action to realize. They are given here in decreased order of priority:

1. Random exploration to discover the environment and the needed sources.
2. Planning to reach the sources in order to satisfy the animat's motivations.
3. Obstacles avoidance allows the animat to follow obstacles until the desired movement becomes possible.
4. When the source is very near, the animat directly sees where the source is located and uses this information to reach it.

3   Model

Place cells have been found in the rat's hippocampus (particularly CA3, CA1 and DG regions) and in the entorhinal cortex (EC) [9]. These cells fire when a rat is at a particular location in its environment. In our model, place cells are coded by a set of couples (landmark,azimuth) from the visual input of the animat. Hence there is no need for a Cartesian map since a particular place is defined by a given set of [landmark, azimuth] pairs. The recognition of the present location is based on the landmark configuration. A place cell P_c responds according to the position of the animat in its environment. The higher this response, the closer the animat is to P_c. After competition between all place cells, the winning cell represents the location where the animat thinks it is [14].

As a place cell still keeps a certain amount of activity even if the animat isn't near the coded place, the neuron response can be quite large: this is the place field. Consequently, we use a rule that controls the recruitment of a new neuron. Hence, a new neuron will code for a place, if all previously learned neurons have an activity lower than the Recognition Threshold (RT). A place cell may be linked with the movement needed to reach a goal. This sensory-motor association may be generalized to the whole environment [14]. However, this simple reactive mechanism is not enough in environment composed of several rooms, or when there are contradictory motivations. A cognitive map will solve these drawbacks.

Transition cells are inspired by a neurobiological model of timing and temporal sequence learning in the hippocampus [15, 16]. Figure 1 shows the hippocampal model and the cognitive map in the prefrontal cortex. Place cells are created in EC by learning the landmark-azimuth configuration. Transition cells are formed in CA3 by the fusion of the current location in EC and the previous one in DG. The transition cell is also coded in CA1. A path integration mechanism computes the mean direction used for going from one place to the other

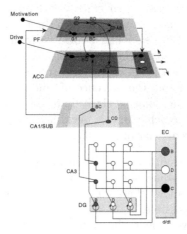

Fig. 1. Planning from transition cells. Place cells recognitions feeds the transition prediction mechanism that provides all possible transitions beginning with the corresponding place cells. In this example, transitions BC and BD are predicted since two different action has been learned from B location. The choice of the correct transition to use is performed (in ACC) by the bias given by the activity of the goal level transitions (PF). As a higher activity means a shorter path to the goal, the BC transition is selected, since the goal is located at location C

[17]. This direction is linked with the transition at the output of the nucleus accumbens (ACC).

The cognitive map is located in our model in the prefrontal cortex (PF). It is built by linking transition cells successively reached during exploration. Learning the cognitive map is performed continuously, until the entire environment is covered densely enough. There is no separation between the learning and planning phase. This map is a graph resulting in a topological representation of the explored environment. The transition cells are the nodes of this graph and arcs link the transitions successively reached. Each source place is associated with a motivation neuron. This allows to define a road to be followed for reaching the goal: the activity diffuses along the links on the map and activates transition cells according to their distance (in number of links) to the goal. Diffusion is achieved by an algorithm similar to the Bellman-Ford one [18]. This activity is sent to ACC layer where it is added to the proposed transitions. A competition mechanism selects one transition, and the corresponding movement is triggered (see fig. 2).

A relevant question is about the growth of the number of transition cells created while exploring the environment. In order to answer this question we first have to underline that this number is intimately linked with the number of place cells, and above all, that the number of place cells created for a fixed R.T value, depends on the complexity of the environment. The degree of complexity

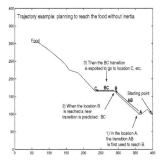

Fig. 2. The animat first follows the direction coded by B and when it comes in B, the transition BC is predicted, and the corresponding direction is used

Table 1. Results of the experiments on the ratio of the number of place cells (nbp) created over the number of transitions created (nbt) according to the number of room in the environment: with one room (top line), with two rooms (middle line) and four rooms (bottom line). Standard deviation is given into brackets. This ratio remains stable. There are five times more transition cells than place cells

Env / RT	0.97
nbp	133.8(2.85)
nbt	735.8(19.80)
ratio	5.49(0.06)
nbp	606.2(6.89)
nbt	3389.2(56.38)
ratio	5.59(0.08)
nbp	643.7(9,88)
nbt	3281,2(48,80)
ratio	5.09(0,04)

of an environment relies mainly on two factors: the number and the location of its landmarks and the number of obstacles found inside.

Hence, we have studied the ratio between created transition cells over created place cells for three environments of increasing complexity according to their obstacle configuration. For these tests, we have chosen to set the number of landmarks at a high value. For each experiment, we have launched a series of animats until 10 survive, and we let them live for 50000 cycles. This number has been chosen high enough to be sure that the animat has learned a complete cognitive map of the environment (40*40). The results shown here are the average on these 10 animat results. We have done these tests for a single, a two and a four room environment. The ratio remains stable around the mean value 5.45 for all environments once the cognitive map of the environment is complete (see table 1). Indeed, only a few transitions can be created, since a transition is a link between "adjacent" place cells. Furthermore the number of a place cell neighbours is necessary limited. So there is no combinatorial explosion on the number of created transitions and they can be memorized for planning purpose.

4   Action Selection Using a Neural Field

Instead of having only one transition win in ACC, we now allow several transitions to be taken into account for the movement. We adopt a dynamical approach in which the action selection and the motor control are obtained by a stable solution of a dynamical system: the neural field [11]. The properties of the neural field have already been successfully experimented to move the robot's arm by imitation using visual tracking of movement [20], or control a robot movement [12, 19].

$$\tau . \frac{f(x,t)}{dt} = -f(x,t) + I(x,t) + h + \int_{z \in V_x} w(z).f(x-z,t)dz \tag{1}$$

Where $f(x,t)$ is the activity of neuron x, at time t. $I(x,t)$ is the input to the system. h is a negative constant. τ is the relaxation rate of the system. w is the interaction kernel in the neural field activation. A difference of Gaussian (DOG) models these lateral interactions that can be excitatory or inhibitory. V_x is the lateral interaction interval that defines the neighborhood. Without inputs the constant h ensures the stability of the neural field homogeneous pattern since $f(x,t) = h$. In the following, the x dimension will by an angle (direction to follow), 0 corresponding to go straight forward.

The properties of this equation allow the computation of attractors corresponding to fixed points of the dynamics and to local maxima of the neural field activity. Repellors may appear too, depending on the inputs. A stable direction to follow is reached when the system is on any of the attractors.

The angle of a candidate transition is used as input. The intensity of this input depends on the corresponding goal transition activity, but also on its origin place cell recognition activity. If only one transition is proposed, there will be only one input with an angle $x_{targ} = x^*$ and it erects only one attractor $x^* = x_{targ}$ on the neural field. If x_c is the current orientation of the animat, the animat rotation speed will be $w = \dot{x} = F(x_c)$ (see fig. 3, bottom).

Fusion of several transition information depends on the distance between them. Indeed the Amari's equation allows cooperation for coherent inputs associated with spatially separated goals (for us different angles proposed). If the inputs are spatially close, the dynamics give rise to a single attractor corresponding to the average of them (see fig. 3). Otherwise, if we progressively amplify the distance between inputs, a bifurcation point appears for a critical distance, and the previous attractor becomes a repellor and two new attractors emerge. An example of two inputs spatially too far to be merged is described in figure 4.

Oscillations between two possible directions are avoided by the hysteresis property of this input competition/cooperation mechanism. It is possible to adjust this distance to a correct value by calibrating the two elements responsible for this effect: spatial filtering is obtained by convoluting the dirac like signal coming from transition information with a Gaussian and taking it as the input to the system. This combined with the lateral interactions allows the fusion of distinct input as a same attractor. The larger the curve, the more fusion there will be.

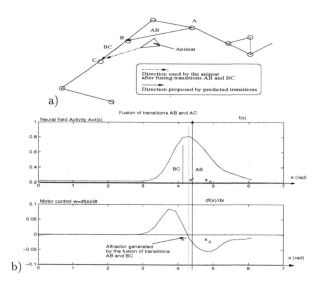

Fig. 3. a) Zoom on the cognitive map. The direction followed by the animat corresponds to the attractor generated from both B and BC. b) Activity of the neural field. The inputs of the system (Gaussian) are centered on the directions coded by transitions B and BC. Since the two inputs are spatially close enough, a unique attractor is created (x*)

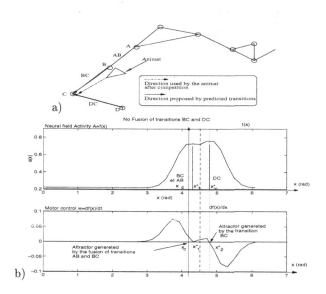

Fig. 4. a) Zoom on the cognitive map. Two transitions are predicted: one from place B that gives BC and another from location D that gives DC. Both come to the same place C, but from a different place and so with a different direction. b) Activity of the neural field. Transitions BC and DC are too spatially distant: two attractors (in x_1^* and x_2^*) are created. The motor control converges to x_1^*, closer to the current direction of the animat

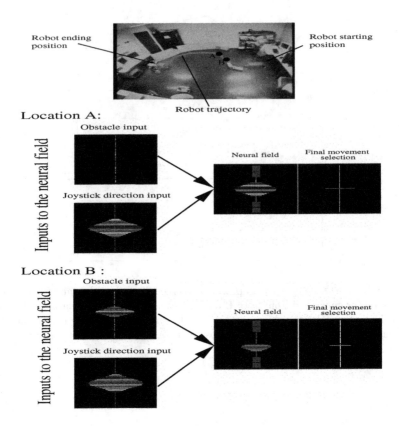

Fig. 5. Top: Trajectory of a Labo3 robot in an open environement with obstacles. The direction to go is given by a joystick input. Middle: Neural field activity without any obstacle. The direction taken corresponds to the joystick input. Bottom: Neural field activity with an obstacle. The obstacle shifts the neural field maximal activity leading to a turning move

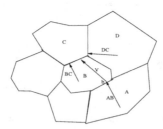

Fig. 6. The merging mechanism allows to get a better direction (V) than the use of the single information obtained from current transition (BC). It takes into account the previous movement performed and the transitions predicted from close enough place cell (D)

The neural field has some interesting properties. First it allows to use multiple entries and to combine them to get something coherent, and this at a very low level (motor control and action selection) in the architecture. For example information from captors to realize obstacle avoidance can be set as negative inputs generating repellors in the direction leading to an obstacle (see fig. 5).

Second, it allows to generalize the decision of the movement and makes it free from noisy input signals. This allows to solve the drawback for an incomplete cognitive map and to get a less suboptimal path plan otherwise. Indeed, the inputs give information about where the animat is located inside the place field, since other transitions are only proposed if close to this place (modulation of the transition activity by the recognition). Moreover, the equation takes into account the time information due to a memory effect. Then the effective transition at time 't-1', and so the direction used to enter this place cell, contributes with the informations of other transitions to the dynamics leading to the effective transition at time 't', even if the entry at time 't-1' is not active anymore. This allows to get an effective transition giving a direction to follow with a better accuracy, by taking into account the location of the animat inside the place field of the current place cell (see fig. 6). This memory effect decreases with time and is an important parameter that must be correctly set. It also allows to smooth the different sequences of direction and also to perform a motor control and an action selection insensible to some discontinuities at its entries (in particular when arriving at different times).

Neural fields are a good alternative to overcome the potential fields local minima problems [21, 22]. The main difference is that we do not have a global minimum corresponding to a goal to reach. Hence we cannot draw a global potential landscape (and use a gradient algorithm). In our system there is always a merging between the external information (obstacle) and the *internal* current direction, leading to follow the attractor which is the nearest from the current direction. Indeed, even though there are several attractors due to different obstacles, they are not added like potentials resulting in high "potential barriers" from which the robot cannot escape, or compensating each other.

5   Conclusions and Perspectives

Transition based models allow to solve several problems in an efficient and very simple way: local minima encountered with place cells based model do not appear and the selection of the action is easier to perform without any ambiguity. From a biological point of view, this model avoids the homunculus problem since the selection is done naturally by exploiting the dynamics of the system used. It may be a possible candidate to explain how the hippocampus uses information.

Neural fields allow to merge multi modal inputs to get a stable result. One can imagine to integrate several other signals such as sound direction and other visual target directions. Note that the updating of these different informations does not need to be performed at the same frequency. The neural field is robust enough to deal with intermittent information or high level signal having a very low frequency due to their computation time.

Some transition directions are only relevant for a specific part of the place field. Updating the direction value each time the transition is used will result in a "mean" direction that will reflect the "best" direction to go from the origin place to the destination one, independently of the place field shapes. Next, integration of a higher level should allow the animat to discover shortcuts, by integrating the informatios of the transitions successively used from the beginning of the planning until reaching the goal. This mechanism should be very useful when the cognitive map is incomplete at the beginning of the exploration, so that the animat could try to use shorter unexperienced paths.

Acknowledgments. This work is supported by two french ACI programs. The first one on the modelling of the interactions between hippocampus, prefontal cortex and basal ganglia in collaboration with B. Poucet (CRNC, Marseille) JP. Banquet (INSERM U483) and R. Chalita (LAAS, Toulouse). The second one on the dynamics of biologically plausible neural networks in collaboration with M. Samuelides (SupAéro, Toulouse), G. Beslon (INSA, Lyon), S. Thorpe (CERCO, Toulouse) and B. Cessac (INLN, Nice). We also thank Patrick Dufrêne for careful reading.

References

1. E.C. Tolman, "Cognitive maps in rats and men," *The Psychological Review*, vol. 55, no. 4, 1948.
2. M.A. Arbib and I. Lieblich, "Motivational learning of spatial behavior," in *Systems Neuroscience*, J. Metzler, Ed. 1977, pp. 221–239, Academic Press.
3. N. A. Schmajuk and A. D. Thieme, "Puposive behavior and cognitive mapping: a neural network model," *Biological Cybernetics*, pp. 165–174, 1997.
4. I. A. Bachelder and A. M. Waxman, "Mobile robot visual mapping and localization: A view-based neurocomputationnal architecture that emulates hippocampal place learning," *Neural Networks*, vol. 7, pp. 1083–1099, 1994.
5. O. Trullier, S. I. Wiener, A. Berthoz, and J. A. Meyer, "Biologically based artificial navigation systems: review and prospects," *Progress in Neurobiology*, vol. 51, pp. 483–544, 1997.
6. B. Schölkopf and H. A. Mallot, "View-based cognitive mapping and path-finding," *Adaptive Behavior*, vol. 3, pp. 311–348, 1995.
7. G. Bugmann, J.G. Taylor, and M.J. Denham, "Route finding by neural nets," in *Neural Networks*, J.G. Taylor, Ed., Henley-on-Thames, 1995, pp. 217–230, Alfred Waller Ltd.
8. J.Y. Donnart and J.A. Meyer, "Learning reactive and planning rules in a motivationnally autonomous animat," *IEEE Transactions on Systems, Man and Cybernetics-Part B*, vol. 26, no. 3, pp. 381–395, 1996.
9. J. O'Keefe and N. Nadel, *The hyppocampus as a cognitive map*, Clarenton Press, Oxford, 1978.
10. A. Arleo and W. Gerstner, "Spatial cognition and neuro-mimetic navigation: A model of hippocampal place cell activity," *Biol. Cybern.*, vol. 83, no. 3, pp. 287–299, 2000.

11. S. Amari, "Dynamics of pattern formation in lateral-inhibition type neural fields," *Biological Cybernetics*, vol. 27, pp. 77–87, 1997.
12. G. Schöner, M. Dose, and C. Engels, "Dynamics of behavior: theory and applications for autonomous robot architectures," *Robotics and Autonomous System*, , no. 2–4, pp. 213–245, 1995.
13. J. A. Meyer and S. W. Wilson, "From animals to animats," in *First International Conference on Simulation of Adaptive Behavior*, Bardford Books2-4, Ed. 1991, MIT Press.
14. P. Gaussier, S. Leprêtre, M. Quoy, A. Revel, C. Joulain, and J.P. Banquet, "Experiments and models about cognitive map learning for motivated navigation," *Robotics and Intelligent Systems Series*, vol. 24, pp. 53–94, 2000.
15. J.P. Banquet, P. Gaussier, J.C. Dreher, C. Joulain, and A. Revel, *Cognitive Science Perpectives on Personality and Emotion*, vol. 124, chapter Space-Time, Order and Hierarchy in Fronto-Hippocampal System: A Neural Basis of Personality, Elsevier Science BV Amsterdam, 1997.
16. P. Gaussier, A. Revel, J.P. Banquet, and V. Babeau, "From view cells and place cells to cognitive map learning: processing stages of the hippocampal system," *Biological Cybernetics*, vol. 86, pp. 15–28, 2002.
17. A. Samsonovich and B. McNaughton, "Path integration and cognitive mapping in a continuous attractor neural network model," *Journal of Neuroscience*, vol. 17, no. 15, pp. 5900–5920, 1997.
18. R. E. Bellman, "On a routing problem," in *Quaterly of Applied Mathematics*, 1958, vol. 16, pp. 87–90.
19. M. Quoy, S. Moga, and P. Gaussier, "Dynamical neural networks for top-down robot control," *IEEE transactions on Man, Systems and Cybernetics, Part A*, vol. 33, no. 4, pp. 523–532, 2003.
20. P. Andry, P. Gaussier, S. Moga, J.P. Banquet, and J. Nadel, "The dynamics of imitation processes: from temporal sequence learning to implicit reward communication," *IEEE Trans. on Man, Systems and Cybernetics Part A: Systems and humans*, vol. 31, no. 5, pp. 431–442, 2001.
21. 0. Khatib, "Real-time obstcle avoidance for manipulators and mobile robots," *Int. Journ. of Rob. Res.*, vol. 5, no. 1, pp. 90–98, 1986.
22. Y. Koren and J. Borenstein, "Potential field methods and their inherent limitations for mobile robot navigation," in *Proc. IEEE Conf. on Rob. and Autom.*, 1991, pp. 1398–1404.

Spatial Navigation Based on Novelty Mediated Autobiographical Memory

Emilia Barakova[1] and Tino Lourens[2]

[1] Brain Science Institute, RIKEN
2-1 Hirosawa, Wako-shi, Saitama, 351-0198, Japan
emilia@brain.riken.jp
[2] Honda Research Institute Japan Co., Ltd,
8-1 Honcho, Wako-shi, Saitama, 351-0114, Japan
tino@jp.honda-ri.com

Abstract. This paper presents a method for spatial navigation performed mainly on past experiences. The past experiences are remembered in their temporal context, i.e. as episodes of events. The learned episodes form an active autobiography that determines the future navigation behavior. The episodic and autobiographical memories are modeled to resemble the memory formation process that takes place in the rat hippocampus. The method implies naturally inferential reasoning in the robotic framework that may make it more flexible for navigation in unseen environments. The relation between novelty and life-long exploratory (latent) learning is shown to be important and therefore is incorporated into the learning process. As a result, active autobiography formation depends on latent learning while individual trials might be reward driven. The experimental results show that learning mediated by novelty provides a flexible and efficient way to encode spatial information in its contextual relatedness and directionality. Therefore, performing a novel task is fast but solution is not optimal. In addition, learning becomes naturally a continuous process - encoding and retrieval phase have the same underlying mechanism, and thus do not need to be separated. Therefore, building a "life long" autobiography is feasible.

1   Introduction

Referring to the memories of experienced events is a usual way to orient in novel situations. Common feature of an embodied agent (that could be animal, human or robot) is that it continuously gathers information about the surrounding world through experiencing sequences of events. Such subjectively experienced sequences are encoded by episodic and autobiographical memory systems in living organisms [27].

Tulving and Markowitsch [28] at present divide memory into five systems. The most complex memory system is the episodic-autobiographical one which requires self-conscious reflection and is embedded in the dimensions of time and locus. Semantic memory, on the other hand, is context-free fact memory. On the more automatic, implicit level, perceptual memory, procedural memory and the priming system constitute the other three long-term memory systems with perceptual memory allowing the assessment of novelty/familiarity of perceptual stimuli, procedural memory referring largely to sensory-motor skills and simple conditioning and priming to an enhanced identification of objects.

Many models in computer science and robotics exploit the characteristics of the semantic (factual) memory. However, memory for events and their relatedness is the way higher organisms build their knowledge. Many spatial navigation tasks in robotics are inspired by navigation behavior of animals. For instance, insects behavior have been simulated in [25,13,12,29]. While insects navigation has mainly reactive nature, the behavior of mammals is memory-driven [1,2,22,24,26]. We aim at memory determined behavior that relies on the neural mechanisms underlying episode formation. Episodic information encoding is related to the hippocampal modeling and with this respect our work is most closely related to [1,2,22,24,26]. We understand episodic memory as including event information within its temporal relatedness and directionality, as modeled in theoretical studies. Moreover, our aim is to obtain behaviors that are mainly driven by old experiences. The experiences are encoded during exploratory learning, a process guided by environmental novelty.

Novelty is a known factor that gates learning in natural and artificial systems [4,11,15,14,5,8]. The relation between novelty and behavior has received much attention by experimental neuroscientists [10,11,15,14,19,5,8]. but there is not enough experimental evidence to build a good computational model. The hippocampus is a brain structure where episodic and perceptual information come together, and where environmental novelty is signaled. Therefore we develop a novelty method that uses the available experimental evidence from hippocampal functioning and optimized it for robotics implementation. Novelty detection is related to experienced episodes rather than a novel place in the environment, as in other robotic studies [3,18].

There is a robotic study that introduces an autobiographical agent as an embodied agent which dynamically reconstructs its individual history (autobiography) during its lifetime [7]. However, they do not consider the neural correlates of autobiographical memory as MTL or prefrontal cortex. Instead, an algebraic model that is not related to the brain processes underlying the autobiographical encoding is used.

This paper is structured as follows: Section 2 proposes a hypothesis based on experimental studies; The computational framework is developed in Section 3. Some results are shown in Section 4. A discussion of the state of the research and its perspectives is made in Section 5.

2   Motivation and Hypothesis

The involvement of the hippocampus in episodic memory encoding has been elaborated extensively (see for instance [6]). This encoding is associated with

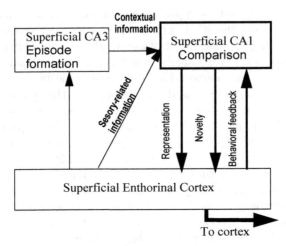

Fig. 1. Computational scheme based on the assumed functionality of the hippocampal formation, accentuating on the comparative role of the CA1 area. CA1-3 denote resemblance with areas 1-3 of cornu ammonis within the hippocampus. The sensory bound and episodic memory related representations are compared to indicate the familiarity. The areas are denoted as Superficial CA1, CA3 or EC to indicate that modeling is not biologically precise

area CA3 of the hippocampus, where due to vast recurrent connectivity the past information is represented in the current representations.

Although the importance of the hippocampus for episodic memory function is undisputed, there is a considerable debate regarding the precise role of this brain region. The hippocampus is reported to be involved in encoding and retrieval of event information, including item and relational information. It is differentially more involved in memory for relational information than in memory for item information.

Recent works report that the hippocampus is selectively engaged in detecting novel ties in the environment. Tulving and Markowitsch [28] argue that perceptual memory allows the assessment of novelty/familiarity of perceptual stimuli. Therefore, our interest is in CA1 area of the hippocampus, to which projections of sensory-bound stimuli come together with episodes of recent memories as formed in area CA3 [9,21]. The same sensory-bound pattern is transferred through both: the direct and the indirect pathway from the enthorinal cortex (EC) to CA1 area, as illustrated in Fig. 1.

Our computational model follows the general structure and functioning of rat hippocampus. We make the following hypothesis: If perceptual memory is involved in novelty detection of places, and the hippocampus is involved in episodic memory formation, the relational information that comes from CA3 to CA1 area interferes with the perceptual information to account for novelty of remembered episodes, rather of novelty of particular places, for which perceptual memory alone is indicative.

3   Computational Model Inspired by the Hippocampal Paradigm

The proposed model aims at novelty driven encoding and recall that facilitates inferential reuse of old memories. Its global functioning is illustrated in Fig. 1. Three structures, resembling EC, CA1 and CA3 areas form the representation, that is further used for navigation. The computations are performed in the superficial CA1 and CA3 areas. The final representation is formed in CA1 area, that is activated through two pathways one directly from EC area and one through the indirect activation pathway going from EC through CA3 areas. The pattern that reaches the CA1 area via the direct pathway is organized on pattern similarity, not on topological principle. Since the same projected pattern from the EC area reaches within a small time interval areas CA1 and CA3, the connection between the current most active neurons in these two areas is strengthened also. This automatically activates the complete episode to which the pattern in CA3 area corresponds, and therefore the contextual information from this episode is transferred to area CA1.

The activation of the individual patterns in the superficial EC area is derived by the established theory [20] that the cells in the rat hippocampus fire when the rat is at particular location of the environment. Because of this feature, these cells are also called place cells. If the rat moves through the environment, at every particular place a number of place cells fire. Cells that code for places in nearest vicinity fire most strongly, while the cells that fire for more distant location fire less. The activity of the place cells can be modeled by a Gaussian for the open environments, where place cells show non-directional firing. Therefore, the movement of a simulated rat at every place of the environment is characterized by a particular pattern of firing, containing the active place cells in its vicinity. The level of activity of every place cell depends on the distance between the rat position and the place fields centers. The mathematical description of this process has been shown in several works before, here we show the outcome of a simulation of two activation patterns formed by simulation of place cells and rat route (Fig. 2). These patterns are external-world related and are further transmitted through the direct pathway.

Activity pattern, coming through the indirect pathway, represents the episodic influence to the representation in CA1 area. It is formed within a network structured as a two layer lattice of neurons, corresponding to the EC and CA3 layers. There are two types of synoptic plasticity, that take place within this network. Between the layers, the afferent connections from superficial EC to the superficial CA3 area are trained through a modified Hebbian rule:

$$\Delta w_{i,j}^{CA-EC} = \alpha_1 g \left(EC_i CA_j - w_{i,j}^{CA-EC} CA_j^2 \right), \tag{1}$$

where α_1 is learning rate, notation $CA - EC$ shows the starting and destination layer of the connection (coming from EC, reaching CA layer) the indices i and j denote neurons on the input and output layer, correspondingly. The CA layer is not denoted as CA1 or CA3, because the learning rule is used for EC-CA1

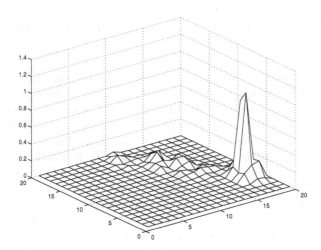

Fig. 2. A learned episode in CA3 area. Activation of a single pattern induces activation within the whole episode

as well as EC-CA3 learning. The term $-w_{ij}^{CA-EC}CA_j^2$ is needed due to internal instability of the Hebbian rule.

The lateral connections within CA3 area are two types: lateral inhibitory and lateral connections that provide the temporal context. The lateral inhibitory connections, denoted as LI have a sharpening effect on the transmitted to CA3 area activations. Eq.(3).

$$\Delta w_j^{LI-CA} = \alpha_2 g \left(LICA_j - \alpha_3 w_j^{LI-CA}CA_j^2\right), \tag{2}$$

where $\alpha2$ and $\alpha3$ are learning rates, and g is a gating factor.

By far, the representation made within the layer denoted as CA3 has not the intrinsic capacity for temporal association. This quality is obtained by applying a Hebbian rule with asymmetric time window over the neurons within the CA3 lattice, since in CA3 layer lateral connections exist between the neurons. Note, that this learning is not affected by the lateral inhibitory connections, denoted as LI in Eq. (3). It concerns the learning due to the excitatory lateral connections only. The asymmetric time window has been simulated to correspond to the experimental measurements as found by [30], see also [17]. The lateral excitatory learning rule is:

$$\Delta v_{i,j}(t) = \sum_{\tau=0}^{bound} LTP(\tau)CA_i(t)CA_j(t-\tau) + LTD(-\tau)CA_i(t-\tau)CA_j(t) \tag{3}$$

where LTP and LTD denote long term potentiation and long term depression, respectively, as found by [30,17]. and adapted to an asymmetrical time window function T_{ep} that shows the length of the temporal interval, considered in episode formation. By introducing the temporal aspect, learning in CA3 is

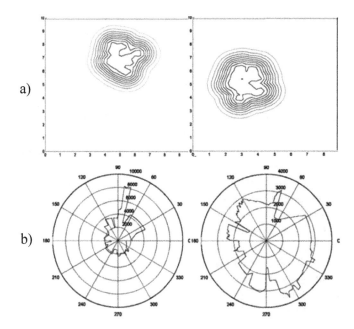

Fig. 3. Samples of sensory patterns. a) Simulation of the place cells formation process. b) Record by omnidirectional range sensor of a robot

episodic. Therefore, when a single pattern in the learned episode is activated the whole episode is activated. Figure 4 shows this effect for a simplified connectivity within CA3 layer. Individual neurons are connected to their immediate neighbors in the simplification needed for visualization process.

The learning in the direct pathway EC-CA1 has the same learning dynamics as described in equations (1) and (2). However, the learning outcome differs substantially, because of the different connectivity between the EC-CA3 and EC-CA1 layers. In the first case, the topological connections are predominant, while in the second, the areas are fully connected. Therefore, the EC-CA1 learning does not preserve the topology of the pattern activation. CA3-CA1 synaptic plasticity has the following dynamics. If k and l are neurons from CA3 and CA1 areas respectively, the connection between them is strengthened if both of them are simultaneously active and weakened if the activation of one decreases. A term that regulates the unbounded growth of the weights is added (Eq. 4). Note that the activation of CA3 neurons might not indicate single neuron activation but the influence of an active episode.

$$\Delta c_{l,k}^{CA3-CA1} = \Theta \left(CA3_l CA1_k - c_{l,k}^{CA1-CA3} CA1_k^2 \right) \tag{4}$$

The sensory bound representation of the direct pathway, and the episodic representation of the indirect pathway, come together in CA1 area, where the comparison (novelty/familiarity judgments) takes place. Note, that at the same time, the CA1 area gets input from the current pattern of EC area and a pattern from CA3 area, which has not been included in an episode yet.

For robotics learning task, several simplifications of the biologically plausible learning algorithm are made. The place cell formation process was replaced by recordings with an omni-directional distance sensor, since the two patterns look similar and uniquely represent a position in space. Fig. 3 shows couples of similar patterns. The patterns shown above are obtained by simulating the place cell formation by an exploration process, while the plots below are recorded by an omnidirectional distance sensor.

The simplification in the learning process are derived from the observation, that topology preserving learning between the EC and CA3 layer does not have substantial contribution to the learning outcome. In contrast, the self-organizing process between EC and CA1, and the temporal association learning within the CA3 layer are essential.

Therefore, the learning process ignores the plasticity of EC to CA3 connections and the inhibitory lateral connections. The Hebbian learning rule between the layers followed by lateral inhibition is replaced by a modification of the competitive Hebbian learning algorithm [2,19], adapted for processing of sequences.

4   Active Autobiography Through Exploration

The mammals, who are able to form episodic memories, and especially humans can remember a particular experience for the whole life span. This fact suggests, that episodic memory encoding is an efficient process, i.e. together with the encoding a decision of the content and importance of the encoded information is taken.

Autobiographical memory overlaps with the concept of episodic memory. It is a memory of one's personal past, but concerns a longer time span. With this respect an analogy can be made with the robotics paradigm of life-long learning. This paper investigates autobiography formed by remembered episodes only. We call it active autobiography to accentuate on its dynamic character.

For building an active autobiography, a latent learning scenario is used. Latent learning is an association of indifferent stimuli or situations with one another without immediate reward. The phenomenon is clearly exemplified in exploratory behavior, and is also known as exploratory learning [20,16,23]. In robotics research the term exploratory learning is preferred, while latent learning is mostly used in animal studies. To compare the results, however, target locations are marked.

To illustrate the active autobiography formation, the episodic encoding when the robot reaches a sample location are shown in Fig. 4a). The learning outcome is shown in Fig. 4b). It indicates, that trajectories (episodes) that are considered similar are not remembered.

Furthermore, the goal reaching by free exploration is shown. To connect the neural representation with the goal reaching navigation behavior, an algorithm

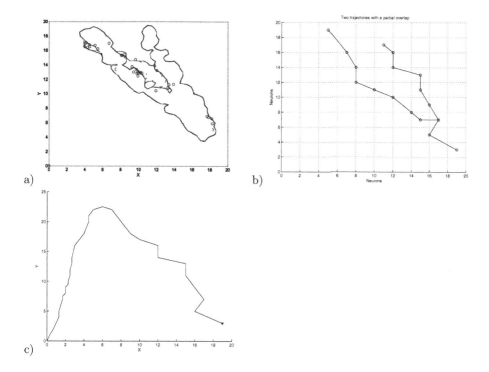

a)   b)   c)

Fig. 4. a) 3 experienced episodes. b) Representation of the encoded episodes. c) Reaching a target location by encountering a known path

we showed in [4] is used. Once a position in the environment is recognized, the robot behavior is guided by the remembered episode. This behavior is illustrated in Fig. 4c. Initially, starting from the beginning of the coordinate system, the reactive behavior is shown. The non-smooth part of the trajectory shows navigation guided by the remembered episode.

5   Discussion

This study shows spatial navigation that is enhanced by robots memories. Formation of the remembered episodes is mediated by novelty/familiarity discrimination method based on the hippocampal modeling for robotics exploration. The embodied nature of an animal and robot makes this parallel useful, and the functional efficiency of the hippocampal encoding, while performing both tasks: episodic encoding and novelty detection, suggests an optimal computational scheme.

The impact of novelty is two-fold: it allows an efficient encoding in the exploration phase and it is a basis for flexible reuse of memories in the recall phase. The same computational paradigm is used in both cases, which makes possible on-line implementation. The paper accentuates on the methodological part and shows simulations of episodic memory encoding and navigation that uses the remembered episodes. A comparison of the method with exploration that uses actor critic learning is made and the advantages of our method for larger environment is obvious. However, the results need good explanation of the method and why this comparison is plausible. Because of that they will be shown elsewhere. The navigation trajectories become more efficient with enriching the autobiography of the stored exploration episodes, since the optimal trajectory is recorded.

References

1. A. Arleo and W. Gerstner. Spatial cognition and neuro-mimetic navigation: A model of hippocampal place cell activity. *Biological Cybernetics*, 83:287–299, 2000.
2. K. Balakrishnan, O. Bousquet, and V. Honavar. Spatial learning and localization in rodents: A computational model of the hippocampus and its implications for mobile robots. *Adaptive Behavior*, 7:173–216, 1999.
3. E. Barakova and U. Zimmer. Dynamic situation and trajectory discrimination by means of clustering and summation of raw range data. In *AISTA*, Australia, 2000.
4. E. I. Barakova. Emergent behaviors based on episodic encoding and familiarity driven retrieval. In *AIMSA*, volume 3192 of *Lecture Notes in Artificial Intelligence*, pages 188–197. Springer-Verlag, 2004.
5. R. A. Bevins and M. T. Bardo. Conditioned increase in place preference by access to novel objects: antagonism by mk-801. *Behav. Brain Res.*, 99:53–60, 1999.
6. M. Bunsey and H. B. Eichenbaum. Conservation of hippocampal memory function in rats and humans. *Nature (London)*, 379(6562):255–257, 1996.
7. K. Dautenhahn and C. Nehaniv. Artificial life and natural stories. In *Proc. of the Third International Symposium on Artificial Life and Robotics (AROB III)*, pages 435–439, 1998.
8. R. Galani, I. Weiss, J.-C. Cassel, and C. Kelche. Spatial memory, habituation, and reactions to spatial and nonspatial changes in rats with selective lesions of the hippocampus, the entorhinal cortex or the subiculum. *Behav. Brain Res.*, 96:1–12, 1998.
9. O. Jensen and J.E. Lisman. Hippocampal ca3 region predicts memory sequences: accounting for the phase precession of place cells. *Learning and Memory*, 3:279–287, 1996.
10. A. Kemp and D. Manahan-Vaughan. Hippocampal long-term depression and long-term potentiation encode different aspects of novelty acquisition. *Proc. Natl. Acad. Sci. U.S.A.*, 2004.
11. R. Knight and T. Nakada. Cortico-limbic circuits and novelty: a review of eeg and blood flow data. *Rev. Neurosci.*, 9:57–70, 1998.
12. D. Lambrinos, R. Muller, T. Labhart, R. Pfeifer, and R. Wehner. A mobile robot employing insect strategies for navigation. *Robotics and Autonomous Systems*, 30:39–64, 2000.
13. L. Leerink, S. Schultz, and M. Jabri. A reinforcement learning exploration strategy based on ant foraging mechanisms. In *Proceedings of the Sixth Australian Conference on Neural Networks*, Sydney, Australia, 1995.
14. S. Li, W. K. Cullen, R. Anwyl, and M. J. Rowan. Dopamine-dependent facilitation of ltp induction in ca1 by exposure to spatial novelty. *Nat. Neurosci.*, 6(5):525–531, 2003.

15. J. E. Lisman and N. Otmakova. Storage, recall, and novelty detection of sequences by the hippocampus: elaborating on the socratic model to account for normal and aberrant effects of dopamine. *Hippocampus*, 11:551–558, 2001.
16. A. Manning. *An Introduction to Animal Behaviour.* Edward Arnold, 1979.
17. H. Markram, H. Lubke, M. Frotscher, and B. Sakmann. Regulation of synaptic efficacy by coincidence of postsynaptic aps and epsps. *Science*, 275:213–215, 1997.
18. S. Marsland. Novelty detection in learning systems. *Neural Comp. Surveys*, 3, 2003.
19. S. Moses, R. Sutherland, and R. McDonald. Differential involvement of amygdala and hippocampus in responding to novel objects and contexts. *Brain Res. Bull.*, 58:517, 2002.
20. J. O'Keefe and L. Nadel. *The Hippocampus as a Cognitive Map.* Clarendon Press, Oxford, 1978.
21. R. C. O'Reilly and J. L. McClelland. Hippocampal conjunctive encoding, storage, and recall: avoiding a trade-off. *Hippocampus*, 4(6):661–682, 1994.
22. T. J. Prescott. Spatial representation for navigation in animats. *Adaptive Behavior*, 4:85–123, 1996.
23. M. J. Renner. Learning during exploration: The role of behavioral topography during exploration in determining subsequent adaptive behavior. *Int. J. Comp. Psychol.*, 2(43), 1988.
24. B. Scholkopf and H. A. Mallot. View-based cognitive mapping and path planning. *Adaptive Behavior*, 3(3):311–348, 1995.
25. J. Svennebring and S. Koenig. Building terrain-covering ant robots: A feasibility study. *Autonomous Robots*, 16(3):313–332, May 2004.
26. O. Trullier and J.-A. Meyer. Animat navigation using a cognitive graph. In R. Pfeifer, B. Blumberg, J.-A. Meyer, and S. W. Wilson, editors, *From Animals to Animats*, pages 213–222. MIT Press, Cambridge, MA, 1998.
27. E. Tulving and W. Donaldson. *Organization of memory.* Academic Press, New York, 1972.
28. E. Tulving and H. J. Markowitsch. Episodic and declarative memory: Role of the hippocampus. *Hippocampus*, 8:198–204, 1998.
29. R. Wehner, B. Michel, and P. Antonsen. Visual navigation in insects: coupling of egocentric and geocentric information. *The Journal of Experimental Biology*, 199:129–140, 1996.
30. L. Zhang, H. Tao anc C. Holt, W. Harris, and M. Poo. A critical window for cooperation and competition among developing retinotectal synapses. *Nature*, 395:37–44, 1998.

Vision and Grasping: Humans vs. Robots

Eris Chinellato and Angel P. del Pobil

Robotic Intelligence Laboratory, Universitat Jaume I, Castellón de la Plana, Spain
{eris, pobil}@icc.uji.es

Abstract. Biomimetic robotics is a rapidly developing field, and the limited literature about biological inspiration in robot grasping at cognitive level suggests that the field has still much to offer.

Neuroscience studies indicate that vision-based reaching and grasping are important to the extent that an entire cortical pathway is dedicated to these skills. Nevertheless, recent findings point out the existence of strict relations between action-oriented (dorsal pathway) and categorization-oriented (ventral pathway) vision.

In this paper, we will compare present day research on vision-based robotic grasping with the above mentioned neuroscience findings. Then, we propose a new approach to vision for grasping in robotics, which aims at improving the emulation of human skills through the integration of the information flows proceeding from the two visual pathways.

1   Introduction

Robotics often aims at reproducing in artificial beings the most relevant skills characterizing intelligent animals in general and humans above all. One of the most distinctive abilities of humans, and in a minor way of other primates, is that of handling every kind of objects in a dexterous way. Indeed, grasping and manipulating skills have been constantly pursued in robotics for their theoretical and practical implications. Technological developments and advancements in neurophysiological research are providing the background on which to build robotic applications closer to the biological reality, and thus more 'intelligent'.

Nevertheless, in spite of the amount of research and technological efforts, the gap between prehension performances of primates and robots is still very large, especially in unstructured, real environments (as is true for many others non-cognitive tasks). In this work, we will analyze visual-based grasping comparing the state of the art of the field in robotics with recent neuroscience findings, revealing useful insights on the way it is achieved in animals, and especially in humans. We will look for similarities and divergences, in order to work out a possible pathway toward a better vision-based robotic grasping, strongly inspired by natural solutions. Our analysis will focus on the visual process prior to the onset of the reaching movement, so we will not consider the action execution side of the problem.

J. Mira and J.R. Álvarez (Eds.): IWINAC 2005, LNCS 3561, pp. 366–375, 2005.

2   Biological Inspiration for Robot Grasping and Manipulation

When trying to develop systems which model or imitate natural skills, either cognitive or practical, two approaches are usually cited as contrary options: bottom-up and top-down. Modern robotics favors the former, which is more plausible from a physiological point of view [1]. According to the paradigm of behavior-based robotics [2], complex behaviors are composed of simpler ones (sometimes called *motor primitives*) in a bottom-up direction, to the extent that some basic components are simple stimulus-effect interactions of the agent with the environment [3]. Grasping and manipulation robotics research following a behavior-based approach is expanding, even though the paradigm may not be as widespread in grasping as it is in other areas of robotics. For all kinds of movements, motor primitives are a type of basic behaviors common to robots and humans. Indeed, they can be drawn from thorough analysis of human motion [4], to form a behavior vocabulary used to produce more complex movements and action sequences. Similar to what happens for biological beings, the composition of simple behaviors can endow robotic systems with notable skills, dexterous manipulation being one of them [5]. Whilst top-down solutions are implemented following knowledge engineering methodologies, bottom-up approaches are often coded with connectionist methods, more or less inspired by biological neural networks. The limited affinity between natural and artificial neural networks nowadays is one of the major handicaps in the quest for intelligent systems.

Recent neuroscience findings show that grasping behaviors in primates are obtained in a way that do not differ too much from the action composition idea described above. Prototypes of simple actions (both cognitive and practical ones) are coded in primate brains, and the way they connect produces higher level behaviors. In the case of grasping and reaching, areas F4 and F5 of the inferior premotor cortex of primates are believed to contain a vocabulary of motor actions of different complexity, duration, significance (e.g. preshape the fingers for a precision grip)[6, 7]. Such actions are selected and combined in different ways according to the task (e.g. push or grasp), the object shape and size, the timing of the action and other relevant issues.

Neuroscience further supports the bottom-up approach as the most suitable to model high-order cognitive processes. In fact, it seems that motor systems strongly contribute in the formation of processes traditionally considered to be "high level" or cognitive, such as action understanding, mental imagery of actions, perceiving and discriminating objects [8]. Mirror neurons are surely the best (and most famous) example of this. They can be found in a purely motor area in monkeys (F5) but show responsiveness to the observation of actions performed by others. They have been related to the ability of social interactions through understanding/prediction of other people's movements [9] and to the explanation of social behavior impairment as in autism [10]. Moreover, the mirror system seems to play a critical role in learning by imitation, a skill that we are only beginning to develop in robots. Broca's area in humans, traditionally related to language production, is the most likely correspondent of F5. This would confirm that complex cognitive processes, as verbal communication, can emerge from simple behaviors which firstly evolved in order to give the organism skills for interacting better with its environment, like action recognition. As regards vision, current approaches and methodologies used in robotics differ very much from the mechanisms of vision in living organisms, and we will describe later with more details what the biggest and most important differences are and their consequences.

Globally, biomimetics, or biologically inspired robotics, is a rapidly developing field [11], and the limited literature about biological inspiration in visual-based robot grasping at cognitive level makes us conclude that the field has still much to offer.

3   Vision for Grasping in Primates: A Dual Mechanism

A most relevant aspect regarding the interaction between vision and grasping, is that visual processes related to grasping actions in primates are different from visual processes with different goals [12]. In fact, looking at an object with grasping purposes activates a neural pathway which is not active when grasping actions are not involved. This activation seems to represent a 'potential grasping action', and is reinforced when the action is actually performed. Some neurons in the anterior intraparietal area (AIP) of posterior parietal cortex in monkeys are found to be active when grasping some particular objects, but also when looking at them with the purpose of grasping (and only in those cases) [13]. Some other neurons of the same zone are sensitive to the size or orientation of objects. Therefore, the AIP area encodes the 3D features of objects in a way suitable to generate grasping actions, which will be performed using the vocabulary of motor actions stored in the premotor cortex. A similar pattern has been found in humans as well, thanks to functional magnetic resonance imaging (fMRI) research [14]. As stated by Jeannerod [15], "object attributes are processed differently according to the task in which a subject is involved. To serve object-oriented action, these attributes are subjected to a 'pragmatic' mode of processing, the function of which is to extract parameters that are relevant to action, and to generate the corresponding motor commands".

Visual data in primates flows from the retina to the lateral geniculate nucleus (LGN) of the thalamus, and then mainly (but not exclusively) to the primary visual cortex (V1) in the occipital lobe. There are two main visual pathways going from V1 to different association areas, the posterior parietal cortex (PPC) and the inferior temporal (IT) cortex (Figure 1). Through the dorsal pathway, object related visual information reaches area AIP in the intraparietal sulcus, which we said is concerned with analyzing visual features in order to organize grasping actions. Area AIP projects mainly to area F5 of the premotor cortex, which contains the motion primitives used to compose grasping actions, and receives most of its input, through the lateral intraparietal area (LIP), from the

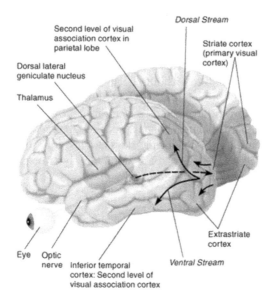

Second level of visual
association cortex in
parietal lobe

Dorsal lateral
geniculate nucleus

Thalamus

Dorsal Stream

Striate cortex
(primary visual
cortex)

Extrastriate
cortex

Eye   Optic
nerve   Inferior temporal
cortex: Second level of
visual association cortex

Ventral Stream

Fig. 1. Dorsal and ventral visual pathways, from http://homepage.psy.utexas.edu

extrastriate cortex V3a [16], which seems to be responsive to stimuli orientation as well as to motion and color [17].

Object information flowing through the ventral pathway passes through V2 and V4 to the lateral occipital (LO) complex, which is related with object recognition [18]. The traditional distinction [19] talks about ventral "what" and dorsal "where/how" visual pathways. Although recent studies confirm that the dorsal stream is more oriented toward action-based vision, whilst the ventral one is more suitable to categorization, such distinction is not completely sharp [12, 8]. New findings [20] suggest that the interaction between the two systems, previously considered nearly independent, is important to allow both of them to function properly. It seems that LO itself is implied in some action-related processing, although the way the two streams communicate is still mostly unknown.

4   Toward Grasp Oriented Vision in Robotics

The most complete attempt to model the sensorimotor mechanisms of visual-based grasping in primates is the FARS (Fagg-Arbib-Rizzolatti-Sakata) model [21], which focuses especially in the final part of the process, more related with action execution (AIP-F5 links). Nevertheless, no robotic applications have been yet developed following this path. In a similar way, even though several different computational models of the human visual systems are available - mainly regarding the ventral, "recognition" path, carrying from the primary visual cortex V1 to the inferior temporal region IT - the integration between the contributions of the two visual pathways is nearly unexplored [22]. Similarly, many robotic approaches follow the scheme of figure 2.a, and consider object recognition or full visual reconstruction previous to the grasp analysis, which is normally based on an object model [23]. Other works exploit visual features in a manner more oriented to grasp purposes, but without the aid of "cognitive" information about objects [24] (see figure 2.b). The first method allows the grasping analysis to be, if not easier, surely more rigorous and less empirical, but it lacks flexibility as it builds on a nearly general-purpose visual processing stage. The second method, more plausible and often very efficient, does not exploit the kind of cognitive information that makes human grasping skills largely transcend a geometric or kinematic analysis.

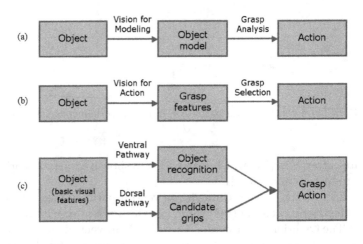

Fig. 2. Different approaches in vision-based grasping

To integrate in a visual-based, robotic grasping system instantaneous visual information gathered in an action oriented-manner (dorsal pathway) with experience-mediated information (ventral pathway) is an interesting goal, and we believe that it is the one to pursue if we want to obtain a robotic grasping system able to interact in the real world. In figure 2.c we can see a representation of this solution, which somehow encompass the two previous ones. Indeed, this integration process is what humans do all the time: we join our knowledge about the thing we are going to grasp and the experience of previous actions with the analysis of the actual, concrete situation we are facing.

Before describing with more details our idea, it is important to stress that, although our approach is inspired on well recognized neuroscience concepts, the following description of the system includes a number of hypothesis that should be confirmed by further neurophysiological and behavioral research.

Before devising our intelligent vision-based grasping system, we need to decide the level of detail of the neural computation stage. This should be neither too rough, to avoid the risk of losing completely the biological plausibility, nor too elaborate, in order to be used in a practical robot application. As our interest is directed mainly toward the integration of different types of information as it is thought to happen through the connections of area AIP with area LO, we leave aside for the moment the steps of visual acquisition and action execution.

In our framework, the visual input is processed in two parallel ways, one more concerned with perceptive information about the object nature, the other oriented to spatial analysis. The products of the visual analysis are, from the dorsal elaboration, precise information about position and geometry of the object, and, from the ventral elaboration, data about expected weight, friction, and previously experienced grasping actions on such object. The dorsal pathway is also responsible for generating possible affordances, or grip candidates. This is accomplished by AIP, in collaboration with the prefrontal cortex F5, which send feedback signals to AIP reinforcing better grips while inhibiting the rivals (this process was already implemented in the FARS model).

The interaction between the two pathways can be realized in two different ways according to the circumstance. For objects which are recognized with a good confidence margin, off-line models of previous affordances can be recovered, but they need to be improved on-line with the information gathered by the "where/how" visual stream. This is because, in our everyday world, there are many types of scissors, or hammers, or chairs, and natural objects (e.g. fruits) are never exactly the same. Thus, to recognize or classify an object does not imply that we know how to grasp it, as we first need to identify plausible surfaces on which to put our fingers. For new objects, or objects whose recognition is considered uncertain or unreliable, a more complex grasping analysis needs to be developed, starting from basic surface extraction, and the generated information can be transmitted to the ventral stream as new memory. More generally, the contribution of the ventral stream on the generation/selection of the final grip can be modulated by the degree of recognition of the target object achieved by LO. A higher confidence in the object recognition reflects in a stronger influence of the past grasping experiences, whilst a more uncertain recognition leads to a more exploratory behavior, giving more importance to the actual observation.

5   Vision-Based Robotic Grasping with Biological Inspiration

A number of issues need to be analyzed accurately in order to plan the development of a system such as the one introduced in the previous section. In figure 3 the system is schematically represented. The blocks outside the dotted line, which are dedicated to the visual acquisition and to the motor output, are not included in this discussion. The blocks inside the continuous line are those in which we are especially interested, as they represent the two visual streams from the moment they separate (after V2) to the moment they join again (in F5).

From a practical point of view, and following a logical sequence, the system will first need an attentional mechanism that allows the robot to focus on the object to grasp and isolate it from the background and possible other objects.

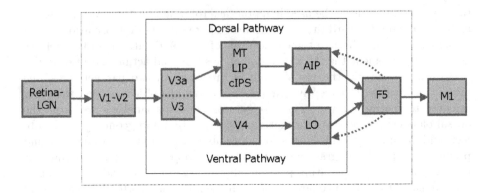

Fig. 3. Block diagram of a grasping system based on human physiology

Once the object is unambiguously identified, the vision elaboration stage comes into action. The visual blocks V1 and V2 provide as output basic features, mainly edges, corners, or simple contours. This features are used by V3 to reconstruct more complex ones. As it can be observed in figure 3, areas V3 and V3a appear slightly separate, as their processing is not exactly the same. In fact, the visual analysis is likely to be performed in different ways by the two pathways starting from V3. If on the one hand achieving a quick and reliable object recognition is probably better done through a volumetric analysis, it seems more plausible for the dorsal stream to look for appropriate surfaces (and not volumes) on which to put the fingers [25]. Thus, V3a can use the basic features obtained by the visual subsystem V1-V2 to reconstruct surfaces, whilst V3 can search for volumetric features. Area V3 is thought to be responsible also for color elaboration. This can be used by the ventral stream to recognize objects more easily, and by the dorsal stream (especially MT) to track objects in motion.

Surface information coming from V3a is the input of areas MT, LIP and cIPS. In humans, whilst MT is more concerned about movements, the intraparietal sulcus is where visual input begin to be coded in a head-based spatial representation. The lateral intraparietal area LIP seems to be responsible of storing and remapping visual memory in eye-centered coordinates [22]. The caudal part of the intraparietal sulcus (cIPS) is believed to code object affordances, as it does not recognize the same object seen from two different view points [18]. In our implementation, cIPS is especially critical, as it is the block in which the search for graspable surfaces is carried out, using the input from V3a. The possible grasping regions found by cIPS are used by AIP, the main grasping area, responsible for finding the candidate grips joining appropriate surfaces.

On the ventral side, the output from area V3 is used by V4 to build a viewpoint invariant simple reconstruction of the object, using volumetric primitives (see e.g. [26]). Then, LO merges spatial and color data with stored information about previously observed objects to finally recognize the target and access its memory on how it has been grasped before. Regarding the last point, it is still unknown if this is what actually happens in our brain, and our solution so far is not supported by neuroscience findings.

The output from AIP and LO is finally used by F5 to choose the most appropriate grip for that object in that situation. Although we do not know exactly in which way the two pathways interact, our idea is that the ventral information can intervene in the grip generation (AIP) and selection (F5) process inhibiting unsuitable grips, or families of grips (e.g. precision grips instead of power ones). The power of the intervention can be modulated according to the degree of confidence of the object recognition process, leaving more or less influence on the dorsal analysis. After the selection process is done, when AIP receives feedback from F5 telling which grip will be finally performed, it forwards the information to LO, in order to memorize the selected action for future reference.

The grasping literature often cite three main factors affecting grasping actions. They are the object, the gripper and the task. We will take now a brief look at the way each of these factors can be taken into account in our framework.

Object. The best and most natural way to consider the object properties is to merge on-line, instantaneous visual information with memory of previously acquired knowledge about the object properties. The on-line, dorsal elaboration would consist of a search for graspable zones. Most often, they can be found looking for reasonably flat or concave surfaces, roughly facing each others, or otherwise for cylindrical and spherical features. In the dorsal elaboration, an estimation of the center of mass of the object should also be included, remembering that such estimation relies on data coming from the ventral pathway, as the expected object composition and density.

Gripper. The geometry, kinematics and strength of the gripper is of extreme importance in the task of grasp planning. Many taxonomies have been created both for the human hand and for robotic grippers. In our opinion, a taxonomy of average complexity is the most appropriate [27]. Human grasps are not extremely precise at first, as possible small misplacements are normally corrected on-line through tactile feedback. This leads to a reduced number of possible grips, as very similar grasping actions would be classified as to be the same grip. In robotic grasping, small differences in finger positioning can lead to big changes in the quality of the grip [24], and thus precision is more critical, unless a haptic system to adjust the grip is available. Arguably, a real-world robotic grasping system can not prescind from good tactile skills, and furthermore, the generation of a big number of grips slightly different from each other is not only very implausible, but also extremely costly for an on-line grasping action. This is why we would opt for a reduced taxonomy for our implementation.

Task. The handling of information about object use or task involves explicit knowledge about the object we are going to grasp, and thus it should use memories stored in the ventral pathway. The task can be a critical factor, as there may be various good ways of grasping an object, but if we want to perform a given action with it, we will probably need to grasp it in an appropriate way that would allow correct execution of that action. The concept of task is thus additional to the normal grip analysis, and considering the task can help during the grip selection process.

6   Conclusions

The analysis and comparison of human and robotic vision-based grasping suggest that to emulate humans grasping abilities with a robotic system we still have much to progress. The way we propose to approach this goal is to mimic the mechanisms of vision for grasping in primates, which is based on the duality between the dorsal and ventral cortical visual streams.

We analyzed the current knowledge on the neurophysiology of vision-based graping in humans and tried to determine the basis for a robotic system, built on a neural computation framework, able to reach high grasping skills in the real world. We claim that this could be achieved integrating on-line, action-oriented visual information (dorsal pathway) with knowledge about the target object and memories of previous grasping experiences (ventral pathway).

Acknowledgments

Research in the Robotic Intelligence Lab is partially supported by Generalitat Valenciana (project CTIDIA/2002/195), by Fundació Caixa-Castelló (project P1-1A2003-10) and by the Ministerio de Educación y Ciencia (DPI2001-3801, DPI2004-01920, FPI grant BES-2002-2565).

We wish to thank Jody Culham for making it possible for the first author to stay at her Lab in the University of Western Ontario and collaborate in her fMRI research, and also for reviewing a draft of this paper. Thanks again to Jody, and to Mel Goodale, for the enlightening discussions.

References

1. Brooks, R.: Cambrian intelligence: the early history of the new AI. MIT Press, Cambridge, Massachusetts (1999)
2. Arkin, R.: Behavior-Based Robotics. The MIT Press (1998)
3. Braitenberg, V.: Vehicles: Experiments in Synthetic Psychology. MIT Press, Cambridge, Massachusetts (1984)
4. Drumwright, E., Jenkins, O., Mataric, M.: Exemplar-based primitives for humanoid movement classification and control. In: IEEE Intl. Conf. on Robot. Automat., New Orleans, USA (2004)
5. Mataric, M.: Getting humanoids to move and imitate. IEEE Intelligent Systems 15 (2000) 18–24
6. Fadiga, L., Fogassi, L., Gallese, V., Rizzolatti, G.: Visuomotor neurons: ambiguity of the discharge or 'motor' perception? Int. J. Psychophysiol. 35 (2000) 165–177
7. Rizzolatti, G., Luppino, G.: The cortical motor system. Neuron 31 (2001) 889–901
8. Gallese, V., Craighero, L., Fadiga, L., Fogassi, L.: Perception through action. Psyche - An interdisciplinary journal of research on consciousness. 5 (1999)

9. Rizzolatti, G., Arbib, M.A.: Language within our grasp. Trends in Neurosciences **21** (1998) 188–194

10. Williams, J., Whiten, A., Suddendorf, T., Perrett, D.: Imitation, mirror neurons and autism. Neuroscience and Biobehavioural Review **25** (2001) 287–295

11. Bar-Cohen, Y., Breazeal, C.: Biologically Inspired Intelligent Robots. SPIE Press (2003)

12. Milner, A., Goodale, M.: The visual brain in action. Oxford University Press (1995)

13. Sakata, H., Taira, M., Kusunoki, M., Murata, A., Tanaka, Y.: The parietal association cortex in depth perception and visual control of the hand action. Trends in Neuroscience (1997)

14. Culham, J.: Human brain imaging reveals a parietal area specialized for grasping. In Kanwisher, N., Duncan, J., eds.: Functional Neuroimaging of Visual Cognition: Attention and Performance XX. Oxford University Press (2004) 417–438

15. Jeannerod, M., Arbib, M., Rizzolatti, G., Sakata, H.: Grasping objects: the cortical mechanisms of visuomotor transformation. Trends in Neuroscience (1995)

16. Nakamura, H., Kuroda, T., Wakita, M., Kusunoki, M., Kato, A., Mikami, A., Sakata, H., Itoh, K.: From three-dimensional space vision to prehensile hand movements: The lateral intraparietal area links the area v3a and the anterior intraparietal area in macaques. The Journal of Neuroscience **21** (2001) 8174–8187

17. Gegenfurtner, K., Kiper, D., Levitt, J.: Functional properties of neurons in macaque area v3. J. Neurophysiology **77** (1997) 1906–1923

18. James, T., Humphrey, G., Gati, J., Menon, R., Goodale, M.: Differential effects of viewpoint on object-driven activation in dorsal and ventral streams. Neuron **35** (2002) 793–801

19. Ungerleider, L., Mishkin, M.: Two cortical visual systems. In Ingle, D., Goodale, M., Mansfield, R., eds.: Analysis of Visual Behavior. MIT Press (1982) 549–586

20. Lee, J.H., Van Donkelaar, P.: Dorsal and ventral visual stream contributions to perception-action interactions during pointing. Experimental Brain Research **143** (2002) 440–446

21. Fagg, A., Arbib, M.: Modeling parietal-premotor interactions in primate control of grasping. Neural Networks **11** (1998) 1277–1303

22. Rolls, E., Deco, G.: Computational Neuroscience of Vision. Oxford University Press, Oxford, UK (2002)

23. Bicchi, A.: Hand for dexterous manipulation and robust grasping: A difficult road towards simplicity. IEEE Trans. Robot. Automat. **16** (2000) 652–662

24. Morales, A., Chinellato, E., Fagg, A., del Pobil, A.: Using experience for assessing grasp reliability. International Journal of Humanoid Robotics **1** (2004) 671–691

25. Moore, C., Engel, S.: Neural response to the perception of volume in the lateral occipital complex. Neuron **29** (2001) 277–286

26. Miller, A., Knopp, S., Christensen, H., Allen, P.: Automatic grasp planning using shape primitives. In: IEEE Intl. Conf. on Robot. Automat., Taipei, Taiwan (2003)

27. Cutkosky, M., Howe, R.: Human grasp choice and robotic grasp analysis. In Venkataraman, S., Iberall, T., eds.: Dextrous Robot Hands. Springer-Verlag (1990) 5–31

Evolved Neural Reflex-Oscillators
for Walking Machines

Arndt von Twickel and Frank Pasemann

Fraunhofer Institute AIS, Sankt Augustin, Germany

Abstract. Legged locomotion has not been understood well enough to
build walking machines that autonomously navigate through rough terrain. The current biological understanding of legged locomotion implies
a highly decentralised and modular control structure. Neurocontrollers
were developed for single, morphological distinct legs of a hexapod walking machine through artificial evolution and physical simulation. The
results showed extremely small reflex-oscillators which inherently relied
on the sensori-motor loop and a hysteresis effect. Relationships with biological findings are shortly discussed.

1   Introduction

The potential of legged machines can be found in their ability to traverse highly
uneven and unstructured terrains. None of the numerous walking machines developed so far seems to keep these promises. This is partly due to the drawbacks
of the hardware; but especially the lack of suitable control mechanisms is apparent. This led to an increased interest in locomotion strategies in animals.

Behavioural as well as electro-physiological studies on model organisms like
stick-insects and cockroaches have shown that the biological mechanisms underlying legged locomotion are quite complex. It involves the control of many degrees
of freedom whereby feed-forward motor patterns and neural and mechanical feedback interact [1]. As was already assumed in [2] the neural locomotion controller
of stick insects *Carausius morosus* is composed of six individual pattern generators, later referred to as single leg controllers [3]. Every single controller consists
of a central neural network and sensors local to the leg and can in turn be decomposed into at least three central rhythm generating networks controlling the
three main leg joints by alternatively exciting and inhibiting the antagonistic
motor neuron pools [4]. Sensory organs of the leg constitute an integral part of
the central rhythm generating networks (for a review see [5]). A set of neural
rules which govern the control of single joints as well the coordination between
joints has been discovered over the last few years (for review see [6]). Contrary to
the coordination of the joints of legs, the coordination of the single legs, and the
influences of leg-external sensory systems and higher brain structures on walking
behaviour is less well understood in terms of neural control.

Despite the constantly increasing knowledge on legged locomotion some important questions concerning the control of legged locomotion with many degrees

J. Mira and J.R. Álvarez (Eds.): IWINAC 2005, LNCS 3561, pp. 376–385, 2005.

of freedom (DOFs) remain unanswered, e.g.: What are the structural and functional dynamic principles of the control of legged locomotion? How do Central Pattern Generators (CPGs) and reflex control work together? To give at least partial answers to these questions the evolution of neural controllers for walking machines should be helpful.

This paper presents some first results derived by applying structure evolution of recurrent neural networks to the control of single 3 DOF legs. Control was applied to the physical simulation of legs belonging to a physical walking machine. The physical walking machine and its physical simulation were developed in parallel. Selection of controllers was with respect to walking distance in given time, and robustness in the sense that equal performance should be achieved under different environmental conditions (obstacles, holes, etc.). The described and analysed neurocontrollers are quite small and equally efficient like much larger ones. The most interesting finding is the conformation of the assumption, that sensory inputs play a major role for pattern generation in locomotion; i.e., rhythmic leg movements are derived by feedback through sensory-motor loops including the environment ([7] and [8]).

The next section describes the methods applied to develop the single leg locomotion controllers. In the "Results" section two example controllers are presented to demonstrate the mechanisms discovered and finally in the last section the results are discussed in the context of the current understanding of locomotion control.

2   Materials and Methods

Inspired by work reported for instance in [6], locomotion controllers were developed and evaluated for single three DOF legs to later develop controllers for the whole walking machine by coupling the single leg controllers (this article will focus on the single leg controllers). This approach has already been successfully employed before ([9], [10], and [11]). Here various tools and techniques have been employed during this process, including physical simulation, structure evolution of (recurrent) neural networks, and analysis of the resulting neurodynamics. These tools and techniques are provided by the integrated structure evolution environment ISEE described in detail in [12].

To be able to simulate just one leg a rail-like structure was developed to which the robot was fixed with a slider joint, altogether enabling the robot to move forward/backward and upwards/downwards but not to move sideways or to turn. A changing environment was provided by randomly placing boxes of different dimensions on the ground, therefore creating gaps and steps. The morphology of the simulated robot was constrained by the physical properties of an actual walking machine. Only two types of sensors were employed (three angle position sensors and one foot contact sensor per leg). The robot was constructed with the idea that different legs (fore-, middle- and hind-leg) have to fulfil different tasks and therefore need to have a distinct morphology. Limited by the constraints, the only morphological differences were the attachment points on the body and the initial orientations at the body as well as the angle ranges of the joints: The fore-legs had a working range in front of the shoulder joint, the middle-legs around the shoulder joint and the hind-legs behind the shoulder joint.

To allow for neural networks with *tanh* as transfer function (see below) to control the robot, motor and sensor signals had to be mapped onto the interval $[-1; 1]$: They were mapped in such a way that either the maximum or the minimum angle possible corresponded to a value of $+1$ and the other to -1. For the contact sensor this was different: A value of zero indicated "no contact" and a value of approx. 0.5 "maximal contact". Mapping conventions, i.e. which sign corresponds to which movement direction can be read from Fig. 1. Since all simulated legs contained three motors and four sensors, all single leg controllers had four input- and three output-neurons. The connections to and from the controller are depicted in Fig. 1.

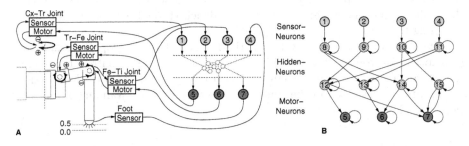

Fig. 1. (A) Linkage of the sensors and motors of the simulator to the sensor and motor neurons of the neural net. The leg is shown from behind. Abbreviations: *Cx-Tr* Coxa-Trochanter, *Tr-Fe* Trochanter-Femur and *Fe-Ti* Femur-Tibia. (B) Interpretation of the finite state mechanism found in [6]

The evolutionary algorithm ENS^3 [12] was employed to develop the controllers. Numerous parameters for the evolutionary process could be set before/during evolution according to some general strategies based on experiences. A summary can be found in [12]. Two types of networks were taken for seed populations: Either a network solely consisting of input- and output-neurons; i.e., without any connections or internal neurons, or a network being our equivalent of the finite state model found in [6] (see Fig. 1).

As fitness function mainly the covered distance in forward direction was taken. Attention was paid by means of cost terms, punishing large networks and high connectivities. Because robustness was one important requirement, controllers were evaluated in seven different environmental scenarios with the fitness obtained in each environment being added to the total fitness. Therefore, the total fitness value was a good measure for the general performance of the controller. Poor or especially high fitness values for single environmental scenarios were good indications for specialisations. After several evolution runs (each about 250 generations) the performances of the best networks were compared with each other and, if available, with some reference controllers. The overall best performing networks or those having a particular interesting structure were subjected to further analysis afterwards.

By utilising this approach several distinct structured and extremely small controllers for each of the three legs were identified. To demonstrate the mechanisms found two example controllers are presented in the following section.

3   Results

A multitude of single leg controllers performing equally well in propelling the body forward were developed. In the following the different motor patterns generated by fore-, middle- and hind-leg-controllers are shortly described. Then two example controllers are presented to demonstrate the principle mechanisms discovered.

A fore-leg movement on even terrain generated by a typical fore-leg controller can be described as follows: At the Anterior Extreme Position (AEP) the fore-leg controller has just completed its swing phase and made foot contact with the ground. The Tr-Fe joint (see Fig. 1 for joint terminology) has already started moving downwards to support the body whereas the Fe-Ti and the Cx-Tr joint only now start moving inwards respectively backwards and exert force on the ground to pull the body forwards. After the leg has accelerated and almost reached the Posterior Extreme Position (PEP) the Tr-Fe joint is activated to move the leg up when it has reached the PEP. Once the leg is in the air, the Cx-Tr and Fe-Ti joints are moved forwards respectively outwards. The Tr-Fe joint is then activated to move downwards so that the foot reaches the ground at AEP. At this point the cycle starts anew. The phase relations between the three joints of the middle-leg are almost identical to those of the fore-leg. The middle-leg exerts its force to the ground more parallel to the body than the fore-leg which rather pulls the body forward with its foot being in front of the body. On the other hand the hind-leg movement differs significantly from those of the fore- and middle-leg because it moves the body forward rather by pushing than by pulling it. During the stance phase the Tr-Fe joint moves downward and the Fe-Ti joint outwards. The Cx-Tr joint supports this backwards movement but starts with the backward directed movement only when the other two joints already started exerting a backwards directed force. Consistent with this observation the hind-leg controllers differed stronger from the fore- and middle-leg controllers than those two in between.

Apart from the differences between the controllers of fore-, middle- and hind-legs similarities could be noted, e.g. the motor-neurons of all controllers approximately acted as toggle switches, either being activated maximal positive or maximal negative. Since the motor output only represents the target value, the actual movements differed significantly from the motor output rather resembling a sine or zig zag curve. Further on two key mechanisms in nets without inherent neural oscillators were found responsible for the oscillatory motor output during walking. The first mechanism involves hysteresis through neural elements which is demonstrated on one of the simplest controllers that was found in the course of the evolution experiments. This fore-leg controller is depicted in Fig. 2 and it consists of one sensory input from the Cx-Tr angle sensor, one self connection (larger than one) and all motor neurons connected in series with one connection being inhibitory. That makes a total of four neurons (including the sensor neuron) and four synapses being involved in the control of the leg. The performance of this controller is comparable to that of more complex controllers. Important is the fact that no neural oscillator can be found in this structure only leaving the possibility of an oscillation via the environmental loop (1-6-7-5-1). The motor

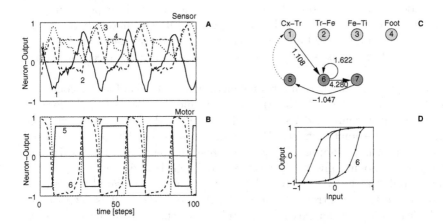

Fig. 2. Single Leg controller (fore-leg) with a neural hysteresis element (self-connection larger than one) and a feedback loop via the environment: (A+B) neuron-outputs during walking on even terrain. (C) structure of the network. Note that only one sensor gives input to the net. The dashed line denotes feedback via the body and the environment. (D) Theoretical hysteresis (iterations $n = 1000$, inner curve) and hysteresis under locomotion conditions (outer curve) of neuron six. Shown is the output of the neuron against its input

signal analysis showed that some kind of a bistable element exists in the controller. One possible realisation is a neuron with a self connection that displays a hysteresis effect. For a hysteresis effect to take place the self-connection has to be larger than plus one [13]. Such single neuron hysteresis elements are found in many of the leg controllers. Its possible role in locomotion control is as follows: In this particular net (see Fig. 2) the central neural element is neuron six which receives the only sensory input, has a self connection greater one and therefore is a hysteresis element. It is also the first element in the chain of all motor neurons. The other motor neurons have the same phase or a phase shifted by 180° compared to neuron six.[1] Neuron seven is in phase with neuron six, neuron five in para-phase. This suggests to take a closer look at the role of the hysteresis

---

[1] Of course they are additionally shifted by either one or two time-steps.

element. Therefore the output of neuron six was plotted against its input (see Fig. 2 D), once for input sequences during normal locomotion and once for a sine-function with a high number of iteration steps as the input. The first thing that may be noted is that the hysteresis element may account for the observation of a bistable element. This is due to the fact that basically two stable fixed points exist in the hysteresis domain either pulling the output of the neuron towards approx. one or approx. minus one depending on the history of the system. As can be seen in Fig. 2 D an important difference exists between the theoretical hysteresis curve and the one under experimental conditions. This is caused by the theoretical hysteresis acting as an attractor which is never reached under real conditions because the input values change too fast. One can therefore regard the frequency of the input signal as an additional parameter determining the behaviour of the system. All motor neurons act as bistable elements, neurons seven and five even stronger than neuron six. This can be explained by the strong connection ($w = 4.28$) from neuron six to neuron seven which amplifies the signal from neuron six therefore pushing it faster to the maximum/minimum of the nonlinear transfer-function $tanh$.

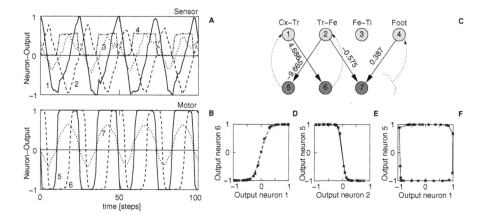

Fig. 3. Single Leg controller without any neural feedback and two environmental feedback loops: (A+B) neuron outputs during walking on even terrain. (C) structure of the network. The dashed lines denote feedback via the body and the environment. (D, E + F) Hysteresis effect without neural feedback. The effect results from the time-delay of the motor-sensor interaction, from the nonlinear transfer-function of the neurons and from the large magnitude synapses

The second mechanism found in the evolved single leg controllers involves hysteresis through interaction with the environment but without neural feedback. A single leg controller, in this case a hind-leg controller, in which no neural feedback occurs at all is exemplified in Fig. 3. Although only four synapses exist at all whereby three of the four sensors, including the foot contact sensor, are involved a total of 6 neurons make up the functional controller. Two loops exist:

The first passes through the environment twice (1-6-2-5-1), the second only once (7-4-7). The motor outputs of the net suggest that bistable elements exist in the controller and therefore not only in controllers with neural hysteresis elements but also for controllers without any self connections, as for the example shown in Fig. 3. Here a combination of the motor-environment-sensor interaction, strong weights and the nonlinear sigmoid transfer function have been found as the underlying mechanisms. In Fig. 3 the loop 5-1-6-2-5 is looked at in detail. Plots $D+E$ show (inverse-)$tanh$-like curves which are steeper than the normal $tanh$ because of the strong connections($w_{61} = 4.69$ and $w_{52} = -9.67$). These steep $tanh$-like curves push the majority of the inputs to values close to plus one respectively minus one. This explains the bi-stability, but not the hysteresis effect. To understand the hysteresis effect depicted in plot F, the properties of the motor-environment-sensor loop have to be considered. In a very simplified point of view, the environment represents a time delay element which, together with the steep $tanh$-like transfer in between the neurons, accounts for the hysteresis effect observed.

4   Discussion

Though in recent years the information available on the neural control of walking has increased rapidly important parts in the "puzzle" on the structure and mechanisms of those controllers still remain unknown [3]. Research on the stomatogastric system for example has shown the complexity of a neural biological system (see e.g. [14]) that in comparison to the control of walking is considered to be "simple". For this reason several researchers have recently begun to take a synthetic approach [15] by means of simulations whereby the goal has mainly been to test hypotheses from biology [16]. Additional to the test of hypotheses the approach taken here considers simulation as a tool to find new hypotheses and therefore alternative perspectives to a problem: Artificial evolution was not only used to perform a parameter optimisation on a given structure, e.g. on a biological inspired controller, but mainly to develop a controller structure from scratch. The only prior knowledge given was the morphology of the simulated robot, the type of neural elements allowed for the controller and the optimisation goal ("move forward as fast as you can"). The evolved structures are distinct to biological systems in that control structures are not optimised for a whole bunch of tasks but for one specific task. The analysis of evolved structures allows to find out about the very principles of a specific control problem because it does not narrow the possible outcome by putting prior knowledge into the system.

Controllers were developed for three legs, each with a distinct morphology which was caused by different operating ranges of the CX-Tr joint when compared to the longitudinal body axis. Consistent with the results of [6] the differences between the movement of the hind-leg and the two other legs (middle- and fore-leg) were much more pronounced than that between the middle- and the fore-leg. This is due to the hind-leg performing a "pushing" movement compared to the other two legs which rather perform a "pull" movement. Consequently the controllers for the hind-leg on the one side and for the fore- and middle-leg on the other side are not compatible. In quite a few cases it was possible to drive the fore-leg with middle-leg controllers and vice versa but no exchange with the hind-leg succeeded. By changing some weights of a fore-leg controller it was possible to make a hind-leg walk backward but without large changes it was not possible to make it walk forward. An interesting correlation can be found in [5]: In experiments where all legs except one are amputated fore-legs always walk forwards, hind-legs always walk backwards and middle-legs do both. A single fore-leg causes the hind-legs to walk forwards, therefore suggesting that walking direction is caused by the coupling of both legs. The question remains whether position dependent influences (morphology), structural differences of the controllers or coupling influences cause the three types of controllers to behave differently.

The diversity of controller structures ranged from structures that were too complex to analyse to extremely small and simple structures. Note that there was an evolutionary pressure (cost term in the fitness function) that favoured smaller controllers. But the speed term was by far the most important term in the fitness function. Therefore only small nets were favoured that performed equally well as the larger ones. Common to all controllers was the existence of a sensori-motor-loop passing through the environment. Also most of the controllers had at least one neuron with a self-connection larger than one. Interestingly, there was no correlation between performance and size of the controller network. Some of the smallest controllers (see e.g. Fig. 2) were also some of the best performing controllers.

The hysteresis effect has been found to be the most important effect in the evolved locomotion controllers. It caused the motor outputs to act roughly like bistable elements, either being fully activated to move to one side or to the other. This effect is also termed as "relaxation oscillator" or "bistable system" and agrees with the literature on stick insect motor activation (i.e. stance-swing transition, see [5]). In the controllers presented in this article the hysteresis effect was caused by two distinct mechanisms which could also work together: First a neural element having a positive self connection larger than one displays a hysteresis effect (see Fig. 2, see also [13]). Second a hysteresis can be caused by strong weights in connection with the bounded $tanh$ function and a time delay in the loop due to the physical characteristics of the body as for example the delay between motor command and actual movement. Theoretical considerations suggest that the total loop has to be positive to allow hysteresis to occur ([17] and specifically for motion control [18] and [19]).

In both cases that caused hysteresis a motor-sensor-loop was needed to drive the oscillation therefore representing a *Reflex-Oscillator*. None of the evolved networks contained a *Central Pattern Generator* (CPG). At least two explanations are conceivable: 1. A reflex oscillator represents the superior solution to the problem. 2. The boundary conditions (evolution parameters, fitness function etc.) either favoured the development of reflex oscillators or they interfered with the formation of a CPG. Since the comparison of the reflex controllers with some parameter optimised constructed CPG controllers showed an approximately equally good performance the first possibility can be ruled out. Assuming that both are equally well suited for the job the reflex controller would have had the advantage of less connections and less internal neurons being necessary to drive the oscillator. In that case the cost function would have favoured the reflex controllers. But in some runs there was no cost function restricting the size and the connectivity of the neurocontrollers and still no oscillators could be found. Therefore it can be stated that under the simulation conditions the reflex controllers performed as well as CPG controllers but with the advantage of a simpler structure. It would be interesting to know if such reflex oscillators exist in biology and if this is true if they are also simpler than the CPG solutions. For several systems it has been established that a CPG can, without any sensory input, generate the rhythm responsible for locomotion (e.g. in lampreys, see [20]) or for the stomatogastric system of crustacea [21]. In other animals such as the stick insect CPGs have been found but their overall contribution to locomotion is still unclear [22]. It is believed that very fast running animals rather rely on CPGs, with sensory inputs acting only modulatory, whereas animals, that walk slowly and possibly on rough terrain, to a stronger extent rely on sensory input to generate the locomotion motor pattern [23]. Further simulations, e.g. with oscillators that incorporate sensory inputs, could be helpful in determining the advantage or disadvantage of certain combinations of CPG and reflex (-oscillator) influence.

The simulation experiments have shown that if the properties of the body and the environment are incorporated into a controller for a complex task, in this case for walking of a 3DOF leg, the controller can itself be extremely simple and small. On the other hand these simple controllers cannot be understood without knowledge about the body, the environment and task, since their main function (oscillation) does not occur without an environmental interaction. It seems promising to continue and expand the artificial evolution experiments to systematically analyse the possibilities of inter-leg coupling and coupling with other, non-locomotory sensory/motor systems. The goal of work in this direction is to generate sensor-driven behaviours in walking machines acting autonomously in rough terrain.

References

1. Dickinson, M.H., Farley, C.T., Full, R.J., Koehl, M.A.R., Kram, R., Lehmann, S.: How animals move: An integrative view. Science **288** (2000) 100–106
2. Wendler, G.: The co-ordination of walking movements in arthropods. Symp. Soc. exp. Biol. **20** (1966) 229–249
3. Orlovsky, G., Deliagina, T., Grillner, S.: Neuronal Control of Locomotion. Oxford University Press (1999)
4. Schmidt, J., Fischer, H., Büschges, A.: Pattern generation for walking and searching movements of a stick insect leg. ii. control of motoneural activity. J. Neurophysiology **85** (2001) 354–361

5. Bässler, U., Büschges, A.: Pattern generation for stick insect walking movements – multisensory control of a locomotor program. Brain Research Reviews **27** (1998) 65–88

6. Ekeberg, O., Blümel, M., Büschges, A.: Dynamic simulation of insect walking. Arthropod Structure & Development **33** (2004) 287–300

7. Brooks, R.A.: Intelligence without representation. Artificial Intelligence (1991) 139–159

8. Brooks, R.A.: New approaches to robotics. Science **253** (1991) 1227–1232

9. Seys, C.W., Beer, R.D.: Evolving walking: The anatomy of an evolutionary search. In: From Animals to Animats 8: Proceedings of the Eighth International Conference on Simulation of Adaptive Behavior, Los Angeles, CA (2004)

10. Brooks, R.A.: A robot that walks: Emergent behaviors from a carefully evolved network. Technical Report AI MEMO 1091, MIT (1989)

11. Schmitz, J., Dean, J., Kindermann, T., Schumm, M., Cruse, H.: A biologically inspired controller for hexapod walking: Simple solutions by exploiting physical properties. Biol. Bull. **200** (2001) 195–200

12. Huelse, M., Wischmann, S., Pasemann, F.: Structure and function of evolved neuro-controllers for autonomous robots. Connection Science **16** (2004) 294–266

13. Pasemann, F.: A simple chaotic neuron. Physica D **104** (1997) 205–211

14. Heinzel, H.G., Weimann, J.M., Marder, E.: The behavioral repertoire of the gastric mill in the crab, *Cancer pagurus*: An in situ endoscopic and electrophysiological examination. The Journal of Neuroscience **13** (1993) 1793–1803

15. Dean, J.: Animats and what they can tell us. Trends in Cognitive Science **2** (1998) 60–67

16. Dean, J., Kindermann, T., Schmitz, J., Schumm, M., Cruse, H.: Control of walking in the stick insect: From behavior and physiology to modeling. Autonomous Robots **7** (1999) 271–288

17. Prochazka, A., Gillard, D., Bennett, D.J.: Implications of positive feedback in the control of movement. The Journal of Neurophysiology **77** (1997) 3237–3251

18. Cruse, H.: What mechanisms coordinate leg movement in walking arthropods? Trends in Neurosciences **13** (1990) 15–21

19. Grillner, S., Deliagina, T., Ekeberg, O., Manira, A.E., Hill, R.H., Lansner, A., Orlovsky, G.N., Wallen, P.: Neural networks that coordinate locomotion and body orientation in lamprey. Trends Neurosci. **18** (1995) 270–279

20. Grillner, S., Ekeberg, O., Manira, A.E., Lansner, A., Parker, D., Tegner, J., Wallen, P.: Intrinsic function of a neuronal network - a vertebrate central pattern generator. Brain Res Rev **26** (1998) 184–197

21. Selverston, A.I., Panchin, Y.V., Arshavsky, Y.I., Orlovsky, G.N.: Shared Features of Invertebrate Central Pattern Generators. In: Neurons, Networks, and Motor Behavior. MIT Press, Cambridge, MA (1999) 105–117

22. Büschges, A., Ludwar, B.C., Bucher, D., Schmidt, J., DiCaprio, R.A.: Synaptic drive contributing to rhythmic activation of motoneurons in the deafferented stick insect walking system. European Journal of Neuroscience **19** (2004) 1856–1862

23. Delcomyn, F.: Walking robots and the central and peripheral control of locomotion in insects. Autonomous Robots **7** (1999) 259–270

A Haptic System for the Lucs Haptic Hand I

Magnus Johnsson[1,2], Robert Pallbo[1], and Christian Balkenius[2]

[1] Dept. of Computer Science, Lund University, Sweden
[2] Lund University Cognitive Science, Sweden

Abstract. This paper describes a system for haptic object categorization. It consists of a robotic hand, the LUCS Haptic Hand I, together with software modules that to some extent simulate the functioning of the primary and the secondary somatosensory cortices. The haptic system is the first one in a project at LUCS aiming at studying haptic perception. In the project, several robotic hands together with cognitive computational models of the corresponding human neurophysiological systems will be built. The haptic system was trained and tested with a set of objects consisting of balls and cubes, and the activation in the modules corresponding to secondary somatosensory cortex was studied. The results suggest that the haptic system is capable of categorization of objects according to size, if the shapes of the objects are restricted to spheres and cubes.

1 Introduction

Identifying materials and objects using the touch in our hands is an ability that we often take for granted, and normally we hardly think about it. But to be possible, this ability demands a hand with a very sophisticated ability to manipulate grasped objects. It also needs receptors for several submodalities, especially cutaneous and proprioceptive mechanoreceptors [9]. In addition, neurophysiological systems are needed, that can actively choose a way to manipulate the object in a beneficial way and then control the execution of these manipulations, while at the same time receiving and categorizing sensory data [7, 8].

One way to learn more about how such an ability works and to find applications for that knowledge by reversed engineering, is to try to build an artificial haptic system with the abilities mentioned above. Such a system should use the human hand and brain as a prototype. This is what we have the ambition to do.

There are several reasons why it should be interesting to build a system capable of haptic object categorization modeled on the corresponding human system. From a pure scientific viewpoint it is interesting because the model can constitute support for the cognitive and neuroscientific theories that it is founded on. It might also provide new insights into the modeled neurophysiological systems. From an applications perspective it is interesting because it might provide new knowledge about robotic haptics and artificial intelligence. Since the system will be founded in the workings of the corresponding human systems, it might also

J. Mira and J.R. Álvarez (Eds.): IWINAC 2005, LNCS 3561, pp. 386–395, 2005.

be used to simulate the effects of nerve injuries between the hand and the brain, and the cortical reorganization that follow these kinds of injuries [12].

In humans, what kinds of procedures are used depend on the age. The haptic procedures used by young children are simpler than those used by adults. In fact, already a few weak old fetuses are sensitive to tactile stimulation, and newborns respond differently depending on how elastic or stiff an object is [20]. A 3-4 year old child has an exploratory procedure for shape discrimination, but it is not optimal [11]. The hand namely stays immobile on the object. Such static contact provides, besides information on temperature, also approximate information on shape, texture, size, and hardness. At adulthood, on the other hand, the explorative procedures have come closer to optimality [11].

Adults use several haptic procedures. The choice of procedure depends on the kind of property explored. When texture is explored, lateral motion is used, and indeed, movements seem to be necessary for the perception of texture [10]. Unsupported holding is used for the estimation of weight, and pressure to explore the hardness of the material [11]. By the aid of contour following, more precise information on shape and size is provided. Unsupported holding is used for the estimation of weight [11]. In weight estimation, it is first and foremost the arms and the shoulders that are most sensitive, while the fingers also have some sensitivity to gravitational constraints [11].

In haptic object identification it is not only information extracted from stimulus that influence the iden-tification, but also expectations based on the context or previous experiences, i.e. there is top-down processing involved [15].

The main focus of the research on robotic hands has been on grasping and object manipulation [6, 21, 3, 18, 5, 19] and surprisingly little work has addressed the problem of haptic perception and only a few haptic perception systems have been built. One example is a system capable of haptic object classification [4].

In the brain, the sensory information that is coded by cutaneous and proprioceptive mechanoreceptors is conveyed to the central nervous system by the dorsal column-medial leminiscal system [7, 8]. This is organized so that the sensory information from the receptors is processed in two subcortical relay regions and the cortex. The receptor neurons are located in ganglions in the dorsal root of a spinal nerve and they have axons that are divided into two branches, whereof one terminates in the skin where its terminals are sensitive to tactile stimuli and the other enters and ascends through the spinal column and terminates in the gracile or the cuneate nuclei in medulla. The axons from medulla terminate in the ventral posterior lateral nucleus of the contralateral thalamus. From the thalamus, the axons go straight to the primary somatosensory cortex (S1), to secondary somaesthetic areas (S2), and to the posterior parietal areas and to the motor cortex [9].

To create a proper model of certain cortical regions we need neural networks that are able to learn without the aid of a supervisor, i.e. an unsupervised neural network. The kind of learning in this category of neural networks is usually called competitive, or self-organizing. A self-organizing map [16, 17], abbreviated SOM, is an instance of this category.

In a competitive learning network each node receives the same input. The nodes influence each other by lateral interactions that are both inhibiting and exciting. The winner, i.e. the node which is most activated is allowed to learn, i.e. update its weights, and in addition the nodes in a vicinity of the winner node are also allowed to update its weights, but to a lesser degree. To what degree the nodes are allowed to update its weights depends on the distance to the winner node. In the long run this will lead to self-organization of the nodes in the network so that every node will be sensitized to a certain domain of the input space, and the network will also conserve the order of the input space.

In a general way, the adaptive SOM processes can explain the organization found in varying structures in the brain and artificial self-organizing maps share many features with brain maps, i.e. the experimentally deduced topographical organization of the brain.

As a beginning to our project to explore haptic perception, we have built a simple haptic system consisting of the LUCS haptic hand I, which is a very simple robotic hand equipped with push sensors, together with a set of software modules that to some extent implements the functioning of SI and SII. In order to model SII we have used Kohonens self-organizing map. The reason for this choice is that it is a simple model of a self-organizing feature map.

The haptic system described in this paper is just the first one in a series of systems with increasing capability. The aim of building it has been to get initial experiences that will enable us to build more advanced and elaborate versions later. Because of this, the technical level of the robotic hand has been kept elementary.

2   LUCS Haptic Hand I

The LUCS haptic hand I (Fig. 1A) has three fingers and one of them, the thumb, is moveable with one degree of freedom. The fingers, that are made from Delrin acetal resin, are straight and rigid and of a rectangular shape (Fig. 1B). The two fixed fingers are mounted so that their superior sides are slanted inwards. The thumb is mounted on a metal joint that in turn is mounted by a RC servo. Besides transmitting torque from the RC servo to the thumb, the metal joint also stabilizes sideway movements, so that the movement of the thumb becomes more accurate. When the thumb moves to close the hand, it ends up right between the two fixed fingers.

Each finger is provided with an array of three force sensitive resistors, attached to the fingers with equal distance in between, i.e. one sensor is placed at the outermost part of the finger, one sensor at the innermost part, and one in between (Fig. 1B). There are tiny plastic plates mounted on top of the sensors to distribute the forces on the fingers. These plastic plates are necessary because otherwise the pressure must be applied right at the push sensor. The size of the plastic plates is such that they fit within the borders of the tactile sensors. Every tactile sensor is, together with a capacitor and a resistor, part of a circuit,

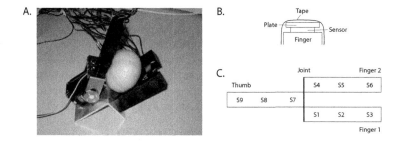

Fig. 1. A. The LUCS Haptic Hand I while grasping a mandarine. (A movie showing the LUCS haptic hand I while grasping an object is available on the web site, see [13]). The mechanical design. V. The design of the tactile sensors with a force sensitive resistor covered with a small plate that distributes the pressure on the finger. C. The placement of the tactile sensors. There are three sensors on each finger

which generates a pulse with a length that depends on the pressure applied to the sensor.

The LUCS haptic hand I communicates with the computer via the serial port, and as an interface a Basic Stamp II is used. The Basic Stamp executes a loop that in every iteration reads a message, coming from the computer, about whether the position of the thumb is going to be changed, and if so to what position. If the position is going to be changed, then a signal is sent to another board, a mini SSC II, which generates a pulse to the RC servo which then moves to the desired position. In every iteration of the loop the pulse length of the signals from each sensor is also read and sent to the computer.

We have tested the LUCS haptic hand I by letting it grasp a number of balls and cubes [14]. This was done in order to study the changes of the signal patterns when the robotic hand grasped different objects.

The results of this test suggest that the signal patterns from the LUCS haptic hand I are, in principle, possible to categorize according to the properties width and height of the curves that describe the signals from the grasping movement over time. These properties of the curves are correlated with the size and the hardness of the objects. Perhaps the slope of the curves are correlated with the shape of the object.

3   The Haptic System

Input from the sensors at time t is given by the vector

$$s(t) = \langle s_0(t), \ldots s_n(t) \rangle .$$

The sensor signals are divided into two channels, the first one $C^{(1)}$ is simply the original signals, while the second channel $C^{(2)}$ consists of the difference between two consecutive inputs

$$C^{(2)} = \langle s_0(t) - s_0(t-1), \ldots s_n(t) - s_n(t-1) \rangle.$$

The two channels are sent to two separate regions, $D^{(1)}$ and $D^{(2)}$, corresponding to primary sensory cortex (S1). In our earlier simulations of the somatosensory system we used self-organizing maps with slowly decaying activation at this level [12], but with the relatively low-dimensional sensory input from the hand we used a simpler mechanism instead based on leaky integrators.

$$D^{(1)}(t+1) = \alpha D^{(1)}(t) + T(C^{(1)}(t)),$$

where, T(x) is a threshold function with $T(x) = 1$ if $x > 0$ and 0 otherwise. The reason for this is that the flow of information through the raw signal channel then will conserve the information concerning the duration of the sensors signal, i.e. the width of the curve describing the signal over time.

An extra channel with a constant value was added to $D^{(1)}$ and $D^{(2)}$. This is necessary since the activity pattern from the sensors often only contained signals from one sensor and the self-organizing maps using cosine similarity measurements can not categorize such signals without reference to a fixed value. For $D^{(1)}$, a fixed value of 1 was used, and for $D^{(2)}$, a value of 10000 was used. In both cases the value was selected to be in the range of the other signals.

The final processing stages consists of two self-organizing maps $S^{(1)}$ and $S^{(2)}$, each receiving input from the corresponding D layer. The net input for calculation for self-organizing map j used the standard cosine measurement

$$I_i^{(j)} = \frac{S^{(j)} w_i^{(j)}}{|S^{(j)}||w_i^{(j)}|},$$

and the activity was calculated using the softmax function

$$a_i^{(j)} = \frac{\left(I_i^{(j)}\right)^m}{\sum_k \left(I_k^{(j)}\right)^m}.$$

where k ranges over the nodes in the self-organizing map and m is the softmax exponent. The learning rate and the Gaussian neighborhood function were gradually reduced in the standard way [17].

4   Implementation

The software part of the haptic system consists of several modules, implemented as Ikaros modules [1, 2]. Ikaros provides a kernel and an infrastructure for computer simulations of the brain and for robot control.

The first module is LucsHapticHandIDriver. This module handles the communication with the robotic hand via the serial port. In addition it orders a grasping movement of the robotic hand and receives information about the status of the sensors. The status of the sensors is sent as output from the module every iterations.

Table 1. The objects used in the categorization test with the haptic system for the LUCS Haptic Hand I

Object	Shape	Size (mm)	Hardness
A	Sphere	Diameter = 60	Rather Hard
B	Sphere	Diameter = 40	Hard
C	Cube	Side = 37	Soft
D	Cube	Side = 55	Soft
E	Sphere	Diameter = 42	Rather Hard
F	Sphere	Diameter = 72	Rather Soft
G	Sphere	Diameter = 62	Medium Hardness
H	Sphere	Diameter = 44	Hard
I	Sphere	Diameter = 66	Hard

During training and testing of the haptic system an alternative module, LucsHapticHandITraining, has been used. This module uses files containing signal patterns generated with the aid of LucsHapticHandIDriver when this was connected to the Lucs Haptic Hand I during the grasping of objects.

LucsHapticHandDriver or LucsHapticHandITraining sends out the raw and differentiated sensory signals. The output from each channel is sent to an instance of a module intended to model the primary somatosensory cortex. This module is called LeakyIntegrator. In our implementation the activity of every unit from the previous iteration is added to the current activation after multiplication with 0.5.

The instance of LeakyIntegrator that receives input from the raw signal channel, first transforms the input vector so that the elements that represent a positive signal are set to one and the other elements are set to zero as described above. Since the Lucs Haptic Hand I only have 9 sensors, we have decided to let the LeakyIntegrator module be hard coded and thus without learning ability.

Each instance of the LeakyIntegrator module conveys its output to an instance of a module that implements Kohonens self-organizing feature map. This module is called SOM and is intended to model the secondary somatosensory cortex. Each instance of this module comprise 2500 neurons. The neighborhood starts out with a radius of 15 nodes and decays gradually to a radius of 4. All weights are initialized to 1.

5   Categorization Tests with Balls and Cubes

To test the haptic system, we have used 9 different objects, 7 balls and 2 cubes (Table 1). The haptic system was exposed to the signal pattern for each object one time in the test.

Before accomplishing the test, the haptic system has been trained with 1000 samples randomly selected from a training set containing 270 training samples generated by the Lucs Haptic Hand I during the grasping of 9 different objects with 30 graspings of each. Every training sample consists of the status of the 9 pressure sensors on the Lucs Haptic Hand I, measured at discrete points in time, during the whole grasping movement.

After the training phase, the haptic system was tested with one sample for each object in the training set.

Studies of the activation in the two SOM modules, that model secondary somatosensory cortex, were made possible partly by writing the best match in every iteration to a file, and partly since graphical representations of the activation were generated during the test. The activation were studied and compared for the different objects, and it were also drawn into depictions (Fig. 2) representing the mapping of the different objects to the two SOM modules.

6   Results

In the grasping tests the raw sensor signals activated three areas in the SOM module map between iterations 26 and 35. The three areas were activated by three different groups of objects. One group comprises the tiniest objects, one

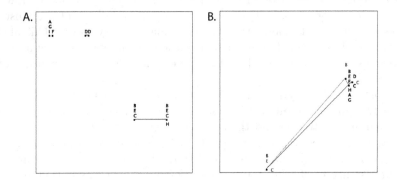

Fig. 2. (A). The mapping of the sensors input from the Lucs Haptic Hand I onto the maps in the two instances of the SOM module in the test of the haptic system with 9 different test objects. The figure shows the mapping over the channel conveying raw sensor signals. Only the best matches in the SOM between iterations 26 and 35 after the start of the grasping movement have been depicted in the figure. If several different co-ordinates are the best match during the sequence of the iterations depicted, then the points have been connected with lines. The grasping of the objects activate 3 areas in the map during the depicted iterations. The objects A, G, I, and F activate one area in the map. These are the largest objects. The objects B, E, C, and H activate another area in the map. These are the smallest objects. Object D activates still another area in the map. The size of object D is between the sizes of the previously described groups of objects. (B). The best matches in the SOM module for the channel that maps the change of sensor signals over time during iterations 26 and 35. Until the two last iterations all objects activate the same area in the map. In the two last iterations, the activation of B, C, and E depart from the activation of the other objects. B, C, and E are the smallest objects

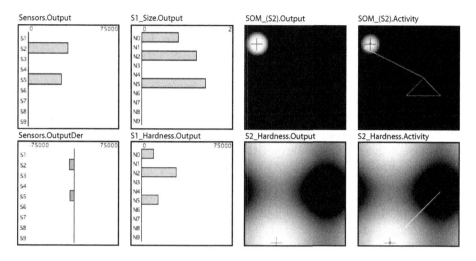

Fig. 3. The status in the haptic system at a moment during a grasping of an object by the Lucs Haptic Hand I. In the upper row, the status in the channel that processes the raw sensor signals from the robotic hand is shown. From left to right: The raw sensor signals, the output from the LeakyIntegrator. The output and activity in the SOM module. The lower row shows the status in the channel that processes the change of the sensor signals over time. From the left to the right: the output from the channel with the change of sensor signals over time, the output from the LeakyIntegrator, the output and activity of the SOM module

group corresponds to the largest objects, and one group corresponds to objects with a size in between the sizes of the objects in the two other groups. In the SOM module that mapped the change of signals over time, only one area was activated before the last two iterations, i.e. before iteration 34. However, during the last two iterations the activation for the objects B, C, and E departed from the activation of the other objects. The objects B, C, And E are the three tiniest objects among all the test objects.

Fig. 2A shows a depiction of the mapping of the different objects on the SOM module that maps the raw sensor signals. In Fig. 2B a similar depiction is shown of the mapping of the different objects on the SOM module that maps change of the sensor signals over time. In Fig. 3 an example of a momentary depiction over the state of the haptic system during the testing of one of the balls is shown.

7   Discussion

An observation that can be done, is that both of the SOM modules categorize the tested objects according to size. In the case of the raw sensor signals the categorization is better than in the case of the change of sensor signals over time. In the latter case the objects are only categorized into two groups.

The current haptic system is not capable to categorize the objects according to hardness and shape. Our idea of having a separate channel for the change of sensor signals over time was that then the objects would be categorized at least according to hardness by the map in the SOM module for this channel, and perhaps even according to shape. A possible explanation to why this did not work is, that the information about the properties hardness and shape may have been overshadowed by information about the objects size that is still present. Information about the objects size is namely partly preserved in the channel with the change of the sensor signals over time, namely in the form of the number of reacting sensors.

With the simplicity of the Lucs Haptic Hand I in mind, it is probably too much hoped for that the haptic system would be able to categorize the objects according to shape. However, hardness should be possible to categorize, since the information about this property is generated by the Lucs Haptic Hand I in the form of the size of the sensor signals.

With our test of the haptic system for the Lucs Haptic Hand I, we have shown that it is capable to categorize objects according to size, to some extent, if the shape of the objects is restricted to be either a sphere or a cube.

This first prototype of a haptic system constitutes the first step towards more advanced haptic systems.

Our future haptic systems will to a larger extent be modelled in accordance with the corresponding human system. Hardness will, for example, be detected in a way similar to that of the human system, namely by relating the pressure applied on the object by the hand to the sizes of the sensor signals [11].

The very next haptic system we build will include a robotic hand with more sensors and more degrees of freedom. The fingers will be jointed so that the grasping of an object provides a better contact between the robotic hand and the object, which will yield a response from more sensors. In addition, we will include an ability to change the grip of the object during the haptic exploration. This will provide the haptic system with more information about the explored object. This will imply a need to develop software modules that take care of motor procedures, and extensive communication between the motor modules and the sensory modules in the system. Taken together, all this should enable the detection of hardness and shape, provided the neural modules are good enough.

References

1. Balkenius, C. (2004). Ikaros (2004-11-24). http://www.lucs.lu.se/IKAROS/.
2. Balkenius, C., and Morén, J. (2003). From isolated components to cognitive systems. *ERCIM News, April 2003, 16.*
3. Dario, P., Guglielmelli, E., & Laschi, C. (2001). Humanoids and personal robots: design and experiments, *Journal of robotic systems, 18, 12,* 673-690.
4. Dario, P., Laschi, C., Carrozza, M.C., Guglielmelli, E., Teti, G., Massa, B., Zecca, M., Taddeucci, D., & Leoni, F. (2000). An integrated approach for the design and development of a grasping and manipulation system in humanoid robotics, *Proceedings of the 2000 IEEE/RSJ international conference on intelligent robots and systems, 1,* 1-7.

5. Dario, P., Laschi, C., Menciassi, A., Guglielmelli, E., Carrozza, M.C., & Micera, S. (2003). Interfacing neural and artificial systems: from neuroengineering to neurorobotics, *Proceedings or the 1st international IEEE EMBS conference on neural engineering*, 418-421.

6. DeLaurentis, K.J., & Mavroidis, C. (2000). Development of a shape memory alloy actuated robotic hand. (2004-10-28). http://citeseer.ist.psu.edu/383951.html

7. Gardner, E.P., & Kandel, E.R. (2000). Touch. In Kandel, E.R., Schwartz, J.H., & Jessell, T.M., (ed.). *Principles of neural science*, 451-471, McGraw-Hill.

8. Gardner, E.P., Martin, J.H., & Jessell, T.M. (2000). The bodily senses. In Kandel, E.R., Schwartz, J.H., & Jessell, T.M., (ed.). *Principles of neural science*, 430-450, McGraw-Hill.

9. Gentaz, E. (2003). General characteristics of the anatomical and functional organization of cutaneous and haptic perceptions. In Hatwell, Y., Streri, A., & Gentaz, E., (ed.). *Touching for knowing*, 17-31, John Benjamins Publishing Company.

10. Gentaz, E., & Hatwell, Y. (2003). Haptic processing of spatial and material object properties. In Hatwell, Y., Streri, A. & Gentaz, E., (ed.). *Touching for knowing*, 123-159, John Benjamins Publishing Company.

11. Hatwell, Y. (2003). Manual exploratory procedures in children and adults. In Hatwell, Y., Streri, A., & Gentaz, E., (ed.). *Touching for knowing*, 67-82, John Benjamins Publishing Company.

12. Johnsson, M. (2004). Cortical Plasticity – A Model of Somatosensory Cortex. http://www.lucs.lu.se/ People/Magnus.Johnsson

13. Johnsson, M. (2005). http://www.lucs.lu.se/People/ Magnus.Johnsson/HapticPerception.html

14. Johnsson, M., Pallbo, R., & Balkenius, C. (2005). Experiments with haptic perception in a robotic hand. http://www.lucs.lu.se/People/Magnus.Johnsson/

15. Klatzky, R., & Lederman, S. (2003). The haptic identification of everyday objects. In Hatwell, Y., Streri, A., & Gentaz, E., (ed.). *Touching for knowing*, 105-121, John Benjamins Publishing Company.

16. Kohonen, T. (1990). The self-organizing map, *Proceedings of the IEEE*, 78, 9, 1464-1480.

17. Kohonen, T. (2001). *Self-organizing maps*, Berlin, Springer-verlag.

18. Laschi, C., Gorce, P., Coronado, J., Leoni, F., Teti, G., Rezzoug, N., Guerrero-Gonzalez, A., Molina, J.L.P., Zollo, L., Guglielmelli, E., Dario, P., & Burnod, Y. (2002). An anthropomorphic robotic platform for ex-perimental validation of biologically-inspired sensorymotor co-ordination in grasping, *Proceedings of the 2002 IEEE/RSJ international conference on intelligent robots and systems*, 2545-2550.

19. Rhee, C., Chung, W., Kim, M., Shim, Y., & Lee, H. (2004). Door opening control using the multi-fingered robotic hand for the indoor service robot, *Proceedings of the 2004 IEEE international conference on robotics & automation, 4*, 4011-4016.

20. Streri, A. (2003). Manual exploration and haptic perception in infants. In Hatwell, Y., Streri, A., & Gentaz, E., (ed.). *Touching for knowing*, 51-66, John Benjamins Publishing Company.

21. Sugiuchi, H., Hasegawa, Y., Watanabe, S., & Nomoto, M. (2000). A control system for multi-fingered robotic hand with distributed touch sensor, Industrial electronics society. *IECON 2000. 26th annual conference of the IEEE, 1*, 434-439.

Action-Based Cognition: How Robots with No Sensory System Orient Themselves in an Open Field Box

Michela Ponticorvo[1,3] and Orazio Miglino[2,3]

[1] Department of Linguistics, University of Calabria, Arcavacata di Rende (CS), Italy
[2] Institute of Cognition Science and Technologies, National Research Council, Rome, Italy
[3] Cognitive Technology Laboratory, Department of Psychology, University of Naples II, Italy

Abstract. This paper shows how spatial cognition and the ability to orient in a closed arena can also emerge in artificial organisms with no sensory apparatus. The control systems (Artificial Neural Networks) of some populations of simulated robots have been evolved in an experimental set-up that is usually used to study spatial cognition of real organisms. Some robots were endowed with a normal perceptive apparatus (infrared sensors and on-board camera), some others had no means of getting information from environment. The control systems of this latter class of robots received stimulation from self-generated input: the feedback of their motor action or the activation of an internal clock. Both kinds of robot learnt some strategies similar to the ones observed in natural organisms. The robots without sensory apparatus displayed a greater amount of micro-behaviour (number of activation patterns of motor output) compared with robots with a normal perceptual system. The amplification of behavioural repertory allowed them to understand the environmental structure even if they couldn't perceive it.

1 Introduction

The success of living creatures in surviving is due to their ability to build a reliable representation of the world they inhabit and to regulate the internal situation of their bodies. It's quite clear that the behaviour of an organism depends both on environmental stimuli it receives through its sensory organs and on internal stimuli the body itself produces (for example visceral stimuli, circadian clocks or motor feedback). Schematically we can state that it's commonly accepted that environmental stimuli are the first step in the construction of a neuro-cognitive representation of the external world while internal stimuli constitute the substrate of the basic functions of life such as keeping body position in the space, regulating internal organs etc. or these stimuli produce internal states as hunger, thirst, sexual impulse that define the meaning of environmental stimulation. This is, in fact, the role of motivation that makes creatures act

J. Mira and J.R. Álvarez (Eds.): IWINAC 2005, LNCS 3561, pp. 396–404, 2005.

differently according to different internal states (for example food produces opposite behavioural effects if perceived by someone who's hungry or sated). Of course, these two systems of stimulation, that we can metaphorically think of as two windows that look onto inside or outside, interact: seeing foodstuffs can rise propensity to eat, but hunger solicits an active behaviour of search (and consequent perception) of food.

These considerations are rather obvious, but they are often disregarded in artificial models of behaviour. For this reason, in this paper, we aim at investigating the possible role of internal stimulation in building a knowledge of the environment an organism lives in. We will explain in detail this question with an hypothetical and concrete example. Let's imagine to have at our disposal an organism provided with an adequate motor apparatus and an internal sensory apparatus, but totally without sensory organs facing the outside. This being is completely closed inside itself, can interact with the world, but cannot get any direct information about it. We can say that it's a model of what Maturana and Varela [6] called an operationally closed system. This concept refers to the exchange of information between the autopoietic system and the outside that allows the system to keep its organization despite perturbations and it's totally different from isolation, from independence from the external world. In this frame knowledge isn't a world representation, but an internal arrangement of system organization. Such a system, as can't react to external stimuli, is forced to create its own interior world on the basis of self-generated stimuli. This system isn't isolated because action warrants a relation between system and environment and the bond with world is not sensory, but behavioural. In nature, we don't know any organism like this, but Artificial Life [4] offers the chance to create and study also impossible creatures. Thanks to methods and techniques developed within its frame we can investigate how an operationally close system knows and adapts to the world with new tools.

One such tool is what has been called Evolutionary Robotics [9], a new discipline belonging to Artificial Life that aims at creating autonomous agents (both physical robot or physically realistic simulation of robot) automatically, without human intervention. In this approach, the agent can develop its own skills interacting with the environment it inhabits, thanks to Evolutionary Robotics methodology that's highly inspired by natural sciences and uses Artificial Neural Networks and Genetic Algorithms. The robots produced by Evolutionary Robotics can have features that we can't find in nature, but can be compared with animals or humans, studied in comparison with living creatures and analyzed more broadly than them because it's possible to manipulate every variable concerning the evolutionary process, the control systems and the sensory and motor apparatus. For these factors Evolutionary Robotics is the right tool for investigating this question.

Even if the main part of robotic models don't take into consideration this problem, a call to pay attention to the internal dynamics comes from Parisi [11], that proposes to build robotic models to explain the emergence of cognition from internal stimulations. In this paper we tried to model a particular behaviour of spatial orientation observed in fresh water little fish (Xenotoca Eiseni) by Sovrano,Bisazza and Vallortigara[12], in rats [5], monkeys [3], chicks [13] and to explain it as a case of behaviour mediated only by internal stimulation.

2   The Open Field Box Task

The task we use is an experimental task often used in the study of spatial behaviour that consists in locating a certain area inside a rectangular arena. At the beginning of the experiment, the subject is placed in a rectangular room, with white walls. In a corner, the experimenter has placed a very inviting object (a reward). The subject is allowed to see the reward, which the experimenter then hides. After a disorienting procedure, designed to eliminate the subject's inertial sense of direction, the experimenter asks the subject to find the reward. Relying on the shape of the room subjects can correctly choose two rotationally-equivalent corners, defined by the geometrical arrangement of the arena. As said before, this experimental set-up has been recently used also to study spatial behaviour of *Xenotoca Eiseni*. As shown in Fig.1 a fish is isolated from its con-specifics putting it in a rectangular compartment of an aquarium. The structure is transparent and in corner A there's a way out. To partially disambiguate the situation, the fish can exploit the geometric cue, provided by the shape of the aquarium and they behave so, because attempt to escape from A (right choice) and C (rotational error). A similar result has been observed in rats [5], monkeys [3], chicks [13]. These results have been interpreted conjecturing that animals have an explicit representation of *rectangularity* of the environment, based on the integration of perceptive signs. This comforts the hypothesis of a neural geometric module inside vertebrates' brain [2]. On the contrary, according to a different thought, animals may have an implicit knowledge of the environment and don't elaborate complex perceptive information, but the active behaviour of searching meaningful proximal stimulation let them reach the target area [7]. This would be an example of Embodied Cognition [10]. Anyway both hypotheses ascribe a fundamental role to environmental stimulation. On the contrary we wonder if this behaviour can be produced by internal stimulation and, if this is the case, how.

3   Method

In our experiments, we used the same experimental setting defined by Sovrano, Bisazza and Vallortigara[12]. We then used the EvoRobot simulator [9] to evolve a population of artificial organisms (software robots) with the ability to solve the task.

3.1   Artificial Organisms

Each artificial organism consisted of a physically accurate simulation of a round robot, with a diameter of 5.5 cm.. Each robot is equipped with 8 infrared prox-

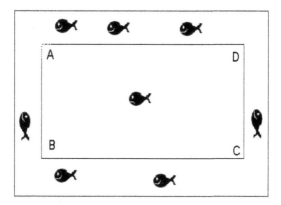

Fig. 1. Schematic representation of experimental set-up used by Sovrano, Bisazza and Vallortigara (2002)

imity sensors (capable of detecting objects within 3 cm of the sensor) and a black/white linear camera with a receptive field of 270 and 6 meter range. The robots moved using 2 wheels (one on each side of the robot) powered by separate, independently controlled motors. The control system was an Artificial Neural Network. In the first simulation the Network is a perceptron whose input layer is formed by 8 input units that codify the activation of Infrared Sensors and receive proximal stimulation, in the second there are 16 input neurons (8 units for the infrared sensors, 8 for the camera) that receive both distal and proximal stimulation. For both simulations in the output layer there are 3 output neurons totally connected to all input units. The first two units control wheels, while the third is a decision unit that signals the choice. In the other simulations the network architecture is different because it receives endogenous stimulation. In fact these networks, have four hidden units with memory [1] and three output units. In the input layer one networks has two nodes that receive motor activation of the previous cycle in input while the second has one unit that codifies the activation of an oscillator that starts at 0, increases its value of 0.01 after each cycle of life of the robot and returns to 0 when it reaches 1.

3.2   Training Environments

In our experiments, we used two distinct training environments: the first one was a rectangular arena (56.8*25.6 cm) with white walls and the second was a rectangular arena (56.8*25.6 cm) with white walls and four additional angular landmarks that can be perceived by the camera. In both environments the target area (the reward) was located in the top-left corner. 1.

3.3   Training Procedures

The robots were trained using a Genetic Algorithm [8].

At the beginning of each experiment, we created 100 simulated robots with random connection weights. We then tested each robot's ability to find the target location. The robot was positioned at the centre of the arena with four face directions (N, S, E, W) and allowed to move around for 200 computation cycles (1ms per cycle). Every time the robot reached and identified the target (activation of the decision unit greater than 0.5) it received a one point reward. Each robot was tested 8 times and assigned a final score consisting of the total number of points received during the tests. At the end of this procedure, the 80 robots with the lowest scores were eliminated (truncation selection). The remaining 20 robots were then cloned (asexual reproduction). Each parent produced five offspring. During cloning, 35 per cent of neural connections were incremented by random values uniformly distributed in the interval [-1, +1]. The testing/selection/cloning cycle was iterated for 100 generations. To analyze the development of the ability to solve the task we ran four different experiments, two with exogenous stimulation and two with endogenous stimulation. The first experiment of exogenous condition was run with agents with both infrared and camera sensors in the arena with walls and angular cues. The second experiment of exogenous condition with robots with only infrared sensors and the two experiments belonging to endogenous stimulation condition were run in the simple rectangular arena. So in the first experiment robots could perceive proximal activation by walls and distal stimulation by angular cues, in the second only proximal activation by walls and in third and fourth self-generated stimulation. For each condition the simulation was repeated twenty times with the same parameters and randomly generated initial connecting patterns.

4   Results

4.1   Behavioural and Neuro-computational Analysis

We focalized our analyses on the individuals with higher fitness measure in each repetition of artificial evolution experiments. In Fig.2 the percentage of correct choices, rotational errors and other choices of fishes (the natural organisms considered as target) and artificial agents are shown. Artificial agents are able to accomplish the task, exploiting the sources of information they perceive and show behavioural indexes similar to the ones observed with animals. The agents that receive both distal and proximal stimulation have reliable information on the shape of the arena and are able to show a perfect behaviour with almost equal choices of the right and the rotationally equivalent corner. The robots endowed with only infrared sensors through an active search of meaningful stimulations are able to reach the geometrically right location. In endogenous stimulation condition, both with motor feedback and internal clock, agents do choose the two corners identified by geometric cues, solving the task. They succeed in coordinating endogenous stimuli with constraints of external environment. Observing their behaviour, we identified some common strategies. The agents initially explore the environment leaving from the central starting position and moving toward the walls of the arena. They approach one of the walls with different trajectories, more probably the long one, and follow it until one of the two rotationally equivalent corners or draw a trajectory that leads them into the right corners. In the simulations without external stimulation, we also find an initial exploratory behaviour that makes agents touch the wall, change direction and follow the wall, but there are some differences. In the Motor Feedback condition we see that, almost always, the first phase consists of drawing a circular trajectory up to a long wall followed by a wall-following behaviour that's quite swinging because the agent doesn't know if it's close to the wall until it bumps into it. Instead, in the Internal Clock condition some agents rotate up to the first wall they meet, go on rotating up to the second wall and then follow it. Some others simple go straight, touch a wall, and follow the perimeter of the arena up to a corner. Other robots start their path going into reverse, impact into a long wall, change direction and follow the wall up to the corners.

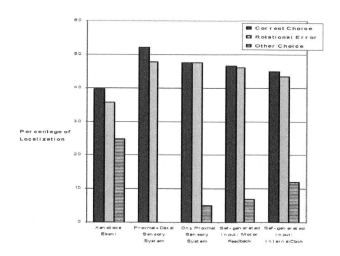

Fig. 2. Percentage of correct choices, rotational errors and other choices of fish (Xenotoca Eiseni) and of the best artificial agents of the last generation. Data on Xenotoca Eiseni are taken from Sovrano, Bisazza and Vallortigara (2002)

Thanks to these strategies the agents of each simulation are able to behave in the best way they can do. In Fig. 3 some prototypic trajectories are represented. In Fig. 4 we present the map of activation (above threshold spikes) of the decision unit, the output neuron that allows localization of the corners during action, in order to clarify how the agents, whose behaviour is shown above, solve the task. The decision unit operates an efficient discrimination between target and non-target zone of the environment when an exogenous stimulation is present, but it's very useful also when no stimulation is perceived. In fact, as we can see in the maps where some spikes are present along arena's perimeter, its activation becomes a signal of wall presence, making the agent change direction or follow the wall, partially substituting external stimulation.

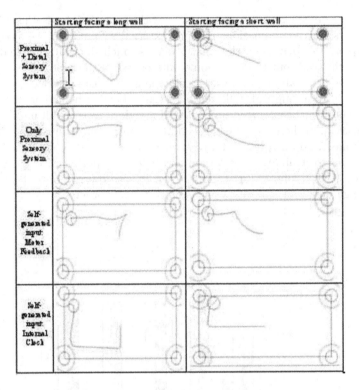

Fig. 3. Trajectories drawn by agents to reach the right corner. In the columns there are two different starting face directions (left column: facing a long; right column: facing a short wall). The trajectories starting from the remaining two face directions are not represented because they are perfectly symmetrical (180 degrees-rotation) and lead to the rotationally equivalent corner). Filled circles represent angular landmarks

4.2   Behavioural Richness

In order to verify the role of endogenous stimuli in producing behavioural sequences we tested the best agents of every repetition and registered the activation of the output layer, a measure of the wideness of micro-behaviour patterns that represents the computational burden of the different stimulation condition.

In Fig.5 the mean number of micro-behaviour for each experimental condition is presented. These data indicate that more and more micro-behaviour (number of different output pattern activation) are necessary to solve the task moving from diversified environmental stimulation toward self-generated stimulation. If we consider data aggregated according to the distinction Endogenous vs. Exogenous Stimulation, we see that the first condition requires a significantly wider repertory of micro-behaviours, in fact probability associated with T test is below 1 per cent, while there's no significant difference inside each condition (probability of T test: Agents with Distal plus proximal Stimulation vs. Agents with Proximal Stimulation above 0.05; probability of T test: Agents with Motor Feedback vs. Agents with Internal Clocks above 0.05). When agents can receive some information (detailed or not) about environment, solving the spatial orientation task requires smaller patterns of behaviour.

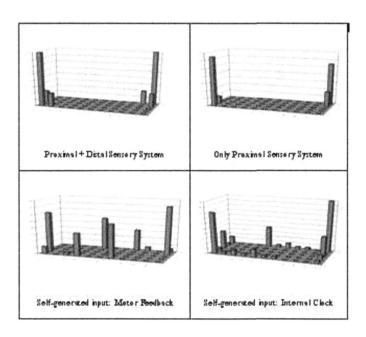

Fig. 4. Maps of activation of Decision unit in the artificial agents. The arena was divided in 5*11 cells and the number of firing of the unit in every cell during action was recorded and reported on z axis

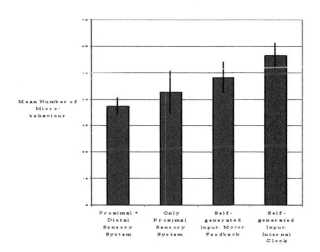

Fig. 5. Mean Number of micro-behaviour in the four experimental conditions

5   Conclusions

The data exposed above suggest that, at least for the artificial organisms we considered, endogenous stimulation can indeed generate a behaviour similar to the one acted by living creatures as fish, rats, chicks and monkeys. Even if they're closed in their own self-generated perceptual world, they do establish a useful relation with the environment around them through action. In fact, exploiting a very precise coordination between self-generated input and produced output, the agent are able to adapt to external constraints, but, of course, this requires a huge resort to action that becomes the vehicle for a representation of the environment it inhabits. Briefly, in this Embodied, but operationally closed system we considered in our experiment, through interaction between organism and environment emerges, after a demanding search, the chance to utilize action to know the environment and the resort to action becomes unavoidable if there is no indication about it.

References

1. Elman, J. L.: Distributed representations, simple recurrent networks, and grammatical structure. Machine Learning 7 (1991) 195–224
2. Gallistel, C.R.: The organization of Learning. Cambridge, MA: MIT Press(1990)
3. Gouteux, S., Thinus-Blanc, C., Vauclair, J.: Rhesus Monkeys use geometric and non-geometric information during a reorientation task. Journal of Experimental Psychology: General130 (2001) 505–519
4. Langton, C.G.: Artificial Life: An Overview.Cambridge, MA: The M. I. T. Press/A Bradford Book (1995)
5. Margules, J.,Gallistel, C.R.: Heading in the rat: Determination by environmental shape. Animal Learning and Behavior16 (1988) 404–410
6. Maturana, H.R.,Varela, F.J.: The tree of knowledge - The biological roots of human understanding. Shambhala, Boston(1992)
7. Miglino O., Lund H.H.: Do rats need euclidean cognitive maps of the environmental shape? Cognitive Processing 4(2001) 1–9
8. Mitchell, M.: An Introduction to Genetic Algorithms. Cambridge, MA: MIT Press(1996)
9. Nolfi S.,Floreano D.: Evolutionary Robotics: The Biology, Intelligence, and Technology of Self-Organizing Machines. Cambridge, MA: MIT Press/Bradford (2000)
10. O'Regan, J.K., No, A.: A sensorimotor account of vision and visual consciousness, Behavioral and Brain Sciences,24(5) (2001) 939–1011
11. Parisi, D.: Internal Robotics. Connection Science (in Press)
12. Sovrano, V.A., Bisazza, A., Vallortigara, G.: Modularity and spatial reorientation in a simple mind: Encoding of geometric and non-geometric properties of spatial environment by fish. Cognition 85 (2002) 51–59
13. Vallortigara, G., Zanforlin, M., Pasti, G.: Geometric modules in animals' spatial representation: A test with chicks (Gallus gallus domesticus). Journal of Comparative Psychology104 (1990) 248–254

A Robotics Inspired Method of Modeling Accessible Open Space to Help Blind People in the Orientation and Traveling Tasks

José R. Álvarez-Sánchez, Félix de la Paz, and José Mira

Dpto. Inteligencia Artificial - UNED - Madrid, Spain
{jras, delapaz, jmira}@dia.uned.es

Abstract. The purpose of this paper is to investigate the viability and potential usefulness of a new method of modeling the open space around a blind person. As an alternative or complement to the usual approaches based on the detection of obstacles, we explore the usefulness of offering information on the temporal evolution of the center of area (CA) of safe navigation in relation to the blind person's position. This approach can be considered ergonomic, egocentric and polar, and easy to learn and use.

1   Introduction and Problem Statement

Blind people need relevant spatial information about the environment around them to construct a suitable mental map for navigating with their objectives safely in this environment.

The general purpose of all electronic aids for blind people's orientation and mobility tasks is to provide them, via the other sensory systems, with the maximum information about the accessible environment around them so that they can get their bearings and move safely. These aids also aim to be ergonomic, robust, and easy to learn and use, and interfere as little as possible with the auditory canal to leave the access free for other sources of information [3, 9, 2].

ONCE (Spanish National Organization for the Blind) establishes these objectives [1–translated]:

> The challenge set concerns technology and knowledge about the minimum information that a person must have to achieve valid, safe and independent orientation and mobility, and also knowledge about the processing capacity of the senses not affected by the information presented. Thus some categories of minimum information about the environment have been determined as a starting point that must be provided for the mobility task: presence, location and nature of the obstacle or obstacles in the way, and the texture, gradient and restrictions of the surface or route.

J. Mira and J.R. Álvarez (Eds.): IWINAC 2005, LNCS 3561, pp. 405–415, 2005.

Conversations with ONCE mobility experts also revealed the following general considerations:

- Blind people always move in familiar or at least potentially familiar environments, as a result of previous experiences in similar environments, since there are common elements that they can identify.
- Anything that blind people cannot touch or hear does not exist in their representation of the world, so they need to have exact but not excessive information.
- Mobile or unexpected objects are also a problem. Very often blind people move too fast to be able to handle dynamic changes in the environment. An electronic aid should therefore prevent them from colliding with anything.
- Given the importance of the auditory canal for blind people, any impediment (e.g. headphones) that we place between their residual senses and the external environment is an additional problem.
- Blind people only need to perceive forward in the direction of their movement. What happens behind is not of immediate interest and may confuse rather than orientate.
- Any aid needs a language to provide useful information for mobility.
- Finally, the information communication interface must be ergonomic, robust and autonomous to be useful for independent and safe movement in partially known environments. It must also be easy to learn, must be guided by the blind person's purpose and give clues for executing the general navigation plan (purpose) in terms of local situated and reactive decisions [7, 8].

There are at least two possible approximations to the design of this kind of aids:

1. Those based on presence, location and the nature of different kinds of obstacles. These proposals [4, 5] aim to transfer the spatial and visual characteristics of the objects in the environment like their position, shape, their three-dimensional axes, their restrictions, their color or texture, etc. to the auditory or somatoesthesic canal.
2. Those based on the quest for open, accessible spaces for safe navigation, as a complement to the white or laser stick. Our proposal fits into this second line.

2   Our Proposal

The interplay between biology and computation is usually posed "from natural to artificial", using biological mechanisms as a source of inspiration in the design of new programming strategies or new sensory and motor systems in robotics, for example. In this work we explore the complementary conjecture that it can be useful to start from the usual robotics methods and techniques for developing potentially useful sensory substitution prostheses for blind people.

The problem of representation of the environment is central to autonomous robotics [6]. Fortunately, the cognitive capacities of any person, irrespective of their degree of eyesight, are fundamentally superior to those of any robot. Moreover, we can use them as a resource to replace the limited computational capacity of any Electronic Traveling Aid (ETA). Much of the visual information about the environment is not necessary for blind people's mobility (i.e. colors of the walls or a painting on them). Furthermore, their capacity for reasoning allows them to construct a cognitive map of the space with a vocabulary of spatial concepts, a detailed knowledge of the areas where they are moving and the verbal orientation that they receive from their instructors. Vocabulary is used that belongs to one of the four kinds of coordinate systems: egocentric, topographic, cartographic and polar-central [3–caps. 2 and 9].

Our proposal is to use a system of reference coordinates with characteristics from the first and fourth kind along with the right discretization.

Starting from the fact that the system must transmit the relevant information to blind people about walking in the accessible space around them without any collisions, we can design the internal representation accordingly. Drawing on our previous experience in autonomous robotics we think that the most useful sensors are the range detectors (sonar, infrared, laser, etc.) and that we can mix this information into a schematic representation (geometric model) of the surrounding space as an open space polygon.

2.1   Open Space Instead of Objects

Instead of focusing the representation on the recognition of objects, we focus on the surrounding accessible open space (gaps) that enable or help blind people to create a "mental representation" or "cognitive map" for their movement and orientation in the environment.

Our approach is complementary but do not sustitute the other approaches based on the recognition of objects, because a description of the open space does not allow certain elements to be distinguished that may be useful for movement in a real environment. For example, it is necessary to distinguish a swing or sliding door from a stationary wall (or a gap in a fence). It may also be necessary to distinguish useful elements from mere obstacles. For example, a counter for attending to the public, a bench to sit on, a bus stop, etc. and even distinguish certain parts or components in the environment (or integrated in the walls) that are necessary to reach or use, like the buttons for calling the lift, an automatic cash point, etc. (see Figure 1). Our proposal is not valid, for example, for very crowded areas although it depends on the speed of processing and detection of mobile objects.

2.2   The Blind's Persons Purpose Guide to the Travel Task

The second characteristic feature of our proposal is that the representation considers the blind person's purpose, which thus becomes a calculus element of the representation. In other words, it is the blind people's proposals that adapt the signals and messages that they want to receive in accordance with what they intend to do, in the same way as blind people move their white stick in the direction in which they want to go.

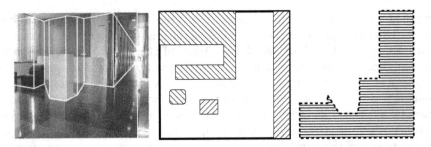

Fig. 1. Illustration comparing the real scenary with, both, the "view" in a representation with objects and that of a representation in a free/open space

3   Solution Method

From the analysis done in the previous sections we propose the general structure of a system to aid mobility that is shown in figure2. In this scheme we see that the two parts or subtasks to be developed are the following:

1. Creation of an internal representation of the external environment in relation to the blind person's position and orientation.
2. User interface.

3.1   Representation

From the observables (sensor signals) an internal representation is constructed that provides information for describing the environment using an open space polygon (vertexes given by the distances in a set of angles distributed around

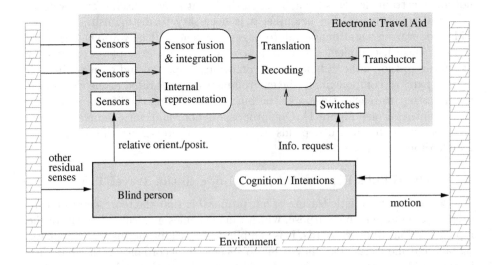

Fig. 2. Interaction scheme of a blind person using an ETA

the current point.) If it were a complete 3D representation it should use a polyhedron with surrounding accessible space, where to find accessible parts it would be necessary to trace the projection of the blind person (cylinder) without colliding with the polyhedron in a horizontal direction. As additional information (constructed from the polygon) the system that builds up this internal representation also obtains the position of the center of area representative of the surrounding space and other representational properties of potential value for the travel task (open or closed areas, etc.).

The spatial representation polygon is constructed increasingly in two phases (from two kinds of information):

1. By spatial-temporal accumulation and fusion (integration) of the information that arrives from the sensors.
2. By transformation of this representation according to the blind person's movement (turns, displacements).

In our experience, the most useful concept derived from representing the open space using a polygon around the blind person's position is the center of area (CA). This virtual magnitude possesses the following useful properties for blind people's mobility:

1. It is a robust reference for getting one's bearings and taking local decisions.
2. Movement between CA's is safe because two adjoining CA's are always connected by open spaces (without any obstacles).
3. The relative displacement speed of the CA allows the blind person to distinguish between different kinds of open spaces (restricted and non-restricted in some direction within the range of the sensors).
4. The system is always going to guide the blind person from the starting point to the nearest CA, which is a safe area without any obstacles. This will give us clues about what we must transmit to the blind person. If the center of area "moves", then probably it is not a place to stop. It is an indication or clue of the entities that will constitute the language that the system will use to communicate with the blind person. Then a grammar will have to be established for this language.
5. It allows mobile objects to be managed as well as dynamical modifications of the accessible space. When a mobile object invades the accessible space around the blind person, the sequence of succesive deviations of the CA (speed of the CA regarding a blind person's movement) can be used to detect in efficient time the modification in the free space, provided that this sequence is followed as a reference. This modification can also be due to a structural change in the environment, as for example doors that close (or open). See an illustration in figure 3.

This utility will be restricted by the speeds of change (or movement of the object) in relation to the speed of detection and processing of the system along with the displacement speed and blind person's reaction (very fast mobile objects are difficult to avoid even for people who can see).

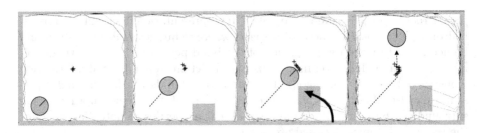

Fig. 3. Illustration of a sequence of change in position of the center of the area when an obstacle invades (square in right bottom half) the surrounding free space. The different surrounding open spaces are represented with fine polygonal lines in a movement sequence of the obstacle. The thick arrows indicate the path of the obstacle (solid line) and center of area (dotted line)

A problem that existed when the robot processed the information was the interpretation task needed to do the modeling of the environment. This problem has now been corrected and no longer exists because the interpretation and the modeling will be done by a blind person who is used to imagining these models from the information that he receives (stick, instructions from an external assistant, guide dog, etc.) Therefore, one part of the high-level interpretation task is no longer so necessary as in the case of the autonomous robot, since the blind person can add his experiences and residual senses (touch, hearing), as well as his cognitive processes, to the information that the sensory substitution system gives him.

3.2   Interface

Let us assume that we have a complete internal representation based on the temporal evolution of the open space polygons and their corresponding centers of area. The following task now is to transmit this information to the blind person in a simple, robust and easy to learn and use manner. In other words, to develop a friendly interface (words, vibrations, sounds, etc.) that help the blind person understand the layout of the space around him in the context of his purposes. To fully describe this interface language we need to specify the semantics, syntax and lexicon.

Semantics. The semantics associated with messages derives from a minimum language for understanding the representation of the surrounding space. The language that we propose uses polar coordinates (angle plus radius) centered on the blind person with reference orientation at the front of his current position, and where distances are measured in entire steps and the angle is a whole number of parts of a turn ($2\pi/N$ rad). This cuantization forms a grid of surfaces with a circular sector shape of rings distributed around the blind person (see figure 4).

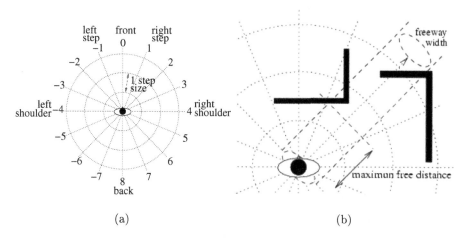

(a)   (b)

Fig. 4. System of centered polar reference coordinates in relation to the blind person's position: a) Seen from above the grid of reference points (intersections between straight lines and circles) for N=16 around the blind person's position. The directions "left-step", "front" and "right-step" must be treated as preferential for receiving information about free space. b) Scheme of how to measure free distance (open space) of a width in relation to the blind person's position (its size plus a margin of safety) in a specific direction in relation to the blind person's orientation

The kind of information (high-level description) that can be transmitted must be useful for mobility and include at least the following points:

1. Warning of obstacle in range of collision (current direction and distance less than safe).
2. Free distance to the detectable obstacle in the direction where the blind person is currently heading (or in a small set of directions near to the current "straight ahead" and "at both sides").
3. Distance or direction to the current center of area (or confirmation of being now in the center of area).
4. Distance or direction to the largest accessible area.
5. Set of ordered distances defining all the accessible area around the blind person.
6. Distance or direction to the nearest obstacle.
7. Change of speed of obstacle in the way.

Syntax. Associated with each item of information there must be a way of right transmission (syntax). If there is a new obstacle that enters the safety area (the obstacle moves towards it or the blind person advances and the safety area moves against the obstacle) and there is danger of a collision an alarm sounds. On the other hand, if the blind person enters an environment by surprise he may want (on his own initiative or request) an extensive description of everything around him (distances in all the surrounding angles).

There can be various ways of transmitting the information to the blind person:

- fixed and continuous: The position information, obstacles, free routes, etc. are translated and transmitted continuously in a predetermined order.
- events or alarms: Information is only transmitted when certain conditions are fulfilled (safety distances, etc.) because of changes that may imply a danger or significant modification in the environment.
- on demand: Only the information requested by the blind person is transmitted (e.g. only free direction or only distance to the nearest obstacle, etc).

To avoid oversaturation of information, transmission "on demand" seems recommendable with the right mixture of some events or alarms.

Lexicon. The final translation (coding) of this information to transmit to blind people depends on what residual functions they have of the other senses and on current or future technological capacities. The simplest medium is auditory one, but for blind people it is very important that this canal is free of interferences since it is used for orientation and total understanding of the environment (echoes, reverberations, etc.) The coding of the information can be made using one or a combination of two or more of the following types:

- by voice, with a small preliminary agreed vocabulary (distance in steps, angles in little/much, etc. or in relation to the "hour hands of a clock").
- by sounds (low-volume beeps):
- distance by pulses (separation or duration) or by distinct frequencies.
- direction by stereophonic sound (balance of volume and delay of tones in both ears).
- by small vibrators on several fingers of a stick or rather on other parts of the body (for example, on the shoulders or waist to indicate directions, on the legs for distances to obstacles, etc.)

4   Simulation

In the initial tests we did experiments substituting the body of the blind person with sensors by an autonomous robot Nomad200 guided remotely by an operator using only the information that would be given to a blind person. We thus simulated the conditions that a blind person would find when walking through an area that we do not know a priori. We have also subtituted the potential actions of the blind person by a language of interaction with the program that represents the environment and produces the advices for safe movement. This program also sends the translation of these messages to the robot simulator (see figure 5).

Usually, sensors will detect the blind person's movement and will use this information to modify the position and relative orientation for internal representation. However, for the initial tests that we are going to do we will use

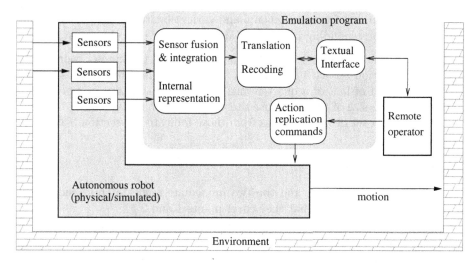

Fig. 5. Interaction schema of the ETA emulated system used in the test. Compare with figure 2 to see the substitution of the blind person body and ETA sensors by the autonomous robot. Transducers, switches and blind person's purposes are also substituted by a textual interface and a remote operator

an extra module to simulate this interaction with the introduction of specific detailed orders corresponding to the actions that the blind person will do.

For the simulated tests we use a system of words that appear on the screen or by voice. The coded version of these words is used to move the robot (emulating the body of the blind person) with simple commands related to individual movements that a blind person would do (without seeing the robot's or simulator's real environment).

The implementation of the blind person emulation program (text interface and robot control) and of the ETA was done in C++ and has several parts that are executed as independent threads (implemented with the GNUPth library) to obtain a better response without dead-time blocks:

* Reading and updating the robot sensors.

* Calculus of an internal representation integrating data from different kinds of sensors accumulated in time and corrected (relative representation) by the robot's movements.

* Extraction of the relevant information (free front distances and relative position of the CA) that would be continuously presented to the blind person (shown on a screen for the human operator controlling the simulation of the blind person's behavior).

* Wait and interpretation of simple commands for individual actions (movements and turns) or requests for extra information that the blind person would do.

* Transformation of the movements and turns that the blind person would do to into the corresponding movements and turns that the robot must to do.

In the information given to the blind person we used the kind of coordinates explained in section 3.2 with the value N=16 resulting in simple angles for understanding (half of half of a quarter turn) and with a step size proportionally equivalent to the size of the robot used to emulate the blind person.

We selected a set of useful values from those indicated in section 3.2 for the message contents:

- Periodically (every 2 sec.):
- the distance ahead data and the two immediate angles at both sides (left and right) are given so that they cover points 1 and 2.
- the distance and orientation data of the CA are given instantaneously, so points 3 and 4 are fulfilled (implicitly).
- Only at the operator's request (emulating the blind person) all the distances are ordered in steps for each of the 16 angles defining all the accessible area around the blind person. This covers point 5.
- In the current version used in the trials, points 6 and 7 were not implemented because they are not as necessary for basic movement.

The possible actions implemented are:

- Advancing one step in the front direction or in the discreet angle direction (1/N of turn) at one of both immediate sides left or right.
- Turning (without advancing) any discreet angle.
- Requesting distance information around discreet angles.

5   Conclusions and Future Works

In this work we have explored the potential validity of an ETA based on the dynamic representation of the open space polygon for a blind person as a traveling aid. This proposal is complementary to other ETAs based on the representation of obstacles and stresses the reactive nature of the sequence of local decisions along a route. The potential benefits of this proposal are:

1. Aid to safe mobility in unknown environments (compared with prefabricated ones or ones requiring infrastructure).

2. Direct, adaptive, dynamic reference of the environment. We do not aim for a totally complete static representation of the entire environment but just an adaptive dynamic representation (in accordance with the blind person's purpose) and local (in space and time).

3. Potentially and in the long term when we expand the scope of the paper, our representation proposal can also be used for creating a map of the environment where the blind person frequently walks.

Acknowledgements. We acknowledge to Antonio Martínez Henarejos and collaborators of the ONCE (Spanish National Organization for the Blind) for the invaluable help offered to us during the conceptualization of this paper.

We also acknowledge the economical support provided by the Spanish Ministerio de Educación y Ciencia under the project TIN2004-07661-C02-01.

References

1. Antonio Bermúdez Cabra, José Antonio Muñoz Sevilla, Antonio Rodríguez Hernández, and Enrique Varela Couceiro. Tecnologías específicas para personas ciegas y deficientes visuales. In *Tecnología y discapacidad visual.* Organización Nacional de Ciegos Españoles (ONCE), 2004.
2. R. M. Blanco Sanz, L. Blanco Zárate, S. Luengo Jusdado, G. Pastor Martínez, M. Rivero Coín, R. Rodriguez de Luengo, and M. J. Vicente Mosquete. *Accesibilidad para personas con ceguera y deficiencia vsiual.* ONCE, 2003.
3. Bruce B. Blasch, William R. Wiener, and Richard L. Welsh. *Foundations of Orientation and Mobility.* American Foundation for the Blind, 1979.
4. Johann Borenstein. The NavBelt - A Computerized Multi-Sensor Travel Aid for Active Guidance of the Blind. In *Proc. of the Fifth Annual CSUN Conference on Technology and Persons With Disabilitie,* pages 107–116, 21 March 1990.
5. Johann Borenstein. The GuideCane - A Computerized Travel Aid for the Active Guidance of Blind Pedestrians. In *Proceedings of the IEEE International Conference on Robotics and Automation,* pages 1283–1288, Albuquerque, NM, 21 April 1997.
6. Johann Borenstein, H.R. Everett, and L. Felng. "Where am I?". Sensors and methods for mobile robot positioning. Technical report, University of Michigan, 1996.
7. Rodney A. Brooks. Intelligence without representation. Number 47 in Artificial Intelligence, pages 139–159. 1991.
8. William J. Clancey. *Situated Cognition: On Human Knowledge and Computer Representations.* Cambridge University Press, 1997.
9. Simon Harper and Peter Green. A Travel Flow and Mobility Framework for Visually Impaired Travellers. In *International Conference on Computers Helping People with Special Needs,* pages 289–296, Germany, 2000. OCG Press. POLI Project (http://proximity.man.ac.uk) - under the auspices of the UNESCO.

A Scientific Point of View on Perceptions

Juan Carlos Herrero

jcherrer@arrakis.es

Abstract. Among the different aspects of natural computation, perception is one of the most amazing phenomena. In order to understand this natural phenomenon, one has to consider not only neuroscience and bio-sciences in general, but also other sciences in the field of physics, like astrophysics and astronomy, or quantum physics. Thus, by suitably widening the scope, we begin to deepen our understanding. Eventually, one retrieves from philosophy some of the most traditional questions, which could well be transferred to science from that moment on. In artificial computation, the question of perceptions appears as soon as we try to build intelligent systems that have to be successfully independent of human assistance, or when one has to approach computational problems, like artificial systems that understand and translate natural language, or in auditive or visual recognition.

1 Introduction

Descartes made one of the best contributions to science when he wrote "I think, therefore I am". In the context of his famous book, this was not a mere conclusion. He started by doubting that anything really existed, and after a while he perceived his own thoughts, besides he was conscious about that circumstance. In this way he implicitly established the figure of the observer (Descartes) and the observed facts (firstly the thoughts he perceived, next his being aware of he was thinking). This was quite a milestone to establish science as based on facts, since the first scientific truth was that he existed (whatever "he" was), this was the first truth among those ones that Descartes was looking for.

Within the observer's domain, the observer can be sure about the perceptions he has, but have those perceptions to do with something else? Since he has perceptions of quite a so called real world, philosophy has sometimes dealt with the affair of the existence of all that we perceive. For instance, Berkeley assured that all our perceptions about a real world are just that, but a real world does not exist at all; in this case, the only existing being would be the observer, and the only existing things would be his perceptions (those of Berkeley in this case; we all know that this is not the real case, because we are not Berkeley, but it could well be each one of us' case). This is really a problem that has not a logical solution, not even a scientific solution, thus it was Bertrand Russell who said that if we think a real world exists, if we think there are other observers besides us, if we think some of our perceptions are caused by events which happen in a real world, this is our choice; however, he said it seems the most suitable choice, according to all our knowledge.

J. Mira and J.R. Álvarez (Eds.): IWINAC 2005, LNCS 3561, pp. 416–426, 2005.

According to all our knowledge, or the knowledge to which we could have access, our choice is to admit that there is a universe in which a lot of events happen and some of them are related to our perceptions. Besides, to be consistent, we have to consider perceptions themselves are facts which happen in nature too, and perceptions as well as ourselves are within that reality, we the observers are part of that universe, facts of the universe.

Thus, as parts of that reality, all those facts should be capable of being studied by science. But the facts known to mankind, really the only ones, are our perceptions. It was Leonardo da Vinci who wrote: "All our knowledge has its origins in our perceptions". However, there are many kinds of perceptions. We perceive colours or sounds, thanks to our senses, but it is not less true that we also perceive that we understand or not what we are told, we perceive fear or pleasure, among a lot of other things, including a wide variety of feelings and thoughts. Those are directly known to us, non-deduced. However, we deduce and infer too, so our knowledge widens. Science is precisely the means we use to do it in a secure way, keeping consistency between all our perceptions –which is not that easy sometimes.

Every measurement instrument used in science is based on our perceptions: most of the time we see images that we later recognise, interpret, and relate to other concepts like distance, volume, weight, voltage, etc. The same thing could be said of your reading this paper, since you are seeing black shades on a more clear piece of paper, and then you are recognising characters, words, and then you are understanding what you are reading.

It was Bertrand Russell who laid the final bridge to cross from Philosophy to Science, when he wrote "physics assures us that the greenness of grass, the hardness of stones, and the coldness of snow, are not the greenness, hardness, and coldness that we know in our experience, but something very different. The observer, when he seems to himself to be observing a stone, is really, if physics is to be believed, observing the effects of the stone upon himself."

Note: in this paper we shall refer to us ("we", "our", etc) as if we were the observer.

2   Perceptions at the End of a Causal Chain of Events

These initial reflections lead us to consider our first facts, not necessarily in the following order. Firstly, we have perceptions like colours, sounds, etc, as mentioned above, whose nature do not know yet, but, secondly, we are conscious about those perceptions.

In the third place, as a result of all our knowledge or the knowledge to which we can access, our choice is to admit that there is a universe, a real world, in which ourselves and our perceptions take place, as well as the rest of the facts that we say we observe.

According to our bio-scientific and medical knowledge, we admit that our brain is the place where perceptions take place, as well as our consciousness– our self– therefore the place of that fact we call "the observer". For instance, the relation between visual perceptions and the brain is strongly supported by medical observations, e.g. those of Justo Gonzalo, taken from patient's testimony during their therapy.

However, that testimony which comes from others is known to us by means of some of our perceptions too, for instance through sounds that we then identify with words, as long as they make sense to us. We certainly have to admit that all our knowledge is based on perceptions, although we can establish different cases: a) direct knowledge coming through our senses, like colours, sounds, heat, etc; b) deduced or inferred knowledge based in our perceptions, this is what we do for instance when we look at a measurement instrument, but whatever this is, it ends with a perception, that of the final deduced knowledge, the result of the deduction; c) deduced or inferred knowledge, based on previously deduced or inferred knowledge, in the end it is the same situation as case b, but c is what we usually do when we elaborate theories, plans, etc; d) direct knowledge coming from inner perceptions, for instance memories, thoughts, feelings; I would ask you: do you understand what I mean? The answer to that question comes from a perception of yours. According to that perception, you may ask "yes", "no", or "I doubt", because if after your answer I ask you: "and how do you know that you understand it, that you don't, or that you doubt?," your only answer can be something like "just because I know it". This was your perception, one of a inner kind, but as perception as a colour can be.

If perceptions are facts which happen in our brain, and there are processes we have called deduction and inference which can generate new perceptions, we have to admit those processes also take place in our brain. This is not opposite to our knowledge about the brain, from a bio-scientific point of view. As we said above, medical and other bio-scientific experiences have shown what parts of the brain and what biochemistry are related to different kinds of perceptions, and this is known by patient's testimony. Those perceptions as we know them (pitch of a sound, red colour, pleasure,...) cannot be "seen" if we look into the brain, for instance looking for colours among the neurons; maybe we could if we connected the patient's brain with ours, or something like that, in a way that we do not know, what seems completely impossible now. Moreover, you have no means to assure, to prove, that the perception that you call "blue" is equal to the perception that I call "blue", since the only thing we can do is to point to the same places and agree that they are blue or not. That is why it is so difficult to agree about inner perceptions, or to explain them.

Due to there being perceptions that are related to inner facts which happen in the brain, as we just have said, and due to there being perceptions which are related to facts that happen outside the brain, for instance which have to do with light coming from a stone, and because we admit our brain exists in the real world or universe, as well as the facts which happen outside our brain, we also think that perceptions are facts which happen in the universe. We just say that the universe in which Descartes' perceptions and Descartes' self existed is the same as the universe in which science works, we just say that the world of our perceptions and our self is the same as the world of the facts that science studies.

Therefore, in the fourth place, we admit there are also causal chains of events which begin in nature and end at our perceptions, chains whose links are not completely known yet, chains which may or may not include other perceptions as links, chains which may begin inside or outside our brain, being our brain a portion of nature.

According to science, an infinity of events happen simultaneously in nature outside our brain, on and on. We know about some of them by direct perception through our senses, but these are just a few, the rest of them remain completely non-perceived. Once again, we have measurement instruments, most of them cause images in our retina, and when we use them we interpret those images in order to deduce facts which cannot be known by means of direct perceptions, but only through deductions or inferences. From this point on, the perceptions, deductions, and inferences which happen in our brain are up to us. Only we steer them all, trying to keep all our perceptions within consistency. This is what scientists do when they elaborate theories: they have to be careful, lest it happen they or anybody else could make them perceive facts which are not consistent with what they have previously perceived, or inconsistent with regard to what they say which can be perceived. For instance, Newton's gravitational theory is consistent with all our measurements (say perceptions set A) as long as we disregard, or do not know sufficiently, things like Mercury orbit about the Sun (say perceptions set B). This orbit is better explained by Einstein's gravitational theory (then perceptions set A and B are consistent with the theory). That's the way of science.

3   Evolution of Perceptions and a Filtering Process

The subject of perceptions is not new to science. It is in the foundations of modern science. It was Darwin who wrote in the 7th edition of his famous book on the origin of species "How it comes that certain colours, sounds, and forms should give pleasure to man and the lower animals, –that is, how the sense of beauty in its simplest form was first acquired,– we do not know any more than how certain odours and flavours were first rendered agreeable".

As we have said, there is an infinity of events happening in the universe, and only some of them are perceived by us. From an evolutional point of view, this has been possible because of the interaction between living creatures' and their environment. If we take the case of light, such light being an instance of those events we are talking about, the living creatures lineages which have been exposed to light may have developed organs which reacted more and more to light, and developed visual cortex structures as well, in order to perceive more and more aspects of light, because this has helped those lineages and successive species to survive.

One of the examples Darwin gave in his book on the origin of species was an explanation on how the eye evolved in living creatures, by showing cases of different stages of eye evolution that one could find in creatures which lived in the 19th century, but creatures that we can still find today. Now, we suppose that a living creature which belongs to a species which has no eyes cannot have perceptions like colours. If we take a living creature which has eyes, for instance a human being, that human being will generally have perceptions like colours. On the evolutional path which leads from species with no eyes to species with eyes, we also travel through a path from no perception of colour to perception of colour. The question would be "when did the perception of colour appear?" The evolutional answer, following Darwin's example about eye evolution, would be something like: it is clear that evolution of perception has taken place, from no perception to perception, so we could expect that in different stages of evolution of the eye, there were different stages of evolution of the perception, accordingly. We could then think of different stages of evolution for perceptions in general, any red colour, pitch of a sound, forms, pleasure, pain, heat, cold, the wide variety of feelings, and so on and so forth, and we could also think of degrees in differentiation among perceptions.

On the other hand, evolution brought about our retinas in order to pick up the kind of light that is more abundant in the environment where those species we come from have evolved. Thanks to physics we know the Sun's surface radiates electromagnetic energy in a continuum spectrum with a maximum around those wavelengths which correspond to the colours we see. Outside the atmosphere, the graph of the irradiation measurements fits that of a black body at Sun's surface temperature. At Earth's surface, that graph's maximum seems to be a step, climbing from ultraviolet zone.

Flowers display their beautiful colours which give pleasure to us, however they are not made for us, but for flying insects. Those insects involuntarily fertilise plants carrying pollen from flower to flower, as we were told at school, and insects are attracted by the colours of the flowers. It seems that insects associate colours to the taste of nectars that they eat or pick. We could expect that insects have perceptions like flavours and colours; even colours that we cannot even imagine, since it seems that insects see colours in the ultraviolet zone: we have no colour perception in that zone (since we do not even see it). So some plants evolved to attract insects and in that way plants reproduce and continue living on planet Earth. So insects evolved to distinguish flowers among the whole electromagnetic radiation which gets to their eyes coming from the Earth's surface, as patches of definite colours.

Thus, eyes have appeared and evolved as a filter for those chains of events that we have talked about. But in order to underline the importance of that filtering, we can observe that the same kind of events may produce chains which end in quite a different kind of perceptions.

For instance, electromagnetic radiations are filtered by eyes, in chains which end at perceptions that we call colours. But if the radiation wavelenght is in the ultraviolet zone, some insects will also see it, but in our case we will not;

however, our skin will produce D-vitamin when that ultraviolet radiation fall on it (no perception at all, here evolution has developed a different function, no less important).

Another very interesting case is that of the movement of air molecules. This can be perceived by us as a sound, if the movement changes periodically in direction, and depending on the speed of change; but if the movement is chaotic, it can be perceived by us as heat, under some circumstances that science measures as temperature, which can also be perceived as cold, among other intermediate or contradictory cases.

4 Computation

As we have said, a huge amount of events happen simultaneously in nature, on and on. For instance, the light reflected by the atoms of any surface consists of a huge amount of photons. So our causal chains could begin there. If we suppose that light comes from the Sun, so it has a wide continuum spectrum, the surface has absorbed some frequencies and reflected some other.

Since we consider a huge amount of events, we have a huge amount of causal chains too. So the final perception is not going to be only one, but a huge set of them. If the surface is a portrait, the huge amount of perceptions could be the whole set of coloured points of the picture that we can distinguish. In this case, the filter has been firstly the retina, then the processing of those colours in our brain, which is known through the theory of colour as well as psychological experiences. But this is the case for a kind of perceptions, those of colour. But we also have simultaneous visual perceptions like contour, distance, etc, among other of different kind, maybe beauty, memories, etc.

Each cell of the retina has been impacted by photons which, due to the lens of the eye, come from the same region of the picture. Taking into account that we have 100 million cells in each retina, which react to either wavelength or intensity of light, we can obtain a filtered image of extremely high resolution. This first filtering can easily be emulated by artificial means, as we can immediately imagine.

When the optical nerve leaves the eye, the number of nerve fibres has been reduced in two orders, and now we have just a million of them coming out from the eye. It is clear that some kind of processing has taken place in the retina. This million fibres get to the visual cortex of the brain.

It is known the visual cortex preserves the geometry of the retinal field, so successive small zones of cortex layers that the signal goes across are related to the same small zone of the image that arrived to the retina. Thus, new filterings take place; for instance in order to locate movements from a zone to the immediate adjacent, or in order to detect contours, among others.

In this way, parallel computation is brought about in our minds as the most natural way to carry out that filtering, since the causal chains are all simultaneous, and geometry is preserved. These are characteristics that we would like to have too in the artificial counterpart.

That filtering can be implemented based on a parallel and distributed computation. We can consider each unit owns a small region of the image, which comes as a signal from a definite pixel. Let us consider the pixels form a bidimensional matrix, whose dimension is not important now.

According to Mira, the calculations that a neural net carries out are performed in parallel and are distributed among the neurons. So let us suppose that each pixel is really processed by one unit neuron, whose receptive field spreads over 8 neighbour units. In order for this ensemble of neurons to work as a filter, we can think of the following matrix:

$$\begin{pmatrix} 0 & -1 & 0 \\ -1 & 4 & -1 \\ 0 & -1 & 0 \end{pmatrix}$$

Figure 1 shows the original image on the left hand side, and the result of the filtering in the middle (on the right hand side the negative image is shown, maybe it can be seen better in the printing of this page). So when the image which is on the left hand side gets to the neural layer $k + 1$, the value of each pixel (i, j), that is $u(i, j, k)$ which comes from layer k is transformed according to the following calculation:

$$u(i, j, k + 1) = -u(i, j - 1, k) - u(i - 1, j, k) + \\ +4u(i, j, k) - u(i + 1, j, k) - u(i, j + 1, k)$$

This kind of calculation is know as convolution, in its discrete form suitable to computation; it results from taking the sum of the bidimensional differentials of the signal at one point (pixel) with respect to the adjacent ones, i.e. the sum of the directional differentials 1 pixel across each time. So

$$f_{k+1}(x, y) = [f_k(x, y) - f_k(x, y - 1)] + [f_k(x, y) - f_k(x, y + 1)] + \\ + [f_k(x, y) - f_k(x - 1, y)] + [f_k(x, y) - f_k(x + 1, y)]$$

which leads to the expression mentioned above.

Fig. 1. Original image, filtered and inverse of filtered

According to the meaning of a differential function, our filter precisely detects changes in the signal distribution (contour, in this case). As Mira wrote: "from the computational perspective, the point is that the whole neural computation is based on the excitation-inhibition game (add, subtract; accumulate, decrease) repeated on and on over signals which detect the spatio-temporal coincidence of events whose semantics increases as we get deeper into the nervous system, coming from the world of sensors." This getting deeper takes place as chains of events which begin at our senses and end at the perceptions, which undergo different filtering processes, so we establish an analogy between natural and artificial computation.

Now the point is that as long as we cannot describe the nature of perceptions in natural computation, we cannot find their analogy in artificial computation, since we do not know if the filtering function is enough to generate the perception– it seems it is not– and perceptions always come together with the observer who perceives, and we do not know the nature of this observer yet, in spite of we are sure that he exists.

5   The 10 Billion Years Experiment

Hubble Space Telescope's heritage currently available on the internet is a magnificent source of information and inspiration. Let us consider now one of the photographs, identified by HST·WFPC2 "Star-Birth Clouds · M16". There we can see an immense cloud, like a column of several cubic light-years of hydrogen; inside that cloud stars are born, that is to say, planetary systems are born.

We can propose an experiment which lasts 10 billion years (i.e., 10 thousand million years, or 10^{10} years). Set up a volume of several light-years of hydrogen, and come back in 10 billion years: you will find starts, planetary systems, living creatures, and perceptions (those living creatures would have), like colours or sounds. You can't believe it? But I am talking about our planetary system. Only in our case we have to add a touch of an early supernova explosion around here, which explains we can find heavy elements like Uranium in our planet.

So if we had hydrogen in the beginning, how can it be that we have perceptions or consciousness in the end? This is a question which has no answer nowadays, but to pose it, is the first step to obtain the answer.

This make us think some things over. It was Dirac who wrote in his interpretation of quantum electrodynamics: "A theory which gives rise to infinite transition probabilities of course cannot be correct. We can infer that there is something wrong with quantum electrodynamics. This result need not surprise us, because quantum electrodynamics *does not provide a complete description of nature.*" (my emphasis).

In this sense, we could say it need not surprise us if Newton's theory of gravitation is not completely correct, since it does not provide a complete description of nature, not even of the phenomena that it approaches; it has been necessary Einstein's theory about gravitation to provide a more complete description. Moreover, we should not reject that this will not be the end; we just do not know.

We have been talking about science during the whole paper, but as a matter of fact science is divided into pieces of ground, as estates in a housing scheme. This is necessary in order for a human being to understand the world, but this must not lead us to ignore each other's findings, to forget that our piece of science is just a part of the whole scheme, which can also grow with new pieces. Thus, as in estates, we have neighbours and meetings of neighbours, or we should. There we discover the problems that we have in common and search for solutions –or we should.

If we ask how it can be that beginning with a huge amount of hydrogen we can obtain consciousness or perceptions after 10 billion years, we have to observe the object of our science more carefully. For instance, we admit that a human body (including the brain) is made of cells. But cells are made of molecules, so we could say with all suitability that a human body is made of molecules, because "cell" is a name that we give to a big set of molecules, under some circumstances and characteristics, certainly with a very complex structure, which works in a way that we do not understand completely yet. But, to be consistent, it is clear that all the properties of a cell derive from the properties of the molecules, i.e. they should be explained by those properties.

However, we know well that molecules are made of atoms. So "molecule" is a name we give to a set of atoms, sometimes bigger, some times smaller, in order to understand our observations in an easier way. We could say that molecules really do not exist, but there are atoms instead. The properties of atoms explain the properties of molecules, if we consider quantum physics as the science which explains both entities' facts, together with quantum electrodynamics.

Nevertheless (please go with me a little bit farther), atoms are made of particles, so "atom" is a name we give in order to understand the behaviour of a set of particles, under some circumstances. Quantum physics also tries to explain atoms, together with quantum electrodynamics, quantum chromodynamics, and the Standard Model of particle physics. We could say with all suitability that atoms do not exist, and there are only particles surrounded by vacuum instead. Electromagnetic field has been present all the time, although we have not mentioned it. At the subatomic levels, more fields come into play, in order to explain the observed interactions between particles, and between particles and fields.

In the current state of the art, some aspects come into play and blur the clear picture of particles and fields in vacuum. It was Dirac himself who was unable to establish an equation for the vacuum state, that he represented by $|V\rangle$. All he could approach is a state $|Q\rangle$ with no particles, so that $H^* |Q\rangle = C |Q\rangle$, but even in that case, he found fluctuations in energy: H^* contains high energy terms which give rise to infinite probability transitions. The vacuum state must contain many particles, which may be pictured as in a state of transient existence with violent fluctuations he said. Even more, it is well known that the Standard Model's particles can be created and annihilated: we can exchange electromagnetic radiation into plenty of those particles and vice-versa. For instance, it is known that $d \rightarrow u + e^- + \overline{\nu_e}$, that is, an elementary particle as a down quark can decay into an up quark, plus an electron, plus an antineutrino, and yet all of them are elementary particles.

It would seem that vacuum is not really empty, so to speak, and it would also seem that elementary particles have a common underlying nature; it could be that vacuum (which really seems not to be empty), and the fields defined in it, and the elementary particles, have a common underlying nature. But this is a new frontier we cannot enter yet. There is no need to go that far for the time being, to say what follows.

When we began our 10 billion years experiment, what had we really? We said we had hydrogen, but we really had particles and fields instead, since hydrogen atoms are made of particles, and these particles have associated fields. And after 10 billion years, we find consciousness and perceptions in living creatures. We could well think that what we call matter+energy and what we call perceptions+consciousness have a common underlying nature, the universe's nature. Within that common underlying nature, evolution takes place, should we considered it in a wider sense: a process with one of its milestones in a huge cloud of hydrogen in the past, and another milestone in perceptions today, certainly with previous milestones, and for sure subsequent ones. Because the current physics' theories do not completely explain all the phenomena related to particles and the so called vacuum which surrounds them, we cannot explain how this underlying nature of the universe is, and we do not know how to map it into our brain's cells, molecules, atoms, particles, or fields. But of course!, since our theories do not provide a complete description of Nature. However, we may think of a common underlying nature for all of them.

If perceptions have that underlying nature of the universe, the fact that they exist now and they did not 10 billion years ago need not surprise us, since evolution really does not work cells, molecules, or atoms, because they properly do not exist, but evolution works eventually the same underlying nature of the universe that we have talked about, which is what properly exists.

In this sense, we get to the old philosophical split and opposition between mind and matter. We have clues which point to the unreality of this split, and we are invited to abandon both concepts and think of a common underlying nature for both. This is not new, since Bertrand Russell already wrote that mind and matter could be the same thing. Now science provides facts enough to consider this.

References

1. Churchland, P.S. and Sejnowski, T.J. *The Computational Brain.* The MIT Press (1992)
2. Cottingham, W.N. and Greenwood, D.A. *An Introduction to the Standard Model of Particle Physics.* Cambridge University Press (1998)
3. Craig, D.P. & Thirunamachandran, T. *Molecular Quantum Electrodynamics.* Dover (1998)
4. Da Vinci, L. *The Notebooks of Leonardo Da Vinci,* Volume 2. (See 1147)

5. Darwin, C. *On the Origin of Species by Means of Natural Selection, or the Preservation of Favoured Races in the Struggle for Life.* 7th edition.
6. Descartes, R. *Le Discours de la Méthode.*
7. Dirac, P.A.M. *The Principles of Quantum Mechanics.* Oxford University Press (1999)
8. DeFelipe, J. *Microcircuits in the Brain.* In Biological and Artificial Computation: From Neuroscience to Technology. Springer (1997) 1–14
9. Gonzalo, I. *Allometry in the Justo Gonzalo's Model of Sensorial Cortex.* In Biological and Artificial Computation: From Neuroscience to Technology. Springer (1997) 169–177
10. Herrero, J.C. *Knowledge and Intelligence.* In Connectionist Models of Neurons, Learning Processes, and Artificial Intelligence. Springer-Verlag (2001) 814–821
11. Herrero, J.C. *Challenges for a real-world information processing by means of real-time neural computation and real-conditions simulation.* In Engineering Applications of Bio-Inspired Artificial Neural Networks. Springer-Verlag (1999) 299–311
12. Herrero, J.C. and Mira, J. *Causality Levels in SCHEMA: A Knowledge Edition Interface.* IEE Proceedings-Software Vol 147, No 6, (1999) 193–200
13. Hester, J. and Scowen, P. *"Star-Birth Clouds · M16" HST · WFPC2. PRC95-44b · ST Slc OPO ·* November 2, 1995. Arizona State University, NASA. Available on the internet HST URL: http://hubblesite.org/newscenter/newsdesk/archive/releases/1995/44/image/b
14. Mira, J. et al. *Aspectos Básicos de la Inteligencia Artificial.* Sanz y Torres (1995)
15. Mira, J. and Delgado, A. *Some Reflections on the Relationships between Neuroscience and Computation.* In Biological and Artificial Computation: From Neuroscience to Technology. Springer (1997) 15–26
16. Mira, J. and Delgado, A. *Reverse Neurophysiology: the Embodiments of Mind Revisited.* In Proceedings of the International Conference on Brain Processes, Theories and Models. The MIT Press (1995) 37–49
17. Moreno-Díaz, R. *Systems Models of Retinal Cells: A Classical Example.* In Biological and Artificial Computation: From Neuroscience to Technology. Springer (1997) 178-194
18. Pitts, W. and McCulloch, W.S. *How we know universals: the perception of auditory and visuals forms.* Bulletin of Mathematical Biophysics, Vol 9. University of Chicago Press (1947) 127–147
19. Russell, B. *Problems of Philosophy.* Oxford University Press (1912)
20. Russell, B. *The Analysis of Matter.* T.J. Press Ltd, Padstow, Cornwall (1927)
21. Russell, B. *History of Western Philosophy.* George Allen & Unwin (1946)
22. Russell, B. *An Inquiry into Meaning and Truth.* The William James lectures for 1940, delivered at Harvard University. George Allen & Unwin (1950)
23. Savage-Rumbaugh, S. and Lewin, R. Kanzi. John Wiley & Sons (1994)
24. Space Telescope Science Institute. http://www.stsci.edu/resources/
25. The Hubble Heritage Project. http://heritage.stsci.edu/
26. Wyszecki, G. and Stiles, W.S. *Color Science: Concepts and Methods, Quantitative Data and Formulae,* 2nd Ed., Wiley (1982)

Reasoning by Assumption: Formalisation and Analysis of Human Reasoning Traces

Tibor Bosse[1], Catholijn M. Jonker[2], and Jan Treur[1]

[1] Vrije Universiteit Amsterdam, Department of Artificial Intelligence,
De Boelelaan 1081a, 1081 HV Amsterdam, The Netherlands
{tbosse, treur}@cs.vu.nl
http://www.cs.vu.nl/~{tbosse, treur}
[2] Nijmegen Institute for Cognition and Information,
Division Cognitive Engineering, Montessorilaan 3, 6525 HR
C.Jonker@nici.ru.nl

Abstract. This paper shows how empirical human reasoning traces can be formalised and automatically analysed against dynamic properties they fulfil. To this end, for the reasoning pattern called 'reasoning by assumption' a variety of dynamic properties have been specified, some of which are considered characteristic for the reasoning pattern, whereas some other properties can be used to discriminate between different approaches to the reasoning. These properties have been automatically checked for the traces acquired in experiments undertaken.

1 Introduction

Practical reasoning processes are often not limited to single reasoning steps, but extend to traces or trajectories of a number of interrelated reasoning steps over time. This paper presents experiments and an analysis for a pattern called 'reasoning by assumption'. This (non-deductive) practical reasoning pattern involves a number of interrelated reasoning steps, and uses in its reasoning states not only content information but also meta-information about the status of content information and about control. For this reasoning pattern human reasoning protocols have been acquired, analysed, formalised, checked on dynamic properties and compared. As a vehicle a temporal technique has been exploited which was already shown to be a useful analysis tool for reasoning processes in (Jonker and Treur, 2002).

Master Mind is a two player game of logic, which was invented in 1970-71 by Mordecai Meirowitz (Nelson). The goal of the game is to discover a secret code of three colored pegs, which can be obtained by making guesses and receiving information about the correctness of the guesses. Because of its protocol, the pattern of reasoning by assumption occurs frequently within this game. Therefore, the game of Master Mind (in a simplified version) will be the main case study within this paper.

Below, in Section 2 the underlying dynamic perspective on reasoning is discussed in some more detail, and focussed on the pattern 'reasoning by assump-

J. Mira and J.R. Álvarez (Eds.): IWINAC 2005, LNCS 3561, pp. 427–436, 2005.
© Springer-Verlag Berlin Heidelberg 2005

tion'. Next, some more details of the temporal language used are described in Section 3. In Section 4 it is shown how think-aloud protocols involving reasoning by assumption in the game of Master Mind can be formalised to reasoning traces. A number of the dynamic properties that have been identified for patterns of reasoning by assumption are shown in Section 5. For the acquired reasoning traces the identified dynamic properties have been (automatically) checked. The results of these checks are provided in Section 6. In addition, it is shown how logical relationships between dynamic properties at different abstraction levels can play a role in the analysis of empirical reasoning processes. Finally, Section 7 is a conclusion.

2   The Dynamics of Reasoning

Analysis of reasoning processes has been addressed from different areas and angles, for example, Cognitive Science, Philosophy and Logic, and AI. For reasoning processes in natural contexts, which are usually not restricted to simple deduction, dynamic aspects play an important role and have to be taken into account, such as dynamic focussing by posing goals for the reasoning, or making (additional) assumptions during the reasoning, thus using a dynamic set of premises within the reasoning process. Also dynamically initiated additional observations or tests to verify assumptions may be part of a reasoning process. Decisions made during the process, for example, on which reasoning goal to pursue, or which assumptions to make, are an inherent part of such a reasoning process. Such reasoning processes or their outcomes cannot be understood, justified or explained without taking into account these dynamic aspects. The approach to the semantical formalisation of the dynamics of reasoning exploited here is based on the concepts reasoning state, transitions and traces.

Reasoning state. A reasoning state formalises an intermediate state of a reasoning process. The set of all reasoning states is denoted by RS.

Transition of reasoning states. A transition of reasoning states or reasoning step is an element $< S, S' >$ of RS x RS. A *reasoning transition relation* is a set of these transitions, or a relation on RS x RS that can be used to specify the allowed transitions.

Reasoning trace. Reasoning dynamics or reasoning behaviour is the result of successive transitions from one reasoning state to another. A time-indexed sequence of reasoning states is constructed over a given time frame (e.g., the natural numbers). Reasoning traces are sequences of reasoning states such that each pair of successive reasoning states in such a trace forms an allowed transition. A trace formalises one specific line of reasoning. A set of reasoning traces is a declarative description of the semantics of the behaviour of a reasoning process; each reasoning trace can be seen as one of the alternatives for the behaviour. In the next section a language is introduced in which it is possible to express dynamic properties of reasoning traces.

The specific reasoning pattern used in this paper to illustrate the approach is 'reasoning by assumption'. This type of reasoning often occurs in practical reasoning; for example, in everyday reasoning, diagnostic reasoning based on causal knowledge, and reasoning based on natural deduction. An example of everyday reasoning by assumption is 'Suppose I do not take my umbrella with me. Then, if it starts raining, I will get wet, which I don't want. Therefore I'd better take my umbrella with me'. An example of reasoning by assumption in the context of a game of Master Mind is: 'Suppose there is a red pin at position 1. Then, guessing the code [red-blue-white] would at least provide one "correct" point. But if I try, it turns out I do not receive any "correct" points. Therefore there is no red pin at position 1.' Examples of reasoning by assumption in natural deduction are as follows. Method of indirect proof: 'If I assume A, then I can derive a contradiction. Therefore I can derive not A'. Reasoning by cases: 'If I assume A, I can derive C. If I assume B, I can also derive C. Therefore I can derive C from A or B'. Notice that in all of these examples, first a reasoning state is entered in which some fact is *assumed*. Next (possibly after some intermediate steps) a reasoning state is entered where *consequences* of this assumption have been *predicted*. Finally, a reasoning state is entered in which an *evaluation* has taken place; possibly in the next state the assumption is retracted, and conclusions of the whole process are added.

3   A Temporal Trace Language

In recent literature on Computer Science and Artificial Intelligence, temporal languages to specify dynamic properties of processes have been put forward; for example, (Dardenne, Lamsweerde and Fickas, 1993; Dubois, Du Bois and Zeipen, 1995; Herlea, Jonker, Treur, and Wijngaards, 1999). To specify properties on the dynamics of *reasoning* processes in particular, the temporal trace language TTL used in (Herlea et al., 1999; Jonker and Treur, 1998) is adopted. This is a language in the family of languages to which also situation calculus (Reiter, 2001) and event calculus (Kowalski and Sergot, 1986) belong, and was also succesfully used to analyse multi-representational reasoning processes in (Jonker and Treur, 2002).

Ontology. An ontology is a specification (in order-sorted logic) of a vocabulary. For the example reasoning pattern 'reasoning by assumption' in a game of Master Mind the state ontology includes unary relations such as assumed and rejected_code on sort INFO_ELEMENT and binary relations such as prediction_for, observation_result_for and holds_in_world_for on INFO_ELEMENT x INFO_ELEMENT. The sort INFO_ELEMENT includes specific domain statements such as at(red, 1), code(red, white, blue), answer(black, black, black).

Reasoning state. A (reasoning) state for ontology Ont is an assignment of truth-values {true, false} to the set of ground atoms At(Ont). The set of all possible states for ontology Ont is denoted by STATES(Ont). A part of the description of an example reasoning state S is:

assumed(code(red, white, blue)) : true
prediction_for(answer(black, empty, empty), code(red, white, blue)) : true
observation_result_for(answer(white), code(red, white, blue)) : true
rejected_code(code(red, white, blue)) : false

RS is the sort of all reasoning states of the agent. For simplicity in the formulation of properties WS is the set of all substates of elements of RS, thus WS is the set of all world states. The standard satisfaction relation $\models$ between states and state properties is used: $S \models p$ means that state property p holds in state S. For example, in the reasoning state S above it holds $S \models$ assumed(code(red, white, blue)).

Reasoning trace. To describe dynamics, explicit reference is made to time in a formal manner. A fixed time frame T is assumed which is linearly ordered. Depending on the application, for example, it may be dense (e.g., the real numbers), or discrete (e.g., the set of integers or natural numbers or a finite initial segment of the natural numbers). A trace γ over an ontology Ont and time frame T is a mapping $\gamma : T \rightarrow$ STATES(Ont), i.e., a sequence of reasoning states γ_t ($t \in T$)in STATES(Ont). The set of all traces over ontology Ont is denoted by Γ(Ont), i.e., Γ(Ont) = STATES(Ont)^T. The set Γ(Ont) is also denoted by Γ if no confusion is expected.

Expressing dynamic properties. States of a trace can be related to state properties via the formally defined satisfaction relation $\models$ between states and formulae. Comparable to the approach in situation calculus, the sorted predicate logic temporal trace language TTL is built on atoms such as state(γ, t) $\models$ p, referring to traces, time and state properties. This expression denotes that state property p is true in the state of trace γ at time point t. Here $\models$ is a predicate symbol in the language (in infix notation), comparable to the Holds-predicate in situation calculus. Temporal formulae are built using the usual logical connectives and quantification (for example, over traces, time and state properties). The set TFOR(Ont) is the set of all temporal formulae that only make use of ontology Ont. We allow additional language elements as abbreviations of formulae of the temporal trace language. The fact that this language is formal allows for precise specification of dynamic properties. Moreover, editors have been developed to support specification of properties. Specified properties can be checked automatically against example traces to detect whether they hold.

4   The Experiment

Participants. Thirty subjects participated in the experiment. They were divided into two groups of 15. Group 1 consisted of 'AI-scientists', all working at the Department of Artificial Intelligence at the Vrije Universiteit Amsterdam. Group 2 consisted of 'non-scientists', a random set of friends and relatives of the authors. Some of them were students, but none of them had any background related to AI. Group 1 included 10 males and 5 females. Group 2 included 9 males and 6 females. The average age of both groups was approximately 28 years.

Method. The subjects were asked to solve a simplified game of Master Mind. Before starting the experiment, they were given the following instructions:

The opponent picks a secret code consisting of three pegs, each peg being one of eight colors. Your goal is to guess the exact positions of the colors in the code in as few guesses as possible. After each guess, the opponent gives you a score of exact and partial matches. For each of the pegs in your guess that is the correct color in the <u>correct</u> *position, the opponent will give you an 'exact' point (represented by a black pin). If you score 3 black pins on a guess, you have guessed the code. For each of the pegs in the guess that is a correct color in an* <u>incorrect</u> *position, the opponent will give you an 'other' point (represented by a white pin). Together, the black and white pins will add up to no more than 3. Notice that the positions of the black and white pins do not necessarily relate to the positions of the colors. Within this specific experiment,* **one initial guess has already been done for you.** *While doing the experiment, please think aloud, explaining each step you perform.*

For each participant, the *solution code* was the same, namely the combination [blue-white-red]. The *initial guess* mentioned above was always the combination [red-white-blue]. Hence, the provided answer corresponding to the initial guess was [black-white-white].

Table 1. Example human reasoning trace

Human transcript	Formalisation
Right? Okay. So, what I'm going to do now. I'm going to... I'm trying to find out which of the colors is in a good place, first. So, let's say I say it's the red one. Maybe.	focus_assumed(at(red, 1))
So, I'm going to put the red here. And then, change these two.	code_extention_for(code(red, blue, white), at(red, 1)) assumed(code(red, blue, white)) prediction_for(answer(black, black, black), code(red, blue, white))
[red-blue-white] *Okay, so this is your guess?* This is my guess.	to_be_observed_for(answer, code(red, blue, white))
Then my answer is like this... [white-white-white] *...two, and three.*	observation_result_for(answer(white, white, white), code(red, blue, white))
Okay, so it wasn't the red. Okay.	rejected_code(code(red, blue, white)) rejected_focus(at(red, 1))
I will always use these ones, apparently. Then, keep the white and exchange red and blue.	focus_assumed(at(white, 2)) code_extention_for(code(blue, white, red), at(white, 2)) assumed(code(blue, white, red)) prediction_for(answer(black, black, black), code(blue, white, red))
[blue-white-red] *Okay, so why do you do this?* I'm testing now if the white one is in the good position.	to_be_observed_for(answer, code(blue, white, red))
Okay. So then my answer is this. Congratulations! [black-black-black]	observation_result_for(answer(black, black, black), code(blue, white, red))

In Table 1 an example trace is shown, and the way in which it was formalised in order to automatically check their properties. The left column contains the human transcript, the right column contains the formal counterpart. The transcripts of all human reasoning traces can be found at:
http://www.cs.vu.nl/~tbosse/mastermind/human-traces.doc.

5   Dynamic Properties

In this section a number of the most relevant of the dynamic properties that have been identified as relevant for patterns of reasoning by assumption are presented. Two categories of dynamic properties exist. The first category is specified by *characterising properties*. These are properties that are expected to hold for all reasoning traces. In contrast, the second category contains *discriminating properties*, properties that distinguish several types of traces from each other. Within each category, *global properties* (GP's, addressing the overall reasoning behaviour) as well as *executable properties* (EP's, addressing the step by step reasoning process) are given.

Characterising Properties

GP2 Correctness of Rejection
Everything that has been rejected does not hold in the world situation.
$\forall\gamma{:}\Gamma$ $\forall t{:}T$ $\forall A{:}$INFO_ELEMENT
state(γ,t) $\models$ rejected_code(A) $\Rightarrow$
state(γ,t) $\not\models$ holds_in_world_for(answer(black, black, black), A)
This property holds for all traces, leading to the conclusion that none of the participants makes the error of rejecting something that is true.

EP5 Observation Initiation Effectiveness
For each prediction an observation will be made.
$\forall\gamma{:}\Gamma$ $\forall t{:}T$ $\forall A,B{:}$INFO_ELEMENT
state(γ,t) $\models$ prediction_for(B,A)
$\Rightarrow$ [$\exists t'{:}T \geq t{:}T$ state(γ,t') $\models$ to_be_observed_for(answer, A)]
This property holds for all traces, leading to the conclusion that in every case that a prediction was made, this was followed by a corresponding observation.

EP6 Observation Result Effectiveness
If an observation is made the appropriate observation result will be received.
$\forall\gamma{:}\Gamma$ $\forall t{:}T$ $\forall A,B{:}$INFO_ELEMENT
state(γ,t) $\models$ to_be_observed_for(answer, A) $\wedge$
state(γ,t) $\models$ holds_in_world_for(B,A)
$\Rightarrow$ [$\exists t'{:}T \geq t{:}T$ state(γ,t') $\models$ observation_result_for(B,A)]
This property holds for all traces. Thus, in all traces, the opponent provided the correct answers.

Discriminating Properties

GP5 Correctness of Assumption

Everything that has been assumed holds in the world situation.

$\forall\gamma:\Gamma$ $\forall$t:T $\forall$A:INFO_ELEMENT

state(γ,t) $\models$ assumed(A) $\Rightarrow$

state(γ,t) $\models$ holds_in_world_for(answer(black, black, black), A)

This property only holds in four of the 30 cases. By checking it, the subjects that made only correct assumptions can be distinguished from those that made some incorrect assumptions during the experiment. Put differently, the subjects that immediately make the right guess are distinguished from those that need more than one guess.

GP7 Observation Effectiveness

For each assumption, the agent eventually obtains the appropriate observation result.

$\forall\gamma:\Gamma$ $\forall$t:T $\forall$A,B:INFO_ELEMENT

state(γ,t) $\models$ assumed(A) $\wedge$ state(γ,t) $\models=$ holds_in_world_for(B,A)

$\Rightarrow$ [$\exists$t':T $\geq$ t:T state(γ,t') $\models$ observation_result_for(B,A)]

This property holds for all but three of the traces. In these three cases people make an assumption that cannot be right, according to the information they have. However, they correct themselves before they decide to observe the answer to this wrong assumption. Thus, the answer to the incorrect assumption is never obtained.

GP9 Initial Assumption

The first focus assumption made was at(red, 1).

$\forall\gamma:\Gamma$ $\exists$t:T

state(γ,t) $\models$ focus_assumed(at(red, 1))

$\wedge$ [$\forall$t':T < t:T $\forall$A:INFO_ELEMENT

state(γ,t') $\models$ focus_assumed(A) $\Rightarrow$ A = at(red, 1)]

This property holds in 18 of the 30 cases. Thus, 18 participants started reasoning by assuming that the red pin was at position 1. Given the fact that they wanted to keep one of the colors at its initial position, and all three options have an equal probability to be the solution, this seems a logical choice, because it is the first pin they encounter when looking from left to right. Nevertheless, there were still 12 participants that started in a different way.

EP4 Prediction Effectiveness

For each assumption that is made a prediction will be made.

$\forall\gamma:\Gamma$ $\forall$t:T $\forall$A:INFO_ELEMENT

state(γ,t) $\models$ assumed(A)

$\Rightarrow$ [$\exists$t':T $\geq$ t:T $\exists$B:INFO_ELEMENT state(γ,t') $\models$ prediction_for(B,A)]

This property holds in 26 of the 30 cases. So in four cases the subjects make an assumption for which no prediction is made. Three of these four traces have already been discussed at GP7. The fourth trace involves a situation where a person has the following reasoning pattern: "...Let's use one of the colors twice. What would happen in that case? Well, I don't know. Let's just see what happens..." Hence, the subject tries a code of which he intuitively thinks that it is an intelligent guess, without really understanding why. Therefore, he does not make a prediction.

6   Results

A special piece of software has been developed that takes a formally specified property and a set of traces as input, and verifies whether the property holds for the traces (see Bosse, Jonker, Schut, and Treur, 2004). By means of this checking software, all specified properties have been checked automatically against all traces to find out whether they hold. In Table 2 an overview of the results is shown. In this table, an X indicates that the property holds for that particular trace. The final row provides the number of guesses needed by each subject to solve the problem.

As can be seen in the table, all characterising properties indeed hold for all traces. The discriminating properties only hold for some of the traces, which allows making a distinction between different classes of reasoners.

In addition to the above, logical relationships have been identified between properties at different abstraction levels. An overview of the logical relationships relevant for overall property GP7 is depicted as an AND-tree in Figure 1.

Table 2. Overview of the results: traces against properties

	1	2	3	4	5	6	7	8	9	10	11	12	13	14	15	16	17	18	19	20	21	22	23	24	25	26	27	28	29	30
GP2	X	X	X	X	X	X	X	X	X	X	X	X	X	X	X	X	X	X	X	X	X	X	X	X	X	X	X	X	X	X
GP5	X	-	-	-	-	-	-	-	-	-	-	-	X	-	-	-	-	X	-	-	-	-	-	X	-	-	-	-	-	-
GP7	X	X	X	X	X	-	X	X	X	X	X	X	X	X	-	X	X	-	X	X	X	X	X	X	X	X	X	X	X	X
GP9	-	-	-	-	-	X	X	X	X	-	X	X	X	-	-	X	X	X	-	X	X	X	X	-	X	-	-	X	X	X
EP5	X	X	X	X	X	X	X	X	X	X	X	X	X	X	X	X	X	X	X	X	X	X	X	X	X	X	X	X	X	X
EP6	X	X	X	X	X	X	X	X	X	X	X	X	X	X	X	X	X	X	X	X	X	X	X	X	X	X	X	X	X	X
EP4	X	X	X	X	X	-	X	X	X	-	X	X	X	X	-	X	X	-	X	X	X	X	X	X	X	X	X	X	X	X
steps	1	2	3	3	3	3	3	2	3	3	2	3	2	1	3	3	3	3	1	3	2	2	3	3	2	1	3	3	3	2

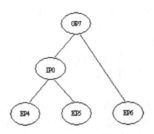

Fig. 1. Logical relationships between dynamic properties

For example, the relationship at the highest level expresses that IP0& EP6 ⇒ GP7 holds. Here, IP0 is an *intermediate property*, expressing the dynamics of the reasoning between two milestones:

IP0 Assumptions lead to Observation Initiation
For each assumption that is made a prediction will be made.

$\forall\gamma{:}\Gamma \quad \forall t{:}T \; \forall A{:}INFO\_ELEMENT$

state$(\gamma,t) \models$ assumed(A)

$\Rightarrow [\; \exists t'{:}T \geq t{:}T \; state(\gamma,t') \models$ to_be_observed_for(answer, A) $]$

Intermediate properties address smaller steps than global properties do, but bigger steps than executable properties do. At a lower level, Figure 1 depicts the relationship EP4 & EP5 ⇒ IP0.

Notice that the results given in Table 2 validate these logical relationships. For instance, in all traces where EP4, EP5 and EP6 hold, also GP7 holds. Such logical relationships between properties can be very useful in the analysis of empirical reasoning processes. For example, if a given person does not obtain the appropriate observation result for her assumption (i.e. property GP7 is not satisfied by the reasoning trace), then by a refutation process it can be concluded that either property IP0, or property EP6 fails (or both). If, after checking these properties, it turns out that IP0 does not hold, then either EP4 or EP5 does not hold. Thus, by this example refutation analysis it can be concluded that the cause of the unsatisfactory reasoning process can be found in either EP4 or EP5. In other words, either the *Observation Initiation* mechanism fails (EP5), or the *Prediction* mechanism fails (EP4).

In this section, only one logical relationship is shown. However, many more global, intermediate, and executable properties for the pattern of reasoning by assumption, as well as the relationships between them can be found at the following URL: http://www.cs.vu.nl/~tbosse/mastermind/properties-and-relationships.doc.

7   Conclusion

This paper shows how given instances of empirical human reasoning traces can be formalised and automatically analysed against dynamic properties they fulfil. To this end a variety of dynamic properties have been specified, some of which are considered characteristic for the reasoning pattern 'reasoning by assumption', whereas some other properties can be used to discriminate between different approaches to the reasoning. For the Master Mind experiments undertaken, properties of the first, characteristic, type indeed hold for the acquired reasoning traces. Properties of the latter, discriminating type hold for some of the traces and do not hold for other traces: they define subsets of traces that collect similar reasoning approaches.

In addition to empirical traces, the analysis method can be applied to traces generated by simulation models. Dynamic properties found relevant for human traces can be used to validate a simulation model, by generating a number of simulation runs and checking the dynamic properties for the resulting traces. This type of validation has been exploited to validate a simulation model for reasoning by assumption to solve the wise men puzzle in (Jonker and Treur, 2003). Moreover, in (Bosse, Jonker and Treur, 2003) a similar analysis approach has been used to validate a simulation model for controlled multi-representational reasoning involving arithmetic, geometric and material representations.

References

1. Bosse, T., Jonker, C.M., and Treur, J., Simulation and analysis of controlled multi-representational reasoning processes. *Proc. of the Fifth International Conference on Cognitive Modelling, ICCM'03*. Universitats-Verlag Bamberg, 2003, pp. 27-32.
2. Bosse, T., Jonker, C.M., Schut, M.C., and Treur, J., Modelling Shared Extended Mind and Collective Representational Content. In: *Proc. of the 24th International Conference on Innovative Techniques and Applications of Artificial Intelligence*. Lecture Notes in AI, Springer Verlag. To appear, 2004..
3. Dardenne, A., Lamsweerde, A. van, and Fickas, S. (1993). Goal-directed Requirements Acquisition. *Science in Computer Programming*, vol. 20, pp. 3-50.
4. Dubois, E., Du Bois, P., and Zeippen, J.M. (1995). A Formal Requirements Engineering Method for Real-Time, Concurrent, and Distributed Systems. In: *Proceedings of the Real-Time Systems Conference, RTS'95*.
5. Herlea, D.E., Jonker, C.M., Treur, J., and Wijngaards, N.J.E. (1999). Specification of Behavioural Requirements within Compositional Multi-Agent System Design. In: F.J. Garijo, M. Boman (eds.), *Multi-Agent System Engineering, Proc. of the 9th European Workshop on Modelling Autonomous Agents in a Multi-Agent World, MAAMAW'99*. Lecture Notes in AI, vol. 1647, Springer Verlag, 1999, pp. 8-27.
6. Jonker, C.M., and Treur, J. (1998). Compositional Verification of Multi-Agent Systems: a Formal Analysis of Pro-activeness and Reactiveness. In: W.P. de Roever, H. Langmaack, A. Pnueli (eds.), *Proceedings of the International Workshop on Compositionality, COMPOS'97*. Lecture Notes in Computer Science, vol. 1536, Springer Verlag, 1998, pp. 350-380. Extended version in: *International Journal of Cooperative Information Systems*, vol. 11, 2002, pp. 51-92.
7. Jonker, C.M., and Treur, J. (2002). Analysis of the Dynamics of Reasoning Using Multiple Representations. In: W.D. Gray and C.D. Schunn (eds.), *Proceedings of the 24th Annual Conference of the Cognitive Science Society, CogSci 2002*. Mahwah, NJ: Lawrence Erlbaum Associates, Inc., 2002, pp. 512-517.
8. Jonker, C.M., and Treur, J. (2003). Modelling the Dynamics of Reasoning Processes: Reasoning by Assumption. *Cognitive Systems Research Journal*. In press, 2003.
9. Kowalski, R., and Sergot, M. (1986). A logic-based calculus of events. *New Generation Computing*, 4:67-95, 1986.
10. Nelson, T. *A Brief History of the Master MindTM Board Game*.
    http://www.tnelson.demon.co.uk/mastermind/history.html
11. Reiter, R. (2001). Knowledge in Action: Logical Foundations for Specifying and Implementing Dynamical Systems. MIT Press, 2001.

Aligning Reference Terminologies and Knowledge Bases in the Health Care Domain

M. Taboada[1], J. Des[2], D. Martínez[3], and J. Mira[4]

[1] Dpto. de Electrónica e Computación, Universidad de Santiago de Compostela,
15782 Santiago de Compostela, Spain
chus@dec.usc.es
http://aiff.usc.es/ elchus/
[2] Servicio de Oftalmología, Hospital Comarcal *Dr. Julián García*,
27400 Monforte de Lemos, Spain
eljjdes@telefonica.net
[3] Dpto. de Física Aplicada, Universidad de Santiago de Compostela,
27002 Lugo, Spain
fadiego@usc.es
http://www.usc.es
[4] Dpto. de Inteligencia Artificial, UNED,
28040 Madrid, Spain
jmira@dia.uned.es
http://www.ia.uned.es/personal/jmira/

Abstract. We study the general question of how knowledge bases can be designed in small domains and scale, importing reference terminologies and taking into account the methodologies and tools available nowadays. For this, we have carried out a case study on a knowledge base oriented to support a diagnosis-aid application in ophthalmology. Our study emphasizes the advantages of extending a knowledge base with a new component that holds both a meta-model representing a very simplified structure of a terminology system and a set of constraints expressed using an axiom language. This set of constraints allows us to check the consistency and coherence of the imported information.

Keywords: knowledge representation, knowledge bases, ontologies and terminology systems.

1 Introduction

In the health care domain and the biomedical sciences, large portions of terminological knowledge are available in electronic form in controlled terminology and classification systems, such as the Unified Medical Language System, UMLS [1]. These terminology systems (TS) supply standard knowledge sources that favor the later share and reuse of the resulting knowledge bases [2]. In many approaches, reference terminologies are embedded in an evolving knowledge base [3, 4]. This is a consequence of transforming large portions of information directly from a TS to the evolving knowledge base. However, applications focused

J. Mira and J.R. Álvarez (Eds.): IWINAC 2005, LNCS 3561, pp. 437–446, 2005.

on designing knowledge bases in small domains and scale, such as [5], require searching the TS previously to import the information. Li et el. [2] emphasized the following activities to be carried out in the design of domain knowledge bases from TS: (1) searching the TS for the information, (2) importing it from the TS, and (3) integrating it with the evolving knowledge base. Firstly, searching the TS mainly is a manual process, where the expert revises and selects the domain terminology. Secondly, in some cases, importing the selected information in the knowledge base can be supported by tools specific to this purpose, such as the UMLS-based extension of Protégé [6] or the one proposed in [5]. Tools like these automatically add the selected information to the knowledge base. Thirdly, the process of integrating the imported information can be carried out by embedding it in the evolving knowledge base or by combining it in a consistent and coherent form, but preserving it separately from the evolving knowledge base. The latter case is known as alignment of knowledge sources.

In this paper, we study the general question of how knowledge bases should be designed in small domains and scale, importing information from a TS and taking into account the methodologies and tools available nowadays. We propose to start the design of knowledge bases from a domain ontology. The latter describes the mainly static information and the knowledge objects in each application domain, using five component types [7]: concepts, relationships, functions, axioms and instances. As it is unlikely to find a domain ontology suitable for each specific application, it will be necessary to build it from merging portions of available ontologies. Later, the enrichment of this core ontology can be achieved by selectively searching small portions of information in TS. For this purpose, we propose to adapt the guide for ontology development created by Noy [8] to cover knowledge reuse. We also suggest to loosely integrate the knowledge base with the imported terminological information, as alignment is usually more suitable when the knowledge bases cover domains that are complementary to each other [9]. In addition, several studies have collected empirical evidence for the lack of logical consistency of some TS, such as the UMLS Metathesaurus [3, 4, 10]. So, we think it is necessary to add a new activity to the set stressed by Li [2]: verification of the imported information.

The structure of the paper is as follows. As our research was application oriented, we firstly describe the purpose and scope of our knowledge base. Then, we present the reuse based methodology we have followed during the development of the knowledge base. Finally, we end with some conclusions.

2   Description of the Knowledge-Based Application

The purpose of the knowledge base is to support a knowledge-based application oriented towards medical diagnosis in the ophthalmologic domain of conjunctivitis. The *level of formality* includes (1) the modeling of the knowledge base using the Protégé-2000 environment for frame-based knowledge acquisition [6] and (2) the implementation of the knowledge base using the development tool

KAPPA-PC from IntelliCorp. In both tools, knowledge representation is based on frames, so the knowledge translation between them is relatively direct.

The scope of the knowledge base is the representation of knowledge about the ophthalmologic problem known as conjunctivitis, which is an inflammation of the conjunctiva. It groups a number of diseases or disorders that mainly affect the conjunctiva. In most of the patients, the conjunctivitis remits on its own, but in some cases, it progresses and can cause serious ocular and extra-ocular complications.

In this work, we have only concentrated on the representation of the static domain knowledge for diagnosis in Protégé-2000. This will allow the knowledge base to be reused to code the operational knowledge by some medical guideline representation language [11] or using specialized algorithms for the diagnostic task [12].

3   Methodology

Until the Nineties, knowledge bases were designed 'ad-hoc'. However, the Artificial Intelligence community now recommends to develop knowledge bases by selecting and integrating previously modeled and formalized knowledge [13]. Ideally, following this approach, if we start with pre-existing descriptions of generic concepts, modeling domain knowledge will simply consist of specifying the precise details of the application [14]. So, nowadays the development of a knowledge base can be viewed as a process consisting of two interrelated stages:

1. Design of a core ontology, starting from other available ontologies.
2. Specialization of the core ontology, by taking into account the specific knowledge contained in available knowledge sources, such as TS.

In the following subsections, we will review the main activities that we have carried out during these two stages. Our first goal has been to obtain an easily reused knowledge base. This has led us to a modular design process. The knowledge base components should be relatively independent, to facilitate their future extraction. Following this strategy, the evolving knowledge base consists of several relatively independent models.

3.1   Development of the Core Ontology

The development of the core ontology has been carried out by reusing portions of knowledge provided by previously designed ontologies:

- The ontology that contains the semantic types defined in the UMLS Semantic Network. This ontology is in the server of DAML+OIL (http://www.daml. org/ontologies/218)
- The EON ontology (http://www.smi.stanford.edu/proyects/eon/), an ontology representing different types of knowledge found in guidelines and medical protocols.

The set of activities performed in this stage includes different types of operations, such as: knowledge extraction, translation of parts of the ontologies to a new framework and the integration of ontologies. For more details, [15] can be consulted.

3.2   Specialization of the Core Ontology

This activity can be seen as a process of core ontology enrichment with selectively imported knowledge from TS. In general, this activity should include the modeling and formalization of the new information, taking into account the structure of the core ontology and following some type of ontology development methodology. We have adhered to the guide for ontology development proposed in [8]. Although this guide is not specifically directed towards knowledge reuse, it can be easily adapted, as we show below. The guide proposes an iterative method to develop the ontology, with seven main stages. Moreover, it discusses a set of decisions that should be taken into account during the development process. We will discuss new types of decisions that appear when the attempt is made to formalize knowledge coming from a TS.

Stage 1: Determine the domain and scope of the ontology. The domain and scope of the ontology has been described in Section 2.

Stage 2: Consider reusing of existing ontologies. This stage corresponds to the development of the core ontology (Section 3.1).

Stage 3: Enumerate important terms in the ontology. The strategy followed to obtain these terms was focused on the revision of a medical text, with the aid of a clinical expert. When the expert underlined a word (name) or a group of words (a name plus some adjective), we searched for them in the UMLS Metathesaurus.

Each concept in the Metathesaurus is mainly represented by a name, a unique identifier (Concept Unique Identifier, CUI), a definition, a set of synonyms and the semantic type which it belongs to. The semantic type is a reference to some concept of the UMLS Semantic Network, which provides a classification of all the concepts of the Metathesaurus. It is structured in two hierarchies, where their root nodes are Entity and Event. Each semantic type is represented by a name, a unique identifier (Type Unique Identifier, TUI) and a definition. Each semantic type is related to at least one other type, through an 'is-a' relationship, except Entity and Event.

The search for concepts in the Metathesaurus is not error free, as many concepts are not defined and, therefore, it is difficult for someone who is not an expert to know whether the search has been successful or not. For this reason, during the search process, we applied the following criteria:

- The search for isolated words within the main text is carried out exactly as it appears in the text (i.e. accompanied by their adjectives). In the case of failure, a search for just the key word is attempted and its definition is compared. If there is no definition in the Metathesaurus for this word (a very frequent case when we model the conjunctivitis guideline), it is checked with the expert.

- The search for words belonging to a list is carried out in the same way, but whether the extracted concepts all belong to the same semantic type is checked afterwards. If the extracted concepts belong to the same type, so the search is considered to be successful. If not, we distinguish two cases:
    - The semantic types of the extracted concepts are nearby subtypes in the hierarchy of the Semantic Network. In this case, we consider the search to have been successful.
    - The semantic types of the extracted concepts are very different (i.e. correspond to types that are poorly related through the semantic hierarchy). In this case, a new search is made of related concepts that allow approximation of the semantic types.

Stage 4: Defining the classes and the hierarchy of classes. The main strategy we followed in defining the classes and hierarchies of the medical domain (Medical Domain Class) was to maintain both hierarchies of the Semantic Network and the 'is-a' relationship of each concept of the Metathesaurus with one or various concepts of the Semantic Network. In this way, we were sure that the class hierarchy was the correct one and thus avoided errors. For example, Fig. 1 shows two concepts in the Medical Domain Class: *Sign or Symptom* and *Symptom*. The first concept is a type of the Semantic Network and the second is the concept of the Metathesaurus Symptom<1>. The latter belongs to the semantic type *Sign or Symptom*, so we have modeled it as a subclass of the *Sign or Symptom* in the Medical Domain Class (see Fig. 1).

Stage 5: Define the properties of the classes (slots). Once the classes have been defined, the following stage consists of describing the internal structure, that is, its slots. With the objective of facilitating the future comprehension and reuse of the KB, we separately maintain the two main types of information that describe a medical domain class: information relative to UMLS (*UMLS Mapping class*) and relative to the KB competence (*Medical Domain Class*). In general, we have distinguished two main types of properties:

- **Properties that define the information about the concept contained in the UMLS.** The information imported from UMLS is stored in the *UMLS Mapping* hierarchy, which is separated from the medical domain class. This hierarchy stores the UMLS information that we have considered relevant. There are three subclasses defined for UMLS Mapping: Metathesaurus Concept, Semantic Type and Semantic Relation. The two first classes can be seen on the right of the Fig. 1. All the imported UMLS information is stored as instances in some subclass of the UMLS Mapping class. For example, the concept symptom<1> imported from the Metathesaurus has been stored as an instance of the Metathesaurus Concept class.
- **Properties that describe the concept from the point of view of the KB competence.** Noy [8] distinguishes different types of properties

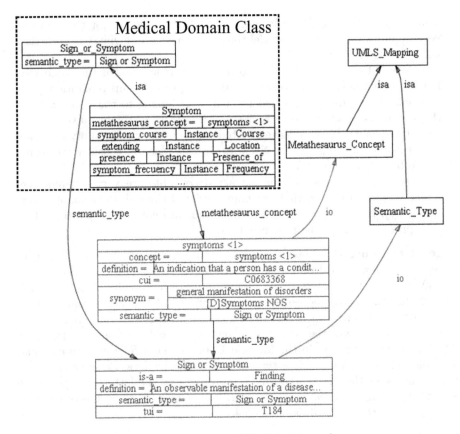

Fig. 1. A screen shot from Protege-2000 (using the Ontoviz plug-in) showing two subclasses of the Medical Domain Class and their relationship with two instances of the UMLS Mapping class

for a class: intrinsic, extrinsic, properties defining parts that form the class and relationships to other classes. To obtain this type of information, many times we have had to resort to the expert. For example, in Fig. 1 we can see some slots that we have defined for symptoms. Many of these were acquired directly from the expert.

When a concept was imported from the UMLS, the following steps were carried out:

1. Addition of a new class to the *Medical Domain Class*. In Fig. 1, the class *Symptom* was added as a subclass of *Sign or Symptom*, thus preserving the 'is-a' relationship of UMLS.
2. Selection of the pattern corresponding to the new added class (Metathesaurus Concept or Semantic Type). In the example, the selected one for *Symptom* was Metathesaurus Concept (Fig. 1).

3. Importing the information from the UMLS and storing it as an instance of some sub-class of UMLS Mapping. In the example, the sub-class is symptom <1>.
4. Relating the new class to the new instance. In the example, we assigned the instance symptom<1> to the slot Metathesaurus Concept of the class *Symptom* (Fig. 1).

Stage 6: Define the attributes. In general, following the criterion of maximum reuse of UMLS information, each time we needed to define an attribute, firstly we searched in the UMLS. For example, in Fig. 1, most attributes of Symptom were defined as instances of concepts imported from the Metathesaurus. When an attribute was found in the Metathesaurus, the following steps were taken:

1. Add a new class to the Medical Domain Class, preserving the relation 'is-a' of the Metathesaurus. In Fig. 1 Course, Frequency, Setting and Location were added.
2. Import information from UMLS and representing it as an instance of some subclass of the UMLS Mapping class.
3. Relate the new class to the new instance of the UMLS Mapping.
4. Add the required instances of the new class. For example, we added the subclass *Eye Symptom Location*, with the objective of specifying different types of symptom location in the eye. Specifically, three instances of Eye Symptom (*Right Eye*, *Left Eye* and *Both Eyes*) collect the types of location of an ocular symptom.
5. Define the attribute as a type of instance of the new class. In Fig. 1, we can see that the possible values of the attribute location are instances of the new class Location.

Stage 7: Create instances. We have distinguished various types of instances, as a function of the knowledge that they model:

- Instances that define expressions of medical knowledge. For example, the expression 'Presence of Discharge' can be modeled through an instance of the class 'Discharge from Eye', where the value of the slot 'Presence' is set to 'yes'.
- Instances of the Medical Domain Class that have been reused from the Metathesaurus. For example, the location of an ocular symptom can be described using three instances: Right Eye, Left Eye and Both Eyes, which are related to concepts in the Metathesaurus.
- Instances that store the information extracted from the UMLS, just as we have seen previously.
- Instances that represent clinical cases. Protégé-2000 generates forms automatically from the definition of the classes. These forms permit direct introduction of data of the clinical cases.

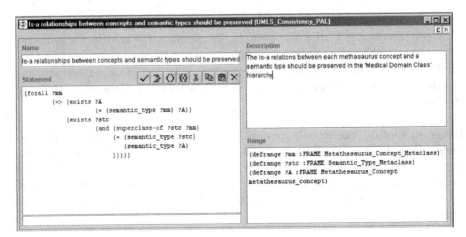

Fig. 2. A PAL constraint expressing that the Medical Domain Class must preserve the relationships between Metathesaurus concepts and semantic types

Revision of the previous steps. Noy [8] proposes general guidelines to keep in mind when defining classes and class hierarchies. These guidelines allow verifying whether the evolving hierarchy has been modeled as intended. In addition, we propose new guidelines focused on the context of reusing knowledge from pre-existing sources. The KB consistency checking can be carried out automatically in two steps: adding constraints to the KB and checking these constraints repeatedly during development time. In Protégé-2000, logical constraints can be expressed using the Protégé Axiom Language (PAL), based on first order logic.

In our KB, the consistency checking has included the following subjects: Has UMLS information been imported correctly? ; Is the medical domain hierarchy defined appropriately? ; Are relationships among the slots defining medical domain classes verified?

As an illustrative example of consistency checking, we can consider the following constraint: *The Medical Domain Class must preserve the relationships between Metathesaurus concepts and Semantic Types.* For example, in Fig. 1 *Symptom* is a concept imported from the Metathesaurus that belongs to the semantic type *Sign or Symptom.* This 'is-a' relationships between Metathesaurus concepts and Semantic Types should be preserved in the Medical Domain Class. The PAL constraint checking the consistency of the Medical Domain Class can be seen in Fig. 2.

4   Conclusions

Using knowledge from unified terminology systems, when they are available, enriches knowledge bases, improving their reuse and share. This process includes importing information selectively and integrating it with the knowledge base. Our application provides a practical scenario of how TS can be used for designing knowledge bases. Many other examples exist in both medical and biological domains, such as the terminological knowledge base designed in the clinical domain of human anatomy and pathology [3] and the knowledge base modeling the Gene Ontology in Protégé [4]. However, the knowledge covered in our application is more sophisticated, as it is oriented to provide sound medical reasoning. In this way, our application has paid attention on searching a TS for smaller portions of information and importing (and reorganizing) them to an evolving knowledge base. Our study provides a set of guidelines oriented to formalize knowledge imported from a TS. Nowadays, searching small portions of information is a very time-consuming task. So, we plan to apply natural language processing-based techniques to reduce the knowledge acquisition time from textual documentation. The set of guidelines presented in this paper could serve as a starting point for developing a new generation of tools focussing on reducing the TS search time.

As a result of our study, we can conclude that, in order to incorporate information from remote knowledge sources (such as UMLS) into a knowledge base, it is necessary to provide a set of logical constraints to support consistency during import of terminological information. Studies, such as [3, 4, 10], have collected empirical evidence for the lack of logical consistency of the UMLS Metathesaurus, so it is always necessary to check the imported information. In our current application, the UMLS structure is represented in a very simplified way, by only checking unique codes, terminological cycles and relationships among concepts. For more details, [16] should be consulted. In the future, we plan to extend it with the aim of supporting a complete and unambiguous description of its structure. For that, we plan to translate the proposal of representation described in [17] to Protégé-2000 and PAL.

Acknowledgements. This work has been funded by the Secretaria Xeral de Investigacion e Desenvolvemento da Xunta de Galicia, through the research project PGIDT01-PXI20608PR.

References

1. Lindberg, D., Humphreys, B. and Mc Cray, A. The Unified Medical Language System. Methods of Information in Medicine, **32** (1993), 281-291.

2. Li, Q., Shilane, P., Noy, N. F. and Musen, M. A.: Ontology Acquisition from Online Knowledge Sources. In: Proc. of the AMIA Annual Symposium, Los Angeles, CA (2000).

3. Schulz, S., Hahn, U. Medical knowledge reengineering- converting major portions of the UMLS into a terminological knowledge base. International Journal of Medical Informatics, **64** (2001), 207-221.

4. Yeh, I. Karp, P. D, Noy, N. F., Altman, R. B. Knowledge Acquisition, Consistency Checking and Concurrency Contol in Gene Ontology. Bioinformatics, **19** (2003), 241-248.

5. Achour, S., Dojat, M., Rieux, C., Bierling, P. and Lepage, E. A UMLS-based knowledge acquisition tool for rule-based clinical decision support system development. Journal of the American Medical Informatics Association, 8 (2001), 351-360.

6. Gennari, J., Musen, M. A., Fergerson, R. W. Grosso, W. E., Crubezy, M., Eriksson, H., Noy, N. F. and Tu, S. W. The Evolution of Protg: An Environment for Knowledge-Based Systems Development. International Journal of Human-Computer Interaction 58 (2002), 89-123. http://protege.stanford.edu

7. Gruber, R. A translation approach to portable ontology specification. Knowledge Acquisition, 5 (1993), 199-220.

8. Noy, N. F. and McGuinness, D.L. Ontology Development 101: a guide for creating your first ontology. SMI Technical Report SMI-2001-0880, In http://www.stanford.edu

9. Noy, N.F. and Musen, M.A. PROMPT: Algorithm and tool for automated Ontology merging and alignment. In: Proc. of the Seventeenth National Conference on Artificial Intelligence (AAAI-2000), Austin, TX (2000).

10. Gu, H., Perl, Y., Geller, J., Halper, M., Liu, L. and Cimino, J. Representing the UMLS as an object-oriented database. The Journal of the American Medical Informatics Association, 7 (2000), 66-80.

11. Elkin, P., Peleg, M., Lacson, R., Berstam, E., Tu, S., Boxwala, A., Greenes, R., Shortliffe, E. Toward Standardization of Electronic Guidelines. MD Computing, 17 (2000), 39-44.

12. Taboada, M., Des, J., Mira, J. and Marín, R. Development of diagnosis systems in medicine with reusable knowledge components. IEEE Intelligent Systems, 16 (2001), 68-73.

13. Neches, R., Fikes, R.E., Finin, T., Gruber, T.R., Senator, T. and Swartout, W.R. Enabling technology for knowledge sharing. AI Magazine, 12 (1991), 36-56.

14. Musen, M.A. Modern architectures for intelligent systems: reusable ontologies and problem-solving methods. In Proc. of the AIMA'98, C.G. Chute (ed.), Orlando, FL (1998) 46-52.

15. Taboada, M., Martínez, D. and Mira, J. Modelling knowledge-bases by reuse: a case study. International Journal of Knowledge-based Intelligent Engineering Systems, 7 (2003), 190-197.

16. Taboada, M., Martínez, D. and Mira, J. Experiences in Reusing Knowledge Sources using Protégé and Prompt. International Journal of Human Compter Studies, in press (2005).

17. de Keizer, N., Abu-Hanna, A. Understanding terminological systems (II): experience with conceptual and formal representation of structure. Methods of Information in Medicine, 39 (2000), 22-29.

Predicting Mortality in the Intensive Care Using Episodes

Tudor Toma[1], Ameen Abu-Hanna[1], and Robert Bosman[2]

[1] Academic Medical Center, Universiteit van Amsterdam,
Department of Medical Informatics, PO Box 22700,
1100 DE Amsterdam, The Netherlands
[2] Department of Intensive Care, Onze Lieve Vrouwe Gasthuis,
1e Oosterparkstraat 279, PO box 10550,
1090 HM Amsterdam, The Netherlands

Abstract. Patient outcome prediction lies at the heart of various medically relevant tasks such as quality assessment and decision support. In the intensive care (IC) there are various prognostic models in use today that predict patient mortality. All of these models are logistic regression models that predict the probability of death of an IC patient based on severity of illness scores. These scores are calculated from information that is collected within the first 24 hours of patient admission. Recently, IC units started collecting sequential organ failure assessment (SOFA) scores that quantify the degree of derangement of organs for each patient on *each day* of the IC stay. Although SOFA scores are primarily meant for recording incidence of organ derangement and failures, the hypothesis is that they contribute to better prediction of mortality. There is virtually no systematic way in the literature to exploit the temporal character of SOFA scores for prediction. This paper adapts ideas from temporal datamining for discovery of sequential episodes and suggests a way to put them into use in the problem of mortality prediction. In particular, we discover frequent temporal patterns, assess their suitability for prediction, and suggest a method for the integration of temporal patterns within the current logistic regression models in use today. Our results show the added value of the new predictive models.

1   Introduction

When making a decision about a treatment, physicians try to *predict* the clinical outcome of each alternative before selecting the one with the best expected outcome. When assessing quality of care, ward managers compare the outcomes of their patients with outcome *predictions* according to some norm. These are two examples of the many uses of medical predictions in clinical practice. These predictions are generated by a *prognostic model* which lies at the heart of various clinical tasks. We define prognosis as *the prediction of the future course and outcome of disease processes, which may either concern their natural course or their outcome after treatment* [1].

J. Mira and J.R. Álvarez (Eds.): IWINAC 2005, LNCS 3561, pp. 447–458, 2005.
© Springer-Verlag Berlin Heidelberg 2005

In the Intensive-Care (IC) there is a long tradition of developing prognostic models to predict mortality of IC patients. Notable examples thereof are the SAPS-II [12] and APACHE-II [9] models. These models are used in quality assessment programs to audit the performance of various IC units (ICUs). The models predict the probability of hospital mortality of each patient in the ICU. Hospital mortality includes deaths in the hospital during or after stay in the ICU. These prognostic models are logistic regression models that use a small number of severity-of-illness scores as co-variates. Each such score is an integer that summarizes the severity of illness of the patient: the higher the score, the more severely ill the patient is, and in turn, the higher the probability of death is. An important characteristic of these scores is that they are based on patient information collected only during the first 24 hours of ICU admission. All information after admission is excluded. This results in relatively simple models and does not require extensive data collection efforts for using the models, which is an important property when Information Technology was still not advanced enough in the ICUs. On the other hand, it is fair to assume that additional temporal information could provide better predictions.

With the advent of Patient Data Management Systems (PDMSs) in the ICUs, routine data collection and capture was facilitated including temporal data. Recently, ICUs started collecting sequential organ failure assessment (SOFA) scores [15] ,[16] that quantify the degree of derangement of organs for each patient on *each day* of the ICU stay. Although SOFA scores are primarily meant for recording incidence of organ derangement and failures, the hypothesis is that SOFA scores contribute to better prediction of mortality. Studies that attempted to incorporate SOFA scores in prognostic models opted to include some summary measures of these scores, and there are virtually no systematic ways in the literature to exploit the temporal character of SOFA scores for prediction.

This paper adapts and applies ideas from temporal datamining to the problem of mortality prediction in the ICU. We develop models that predict mortality after one, two, three, four and five days of admission to the ICU based on the available information up to the respective day. Our approach is characterized by discovering frequent sequential episodes, which are temporal patterns, and integrating them as dummy variables (also refereed to as design variables) in the current logistic regression framework. More specifically, the dummy variables are added to a reference model based on the popular SAPS score [12]. This allows for assessing the added value of the patterns in comparison to this static-based model. From a temporal datamining point of view, our approach does not only extend existing models with additional temporal information, but also takes account of the interdependency between patterns and allows an intuitive assessment of their importance in prediction.

The paper is organized as follows. The next section describes the data, methods and results used, from preprocessing the data through the strategy for developing the models, till performance evaluation. Section 3 discusses and concludes this paper and delineates future research directions.

2   Materials, Methods and Results

2.1   Data

The data has been collected in the adult ICU at the OLVG teaching hospital in Amsterdam from July 1998 until December 2004. It contains the information about all 5160 patients admitted to the ICU in that period. The data includes two types of information: static (or *a-temporal*) and *temporal* patient information.

The *a-temporal* data consists of more than 100 attributes for each patient. The data contains: demographics, like age and sex; reason for admission to the ICU, like surgery; and physiology like body temperature and heart rate. The physiological variables are all collected within the first 24 hours of IC admission. In addition, and of special importance, there are the severity-of-illness scores like the SAPS score, which is used in this paper. This score is an integer that takes into consideration the demography, reason of admission and the physiology of the patient in order to quantify the severity of illness of the patient in the first 24 hours of admission.

The *temporal* data contains information about the sequential organ failure assessment (SOFA) scores. These are scores computed on a *daily basis* for each patient to quantify the severity of derangement of the organ systems (cardiovascular, respiratory, hematologic, renal, neurological, and hepatic). Each of these organ systems contributes as a sub-score from 0 to 4 points to the SOFA score which, hence, may theoretically vary between 0 to 24 points. Greater values of the SOFA score imply worse condition of the patient's organ systems. For each patient there will be a SOFA sequence with length equal to the number of days that the patient stayed in the ICU. An example of a SOFA sequence for a patient that stayed for 4 days and steadily recovered is 7-4-2-1.

For patients that were readmitted to the ICU, e.g. because of complications after discharge to another ward, only the last admission to the ICU was included. This is because only the last readmission data is relevant to the in-hospital mortality, and earlier re-admissions may bias the analysis as they have, by definition, the same outcome as the last admission. Excluding the earlier re-admissions resulted in a dataset of 4771 patients.

This dataset has been randomly split into a training set, (two thirds or 3181 patients) for model development, and a test set (one third or 1590 patients) for model validation. Table 1 shows important characteristics of these datasets.

2.2   Methods

In dealing with temporal data, such as the SOFA sequences, there are two main approaches followed in the literature. In the first approach, temporal data is reduced by some summary statistic. Examples thereof are the mean value, or time-until-peak value in the series or sequence. This approach is convenient because it allows for using analytic methods that work for static data. It is appropriate when one is focused on and knows apriori what the summary measure is, like time-to-event. It, however, loses information about the temporal *course* of the sequence. The other approach, which we adopt in this paper, preserves the

Table 1. Descriptive statistics of whole data, training data and testing data

	Data set	Training set	Test set
No. of patients	4771	3181	1590
Male/Female %	65.56	65.14	66.42
Age mean ± sd	64.4 ± 14.41	64.7 ± 14.21	63.7 ± 14.79
Age median	67	68	66
SAPS mean± sd	34.2 ± 15.6	34 ± 15.3	34.4 ± 16.3
Hosp. mortality %	10.7	10.3	11.6
Length of stay median	0.92	0.92	0.92
Length of stay mean	2.11	2.13	2.08

inter-relationship between the values in the sequence or series. More specifically, we are interested in the discovery of SOFA patterns that influence mortality and then use these patterns for mortality prediction. One important merit of this approach is that the patterns can be scrutinized qualitatively by asking whether they make sense, and quantitatively by assessing their significance on the predictive performance. The formalism that we use for the predictive models is logistic regression due to the properties discussed later on in this section. Our approach consists of the following steps:

1. Data preprocessing for categorizing SOFA scores
2. Frequent episode discovery tailored to sequences with different lengths
3. Development of logistic regression models that use episodes
4. Model validation

Data Preprocessing. The SOFA score is an ordinal variable that ranges from 0 to 24. We categorize its values in few qualitative categories. The reason is two fold. First, it is more convenient to interpret and think about qualitative categories such as "low" and "high" than numbers, for both physicians and analysts. Second, lumping up the scores in few categories enhances finding enough cases for supporting patterns, otherwise the discovered patterns will have little support and the inter-relationship between the patterns will be lost.

One way to categorize the SOFA score is to ask a medical expert (intensivist, in our case) to provide the categories. Due to the subjectivity of the categories, they might not be acceptable to others. Hence we categorize the SOFA scores automatically. Because the ultimate goal is to predict mortality, the categorization will be performed with respect to it. We use entropy-based categorization whose merits are described in [10]. In other words, the SOFA score will be categorized in groups that minimize entropy with respect to mortality. Because this method assumes only one value of a SOFA score per patient and because we are looking for categories correlated with mortality, we used the maximum SOFA score during the ICU stay for each patient. The use of the maximum value was based on expert advice to capture the worst condition of the patient during ICU stay. We implemented the method by fitting binary classification trees with

Table 2. Distribution of the frequent episodes' length

Length	1	2	3	4	5	6	7	Total
Frequency	3	7	13	18	31	19	10	101

maximum SOFA score as predictor and mortality as the decision variable. We first over-fitted the tree and then pruned it according to the minimum 10-fold cross-validation error. The resulting splits of this tree provided the cut-off points to categorize the SOFA score. Three SOFA categories were obtained: SOFA $\leq$ 10 , 11 $\leq$ SOFA $\leq$ 13 and SOFA $\geq$ 14. For the sake of interpretability we denote these categories, respectively, as L (LOW), M (MEDIUM) and H (HIGH). It is important to note that categorization has been performed only on the training set, the test set will use these categories.

Discovery of Frequent Episodes. In this section we assume the reader is familiar with the notions pertaining to the Apriori algorithm [3]. We implemented an algorithm for the discovery of sequential episodes based on an adaptation of the Apriori-like algorithm described in [13]. A sequential episode is a temporal pattern indicating the *relative* order between values, called events, of a sequence. The events do not have to be consequent in the sequence. For example the episode L-H will not only match the sequence L-H but also the sequences L-M-H and L-L-M-L-H. This choice allows for capturing patterns regardless of when they happened in the sequence but it is also less specific. We will say more in the discussion about this choice and ways to improve it, but it forms a reasonable good start. In practice one constrains the allowable distance between episode events by limiting the search within a fixed-sized window. This window implies an upper limit for the episodes' length. We have used a window size of 7. This does not constrain sequences of those who stayed for one or less than a week (this includes the majority of patients).

Our main adaptation to the algorithm in [13] is related to the way the support of episodes is calculated. In the ICU, patients have markedly different length of stay (regardless of their vital status at discharge), and hence, different lengths of SOFA sequences. If we calculate support of an episode based on its occurrence in the whole patient population, then the longer episodes will have much less support because the longer sequences are much less frequent. Because we want to make predictions for five different cohorts, for those that stayed for at least 1, 2, 3, 4, and 5 days, we must adjust the support based on the respective cohort. For example, when calculating the support for an episode of length 3 we should seek support within sequences of at least length 3. In other words, only patients that stayed for at least 3 days are eligible to be counted in the denominator of the support. This allows longer sequences to emerge in the set of frequent episodes without being negatively biased by the existence of shorter sequences. We require a minimum support of 5% for an episode to be considered frequent. Applying the frequent episodes discovery algorithm on the data generated 101 frequent episodes. Table 2 provides the distribution of the episodes' length. Note

Table 3. A selection of dummy variables representing frequent episodes. The table shows, per day, the frequency of matching sequences in the training set and the respective mortality counts

		M1		**M2**		**M3**		**M4**		**M5**	
Dummy Id.	**Episode**	Freq	Died	Freq	Died	Freq	Died	Freq	Died	Freq	Died
S19	L	2958	186	905	139	524	118	409	104	319	89
S86	M	142	74	147	62	140	71	129	65	115	57
S34	L L			804	101	490	103	393	95	309	84
S61	H H			26	24	25	22	20	18	17	13
S33	L L L					412	67	353	76	289	73
S3	H H H					12	12	13	11	9	7
S40	M M L					12	3	18	5	21	11
S17	L L M L							9	6	18	8
S12	L L L L							297	55	260	56
S7	M L M					6	4	18	9	23	9
S46	M L L L L									9	5
		3181 pat.		990 pat.		573 pat.		440 pat.		338 pat.	

that short episodes have less realizations but enjoy high support. The first two columns is Table 3 provide examples of frequent episodes, which will be also encountered later. The other columns provide their frequency counts in the five cohorts, defined based on the first five days.

Development of the Logistic Regression Models. A logistic regression model [5] is a parametric model that specifies the conditional probability of a binary variable Y ($\{0, 1\}$) to have the value 1, given the values of the covariates of the model. $Y = 1$ indicates the occurrence of an event such as death in our concrete case. The logistic model has the following form:

$$p(Y = 1 \mid \mathbf{x}) = \frac{e^{g(\mathbf{x})}}{1 + e^{g(\mathbf{x})}} \tag{1}$$

where $\mathbf{x} = (x_1, ..., x_m)$ is the covariate vector. For m variables (also called predictors) the *logit function* $g(\mathbf{x})$ has the following form:

$$g(\mathbf{x}) = \beta_0 + \sum_{i=1}^{m} \beta_i x_i \tag{2}$$

where β_i, $i = 1, ..., m$, denote the coefficients of the m predictors. One reason for the popularity of the model is the interpretation that is given to each β_i in terms of an *odds ratio*. Suppose the logit function is $\beta_0 + \beta_1 sex + \beta_2 age$ where $sex = 1$ for males and 0 for females, and age is calculated in years. The odds of dying for males, $odds(sex = 1)$, is $P(Y = 1|sex = 1)/P(Y = 0|sex = 1)$ and for females, $odds(sex = 0)$, is $P(Y = 1|sex = 0)/P(Y = 0|sex = 0)$. The quantity e^{β_1} turns out to be equal to the odds ratio $odds(sex = 1)/odds(sex = 0)$. If there is no difference between the odds for dying for males compared to females, assuming all other variables in this case only age have the same values, the odds ratio will be 1. A higher value indicates higher risk to die for males, and a lower value than 1 indicates higher risk for females. The interpretation of e^{β_2} is similar, it indicates the odds ratio between a group of patients who are one unit, here one year, older than the other group.

The motivation for using logistic regression as our formalism of choice is two-fold. First, it is the most popular model used in medicine for binary classification problems, and is also the *de facto* model used in quality assessment programs in the ICU. Second, the β_is take into account the dependence between the variables used, so the value of any β_i is adjusted for all other variables in the model and there is no assumption that the variables are independent.

Our idea in developing logistic-based prognostic models it to use a dummy (sometimes called design) variable for each episode. Each patient will have a vector of dummy variables, one for each episode. In our case every dummy variable will only have two values: 0, indicating the episode does not match the SOFA sequence of a patient, and 1 if it does match. When predicting mortality for patients on the kth day, one may only use episodes with maximum length of k. For example, to predict mortality for patients on the third day of their stay, we only consider patients who stayed for at least three days and we use episodes of a maximum length of 3 which are matched against only the SOFA sequences of the first three days. In total we will hence have 5 models to predict mortality on each of the first five days. Prediction in these days is relevant for the ICU. Note that the number of patients considered for prediction on day 1 includes all patients, and that this number decreases with each day due to patient discharge (regardless of mortality status). To assess the *added* value of the temporal patterns to current logistic models, we develop for each day a model with only the SAPS score as a covariate, and we compare this model with a temporal one in which also dummy variables representing the existence of the episodes are included.

The strategy for fitting the temporal models is as follows. For any given day for prediction, the best dummy variables are included by a hill-climbing search process. We first fit a model including only the intercept, β_0, and the term for the SAPS score, $\beta_1 SAPS$. This is the reference model. We then assess the inclusion of each dummy variable to this model by the log-likelihood test [5]. The dummy variable with the most significant p-value, as long as it is ≤ 0.05, is included in the model. We reiterate this process till a maximum of 4 covariates. This restriction is meant to keep the models manageable and also to combat over-fitting. We excluded a dummy variable if its corresponding β_i, and in turn the odds ratio, is not in line with expert expectation. For example, consider a dummy variable for an episode that ends with M-L, which indicates going from medium severity to low severity at some later time. We would expect, on average, that this pattern should not be associated with an odds ratio greater than 1 because we do not expect those patients matching the episode to be associated with a higher odds of mortality compared to the odds of those who do not match this

Table 4. Comparative prediction performance - reference SAPS based models vs. temporal models combining SAPS with episodes represented by dummy variables

Day	Test set	Died	SAPS-only based models		Dummy based models		Model structure	β	e^{β}
	# patients	Count	Log	Brier	Log	Brier			
1	1576	184	356.4333	0.1315	348.2468	0.1272	intercept+	-5.04	0.006
							SAPS+	0.09	1.10
							S19+	-1.47	0.23
2	504	110	222.2785	0.2838	215.7803	0.2730	intercept+	-3.89	0.02
							SAPS+	0.06	1.06
							S34+	-0.85	0.42
							S61	2.22	9.26
3	299	86	164.9534	0.3708	160.6943	0.3580	intercept+	-2.42	0.09
							SAPS+	0.04	1.04
							S33+	-1.21	0.29
							S3	6.55	705.45
4	206	65	123.6431	0.4101	122.098	0.4025	intercept+	-1.49	0.22
							SAPS+	0.03	1.03
							S33+	-1.41	0.24
							S40+	-1.39	0.24
							S61	1.44	4.24
5	155	46	89.9248	0.3922	88.1640	0.3820	intercept+	-1.13	0.32
							SAPS+	0.03	1.02
							S12	-1.47	0.22

episode. Hence if the respective β is positive, which is equivalent to an odds-ratio greater than 1, then we exclude the dummy variable, refit the model without it, and repeat the procedure.

The last three columns in Table 4 show, for each one of the five temporal models, the frequent episodes whose dummy variables were selected in the logistic regression; the corresponding β_i; and the respective odds ratio. For example the first temporal model for predicting mortality after observing the first day is:

$$P(Y = 1 \mid SAPS, \ S19) = \frac{e^{-5.04+0.09SAPS-1.47S19}}{1 + e^{-5.04+0.09SAPS-1.47S19}} \qquad (3)$$

When inspecting the selected episodes we note that the models include episodes with consistently high or low values like *H-H* or *L-L-L*. Also the models include episodes that vary over time like *M-L-M* and *L-L-M-L* which apparently capture trends influential to mortality. However, Table 3 includes episodes not selected in the models. They either did not have enough discriminating power or were removed in the model selection stage. Episodes which were statistically significant but their β coefficients were not in line with our clinical expectations were *L-L-M-L*, *M-L-M*, *M-L-L-L-L* represented by dummy variables S17, S7 and S46. This phenomenon happened in the models for the fourth and fifth days.

Validation. After model selection, we validated the models on the test set. Our performance measures were the Brier score, which is:

$$1/N \sum_{i=1}^{N} (P(Y_i = 1 \mid x_i) - y_i)^2$$

where N denotes the number of patients, y_i denotes the actual outcome for patient i, and the logarithmic score which is:

$$\sum_{i=1}^{N} LS_i \text{ where } LS_i = -\ln P(Y_i = y_i \mid x_i).$$

Lower values mean better performance for both scores. These performance scores penalize models when they do not provide the true probability, and are more appropriate than purely discriminating measures such as error rate and the area under the ROC curve which might mask under- and over-prediction (see discussion in [2]). The performance of each temporal model is then compared to its respective reference model on the same part of the test set.

The validation results are presented in the columns labeled "Log" and "Brier" in Table 4. Similar to partitioning the training set, five test sets were used having patients staying at least one day for the first data set until at least five days for the fifth data set. All the temporal models 1, 2, 3, 4 and 5 outperform the reference model based on SAPS alone.

3   Discussion, Conclusions and Future Work

In this section we reflect critically on our approach and results, draw conclusions, and provide context and an outlook for further research.

The entropy-based categorization method using maximum SOFA score per patient resulted in three categories. Inspection of the categories showed that these make clinical sense as they correspond to three groups of patients with distinctively different number of multiple organ failures as can be calculated from the mean of the 6 sub-scores (an organ is failing when its sub-score, which ranges from 0 to 4, is 3 or 4). Alternatives to using the maximum SOFA score include using the last SOFA value in a patient's sequence, or the mean in the last 3 days but further analysis showed that our choice for the maximum value is quite robust as the cut-off points hardly changed.

The frequent episodes that were discovered have a clear clinical meaning in terms of improvement or worsening of a patient's condition. When an episode appears statistically significant to be included in a temporal model, the analyst can still judge whether it makes clinical sense by inspecting whether its corresponding β (either positive or negative) is in accordance to its clinical meaning. In this way we have been able to exclude some episodes especially in the model for the fourth day. Further analysis showed that including them would have indeed resulted in a degradation in performance on the test sets.

There seems to be a preference for selecting episodes that are as long as the number of days under consideration. It seems that the last couple of days before the prediction are probably the most important ones but our episodes cannot capture this, because they are not bound and can appear anywhere between the first day and the day of prediction. Our hypothesis is that the choice for the longest possible episodes somehow compensates for this situation because it is the only way now to include SOFA scores at the last days before prediction. Also we allow nonconsecutive episodes while for such short series it is perhaps better to use only consecutive ones. These two points form a possible weakness of our approach.

The results obtained are a proof of concept that SOFA score episodes, based on our categorization, is beneficial indeed and has an added value compared to static models. Note that this added value is inherent in the patterns themselves and not because the models are developed on cohorts with different lengths of stay, because the static ones have also been fitted separately for each of these cohorts as well. Although better than the reference model, the model for day four has turned out to be overfit because excluding some of these episodes resulted in improved performance. In fact, the model with the SAPS score and episodes S33 plus S61, which stands for *L-L-L* and *H-H* turned out to be superior to the current temporal model for day four. This is an indication that our strategy for developing the temporal models should be refined to allow for less covariates when the number of patients, and deaths, shrinks.

3.1   Related and Future Work

Frequent episode discovery and prediction tasks are popular research topics. However, their combination has not received much attention, let alone for solving a complex real world problem. In the IC, most prognostic models that made use of temporal information has opted to reducing the temporal information into summary statistics, as typified e.g. by [8]. In other domains, temporal abstractions have been used such as in [11] but this requires defining the abstractions to be known in advance. We are interested in discovered sequential episodes that keep the temporal relationship between events intact.

The notion of using indicator variables, like our dummy variables, for episodes to denote the existence of a matching sequence is not new, for example it has been applied in [4]. However, the episodes were used for further clustering and not for predictions. Perhaps the most similar work to ours, in its general approach and aim, is that described in [6] and [7]. The work in [6] considers the notion of stationarity and non-stationarity. From this perspective we assume stationarity of the temporal process in the sense that the frequency of an episode is calculated independently on when, in time, it occurred. In [7] temporal patterns using SOFA sub-scores are calculated from the data, then integrated in a Naive-Bayes framework. Apart from the fact that the data set in that study is smaller and the value categories were created subjectively, there are other important differences. In our approach, the discovered frequent sequential episodes, based on an adaptation of the algorithm in [13], are integrated in a logistic regression framework. This not only allows for integrating new methods into the established framework in IC prediction, but also takes into account the interdependence between the sequential episodes, unlike the Naive Bayes approach which assumes conditional independence of patterns. In addition, our approach provides an intuitive interpretation of significance of these episodes, by means of the βs, by which the episodes can be judged. Another difference is the way models are validated. We use the Brier score and the logarithmic score instead of the area under the ROC curve used there because, in quality assessment programs, it is important to measure discrimination and precision in combination, instead of relying on only the discrimination power of the model.

Unlike in our approach, in [7] patterns specify *consecutive* events and they are always considered backwards from the day of the prediction. These properties might be beneficial and, as future research, we plan to investigate whether they would lead to improvements of our temporal models. Another future research is investigating new temporal event types. The intensivists hypothesize that dealing with the notion of recovery, instead of the SOFA scores themselves, might be beneficial. For example the sequence could reflect the recovery, or recovery rate, in time.

Acknowledgment. This research was supported by the Netherlands Organization for Scientific Research (NWO) under the I-Catcher project number 634.000.020. We thank Arno Siebes and Manuel Campos for their helpful discussions on the topics described in this paper.

References

1. Abu-Hanna, A. and Lucas, P.J.F., Editorial: Prognostic models in medicine - AI and statistical approaches, Methods of Information in Medicine **40** (2001) 1–5
2. Abu-Hanna, A. and Keizer, N. de, Integrating classification trees with local logistic regression in Intensive Care prognosis, Artificial Intelligence in Medicine **29(1-2)** (2003) 5–23
3. Agrawal, R. and Srikant, S., Fast algorithms for mining association rules, In Proc. of the 20th VLDB Conf., (1994) 487–499
4. Bathoorn, R. and Siebes, A., Constructing (almost) phylogenetic trees from developmental sequences data, PKDD (2004) 500–502
5. Hosmer, D.W. and Lemeshow, S., Applied logistic regression. New York: John Wiley & Sons, Inc. (1989)
6. Kayaalp, M. and Cooper, G. Clermont, G., Predicting ICU mortality: a comparison of stationary and nonstationary temporal models, Proc. of AMIA, (2000) 418-422
7. Kayaalp, M.and Cooper, G. and Clermont, G., Predicting with variables constructed from temporal sequences, Proceedings of the Eighth International Workshop on Artificial Intelligence and Statistics. (2001) 220-225.
8. Kajdacsy-Balla Amaral, A.C. and Andrade, F.M. and Moreno, R. and Artigas, A. and Cantraine, F. and Vincent, J.L., Use of Sequential Organ Failure Assesment score as a severity score, Intensive Care Med **31** (2005) 243–249
9. Knaus W, and Draper E, and Wagner D, Zimmerman J. APACHE II: a Severity of Disease Classification System. *Crit Care Med* **13** (1985) 818–829.
10. Kohavi, R. and Sahami, M., Error-Based and entropy-based discretization of continuous features, KDD (1996) 114–119

11. Larizza, C. and Bellazzi, R. and Riva, A., Temporal abstractions for diabetic patients management , AIME, (1997) 319–330
12. Le Gall, J. and Lemeshow, S. and Saulnier, F., A new Simplified Acute Physiology Score (SAPS-II) based on a European/North American multicenter study. *JAMA* **270** (1993) 2957–2963.
13. Mannila, H. and Toivonen, H. and Verkamo, A.I., Discovering frequent episodes in sequences, Data Min. Knowl. Discov., **1(3)** (1997) 259–289
14. Mannila, H. and Pavlov, D. and Smyth, P., Predictions with local patterns using cross-entropy, KDD, (1999) 357–361
15. Vincen,J.-L. and Ferreira, F.L., Evaluation of organ failure: we are making progress, Intensive Care Med **26**(2000) 1023–1024
16. Vincent, J.L. and Mendonca, A. de and Cantraine, F. and Moreno, R. and Takala, J. and Suter, P. and Sprung, C. and Colardyn, F.C. and Blecher, S., Use of the SOFA score to assess the incidence of organ dysfunction/failure in intensive care units: results of a multicentric, prospective study. Crit Care Med **26** (1998) 1793-1800

A Fuzzy Temporal Diagnosis Algorithm and a Hypothesis Discrimination Proposal*

J. Palma, J.M. Juárez, M. Campos, and R. Marín

Artificial Intelligence and Knowledge Engineering Group,
Departamento de Ingeniería de la Información y las Comunicaciones,
Facultad de Informática. Campus de Espinardo,
Universidad de Murcia, Murcia 30071, Spain
Tel: +34 968 364631; Fax: +34 968 363151
jpalma@dif.um.es

Abstract. Over the last decades, Artificial Intelligence has approached the decision support system design in medical domains by capturing the knowledge and configuring it in knowledge intensive software systems. Model-based diagnosis is one of the techniques which has produced the best results, such as diagnosis intelligent systems in the realm of medicine. In this domain, one of the key factors is the temporal dimension. This variable enormously complicates the design of such systems, and in particular the process of getting a reliable diagnosis solution. This paper presents a Diagnosis Abductive Algorithm based on Fuzzy Temporal Abnormal Model. This algorithm provides a solution for the above problem by the description of its dianosis explanation, allowing an approach based on the Possibility Theory for the evaluation of the diagnosis hypotheses.

Keywords: Model-based Diagnosis, Temporal Reasoning, Possibility Theory.

1   Introduction

In recent years, medical knowledge has ben subjected to a constant gorwth, achieving high levels of specialization, which makes the desgn of models more difficult. The use of deep causal models has rapidly showed the advantages of modelling domain knowledge in this way, in contrast with the classical rule-based systems [1]. Recent research in this area has paid increasing attention to the use of deep causal models, especially if they are considered as integrated in Model Based Diagnosis (MBD) techniques.

A good analysis of patient evolution over time lies in an efficient diagnosis process. The ICUs domain reveals the importance of the temporal component modelling in capturing the temporal information associated to patient evolution [2]. However, the inclusion of temporal representation techniques in MBD has increased the complexity

* This work was supported by the Spanish MEC under project MEDICI (TIC2003-09400-C04-01), the Murcia Regional Government under project PB/46/FS/02, the Spanish MEC under the FPU national plan (grant ref. AP2003-4476), and the SENECA Foundation of the Murcia Reginal Government (grant ref. FPI00911CV02).

J. Mira and J.R. Álvarez (Eds.): IWINAC 2005, LNCS 3561, pp. 459–468, 2005.

of the diagnosis process. A serious attempt to provide a general framework for temporal MBD can be found in [3], which presents a general characterization of temporal MBD at knowledge level.

Our goal, therefore, is to present a diagnosis algorithm based on a Diagnosis Fuzzy Temporal Model. Our proposal deals with temporal uncertainty using the Fuzzy Temporal Constraints Network formalism ($FTCN$). Furthermore, this paper present a hypothes discrimination process in order to achieve a relief diagnosis.

The structure of the paper is as follows: the underlying temporal framework is described in a concise manner in Section 2. Section 3 presents the temporal behavioral model. The hypotheses discrimination is analysed in Section 5. Finally, we provide some conclusions.

2   Temporal Framework

In some proposals for Temporal MBD, the temporal dimension is modelled by means of the so-called *Fuzzy Temporal Constraint Network* ($FTCN$) formalism. A $FTCN$ is a pair $\mathcal{N} =< \mathcal{T}, \mathcal{L} >$ consisting of a finite set of temporal variables, $\mathcal{T} = \{T_0, T_1, ..., T_n\}$, and a finite set of binary temporal constraints, $\mathcal{L} = \{L_{ij}, 0 \leq i, j \leq n\}$ defined on the variables of $\mathcal{T}$. A $FTCN$ can be represented by means of a directed constraint graph, where nodes represent temporal variables and arcs represent binary temporal constraints.

Each binary constraint L_{ij} on two temporal variables T_i and T_j is defined by means of a convex possibility distribution $\pi_{L_{ij}}$ ($\pi(v') \geq min\{\pi(v), \pi(v')\}; v \leq v' \leq v''$), whose discourse universe is $\mathbb{Z}$, and which restricts the possible values of the time elapsed between both temporal variables. In the absence of other constraints, the assignments $T_i = t_i$ and $T_j = t_j$ are possible if $\pi_{L_{ij}}(t_j - t_i) > 0$ is satisfied.

An n-tuple $S = (t_1, ..., t_n) \in \tau^n$ is a σ-*possible solution* of a $FTCN$ network $\mathcal{N}$ if $\pi^{S_{\mathcal{N}}} = \sigma$, where $\pi^{S_{\mathcal{N}}} = min\{\pi_{L_{ij}}(t_j - t_i), 0 \leq i, j \leq n\}$. The possibility distribution $\pi^{S_{\mathcal{N}}}$ defines the fuzzy set $S_{\mathcal{N}}$ of the σ-possible solutions of the network, with $\sigma \geq 0$. A $FTCN$ network $\mathcal{N}$ is *consistent* if and only if $S_{\mathcal{N}}$ is greater than a previously established threshold α, where $\alpha \in [0, 1]$, with $\alpha = 1$ being equivalent to the crisp case. The value of α is conditioned by the context and is set up arbitrarily by the user.

This model has been implemented and extended in FuzzyTIME [4], a general purpose temporal reasoner that provides high level language and reasonings capabilities on fuzzy temporal constraints between temporal variables which can represent intervals or time instants.

3   Temporal Behavioral Model

In this proposal, we opt for a Temporal Behavioral Model, TBM, an abnormal behavioral model in which only the causal and temporal relations between hypotheses (diseases) and abnormal observations caused by them are represented. These relations are defined by *Diagnostic Fuzzy Temporal Patterns* ($DFTPs$). Hence,

$TBM = \{DFTP_k\}$. Each $DFTP$ can be formally defined by the tuple $DFTP = \langle H, IM, IH, R^{dftp}, \rangle$ where:

- H is the diagnostic hypothesis described by $DFTP$.
- $IM = \{im_k | k = 1, \ldots, n_{im}\}$, is the set of abnormal manifestations implied by the hypothesis H.
- $IH = \{ih_k | k = 1, \ldots, n_{ih}\}$ is the set of hypotheses implied by H (in medical domains, ih_k is a disease caused by H).
- $R^{dftp} = \langle \mathcal{T}^{dftp}, \mathcal{L}^{dftp} \rangle$ is a consistent $FTCN$, where temporal variables in $\mathcal{T}^{dftp}$ are associated to H, IM and IH, $\mathcal{T}^{dftp} = \{t^H, t_1^{im}, \cdots, t_{n_{im}}^{im}, t_1^{ih}, \cdots, t_{n_{ih}}^{ih}\}$ and the temporal constraints between them are defined in $\mathcal{L}^{dftp}$, where $\mathcal{L}^{dftp} = C(t^H, t_1^{im}, \cdots, t_{n_{im}}^{im}, t_1^{ih}, \cdots, t_{n_{ih}}^{ih})$. Furthermore, only those constraints defined by the expert are instantiated, and a subsequent process computes the minimal network of constraints between all temporal variables.

Other elements of the model could be taken into accont. In [5], we suggest the importance of temporal contexts which describes how the $DFTP$ definition is modified when a context factor occurs (temporal or atemporal concepts).

4 The Diagnosis Process

In order to provide a solution, the diagnosis process requires as inputs (apart from the TBM) the patient's observations ($EVT = \{evt_i | i = 1, \ldots, n_{obs}\}$), the consistent temporal network (R^{input}), whose temporal variables are associated to elements in EVT. In most cases, these temporal variables are specified as absolute time instants, which makes the reasoning process more efficient.

In our proposal, the diagnostic process output (i. e., the explanation provided) is composed by all the elements that conform the final instantiated causal network (physiopathological and ethiological diagnosis, in medical domains). This kind of diagnosis explanation is necessary from the point of view of decision support system development. Therefore, the diagnosis algorithm output can be formally defined as the tuple $EXP = \langle CN_{exp}, R^{exp}, AB_{exp} \rangle$ where:

- CN_{exp} represents a directed graph describing the final causal network, where nodes represent observables and hypotheses in the final explanation.
- R^{exp} is a $FTCN$ where the temporal variables are associated to the CN_{exp} nodes.
- $AB \subset DFTP_{exp}$ is the set of abducibles generated by the diagnosis process.

In MBD, different interpretations of temporal diagnosis explanation have been proposed. On the one hand, there is totally consistency-based diagnosis [6], in which the explanation provided should be consistent with all observations. On the other hand, there is totally abduction-based diagnosis [7, 2], in which the explanation should logically entail all the observations.

In our proposal we opt for a model in which an abductive component is applied to the set of events (EVT). The diagnosis process tries to subsume these events in order to find a diagnosis explanation. Due to the abnormal nature of $DFTPs$, all events that can be explained are labelled as abnormal events(EVT^-), and the rest are considered normal (EVT^+).

We consider that a diagnosis explanation (EXP) is a solution of a diagnosis problem if the set of abnormal events (a subset of EVT) is a logical consequence of the set of the diagnosis temporal patterns (causal consistency); and if the temporal contraint network of the explanation and the one of the imput are temporally consistent.

4.1   The Diagnosis Process Description

The diagnosis process in this work is described by an algorithm based on the TBM described in Section 3. Our approach follows two assumptions. Firstly, we consider a **multiple cardinality solution**. Several hypotheses may befound in a solution, which represent alternative or complementary solutions. Furthermore, different instances of the same hypothesis (the same hypothesis located at different time instants) are possible in a solution. However, all hypotheses should be consistent with the context information.Secondly, the diagnosis process is based on **Parsimonious covering**. The proposed algorithm explains the abnormal event set EVT^-_{new} through parsimonious covering. New hypotheses are included in the final explanation if and only if events cannot be explained by the hypotheses already instantiated. Of course, the solutions provided do not contradict either temporal or atemporal contextual concepts.

4.2   Subsumption

The aim of the subsumption process is to avoid an excessive proliferation of temporally nearby hypotheses. Thus, before creating a new instantiated pattern to explain a given event evt_i, the subsumption process tries to include it in one of the already instantiated patterns, particularly those patterns in $DFTP_{exp}$ which match with the patterns in TBM and which explain evt_i.

In order to subsume a given evt_i, with $dftp_k = evoke(evt_i)$, in $DFTP \in DFTP_{exp}$, the subsumption process checks if the temporal constraints defined in $dftp_k$ (in which evt_i takes part) are consistent with the temporal constraints of R^{exp} (the temporal constraint network of the solution). This process is carried out by a temporal query to the temporal reasoner using local propagation of the fuzzy constraints, similar to the technique defined in [8].

However, subsumption usefulness refers to the time execution factor. The subsumption process slows down the growth of instantiated hypotheses, which is exponential. Subsumption allows events to be explained by instantiated hypotheses of the solution, avoiding temporal nearby instances of hypotheses.

4.3   Temporal Shifting

When subsumption is not possible, is a new pattern instantiation enough? The answer is no. When a given event cannot be subsumed, it is due to temporal inconsistencies in the instantiated pattern, $DFTP$. However, $DFTP$ could possibly explain the new event if temporal conditions were different.

We, therefore, propose to include a new instance of the same pattern and to associate the event to it. If we reconsider the failed subsumption, we will notice that only a few of the associated events subsumed into it do not allow the new subsumption. According to

---

Algorithm 1 The Temporal Covering Process

Function *TemporalCovering*(TBM, EVT) **return** EXP

    Mark all events in EVT as event to be explained

2: **while** Unexplained events exists **do**

      for each $evt_i \in EVT$ **do**

4:      $\mathcal{D} = evoke(evt_i, TDM)$

        if $\mathcal{D} \neq \emptyset$ **then**

6:        Mark evt_i as belonging to EVT^-

          for each $dftp_i \in \mathcal{D}$ **do**

8:          **if** $contextualizable(dftp_i, EVT, EXP)$ **then**

             $DFTP_i^c = contextualize(dftp_i, EVT, EXP)$

10:           **for each** $dftp_i^{c_j} \in DFTP_i^c$ **do**

              if $explains(evt_i, dftp_i^{c_j})$ **then**

12:             $DFTP_{exp} = DFTP_{exp} \cup \{dftp_i^{c_j}\}$

              end if

14:          **end for**

         else

16:         $DFTP^{evt_i} = evoke(evt_i, DFTP_{exp})$

          if $DFTP^{evt_i} \neq \emptyset$ **then**

18:           **for each** $dftp_j \in DFTP^{evt_i}$ **do**

             if $subsumible(evt_i, dftp_j, EXP)$ **then**

20:             $subsume(evt_i, dftp_j)$

             else

22:              **if** $LAST(\mathcal{D})$ **then**

                $dftp_k = temporally\_shift(evt_i, dftp_j, EXP)$

24:                $DFTP_{exp} = DFTP_{exp} \cup \{dftp_k)\}$

             end if

26:            **end if**

          end for

28:          **else**

          $dftp_new = generate\_new(evt_i, dftp_i, EXP)$

30:          Generate a new event evt_j associated to the new hypothesis

          Mark evt_j as an event to be explained

32:          **end if**

         end if

34:        **end for**

       Mark evt_i as belonging to EVT^-

36:     **else**

       Mark evt_i as belonging to EVT^+

38:     **end if**

      end for

40: **end while**

---

this, some of these events (already subsumed) can be subsumed by the new instance of the pattern. In conclusion, a temporal shifting process will produce two instances of the same pattern (at different time instants), whose hypotheses explain at least one different event and, perhaps, some common events.

The temporal shifting process is used when subsumption is not possible. However, this process reduces the algorithm's efficiency because of the large amount of calculi for temporal consistence checking, in spite of the local propagation process. Furthermore, this process could imply new subsumptions. In our opinion, this problem could be partially reduced using some heuristics, which determine whether the hypothesis must be shifted or not. In this work, we suggest the application of this shifting technique only with the latest temporal instance of the pattern ($last(D)$, line 22 of Algorithm 1).

4.4   The Diagnosis Algorithm

Once the selected event is explained and removed from EVT_{new} , its explaining hypothesis (hypotheses) will be a new event to be explained, and therefore will be included in EVT_{new}. The algorithm finishes when it is not possible to find a higher level hypothesis, abducibles of the solution, that can explain any of the EVT_{new} events. The diagnosis process can be described, as follows:

1. An event e is selected from EVT_{new}. The event e is possibly associated to an evidence or a hypothesis.
2. The algorithm searches ($evoke()$, line 5) all possible patterns D from TDM that can explain event e.
3. Finally, the algorithm tries to include each pattern of D found in the solution as follows: **A** The algorithm considers temporal and atemporal concepts from the context information input (CTX_{obs}), then the algorithm tries to contextualize ($contextualized()$, line 9) the pattern. **B** If contextualization is not possible, the diagnosis process tries to subsume ($subsume()$, line 20) the event in any of the already instantiated patterns that exist in the partial solution (see Section 4.2). **C** When subsumption is not possible either, the temporal shifting process (see Section 4.3) is applied ($temporal\_shifting()$, line 23). Due to the computational cost of this procedure, this process must fulfil some heuristic conditions like that proposed in section 4.3. **D** If non previous actions are possible, the diagnosis process will generate an instance of the new pattern in the solution ($generate\_new()$, line 29).

5   Hypothesis Discrimination

The Temporal Covering Process (TCP), described above (Section 4.4), can generate hypotheses which are not sufficiently supplied by the evidence or which may be inconsistent with some of the observations not explained in the previous process. In order to eliminate such hypotheses it is necessary to establish some measure that will indicate the degree of credibility associated to a hypothesis. The approach we propose is based on Possibility Theory [9], by means of which we can calculate the degree of possibility and of necessity associated to each hypothesis. Such degrees allow the Hypothesis discrimination process to eliminate from the final explanation all those hypotheses which should not be included therein.

In order to analyse the calculus process of the degrees of necessity, N, and possibility, Π, associated to a hypothesis, we need to resort to a logic approach to TBM. A $DFTP$ with several manifestations and implied hypotheses can be rewritten as:

$$(\hbar, t^\hbar) \rightarrow (m_1, t_{m_1}) \wedge \ldots \wedge (m_n, t_{m_n}) \wedge (h_1, t_{h_1}) \wedge \ldots \wedge (h_r, t_{h_r}) \wedge \\ C(t^\hbar, t_{m_1}, \cdots, t_{m_n}, t_{h_1}, \cdots, t_{h_n}) \tag{1}$$

As can be deduced from temporal formula 1, the TCP process seeks to construct an explanation which satisfies the backward meaning of the temporal formula. The hypothesis discrimination process based on the calculus of the degrees of necessity and possibility for the hypotheses will take charge of satisfying the forward meaning of the temporal formula.

For shake of simplicity, let us suppose that we have a $DFTP$ which only associates one hypothesis, $\hbar$, with one manifestation, m. In such a case, temporal Formula 1 reduces to

$$\hbar \rightarrow m \wedge C(_\hbar^t, t_m) \tag{2}$$

The degrees of necessity and possibility associated to $\hbar$, $N(\hbar)$ and $\Pi(\hbar)$ can be calculated by a combined form of *Modus Tollens* and *Modus Ponens*, used in the Possibility Theory [9], and which in our case is reduced to Expression 3.

$$\begin{array}{ll} N(\hbar \rightarrow m) = S_1(m) \\ N(m \rightarrow \hbar) = S_2(m) \\ \dfrac{N(m) = b \qquad\qquad \Pi(m) = c}{N(\hbar) = \max(S_2, b) \quad \Pi(\hbar) = \max(1 - S_1, c)} \end{array} \tag{3}$$

In the calculus of $N(m)$ and $\Pi(m)$, it has to be taken into account that for an implied manifestation to be considered as a necessary condition of $\hbar$, the stimated appearance time of the pattern must occur before τ_{now}, the time at which the diagnosis process began. This means that the absence of a manifestation −but one which according to the information in the temporal pattern may, nevertheless, still occur− may contribute positively to the consistency of the hypothesis. Hence, a new, disjunctive term in the consequent of equation 1 must be added:

$$\hbar \rightarrow (m \wedge C(_\hbar^t, t_m)) \vee before(\tau_{now}, t_m) \tag{4}$$

Therefore, the possibility degree calculus of a given manifestation can be formulated as follows:

$$\begin{aligned} \Pi(m) &= \Pi(m \wedge C(t^\hbar, t_m) \vee before(\tau_{now}, t_m)) \\ &= \max(\Pi(m \wedge C(t^\hbar, t_m)), \Pi(before(\tau_{now}, t_m))) \end{aligned} \tag{5}$$

It needs to be taken into account that the temporal query $\Pi(m \wedge C(t^\hbar, t_m))$ (and its extension to all the constraints defined in the $DFTP$ in the case of there existing more than one manifestation) is implicit in the query on consistency. In the case of there existing an event associated to the manifestation in the explanation, a high possibility value will be returned. If the associated event does not exist or another event exists, the possibility value will be low or even 0. In the same way, $N(im_i)$ can be defined as follows:

Algorithm 2 Calculate Possibility and Necessity function.

Function $CalculateΠN$(d,EXP) **return** $Π(h_d), N(h_d)$

1: Let $d = \{H_d, IM_d, IH_d, R^d, CTX_d, N1^d, N2^d\}$
2: Assert R_{im}^d in R^{exp}
3: $Π_{hd} = 1, N_{hd} = 0, Π_{ih} = 1, N_{ih} = 0$
4: $Π_{im1} = Π(EXP, d), N_{im1} = N(EXP, d)$
5: **for each** $im \in IM_d$ **do**
6:     Let $im = (p, s, v, t^{im})$
7:     $Π_{im3} = Π^{R^{exp}}(C_b(t_{now}, t^{im}))$
8:     $Π_{im} = max(Π_{im1}, Π_{im3})$
9:     $N_{IM} = max(N_{im1}, N_{im3})$
10:    $N1$ =get N1 of $im_d \in DFTP_d$, $N2$ =get N2 of $im_d \in DFTP_d$
11:    $Π_{hd} = min(max(1 - N1, Π_{im}), Π_{hd})$
12:    $N_{hd} = max(max(N2, N_{im}), N_{hd})$
13: **end for**
14: **for each** $ih \in IH_d$ **do**
15:    $Π_{ih}, N_{ih} = CalculateΠN(ih, EXP)$
16:    $N1$ =get N1 of $ih_d \in DFTP_d$, $N2$ =get N2 of $ih_d \in DFTP_d$
17:    $Π_{hd} = min(max(N1, Π_{ih}), Π_{hd})$
18:    $N_{hd} = max(max(N2, N_{ih}), N_{hd})$
19: **end for**
20: $return(Π_{hd}, N_{hd})$

$$N(m) > max(N(m \wedge C(t^{ℏ}, t_m)) \vee N(before(\tau_{now}, t_m))) \qquad (6)$$

Hereinafter, we will assume that $N(im_i)$ is represented by its upper bound. For the evaluation of implied hypotheses we must take into account two different situations. Firstly, the implied hypothesis has been explained by the TCP process, Expression 3 is applied in a recursive way. Secondly, if the implied hypothesis has not been explained, nothing can be said about the hypothesis, and therefore it is completely possible that it occurs, $Π(ih_i) = 1$ (no evidences to refute it), but there is no certainty of its occurrence, $N(ih_i) = 0$.

Finally, it has to be taken into account that several implied hypotheses and manifestations can contribute to the evaluation of $Π(ℏ)$ and $N(ℏ)$. In this case, they can be combined by means of expressions 7 and 8 [9].

Algorithm 3 The Hypotheses Discrimination function.

Function $Hypotheses\_Discrimination(EXP)$ **return** EXP

1: **for each** $ab \in abducibles(EXP)$ **do**
2:     $Π(ℏ_{ab}), N(ℏ_{ab}) = CalculateΠN(ab, EXP)$
3: **end for**
4: **if** $Π(ℏ_{ab}) < Π_{th}$ **then**
5:     $Conservative\_pruning(ab, EXP)$
6: **end if**

$$N(\hbar) = \max_k(N^{\hbar}_{mh_k}) \tag{7}$$

$$\Pi(\hbar) = \min_k(\Pi^{\hbar}_{mh_k}) \tag{8}$$

where mh stands for all the manifestations and implied hypotheses included in the temporal pattern; $N^{\hbar}_{mh_k}$ and $\Pi^{\hbar}_{mh_k}$ stands for the contribution to the global pattern's necessity and possibility degree due to implied manifestation or hypotheses mh_k.

5.1 Hypothesis Discrimination Process

The hypotheses discrimination (Algorithm 3) starts with the evaluation of the possibility, $\Pi(\hbar)$, and necessity, $N(\hbar)$, degrees of the abducibles generated by the TCP process ($abducibles(EXP)$). This evaluation triggers the evaluation downwards through the causal network of all the hypotheses. The calculus of possibility/necessity of each node of the causal network is stablished by a recursive function, named $Calculate\Pi N$. This function is described in Algorithm 2 where possibility and necessity values are determined by the maximum/minimum of the possibility/necessity degrees associated to the different implied manifestations and hypotheses.

In the hypotheses discrimination (algorithm 3), the possibility degree for each hypothesis is compared to the given threshold, Π_{th}. If $\Pi(\hbar)$ is less than Π_{th}, there is no enough evidence to confirm the hypothesis, so it is removed from explanation causal network. This hypotheses pruning is propagated downwards the causal network through implied hypotheses applying the $Conservative\_pruning$ process (line 5 in Algorithm 3). $Conservative\_pruning$ only removes from the explanation those elements with no alternative explanation (an alternative path to another abducible exists). In the case that an alternative explanation exists, the pruning process is stopped and the link with the explanation path pruned is removed.

6 Conclusions

This paper describes a general abductive diagnosis algorithm which tackles the problem of managing complex interaction between deep causal models and context knowledge, evaluation of hypotheses possibility degree, and the structure and the expressiveness of the solution provided. The proposed abnormal behavioural model demonstrates the suitability of FTCNs for time management.

One of the main contributions of our proposal is the evaluation of hypotheses, which is more complete since it is based on inference mechanisms based on Possibility Theory. This approach is related to the possibility of modelling complex contextual interactions, such as those found in medical domains. In classical TMBD models [3, 10] contextual information is modelled as atoms, which are not included in the observations sets and can be added in the antecedent of diagnostic rules. Our model allows elements from the observation set to be included in the description of diagnostic rules modifications, making it possible to capture complex contextual situations in a more compact way. Another advantage of this proposal is the causal network. Hence, the integration with medical decision making processes is more simple.

References

[1] P. Torasso. Multiple representations and multi-modal reasoning in medical diagnostic systems. *Artificial Intelligence in Medicine*, 23:46–69, 2001.

[2] W. Long. Temporal reasoning for diagnosis in causal probabilistic knowledge base. *Artificial Intelligence in Medicine*, 8:193–215, 1996.

[3] V. Brusoni, L. Console, P. Terenziani, and D. T. Dupré. A spectrum of definitions for temporal model-based diagnosis. *Artificial Intelligence*, 102:39–79, 1998.

[4] M. Campos, A. Cárceles, J. Palma, and R. Marín. A general purporse fuzzy temporal information management engine. In *EurAsia-ICT 2002. Advances in information and communication technology*, pages 93–97, 2002.

[5] J. Palma, J. M. Juárez, M. Campos, and R. Marín. A fuzzy approach to temporal model-based diagnosis for intensive care units. In *Proccedings of 16th ECAI*, pages 868–872, 2004.

[6] W. Hamscher, L. Console, and J. de Kleer. *Readings in Model-Based Diagnosis*. Morgan Kauffman, San Mateo, 1992.

[7] L. Console, L. Protinale, and D. T. Dupré. Using compiled knowledge to guide focus abductive diagnosis. *IEEE Transactions on Knowledge and Data Engineering*, 8(5):690–706, 1996.

[8] V. Brusoni, L. Console, and P. Terenziani. Efficient query answering in LaTeR. In *TIME-95 International Workshop on Temporal Representation and Reasoning*, pages 121–128, 1995.

[9] D. Dubois and H. Prade. *Possibilistic Theory: An Approach to Computerized Processing of Uncertainty*. Plenum Press, New York and London, 1988.

[10] L. Portinale, D. Magro, and P. Torasso. Multi-modal diagnosis combining case-based and model-based reasoning: a formal and experimental analysis. *Artificial Intelligence*, (158):109–153, 2004.

Spatial Reasoning Based on Rules

Haibin Sun and Wenhui Li

College of Computer Science and Technology, Jilin University,
Changchun 130012, China
Offer_sun@hotmail.com

Abstract. In this article, we investigate the problem of checking consistency in a hybrid formalism, which combines two essential formalisms in qualitative spatial reasoning: topological formalism and cardinal direction formalism. Although much work has been done in developing composition tables for these formalisms, the previous research for integrating heterogeneous formalisms was not sufficient. Instead of using conventional composition tables, we investigate the interactions between topological and cardinal directional relations with the aid of rules that are used efficiently in many research fields such as content-based image retrieval. These rules are shown to be sound, i.e. the deductions are logically correct. Based on these rules, an improved constraint propagation algorithm is introduced to enforce the path consistency. The results of computational complexity of checking consistency for constraint satisfaction problems based on various subsets of this hybrid formalism are presented at the end of this article.

1 Introduction

Combining and integrating different kinds of knowledge is an emerging and challenging issue in Qualitative Spatial Reasoning (QSR), content-based image retrieval and computer vision, etc. Gerevini and Renz [1] has dealt with the combination of topological knowledge and metric size knowledge in QSR, and Isli et al. [2] has combined the cardinal direction knowledge and the relative orientation knowledge. Assumed that we are given two map layers, we know that region A is north of region B and B overlaps with region C from one layer, and that region C is north of region D from another. What are the cardinal direction relation and topological relation between regions A and D? Obviously, we need to investigate the reasoning problem with the hybrid system combining topological and directional relations between spatial regions.

To combine topological and directional relations, Sharma [3] represented topological and cardinal relations as interval relations along two axes, e.g., horizontal and vertical axes. Based on Allen's composition table [4] for temporal interval relations, Sharma identifies all of the composition tables combining topological and directional relations. But his model approximated regions with Minimal Boundary Rectangles (MBRs), and if a more precise model (e.g., in this paper) is used, his composition tables will not be correct. We base our work on

J. Mira and J.R. Álvarez (Eds.): IWINAC 2005, LNCS 3561, pp. 469–480, 2005.

the same topological model as Sharma's, and a different directional model from his, which is more general and thereby, is more practical.

In this paper, we detail various interaction rules between two formalisms and we are also devoted to investigating the computational problems in the formalism combining topological and cardinal directional relations.

In the next section, we give the background for this paper. The interaction rules are introduced in section 3, which are used to implement our new path consistency algorithm in section 5 after some definitions and terminologies are prepared in section 4. In section 6, the computational complexity of consistency checking is analyzed, which is followed by the conclusion section.

2   Background

We first introduce the two formalisms of topological and cardinal directional relations, respectively. The region considered in this paper is a point-set homeomorphic to a unit disk in Euclidean space $\mathbb{R}^2$.

2.1   Topology Formalism

Topology is perhaps the most fundamental aspect of space. Topological relations are invariant under topological transformations, such as translation, scaling, and rotation. Examples are terms like *neighbor* and *disjoint* [6]. RCC8 is a formalism dealing with a set of eight jointly exhaustive and pairwise disjoint (JEPD) relations, called basic relations, denoted as DC, EC, PO, EQ, TPP, $NTPP$, $TPPi$, $NTPPi$, with the meaning of DisConnected, Extensionally Connected, Partial Overlap, EQual, Tangential Proper Part, Non-Tangential Proper Part, and their converses (see Fig.1). Exactly one of these relations holds between any two spatial regions. In this paper, we will focus on RCC8 formalism.

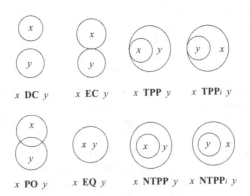

Fig. 1. Two-dimensional examples for the eight basic relations of RCC8

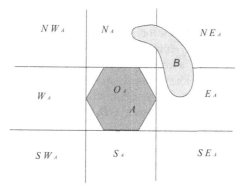

Fig. 2. Capturing the cardinal direction relation between two polygons, A and B, through the projection-based partitions around A as the reference object

2.2   Cardinal Direction Formalism

Goyal and Egenhofer [8] introduced a direction-relation model for extended spatial objects that considers the influence of the objects' shapes. It uses the projection-based direction partitions and an extrinsic reference system, and considers the exact representation of the target object with respect to the reference frame. The reference frame with a polygon as reference object has nine direction tiles: north (N_A), northeast (NE_A), east (E_A), southeast (SE_A), south (S_A), southwest (SW_A), west (W_A), northwest (NW_A), and same (O_A, i.e., the minimum bounding rectangle) (see Fig.2). The cardinal direction from the reference object to a target is described by recording those tiles into which at least one part of the target object falls. We call the relations where the target object occupies one tile of the reference object *single-tile* relations, and others *multi-tile* relations. We denote this formalism by CDF(Cardinal Direction Formalism) for brevity. It should be noted that Sharma [3] did not consider the kind of *multi-tile* relation and the intermediate relations, i.e., *NW, NE, SE* and *SW*.

3   Interaction Rules Between RCC8 and CDF

The internal operations, including converse and composition, on RCC8 can be found in [10]. The internal operations on CDF have been investigated in [9] and [11]. In order to integrate these two formalisms, we must investigate interaction rules between them. These rules are very useful to improve the spatial reasoning and can be the complement of the present composition tables. The spatial reasoning based on rules is more efficient and extended easily in the future as Sistla et al. [5] indicated.

The notation and representation of these rules are similar to [5], i.e. each rule will be written as $r :: r_1, r_2, \cdots, r_k$, where r is called the head of the rule, which is deduced by the list $r_1, r_2, \cdots, r_k$ called the body of the rule.

To facilitate the representation of the interaction rules, we denote a basic cardinal direction (i.e., single-tile or multi-tile relation) relation by a set SB, which includes at most nine elements, i.e. the nine single-tile cardinal direction relations. For example, a relation $O{:}S{:}SE{:}SN$ (multi-tile relation) can be denoted by $\{O,S,SE,SN\}$. The general cardinal direction relation (i.e., a basic cardinal direction relation or the disjunction of basic cardinal direction relations) can be regarded as a superset GB, whose element is the kind of set SB. So we have the relation: $SB \in GB$. The universal relation is the set $BIN = \{O, N, NE, E, SE, S, SW, W, NW\}$, and the universe, i.e. the set of all possible cardinal relations, is denoted by U.

Let A be a region. The *greatest lower bound* of the projection of region A on the x-axis (respectively y-axis) is denoted by $inf_x(A)$ (respectively $inf_y(A)$). The *least upper bound* of the projection of region A on the x-axis (respectively y-axis) is denoted by $sup_x(A)$ (respectively $sup_y(A)$). The *minimum bounding box* of a region A, denoted by $MBB(A)$, is the box formed by the straight lines $x = inf_x(A)$, $x = sup_x(A)$, $y = inf_y(A)$ and $y = sup_y(A)$. Based on these symbols, Skiadopoulos and Koubarakis [9] formally defined the cardinal directional relations.

Now, we present a system of rules for deducing new spatial relations from existing ones.

3.1   Rules for Deducing CDF Relations from RCC8 Relations (RCC8 → CDF)

Assuming that there exists some RCC8 relation between two regions A and B and we want to know the potential cardinal direction relations between them, we show the deduction rules in three cases and give their proofs if necessary.

Case 1. From the RCC8 relation A DC B, we can not specify the CDF relation between them, i.e.,

$$\text{A } U \text{ B} :: \text{A } DC \text{ B,} \tag{1}$$

where U is the universe of possible CDF relations between two non-empty and connected regions.

This rule is obvious, because the DC relation is the least restricted relation between two regions.

Case 2. Let x denote any relation symbol in $\{EC, PO, TPPi, NTPPi\}$. We have the following rule for each x. Because this rule is difficult to represent, we adopt first-order logic and the notations for CDF.

$$\forall SB \in GB(A, B), O \in SB \; :: \text{A } x \text{ B} \tag{2}$$

Proof. According to definitions for $EC, PO, TPPi$ and $NTPPi$ [7], A and B must have a common part. From B $\subseteq MBB(B)$, it follows that A and $MBB(B)$ must have a common part (i.e., $A \cap MBB(B) \neq \emptyset$). According to the definitions for relation O and multi-tile relation [9], region A must have a part which satisfies the relation O with respect to B.   □

Case 3. Let x denote any of the relation symbols in $\{TPP, NTPP, EQ\}$. We have the following rule for each such x.

$$A \; O \; B :: A \; x \; B \tag{3}$$

Proof. From the relation A x B, we have A$\subseteq$B. Hence A$\subseteq MBB$(B). According to the definition for CDF relation O [9], we conclude that the relation A O B holds. □

3.2   Rules for Deducing RCC8 Relations from CDF Relations (CDF → RCC8)

In this section, we will investigate the rules deducing RCC8 relation between any two regions A and B from the CDF relation between them in three cases.

Case 1. Let y denote any relation symbol in $\{DC, EC, PO, TPP, NTPP, EQ, TPPi\}$ (i.e., $\overline{NTPPi}$). We have the following rule.

$$A \; y \; B :: A \; O \; B \tag{4}$$

Proof. From the relation A O B and the definition in [9], we have A$\subseteq MBB$(B). we can construct a scenario where A$\subseteq MBB$(B) and A y B are simultaneously satisfied. We now prove the relation A $NTPPi$ B is impossible if A O B holds. According to definition for $NTPPi$ in [7], it is clear that there must be a part belonging to A which is outside of MBB(B). Hence the CDF relation between A and B must be a multi-tile one according to definition for multi-tile relation in [9]. So there is a contradiction. □

Case 2. Let x denote a cardinal direction relation which is a multi-tile relation at least including O and another single-tile relation, for example $\{O:N:NE\}$. Let y denote the relation set $\{DC, EC, PO, TPPi, NTPPi\}$, which means y can be anyone of these relations. We have the rule below.

$$A \; y \; B :: A \; x \; B \tag{5}$$

Proof. From the relation x, we know there must be a part of A in MBB(B), and another outside it. So any of the RCC8 relations $\{TPP, NTPP, EQ\}$ is impossible, because, if so, A will be contained in MBB(B). □

Case 3. Let x denote any of the cardinal direction relations which do not contain O. Another rule can be described as follows.

$$A \; DC \; B :: A \; x \; B \tag{6}$$

Proof. This rule is obvious. Because x does not contain relation O, we have A$\cap MBB$(B)=∅. Hence A$\cap$B=∅, it follows A DC B according to definition for RCC8 relation DC [7]. □

3.3   Rules for Deducing Relations from the Composition of RCC8 and CDF Relations (RCC8 ∘ CDF)

We will discuss these rules in three cases.

Case 1. Let x denote any of the relation symbols in $\{TPP, NTPP\}$, y any CDF relation and z the induced CDF relation. The rule is described as follows.

$$A \; z \; C :: A \; x \; B, B \; y \; C, \tag{7}$$

Where, if y is a single-tile CDF relation, z equals y, and if y is a multi-tile CDF relation, z is any subset of y.

Proof. From A$\{TPP, NTPP\}$ B, We know A⊆B. Hence, if B satisfies a single-tile CDF relation with respect to C, A must also satisfy it. Then it follows that A y C holds. We now consider the situation where y is a multi-tile CDF relation. According to definition for multi-tile relations [9], B can be regarded as consisting of several subregions which satisfy single-tile relations in y with respect to C, respectively. So region A can be one of, or consist several of these subregions. It follows that the relation z can be any subset of y.   □

For example, if there exists A *TPP* B and B *N:NE:E* C, we can deduce the new relation A N C, or A NE C, or A E C or A *N:NE* C, or A *NE:E* C. It should be noted that the relation A $N : E$ C does not exist, because the region considered in this paper is non-empty and *connected*.

Case 2. This rule is similar to the above except that x is anyone of the relation symbols in $\{TPPi, NTPPi\}$. So we have the relation A⊇B. It follows that the rule can be described as follows.

$$A \; z \; C :: A \; x \; B, B \; y \; C, \tag{8}$$

where z is any superset of y, i.e. y is the subset of z.

Case 3. This rule is obvious, so we present it directly.

$$A \; y \; C :: A \; EQ \; B, B \; y \; C \tag{9}$$

The rules for deducing RCC8 relations from the composition of RCC8 and CDF relations can be derived by combining the above rules (7)-(9) and rules (4)-(6).

3.4   Rules for Deducing Relations from the Composition of CDF and RCC8 Relations (CDF ∘ RCC8)

The rules are presented in three cases as follows.

Case 1. Let x denote any single-tile CDF relation and y denote the deduced CDF relation. The rule is described as follows.

$$A \; y \; C :: A \; x \; B, C \; \{TPP, NTPP\} \; B, \tag{10}$$

Where, if x is any of the relation symbols in {NW, NE, SE, SW}, y equals x, and if x is N (respectively S, E or W), y is any subset of {NW, N, NE} (respectively {SW, S, SE}, {NE, E, SE} or {SW, W, NW}).

Proof. To prove the first case, we take the relation NW for example. From the relation C {*TPP, NTPP*} B and definitions in [7], we have the following ordering relations: $sup_x(C) \leq sup_x(B)$, $inf_x(B) \leq inf_x(C)$, $sup_y(C) \leq sup_y(B)$ and $inf_y(B) \leq inf_y(C)$. From the relation A NW B, we can list the following ordering relations according to its definition [9]: $sup_x(A) \leq inf_x(B)$ and $sup_y(B) \leq inf_y(A)$.

From the above ordering relations and transitivity of $\leq$, we see that $sup_x(A) \leq inf_x(C)$ and $sup_y(C) \leq inf_y(A)$, which corresponds to the definition for relation A NW C [9]. The proof for NE, SE or SW is similar.

To prove the second case, we take the relation N for example. From the relation A N B, we have the following ordering relations according to its definition [9]: $sup_y(B) \leq inf_y(A)$, $inf_x(B) \leq inf_x(A)$ and $sup_x(A) \leq sup_x(B)$. From the above relations and transitivity of $\leq$, we see that $sup_y(C) \leq inf_y(A)$, which restricts the CDF relation between A and C to be any subset of {*NW, N, NE*} (i.e., N, or NW, or NE, or NW:N, or N:NE). The proof for S, W or E is similar. $\square$

Case 2. Using the above methods, we can also verify the following rule.

$$\text{A } y \text{ C} :: \text{A } x \text{ B, C } \{TPPi, NTPPi\} \text{ B,} \tag{11}$$

Where, if x is SW (respectively NW, NE or SE), y is any subset of {W, SW, S, O} (respectively {N, NW, W, O}, {N, NE, E, O}, or {E, SE, S, O}), and if x is N (respectively S, E or W), y is any subset of {N, O} (respectively {S, O}, {E, O} or {W, O}).

For example, if we have the relations A N B and C $TPPi$ B, we can conclude that the possible relation between A and C is A N B, A O B or A $N : O$ B, i.e., all the subsets of {N, O}.

Case 3. Let x denote any CDF relation. This rule is obvious. We just describe it directly as follows.

$$\text{A } x \text{ C} :: \text{A } x \text{ B, B } EQ \text{ C} \tag{12}$$

The rules for deducing RCC8 relations from the composition of CDF and RCC8 relations can be derived by combining the above rules (10)-(12) and rules (4)-(6).

3.5   Composite Rules

The advocation of the rules in this section is motivated by such situations where given the relations A N B, B PO C, C N D, what is the relation between A and D? We can not find the answer using the above rules and we should find more powerful rules.

Sharma [3] verified and extended [12]'s inference rule:

$$\text{A } x \text{ D} :: \text{A } x \text{ B, B } y \text{ C, C } x \text{ D} \ .$$

In this paper, we adapt this rule to our model and investigate its properties. Let R denote any of the RCC8 relation symbols in $\{EC, PO, TPP, NTPP, TPPi, NTPPi, EQ\}$, x and y denote any single-tile CDF relation and z denote the deduced CDF relation, respectively. These rules are discussed in three cases.

Case 1.

$$A \; z \; D :: A \; x \; B, B \; R \; C, C \; y \; D, \tag{13}$$

where x is N (respectively S, W, or E), y is any of the relation symbols in $\{NW, N, NE\}$ (respectively $\{SW, S, SE\}$, $\{NW, W, SW\}$, or $\{NE, E, SE\}$) and then z is any subset of $\{NW, N, NE\}$ (respectively $\{SW, S, SE\}$, $\{NW, W, SW\}$, or $\{NE, E, SE\}$).

Proof. When x is *NW* and y is *NW*, we have the relations A *N* B, B *R* C and C *NW* D. From A *N* B and the definition for relation N [9], we have the following ordering relations: $sup_y(B) \leq inf_y(A)$, $inf_x(B) \leq inf_x(A)$ and $sup_x(A) \leq sup_x(B)$.

From C *NW* D and the definition for relation *NW* [9], we have the following ordering relations: $sup_x(C) \leq inf_x(D)$ and $sup_y(D) \leq inf_y(C)$.

From B *R* C, we know that $B \cap C \neq \emptyset$. So let p be an arbitrary point in $B \cap C$. p_x is its x-coordinate and p_y its y-coordinate, respectively. So, p satisfies the following ordering relations. $inf_x(B) \leq p_x \leq sup_x(B)$, $inf_x(C) \leq p_x \leq sup_x(C)$, $inf_y(B) \leq p_y \leq sup_y(B)$ and $inf_y(C) \leq p_y \leq sup_y(C)$.

From the above ordering relations and transitivity of $\leq$, we have the resulting ordering relation $sup_y(D) \leq inf_y(A)$, which means the possible relations between A and D can be A *N* D, A *NW* D, A *NE* D, A *N:NW* D or A *N:NE* D, i.e., all the subsets of $\{NW, N, NE\}$. When y is *N* or *NE*, the same result can be derived. Other cases can be proved similarly. □

Using the above methods, we can validate the following two rules.

Case 2.

$$A \; z \; D :: A \; x \; B, B \; R \; C, C \; y \; D, \tag{14}$$

where x is any of the relation symbols in $\{NW, NE\}$ (respectively $\{SW, SE\}$, $\{NW, SW\}$, or $\{NE, SE\}$), y is N (respectively S, W, or E) and then z is any subset of $\{x, N\}$ (respectively $\{x, S\}$, $\{x, W\}$, or $\{x, E\}$), i.e., when x is NE and y is N, then z is any subset of $\{NE, N\}$.

Case 3.

$$A \; z \; D :: A \; x \; B, B \; R \; C, C \; y \; D, \tag{15}$$

where x is NW (respectively SW, NE, or SE), y equals x, and then z is NW (respectively SW, NE, or SE).

4   Terminologies and Definitions

Definition 1. *Binary Constraint Satisfaction Problem (BCSP)*
If every one of the constraints in a Constraint Satisfaction Problem (CSP) involves two variables (possibly the same) and asserts that the pair of values assigned to those variables must lie in a certain binary relation, then the constraint satisfaction problem is called Binary Constraint Satisfaction Problem.

Definition 2. *We define an RCC8-BCSP as a BCSP of which the constraints are RCC8 relations on pairs of the variables. The universe of a RCC8-BCSP is the set $\mathbb{R}^2$ of regions anyone of which is a point-set homeomorphic to a unit disk. Similarly we can define CDF-BCSP as a BCSP of which the constraints are CDF relations on pairs of the variables and the universe is the set $\mathbb{R}^2$ of regions anyone of which is a point-set homeomorphic to a unit disk, and RDF-BCSP as a BCSP of which the constraints consist of a conjunction of RCC8 relations and CDF relations on pairs of the variables and the universe is the set $\mathbb{R}^2$ of regions anyone of which is a point-set homeomorphic to a unit disk.*

A binary constraint problem with n variables and universe U can be simply viewed as an n-by-n matrix M of binary relations over U: the relation M_{ij} (in row i, column j) is the constraint on $< x_i, x_j >$.

Let M and N be n-by-n matrices of binary relations. We have definitions as follows:

Definition 3.

$$(M \circ N)_{ij} = (M_{i0} \circ N_{0j}) \cap (M_{i1} \circ N_{1j}) \cap ... \cap (M_{in-1} \circ N_{n-1j}) = \underset{k<n}{\cap} M_{ik} \circ N_{kj} \quad .$$

Let $M^2 = M \circ M$.

Definition 4. *An n-by-n constraint matrix M is path-consistent if $M \leq M^2$.*

M is path-consistent just in case $M_{ij} \subseteq M_{ik} \circ M_{kj}$. We must note that path consistency is the necessary, but not sufficient, condition for the consistency of a BCSP.

5   Path Consistency in RDF-BCSP

To enforce the path consistency in RDF-BCSP, we must consider the interactions between the RCC8 component and CDF component in RDF-BCSP in addition to the internal path consistency in RCC8-BCSP and CDF-BCSP, respectively.

We devise a constraint propagation procedure *Dpc()* for enforcing path consistency in RDF-BCSP, which is adapted from the path consistency algorithm described in [4]. Our algorithm employs two queues RCC8-Queue and CDF-Queue, which are initialized to all pairs (x, y) of the RCC8-BCSP and CDF-BCSP variables, respectively, verifying $x \leq y$ (the variables are supposed to be ordered). The algorithm removes pairs of variables from the two queues in parallel or in turn. When a pair $\langle X, Y \rangle$ of variables of RCC8-BCSP (respectively CDF-BCSP) is removed from RCC8-Queue (respectively CDF-Queue), firstly the RCC8 (respectively CDF) relation on $\langle X, Y \rangle$ is converted to the CDF (respectively RCC8) relation on $\langle X, Y \rangle$ according to the rules (1)-(3) (respectively (4)-(6)). If the resulting CDF (respectively RCC8) relation on $\langle X, Y \rangle$ is different from the original relation on $\langle X, Y \rangle$, the pair of variables will be entered to the CDF-Queue (respectively RCC8-Queue); Then this CDF (respectively RCC8) relation on the pair $\langle X, Y \rangle$ is used to update the CDF (respectively RCC8) relations on the neighboring pairs of variables (pairs sharing at least one variable) according to the prerequisites in the rules provided by section 3. If a pair is successfully updated, it is entered into RCC8-Queue (respectively CDF-Queue), if it is not already there, in order to be considered at a future stage for propagation. This propagation procedure is common with Allen's algorithm, what's different is that the RCC8 (respectively CDF) relation on every pair of variables will be used to refine the relevant relations according to these rules provide by section 3.

The algorithm loops until it terminates if the empty relation, indicating inconsistency, is detected, or if RCC8-Queue and CDF-Queue become empty, indicating that a fixed point has been reached and the input RDF-BCSP is made path consistent.

Theorem 1. *The constraint propagation procedure Dpc() runs into completion in $O(n^3)$ time, where n is the number of variables of the input RDF-BCSP.*

Proof. The number of variable pairs is $O(n^2)$. A pair of variables may be placed in queue at most a constant number of times (8 for a pair of RCC8 variables, which is the total number of RCC8 atoms; and 218 for a pair of CDF variables, which is the total number of CDF basic cardinal direction relations. Every time a pair is removed from queue for propagation, the procedure performs $O(n)$ operations. □

6   Complexity of Consistency Checking in RDF-BCSP

We use T to denote the set of general RCC8 relations, T_b the set of basic RCC8 relations including universal relation, C the set of general CDF relations and C_b the set of basic CDF relations in *RDF-BCSP*.

The work by Renz and Nebel [13] and Renz [14] identified three maximal tractable subsets, i.e. $\mathcal{C}_8$, $\mathcal{Q}_8$ and $\hat{\mathcal{H}}_8$, of the relations in RCC8 that contains all basic relations and showed that path-consistency is sufficient for deciding consistency for BCSPs based on these subsets. Skiadopoulos and Koubarakis [15] has presented the first algorithm for checking the consistency of a set of cardinal direction constraints and proved that the consistency checking of a set of basic cardinal direction constraints can be performed in $O(n^5)$ time while the consistency checking of an unrestricted set of cardinal direction constraints is NP-complete.

Theorem 2. *The complexity of checking the consistency of RDF-BCSP based on set $S = T_b \cup C_b$ is $O(n^5)$.*

Proof. Checking the consistency of RCC8-BCSP based on T_b is polynomial [13], and checking the consistency of CDF-BCSP based on C_b has been shown to be $O(n^5)$[15]. From the rules (4), (5) and (6), all possible CDF basic relations can only entail $DC \vee EC \vee PO \vee TPP \vee NTPP \vee EQ \vee TPPi$,

$DC \lor EC \lor PO \lor TPPi \lor NTPPi$ or DC relations, which belong to the maximal tractable subset $\hat{\mathcal{H}}_8$ (see Appendix B of [13]) of RCC8. So checking the consistency of RDF-BCSP based on the union of T_b and C_b is polynomial. First we can run the improved constraint propagation algorithm to enforce the path consistency, and then the algorithm in [15] will be employed to check the consistency for CDF-BCSP component. Obviously the complexity is $O(n^5)$.   □

Because checking the consistency of RCC8-BCSP based on the set T is **NP**-Complete (see theorem 22 of [13]) and checking the consistency of CDF-BCSP based on the set C is **NP**-Complete (see theorem 3 of [15]), we have the following corollary:

Corollary 1. *Checking the consistency of RDF-BCSP based on the set $S = T \cup C_b$ or $T_b \cup C$ or $T \cup C$ is **NP**-Complete.*

7   Conclusions

In this paper, we have combined two essential formalisms in qualitative spatial reasoning, i.e., RCC8 and cardinal direction formalism. The interaction rules have been given and they can be embedded into the propagation algorithm to enforce the consistency of BCSP based on the new hybrid formalism, and then the results for the complexity of checking consistency based on various subsets of this new formalism are given. The complexities for other combinations of formalisms in QSR should be investigated in the future, and the modeling and computational problems in Fuzzy QSR should be also interesting.

References

1. A. Gerevini and J. Renz. Combining Topological and Size Constraints for Spatial Reasoning. *Artificial Intelligence (AIJ)*, vol. 137(1-2): 1-42, 2002
2. A Isli, V Haarslev and R Moller. Combining cardinal direction relations and relative orientation relations in Qualitative Spatial Reasoning. Fachbereich Informatik, University Hamburg, Technical report FBI-HH-M-304/01, 2001
3. J. Sharma. Integrated spatial reasoning in geographic information systems: combining topology and direction. Ph.D. Thesis, Department of Spatial Information Science and Engineering, University of Maine, Orono, ME, 1996
4. James F. Allen. Maintaining knowledge about temporal intervals. *Communications of the ACM*, vol. 26(11): 832-843, November, 1983
5. A. Prasad Sistla, Clement T. Yu and R. Haddad. Reasoning About Spatial Relations in Picture Retrieval Systems. In: *20th International Conference on Very Large Data Bases*, pp. 570-581, Morgan Kaufmann, 1994
6. M. Egenhofer. A Formal Definition of Binary Topological Relations. In *Third International Conference on Foundations of Data Organization and Algorithms (FODO)*, Vol. 367, pp. 457-472, Paris, France: Lecture Notes in Computer Science, Springer-Verlag, 1989
7. Randell D, Cui Z and Cohn A. A spatial logic based on regions and connection. In: Nebel B, Rich C, Swartout W, (eds.) *Proc. of the Knowledge Representation and Reasoning*, pp. 165~176, San Mateo: Morgan Kaufmann, 1992

8. R. Goyal and M. Egenhofer. Cardinal Directions between Extended Spatial Objects. *IEEE Transactions on Knowledge and Data Engineering* (to be published), 2000

9. S. Skiadopoulos and M. Koubarakis. Composing cardinal direction relations. *Artificial Intelligence*, vol. 152(2): 143—171, 2004

10. D. A. Randell, A. G. Cohn and Z. Cui. Computing Transitivity Tables: A Challenge For Automated Theorem Provers. In *11th International Conference on Automated Deduction*, pp.786-790, Berlin: Springer Verlag, 1992

11. Serafino Cicerone and Paolino Di Felice. Cardinal directions between spatial objects: the pairwise-consistency problem. *Information Sciences*, vol. 164: 165-188, 2004

12. A. Prasad Sistla, Clement T. Yu, and R. Haddad. Reasoning About Spatial Relations in Picture Retrieval Systems. *International Journal on Very Large Databases (VLDB)*, vol. 3(4): 570-581, 1994

13. J. Renz and B. Nebel. On the Complexity of Qualitative Spatial Reasoning: A Maximal Tractable Fragment of the Region Connection Calculus. *Artificial Intelligence (AIJ)*, vol.108(1-2): 69-123, 1999

14. J. Renz. Maximal Tractable Fragments of the Region Connection Calculus: A Complete Analysis. In: *16th International Joint Conference on Artificial Intelligence (IJCAI'99)*, pp. 448-455, Stockholm, Sweden, August, 1999

15. S. Skiadopoulos and M. Koubarakis. Qualitative spatial reasoning with cardinal directions. In: *7th International Conference on Principles and Practice of Constraint Programing (CP'02)*, Vol.2470, pp. 341-355, in Lecture Notes in Comput. Sci., Springer, Berlin, 2002

Key Aspects of the Diagen Conceptual Model for Medical Diagnosis

Rafael Martínez, José Ramón Álvarez Sánchez, and José Mira

Dpto. Inteligencia Artificial. ETSI Informática,
Universidad Nacional de Educación a Distancia, Madrid, Spain
{rmtomas, jras, jmira}@dia.uned.es

Abstract. As a result of the search for recurring elements in medical diagnosis, components of an ontology aspiring for generic medical diagnosis have been presented in earlier works and in this work, which allows semiautomatic translation of the natural language expert's description to its implementation. This article presents the most important aspects of the conceptual model, in particular the suggestion and confirmation relations are described that are considered central to the diagnosis. The characteristics of these relations have been exemplified in the diagnosis of colon cancer.

1 Introduction

In different works [4, 5, 11], we have looked for structured natural language segments for establishing correspondences with the primitive inferences of the different problem solving methods (PSMs) ("establish and refine", "propose-act-modify", "cover and differentiate", ...). As a continuation of these works and following a complementary line, in [6, 8], and within the Diagen project [7, 9], we have presented a simple relational model based on the natural language description of the tasks, which looks for recurrent abstractions in the magnitudes and relations (nouns and verbs) that the expert uses in the causal chain of his/her reasoning (formal implication in the model). We have explored the alternative of looking for reusable components in the higher part of the domain knowledge layer (primitives representing the relational part of the domain model), linking knowledge elicitation with implementation. This model, from the most abstract point of view, consists of two main kinds of reusable components: *magnitudes* and *relations*. *Magnitudes* play origin and/or destination *roles* in *relations*. *Relations* define differentiated conditions on origin and destination *magnitudes* and action in this model.

Magnitudes represent separate concepts, which can be measured or evaluated, and which the expert considers necessary and sufficient to describe his/her knowledge. They constitute the universe of discourse for relations and are associated with names whose referent can be physical or mental. *Magnitudes* play the same role as the physical magnitudes in an analytic model.

A *fact* defines the knowledge about a temporal state of a magnitude, thus it can be represented as a magnitude instance. Both concepts are related by an

J. Mira and J.R. Álvarez (Eds.): IWINAC 2005, LNCS 3561, pp. 481–488, 2005.

instantiation association (but conceptually, this association is not exactly the instantiation association as could be interpreted in object orientation).

Magnitudes represent the facts, while relations are concerned with the links in the chain of causal reasoning that the expert uses and they are generally represented by verbs like "suggests", "is-confirmed-by", "is-composed-of", etc. These entities and relations form a relational network between concepts or a semantic network which gives rise to two possible forms of PSM behavior: from a domain structure prepared for a specific PSM or design of the structure from scratch where the expert does not need to know that he/she is using a specific PSM in his/her daily clinical practice. Thus instead of a PSM library, we have a library of relations, roles and magnitudes. Thanks to the preservation of the structure [1] of the different ontology elements [3, 2], only their semantic table changes in the reduction process.

Later works followed the line of exemplifying the proposals and developing a software environment for the elicitation and reduction of medical knowledge models in diagnostic tasks and therapy planning. The software environment has a library of these components so that the expert can design his/her own network in the domain layer and it is from this network that the right PSM behavior emerges. It makes it easier for the medical expert to first construct the model of the task from the reusable modelling components independently of the implementation. At the same time representation and implementation elements are provided within the same software environment, corresponding to the modelling elements, so that the medical expert also makes the model operational and fully develops his/her own diagnostic tool.

This article thus presents the most important aspects of the conceptual model defined on the ontology of reusable components from the Diagen project. It has been exemplified in the diagnosis of colon cancer, thanks to the works with the oncology department at *Hospital Clínico Universitario de Madrid*.

2   Key Relations in the DIAGEN Model

Both in earlier works and works for this article, the key relations of our diagnosis proposal are suggestion and confirmation. The suggestion relation links a superficial alteration with a deeper anomaly which can be its origin (or participate in its origin) (Fig. 1). According to the ontology of the DIAGEN medical domain, alterations from three planes play a role in this relation: findings (observations whether from anamnesis, clinical exploration or complementary tests, which are significant), whether individual or syndrome (the introduction of syndromes as specific accumulations or conjunctions of findings is usual), physiopathological processes and pathologies (like the differentiation established in CASNET [12]). The basic structure of the *suggests* chain would only present interplanes, i.e. a finding (simple or syndrome) suggests some physiopathological processes and a physiopathological process suggests in turn some specific pathologies. Although the colon cancer domain adapts perfectly to this structure, in order to generalize, we must put a *suggests* chain within each of the planes, so that a physiopathological process can suggest another physiopathological process. Within the plane of physiopathological processes and a pathology another pathology can be suggested, i.e. from a level of higher causal depth. The *suggests* chain ends in a suggested pathology, whose confirmation would allow a therapeutic action to be determined.

Figure 1 shows the usual schematic representation in CommonKads [10] modelling of the chain of the origin and destination roles of the suggestion relation, labelled with the verb *suggest* on the left. An example of a diagnosis of cancer colon is shown; an acute abdominal pain (*finding*) suggests an ulcerated injury in the colon and suggests an ulcerated injury in the rectum (*physiopathological process*).

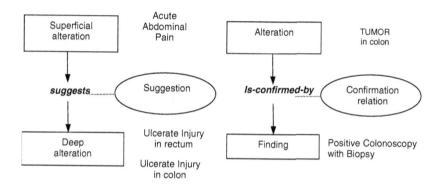

Fig. 1. Left. Chain of the origin and destination roles of the suggestion relation, labelled with the verb *suggest*. We term the origin role *superficial alteration* and the destination role *deep alteration*. As an example in the colon cancer domain, an acute abdominal pain suggests an ulcerated injury in the colon and an ulcerated injury in the rectum. On the right is the schematic representation for describing the confirmation relation between the origin role, *alteration*, and destination role, *finding*, labelled as *is-confirmed-by*. As an example, the alteration role is played by the diagnosis *tumor in colon* and the finding role is played by the conjunction of *positive colonoscopy with biopsy*

It is obviously a causal relation expressed inversely, prepared for a direct monitoring of the relation, from the effect to the cause, not for an abductive monitoring from the cause to the effect, as is usually done.

The suggestion relation of more or less deep alterations connects with the confirmation relation, between alterations and findings. On the right of Figure 1, the connection between an origin role, *alteration*, and a destination role, *finding*, with the confirmation relation labelled as *is-confirmed-by* is shown. It is exemplified with the connection between the alteration *tumor in colon* with the finding *Positive Colonoscopy with Biopsy.*

Figure 2 shows a chain *suggests* where from a finding at a superficial level (*acute abdominal pain*) the suggestion of some pathologies is reached (*tumor*

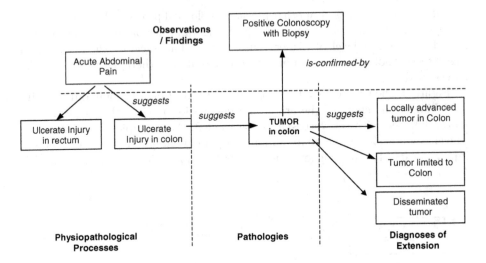

Fig. 2. Example of partial relation network in the colon cancer diagnostic domain. The chain of suggestions is shown from a more superficial level (findings) to a level of diagnostic depth (diagnoses of extension) whose confirmation allows determination of the therapeutic action

locally advanced in colon, tumor limited to colon, disseminated tumor) of the last and lowest level of diagnosis, i.e. the level that determines a therapy.

The alteration in the origin role of a *suggests* must have been suggested previously. Moreover, according to the chain of the two relations *suggests* and *is-confirmed-by*, we find two kinds of *suggests*: which require prior confirmation of the superficial alteration (origin role) or not. Thus, while in some cases the suggestion connection requires the alteration origin to be confirmed, there are other cases where this is not so, since *alterations* are introduced into the chain exclusively for explanatory purposes, but not because of a strict diagnostic need. This occurs especially with the physiopathological processes that play the role of *alterations*, although also in some intermediary diagnoses. It is the case, for example, in Figure 2, since the suggestion of tumor in colon from *ulcerate injury in colon* requires immediate confirmation (before continuing to study in detail at lower causal levels), there must be a connection between this alteration and the findings confirming it. In the example in Figure 3, on the other hand, suggestions are linked without a confirmation, from the *obstruction in the lower cavity* via the *tumor in colon* to a *disseminated tumor*, which does require its confirmation via *CAT of abdomen shows metastasis* or *X-ray of thorax shows metastasis*.

If we wish to generalize further (Fig. 4 left), the connection of an *alteration* with the corresponding *finding* using the connection *is-confirmed-by* depends on the linked *alteration* using *suggests* from the former level (Al-). This occurs particularly more at the stage of principal diagnosis, where the alterations are played by physiological causes in the therapeutic monitoring which includes a diagnosis-assessment of the state of evolution of the illness. Yet this state is an

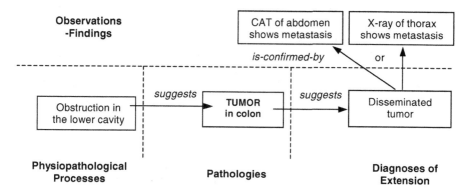

Fig. 3. In this example, suggestions are linked via the *obstruction in the lower cavity* and *tumor in colon,* without any confirmation, before reaching a *disseminated tumor.* This pathology is confirmed with the existence of at least one of the findings: *CAT of abdomen shows metastasis* and *X-ray of thorax shows metastasis*

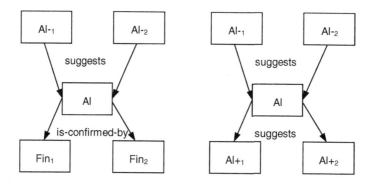

Fig. 4. Left: The connection of an *alteration* (Al) with the corresponding *finding* (Fin) via the connection *is-confirmed-by* depends in turn on the linked *alteration* via *suggests* in the previous level (Al-). Right: The connection *suggests* of an *alteration* (Al) with the corresponding alteration in the following lower level (Al+) depends in turn on the linked alteration of the previous level (Al-)

assessment within a typology and its confirmation depends on what the evolution was before the illness. For example, an Al state, suggested from a limited tumor as a principal diagnosis, is going to require confirmation from a surgery controlling it with a positive result However, if it has been suggested from a relapse it will have to be confirmed with different complementary tests with a normal result (Scan or abdomen CAT, analytical, radiography of thorax and colonoscopy).

Another characteristic of *suggests* is that the connection of an *alteration* (Al) with the corresponding alteration in the following lower level (Al+) depends on the alteration linked with the former level (Al-). This characteristic is shown on the right of Figure 4 and therefore affects the internal structure of the chain. The example that we give is the connection between the constitutional syndrome (Al_{-1}) with the existence of a tumor in the colon (Al), within the framework of other findings that have directed the patient to intestinal diagnosis, and more particularly, the existence of distance metastasis (Al_{+1}). Conversely, the non-existence of the constitutional syndrome (Al_{-2}), together for example with an ulcerated injury in the colon, suggests a cancer in the colon, but this will require immediate confirmation and also suggests the possibility of the tumor being located (Al_{+2}).

3   Inference as a Selection on DIAGEN Relations

In the previous section the key aspects of the nucleus of the Diagen relational network were presented for diagnosis and therapeutic monitoring. In this section we will show the basic navigation elements, the inferences, on these relations. Inferences are defined as very simple, a selection of the associated entities or rather a navigation on the chains.

Thus, key selection inferences navigate on the relations *suggests* and *is-confirmed-by*. In the first case, *select* links the domain role *claim* with the domain role *hypotheses*, while in the second case, *select* links *hypotheses* with *confirmation-findings*.

The input role *claim* (Fig. 5) to *select-1* is played by alterations, whether they are findings, physiopathological processes or pathologies. In any case the confirmed origin of *suggests* relations or in specific cases merely suggested, as seen in the previous section. The selection obtains *hypothesis*, the role played by the suggested alterations, and therefore at a greater level of diagnostic depth, which requires confirmation. *Select-2* selects from these hypotheses the confirmation findings on the relation *is-confirmed-by*. Obviously, from the selected findings the necessary observables are presented to the user: clinical or complementary tests.

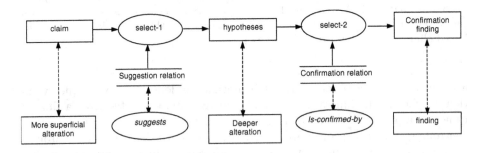

Fig. 5. At the inferential level, a first inference *select* obtains the hypotheses from a claim requiring confirmation. This inference is based on the suggestion relation between alterations. From the hypotheses selected a second *select* obtains the confirmation findings. In this instance the inference is based on the confirmation relation between the alterations of a specific depth level requiring confirmation and the relevant findings

4   Conclusion

From the work developed in the Diagen project with different medical experts, especially in the area of diagnosis and treatment of colon cancer, a model has been defined that can be applied to other medical specialties. In other works aspects of the Diagen results have been shown to be closer to the software implementation. Here aspects of the most abstract model are presented, focusing on the suggestion relations, between more superficial alterations and deeper alterations, and confirmation relations between the suggested alterations and the findings confirming them.

The fundamental result is that there is an ontology with immediate computational translation. Either the knowledge engineer from the medical expert's description in natural language, or the medical expert, can represent domain knowledge as the composition of a network between a limited number of relational verbs (arcs), of representative terms of physiopathological alterations of different levels, from the symptom to the deep pathology, and of possible results of confirmation tests (nodes). Executability is guaranteed in this network, since another of the project's results has been the implementation of the corresponding inferential level as a motor for diagnosis, which attaches a relational network edition module.

References

1. R. Benjamins and M. Aben. Structure-Preserving KBS Development through Reusable Libraries: a Case-Study in Diagnosis. *International Journal of Human-Computer Studies*, 47:259–288, 1997.
2. N. Fridman Noy and C. Hafner. The State of The Art in Ontology Design. *AI Magazine*, 4:53–74, 1997.
3. N. Guarino. Understanding, building and using ontologies. *International Journal of Human-Computer Studies*, 46:293–310, 1997.
4. J.C. Herrero and J. Mira. In Search of a Common Structure Underlying a Representative Set of Generic Tasks and Methods: The Hierarchical Classification and Therapy Planning Cases Study. In J. Mira, A. P. del Pobil, and A. Moonis, editors, *Methodology and Tools in Knowledge-Based Systems*, volume 1415 of *LNAI*. Springer-Verlag, Berlin, 1998.
5. J.C. Herrero and J. Mira. SCHEMA: A knowledge edition interface for obtaining program code from structure descriptions of PSMs: Two case studies. *Applied Intelligence*, 10:139–153, 1999.
6. R. Martínez-Tomás. *Modelado y reducción del Conocimiento basados en la Teoría de Sistemas: Aplicaciones al diagnóstico Médico*. PhD thesis, Facultad de Ciencias. UNED, 2000.
7. J. Mira. Report of project Diagen. Technical Report TIC 97-0604, CICYT, 1997.
8. J. Mira, J. R. Alvarez, and R. Martínez. Knowledge Edition and reuse in DIAGEN: A relational approach. *IEE Proceeding-Software*, 147(5):151–162, 2000.
9. J. Mira, R. Martínez, J. R Álvarez, and M. J. Taboada. DIAGEN: A Software Environment for the Reduction of Generic Models of Medical Knowledge in Diagnosis and Therapy Planning. In *Jornadas de Seguimiento CICYT-TIC-Info*, Valladolid (Spain), 10 November 2000.

10. G. Schreiber, H. Akkermans, A. Anjewierden, R. de Hoog, N. Shadbolt, W. Van de Velde, and B. Wielinga. *Knowledge engineering and Management: The CommonKADS Methodology*. MIT Press, 1999.
11. M. Taboada, M. Lama, S. Barro, R. Marín, J. Mira, and F. Palacios. A Problem-Solving Method for "Unprotocolised" Therapy Administration Task in medicine. *Artificial intelligence in medicine*, 17:157–180, 1999.
12. S. M. Weiss, C. A. Kulikowski, S. Amarel, and A. Safir. A model-based method for computer-aided medical decision-making. *Artificial intelligence*, 11:145–172, 1978.

Connectionist Contribution to Building Real-World Ontologies

Miłosław L. Frey

FGAN — FKIE,
Neuenahrer Straße 20,
53343 Wachtberg, Germany
m.frey@fgan.de

Abstract. The paper at hand presents an unsupervised connectionist network using spreading activation mechanism. By means of self-organization, the network is capable of creating a taxonomy of concepts which serves as a backbone for a respective ontology. The system is a biologically inspired constructivist hybrid between connectionist networks using distributed and localist data representation. Unlike most currently developed models it is capable to deal with analog signals and displays cognitive properties of categorization process. The paper at hand presents the general overview over the system's architecture and method of network build-up and shows results of several experiments exploring the nature of categorization performed with the use of the described network.

1   Introduction

This paper presents a semantic network that implements the spreading activation mechanism. The novel feature is the use of a data representation which is a mix of localist and distributed ones. Additionally, a biologically inspired internal architecture of a single network node allows for classification of non-binary features.

Beside classification task ability the network has features of auto associative networks in the sense that it can perform a completion of a concept's description and reduce noise in the input pattern. In some cases of vague or not clear enough definitions, it is able to generalize to superclasses even if no object belonging to them was previously presented. Due to the capacity for classification of non-binary stimuli the network exposes also properties of fuzzy classification as well as priming and asymmetric category learning.

The paper is organized as follows. Section 2 outlines the network's architecture, components and working principle. Section 3 describes in more details the learning and taxonomy creation procedure, and section 4 deals with cognitive features of the presented model. The concluding section 5 provides more comments and summarizes the paper.

J. Mira and J.R. Álvarez (Eds.): IWINAC 2005, LNCS 3561, pp. 489–497, 2005.

2   Network's Architecture and Data Representation

The network to be discussed consists of two kinds of nodes and two kinds of connections. Nodes can be divided into feature nodes and class nodes while the connections are either excitatory or inhibitory ones.

Data presented to the system consists of "definitions". These definitions are sets of features accompanied by the degree (expressed by a real number) in which the respective feature is present in the described class. These definition sets constitute descriptions of classes which have their characteristics known for example from experience or other data source in contrast to classes created by the system in further processing phases. The data presented to the network is stored in its structure and serves as a source for further operations.

2.1   A Node

The operation principle of single nodes bases on the finding that neurons are divided into subregions which are able to perform complex computations on incoming signals (cf. Koch et al. [1] and [2]; Mel [7]). While a single real neuron can be seen as an analogue to the whole "classical" distributed connectionist network, it is justified to assume that it can perform virtually any operation on the input signal. Thus in the presented network the node processes signals coming from its parent nodes in a different way than signals received from child nodes. The general internal structure of a node is sketched on the Fig. 1. In the following the signal processing within one node is shortly outlined.

Signals from Parent Nodes. Signals coming from parent nodes are processed in a way similar to calculating a distance in the multi dimensional space. The dimensions' number is defined by the number of incoming connections. Additionally, the weights of those connections set up a point in this space. The node is then able to calculate a function of an Euclidean distance between the point representing the incoming signal (defined by activations of parent nodes connected to it) and the point set up by the weight values. Finally, an activation function is applied which calculates the activation coming from parent nodes basing on current input signal and the node's previous activation (cf. the reciprocal dotted link from activation buffer in the Fig. 1.)

The resulting part of activation coming from parent nodes expresses the difference between the incoming signal and a signal to which a node is most sensitive as well as a kind of history of previous input signals.

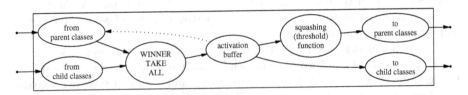

Fig. 1. Internal structure of a node

Signals from Child Nodes. Signals from child nodes are calculated simply as a mean of child nodes' activations weighted by connections' strengths.

Final Activation Function. Activation parts coming from parent and child nodes are finally subject to a winner-take-all process. Further calculations differ between connections going in the direction of feature nodes or of class nodes. For connections going to class nodes a squashing function is applied which keeps the final value in a range between 0.0 and 1.0. On the other hand, for connections going in the direction of feature nodes, a threshold function is used.

2.2 Connections

Nodes are connected by links with symmetric weights. There are both excitatory and inhibitory ones. Excitatory connections form a structure of taxonomy while inhibitory ones take care of increasing differences between different classes.

3 Learning

The presented network incorporates three kinds of learning [12].

1. Rote learning (storing data in network's structure). This kind of learning is used to store input data in the network. It is comparable to the long-term memory.
2. Connection weight changes. Changes of connection weights are means to create the working structure of the network; the taxonomy itself.
3. Restructuring (by creating nodes and connections) takes place also in the development of the taxonomy. The newly added nodes denote taxonomy classes.

3.1 Creating a Taxonomy

Step 1: Remembering Data. In the first step the presented data is remembered only (stored). For each dataset feature nodes are created when necessary. Additionally, class nodes are created denoting the respective set of all co-occurring features. Because the features are characterized not only by their presence but also by a degree of this presence (value), the features are said to co-occur only when their values equal with a given precision. Between class nodes and feature nodes excitatory connections are created with weights corresponding to the values of respective features.

Step 2: Creating a Raw Taxonomy. Based on data stored in already created feature and class nodes the hierarchy is created. For each pair of class nodes both nodes are subsequentially activated, the activation is spread to the feature nodes layer and the activation patterns are compared. If one of the nodes generates an activation pattern comprised in the other one's pattern, it is assumed to be its' superclass. This principle bases on the simple assumption that a subclass contains all features of its superclass and at least one more, a distinctive one. The comparison of patterns is performed with a given precision in order to gather classes characterized by features which values do not differ significantly. At this stage, a network is formed which contains many superfluous excitatory connections that do not represent direct class — superclass relations. Those connections are removed by an introspective process. This process analyzes the activation flow between two nodes. Based on activation value comparison the decision is made whether two nodes remain in direct class — superclass relation or not. Subsequently inhibitory connections are introduced to enhance differences between exemplars presented to the system.

Step 3: Discovering Similarities. The network constitutes representation for raw facts known from input data. This representation is structured only as far as it is provided by this data. That means that relations between classes are known only if they result from definitions.

The next step toward a "better" taxonomy is to discover parts of the hierarchy which were not provided explicitly. This is achieved by analyzing pairs of exemplars. Again pairs of class nodes are analyzed with respect to the featural patterns. Those which are common to them, form new descriptions.

4   Properties of the Network

This section presents the overview over the properties of the presented network. It starts from simple properties of the network itself and leads to the cognitive properties of the network's performance.

4.1   Description Autocompletion and Noise Reduction

The network manifests autoassociation properties. Given a part of a description (an incomplete set of features) it attempts to complete it in the best possible way. That means it recalls the features set (in a way that it activates corresponding nodes) for which the given features are most distinctive. This is possible even if only one feature is given as long as this feature is distinguishing.

The noise reduction property is displayed in a case when the network is fed with data containing features that should not co-occur with other ones. In case the other features define the class to a sufficient extent, the corrected pattern is reproduced and the activation of the "noise" features is canceled.

4.2   Generalization

Generalization takes place when the given data contain features that contradict each other on some level but correspond to nodes in the same taxonomy branch. In this situation a network converges to the feature pattern describing the superclass which is common for all nodes partially described by given data. In other words the network finds the more general term to describe the presented set of features.

For this mechanism, the discovery of common features is crucial. It allows for correct classification of novel datasets. Moreover, the system may enhance the taxonomy by the newly encountered object in case its featural description is verified.

4.3   Family Resemblance

In the resulting network not all members of a category must have features common to the other members. Membership is based on family resemblance (e.g. Rosch [11]). The family resemblance can be measured in presented model in the terms of activation values.

4.4   Fuzzy Categorization

Based on the famous experiment by Labov [4], a test was conducted in order to discover the behavior of the network in categorization tasks. The network was trained to categorize 4 cup-like objects. The categorization task was performed with respect to the ratios of upper and lower diameter as well as height. Additionally a context was taken into account: neutral context or food context. The network trained on cup-like objects was then multiplied 500 times with random changes in weights in order to reflect different experiences of different people.

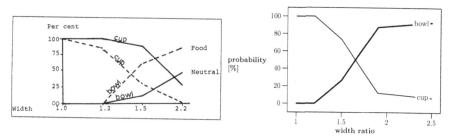

Fig. 2. Results of the original Labov experiment (left) and of the simulation (right)

In the test phase, all 500 networks performed categorization of the above set of objects. The sum of activation of nodes assigned to a given category was calculated and the category with the highest value was chosen as a categorization result. The probability to assign an object to a given category was calculated. As in the original experiment, the results show that categorization is an ambiguous process (cf. Figure 2). There is no clear-cut border between investigated objects, and those which are categorized as cups by some people (networks) can be treated as bowls by others.

4.5   Priming

Context-based priming was observed in the experiment described above. Depending on the context (neutral or "food") the probability to classify a given object as a cup or a bowl differs. Also priming driven by previous categorization was simulated.

As in experiment presented above, 500 networks with randomly modified connections were used. The networks underwent the following procedure. Firstly, they did a categorization of an item and after some delay, categorization of another one. Activation decay in the nodes was time dependent. Two effects were measured: dependency of categorization probability on time between categorization acts and dependency of categorization on previously categorized item.

The results of the priming experiment clearly showed the expected phenomenon: the increase in probability to categorize an object as a cup under the system's prior exposure to information in the decision context, in this case on previously performed categorization. This dependency decays with time as expected.

4.6   Lexical Items

The connectionist network created with the method described in this paper contributes also into models of representation of lexical items. "[T]he lexical item serves a central controlling and stabilizing role in language learning and processing" (MacWhinney [5]). This means that lexical items associate different semantic, phonological, perceptual and possibly also other features. A class node in the network plays exactly the same role. It creates a link between different feature nodes. The lexical item representation has its origin in autoassociation property of the network as well as in its highly localist nature.

The advantage over MacWhinney's solution using self-organizing maps (Kohonen [3]; Miikkulainen [8]) is that this network is able not only to tie different types of features together but also delivers a structure of the lexicon: that means items sharing similar features are not only close to each other but the hyponymy relation is preserved as well.

4.7   Asymmetric Category Learning

This simulation explores the properties of category learning by infants. The investigation by Quinn et al. [10] showed that in case of dogs and cats categories, the infants' categorization displays an asymmetry property. That means, they more frequently categorize unknown cats as dogs than the other way round. In the simulation made, the dataset used by Mareschal et al. [6] was reused. The novelties discovery was measured in terms of mean square error (MSE) between network's output and expected values.

The Fig. 3 shows result of the experiment. The dark bars correspond to the network's answers on cats while the lighter ones correspond to dogs. In the first setup the network was trained on cats only and during the test phase the mean square error was considerably higher for novel dogs than for novel cats. In the second setup, the network was trained on dogs only, but this time novel cats caused only little more errors than novel dogs. Thus, the network developed the representation of dog and cat categories with asymmetric exclusivity. This

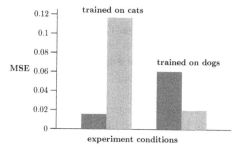

Fig. 3. Mean square error for new items in asymmetric category learning simulation

finding is consistent with results of the Quinn et al.'s experiment performed on infants.

The further investigation on the reasons of the assymetric category learning phenomenon in this experiment can be found in [6].

5   Conclusions

In this final section further issues concerning the network are discussed and a summary is given.

5.1   Localism and Distributionism

The network presented here is neither purely localist nor purely distributed. It is localist in the sense, that each node can have its own interpretation independent from the state of the whole network. It is also localist in that class nodes represent single entities which can be referred to by name. It is however also of distributed nature because in most network architectures the representation of any class would consist of multiple nodes activated on a given level of network. I use the word "level" but "level" cannot be defined clearly. The network is not structured in layers, as most classical (localist and distributed) networks are. As a level a set of nodes described by an equal number of co-occurring features can be taken. However, in this case inhibitory connections exist not only between nodes on the same level but also between nodes on different levels.

5.2   Biological Inspiration

As for each network having localist traces, this one's biological plausibility can be questioned. That is why I would rather say about biological inspiration than plausibility. It is clear that in the brain there are no single neurons corresponding to single concepts, or combinations of concepts. However, Page [9] states that distinct populations of neurons (e.g. cortical minicolumns) can have similar representation properties as nodes of localist network. In this view, the use of semi-localist representation can be justified.

What makes the network even more biological inspired, is the structure of a node. Unlike in usually used both distributed and localist models the node has its internal structure. The neuron is not an object capable for simple single operation on the incoming signal (like summing or integrating) only but can perform much more complex calculations, and so does a node in the presented model.

With respect to connections between nodes, it should be noticed that although they are symmetric in their weights, the signal flowing from node A to node B should not be processed exactly the same as the one flowing from node B to node A. This has its origin in the fact that even if two neurons are mutually connected, two different axons and two different sets of dendrites are engaged.

5.3 Summary

A spreading activation network, which is capable to represent a taxonomy of concepts described by non-binary features, has been presented. This taxonomy forms a skeleton for an ontology valid in the domain defined by concepts in question. This network undergoes self-organization to create the taxonomy containing objects presented as input data as well as attempts to enrich this taxonomy with new terms which are discovered by the system itself.

Beside pure representational power the network is capable to model some cognitive phenomena like categorization, fuzzy categories borders, priming and asymmetry in category learning. It can also be used to model the structure of a lexicon, because of its autoassociative properties. The generalization capability allows for expanding the taxonomy structure as well as finding similarities between objects and the description of some entities that had never been presented to the network before.

References

1. C. Koch, T. Poggio, and V. Torre. Retinal ganglion cells: A functional interpretation of dendritic morphology. *Philosophical Transactions of the Royal Society of London. Series B: Biological Sciences*, 298:227–263, 1982.
2. C. Koch, T. Poggio, and V. Torre. Nonlinear interactions in a dendritic tree: Localization, timing and role in information processing. *Proceedings of the National Academy of Sciences of the United States of America*, 80:2799–2802, 1983.
3. Teuvo Kohonen. Self–organized formation of topologically correct feature maps. *Biological Cybernetics*, 43:59–69, 1982.
4. William Labov. The boundaries of words and their meanings. In C.-J. Bailey and R. Shuy, editors, *New Ways of Analyzing Variation in English.*, pages 340–373. Georgetown U. Press., Washington, DC, 1974.
5. Brian MacWhinney. Connectionism and language learning. In Michael Barlow and Suzanne Kemmer, editors, *Usage–Based Models of Language*, pages 121–149. CSLI Publications, Stanford, California, 2000.
6. Denis Mareschal, Robert M. French, and Paul Quinn. A connectionist account of asymmetric category learning in early infancy. *Developmental Psychology*, 36(5): 635–645, 2000.

7. B. W. Mel. Information processing in dendritic trees. *Neural Computation*, 6: 1031–1085, 1994.
8. Risto Miikkulainen. A distributed feature map model of the lexicon. In *Proceedings of the 12th Annual Conference of the Cognitive Science Society*, pages 447–454, Hillsdale, NJ, 1990. Lawrence Erlbaum.
9. Mike Page. Connectionist modelling in psychology: A localist manifesto. *Behavioral and Brain Sciences*, 23:443–512, 2000.
10. Paul Quinn, Peter D. Eimas, and Stacey L. Rosenkrantz. Evidence for representations of perceptually similar natural categories by 3-month-old and 4-month-old infants. *Perception*, 22:463–475, 1993.
11. Eleanor Rosch. Family resemblance: Studies in the internal structure of categories. *Cognitive Psychology*, 7:573–605, 1975.
12. John. F. Sowa. *Semantic Networks.* http://jfsowa.com/pubs/semnet.htm, 2002.

Self Assembling Graphs

Vincent Danos[1] and Fabien Tarissan[2]

[1] Équipe PPS, CNRS & Université Paris VII
[2] Équipe PPS, Université Paris VII

Abstract. A self-assembly algorithm for synchronising agents and have them arrange according to a particular graph is given. This algorithm, expressed using an ad hoc rule-based process algebra, extends Klavins' original proposal [1], in that it relies only on point-to-point communication, and can deal with any assembly graph whereas Klavins' method dealt only with trees.

1 Introduction

In a number of different subject areas, nanotechnologies [2], amorphous computations [3], molecular biology [4], one commonly finds a debate about whether and how complex shapes, structures and functions can be generated by local interactions between simple components. Klavins adressed this question in the field of robotics [1]. The problem is that of synchronising a population of autonomous agents and have them achieve a particular disposition in space specified as a tree. The aim of the present paper is to extend the solution given by Klavins to the case of arbitrary graphs, and to provide a formalization of the self-assembly algorithm that takes complete care of the subtler part of building a distributed consensus among agents.

The idea of the algorithm is to circulate between agents belonging to a same connected component a single copy of a mapping of their component. Whoever possesses this mapping can either pass it over to a neighbour, or decide to create a new connection, based on a successful point-to-point communication with another agent. Note that since agents are building a potentially cyclic graph, they may have to create edges to their own component. Necessary updates after a growth decision are shipped along a tree spanning the current component. Both the component and the tree are dynamically created. Interestingly, the algorithm is parameterized by the choice of a *growth scenario* specifying when an edge can be created. Thus, the solution we propose naturally supports additional constraints pertaining to which intermediate graphs are allowed during the growth of the graph.

The solution and the problem itself are laid down in the language of concurrency theory, and the algorithm is written in a rule-based process algebra that one could view as a simplified version of Milner's π-calculus [5]. Although the self-assembly algorithm we present is independent of this particular choice, there is a good reason for such a formal approach. More often than not, one

J. Mira and J.R. Álvarez (Eds.): IWINAC 2005, LNCS 3561, pp. 498–507, 2005.

can go wrong in the description of such synchronisation procedures, and the use of formal methods seems legitimate in this context, since they allow for a clear statement of correctness, and a correctness proof based on a well-established notion of equivalence known as *barbed bisimulation* [6].

Our formal treatment is made relative to abstract or logical space. Including true space and explicit motorization in the agents supposes a significant extension of the usual concurrency models and as such represents an interesting challenge to formal methods. Such an extension would in particular allow a refined description of the agents behaviour in the case of a group being dislocated. This is a matter to which we plan to return in a further work. For now, we provide a crude treatment of such "crashes" by introducing non deterministic alarms. The correctness of the algorithm enriched with alarms is also proved. A demo illustrating the algorithm is available on line.[1]

The self-assembly question we adress here was inspired by similar questions raised in the context of formal molecular biology [7, 8]. Indeed, a strong structural property that one might look for when defining a formal language for protein-protein interaction is precisely whether the formation of complexes (assemblies of proteins) can be explained in terms of only local interactions. In the context of biology, there is an additional constraint, namely that the self-assembly algorithm doesn't build in the agents unrealistic computational prowess. With robots however, agents can be taken to be computationally strong and no such objection stays on the way of a completely satisfying result.

2   Graph Rewriting

2.1   Agents and Networks

In order to handle graphs and the kind of local graph rewriting our agents will perform, we introduce first a notation for graphs inspired by π-calculus, where nodes are agents, and edges are represented by name-sharing. Let $\mathcal{C}$ be a countable set of *names* ranged over by $x, y, z, \ldots$, one defines an *agent* as a finite set $C \subset \mathcal{C}$, written $\langle C \rangle$, where the set C itself is refered to as the agent *interface*. Agents can be arranged in *networks* according to the following grammar:

$$G := \varnothing \mid \langle C \rangle \mid G, G \mid (\nu x)G$$

where $\varnothing$ is the empty network, G_1, G_2 stands for the juxtaposition of G_1 and G_2, and $(\nu x)G$ stands for G where the name x has been made private to G.

Here is an example:

becomes $(\nu x)(\nu y)(\langle x \rangle, \langle x, y \rangle, \langle y \rangle)$

Our algebraic notation is redundant in that there are many distinct ways to represent the same graph. The notion of *structural congruence* below will take care of this redundancy.

---

[1] http://www.pps.jussieu.fr/~tarissan/self

The "new" operator, written in symbols ν, is a binder for names in $\mathcal{C}$ and allows for a smooth treatment of name creation. It comes along with the usual inductive definition of *free names*:

$$
\begin{aligned}
\mathsf{fn}(\varnothing) &= \varnothing \\
\mathsf{fn}(\langle C \rangle) &= C \\
\mathsf{fn}(G, G') &= \mathsf{fn}(G) \cup \mathsf{fn}(G') \\
\mathsf{fn}((\nu x)G) &= \mathsf{fn}(G) \smallsetminus \{x\}
\end{aligned}
$$

An occurrence of name is said to be bound if not free. The operation of renaming bound variables is often called α-*conversion*.

Definition 1. *Structural congruence, written* $\equiv$, *is the smallest congruence relation closed under* α-*conversion and such that:*

1. $(\mathcal{N}/\equiv, \, ',', \, \varnothing)$ *is a symmetric monoid*
2. $(\nu x)(\nu y)G \equiv (\nu y)(\nu x)\ G$
3. $(\nu x)G \equiv G$ *if* $x \notin \mathsf{fn}(G)$
4. $(\nu x)G, G' \equiv (\nu x)(G, G')$ *if* $x \notin \mathsf{fn}(G')$

There is a unique network, up to structural congruence, representing a given isomorphism class of graphs, which is what we wanted. Based on this, we will now consider graphs as networks and don't distinguish them notationally. We also observe in passing that our notation also accomodates the description of hypergraphs.

Working up to structural congruence, and using the last three clauses, one can use any of the equivalent notations $(\nu\{x, y\})$, (νxy) or $(\nu x)(\nu y)$. Sets of names will be sometimes denoted succinctly by $\tilde{x}$.

2.2   Reactions and Transition Systems

Now that we have our notation for graphs in place, we turn to the definition of a notion of graph rewriting which will be expressive enough for our needs.

Definition 2. *A reaction is a pair* L, $(\nu\tilde{x})R$, *also written* $L \to (\nu\tilde{x})R$, *where* $L = \langle L_1 \rangle, \ldots, \langle L_n \rangle$, $R = \langle R_1 \rangle, \ldots, \langle R_n \rangle$, *and* $\mathsf{fn}((\nu\tilde{x})R) \subseteq \mathsf{fn}(L)$.

When in addition, $n \le 2$, one will say the reaction is *local*. Local reactions express point-to-point communications and will be used to state the self-assembly problem.

Names occurring in a reaction fall naturally in three classes: the names *created* by the reaction $\tilde{x}$, the names *erased* by the reaction $\mathsf{fn}(L) \smallsetminus \mathsf{fn}((\nu\tilde{x})R)$, and the rest $\mathsf{fn}(L) \cap \mathsf{fn}((\nu\tilde{x})R)$. The condition in the definition above makes sure that any name occurring in R is either created, or already occurs in L.

To fire a reaction in a network G, one looks for an instance of L in G and then replaces it with the right hand side $(\nu\tilde{x})R$. More precisely, given a set of reactions $\mathfrak{R}$, one defines inductively a binary relation $\rightarrow_{\mathfrak{R}}$ as follows:

$$\text{(DIR)} \quad \frac{G \rightarrow_{\mathfrak{r}} G' \quad \mathfrak{r} \in \mathfrak{R}}{G \rightarrow_{\mathfrak{R}} G'} \qquad\qquad \text{(GROUP)} \quad \frac{G_1 \rightarrow_{\mathfrak{R}} G_2}{G_1, G \rightarrow_{\mathfrak{R}} G_2, G}$$

$$\text{(NEW)} \quad \frac{G_1 \rightarrow_{\mathfrak{R}} G_2}{(\nu x)\, G_1 \rightarrow_{\mathfrak{R}} (\nu x)\, G_2} \qquad \text{(STRUCT)} \quad \frac{G_1 \equiv G_1' \quad G_1' \rightarrow_{\mathfrak{R}} G_2' \quad G_2' \equiv G_2}{G_1 \rightarrow_{\mathfrak{R}} G_2}$$

with $G \rightarrow_{\mathfrak{r}} G'$, if $\mathfrak{r} = \langle L_1\rangle, \ldots, \langle L_n\rangle \rightarrow (\nu\tilde{x})(\langle R_1\rangle, \ldots, \langle R_n\rangle)$, and there exists an injection $\mathfrak{i}$ from $\mathsf{fn}(L) \cup \tilde{x}$ to $\mathcal{C}$ such that:

$$\begin{aligned} G &= \langle \mathfrak{i}(L_1)\rangle, \ldots, \langle \mathfrak{i}(L_n)\rangle \\ G' &= (\nu\mathfrak{i}(\tilde{x}))(\langle \mathfrak{i}(R_1)\rangle, \ldots, \langle \mathfrak{i}(R_n)\rangle) \end{aligned}$$

One also writes $G \rightarrow_{\mathfrak{R}}^* G'$, whenever G' can be obtained from G by repeatedly firing reactions in $\mathfrak{R}$. This includes the case when no reaction is used and $G' = G$. A *transition system* is a pair G, $\mathfrak{R}$, where G is a graph, called the *initial state*, and $\mathfrak{R}$ is a set of reactions. A transition system G, $\mathfrak{R}$ is said to be *local* when each reaction in $\mathfrak{R}$ is local.

2.3   Self-assembly

Suppose now we want our agents to self-assemble according to a given connected graph $G = (V, E)$. Write $|V|$ for the number of nodes in G, $\langle\rangle_n$ for the network $\langle\rangle, \ldots, \langle\rangle$ consisting of n empty agents, and $(\langle\rangle_n, \{r_G\})$ for the transition systems associated to G, with initial states $\langle\rangle_n$ and only reaction $r_G := \langle\rangle_{|V|} \rightarrow G$.

The *self-assembly* problem for G is to find a set of *local* reactions $\mathfrak{R}_G$ and a map θ from agents to some suitable notion of enriched agents, such that for all n, the transition system $(\theta(\langle\rangle_n), \mathfrak{R}_G)$ simulates —in a sense yet to be defined— the original system $(\langle\rangle_n, \{r_G\})$.

Two points need to be clarified here. First we need to explain how to enrich agents. An enriched agent will no longer be a mere set of names but a list of sets of names or integers. The θ function above will take care of structuring the initially empty agents according to the enriched agent format. Second, we need a definite statement about what we mean when we say that $(\theta(\langle\rangle_n), \mathfrak{R}_G)$ simulates $(\langle\rangle_n, \{r\})$, and this is where bisimulation comes in the picture.

The first point will be adressed in the course of the construction of the local transition system $(\theta(\langle\rangle_n), \mathfrak{R}_G)$ to which we turn now, while the second will be the object of the next section devoted to correctness.

3   The Construction

Transition systems over enriched agents are similar to the ones defined before and we don't go over all the definitions of the preceding section.

We proceed to our construction in two steps. First we define the notion of a *growth scenario*, in essence a transition system describing which intermediate graphs one should seek for. Then we obtain the corresponding local transition system.

3.1   Growth Scenarios

Agents manipulate concrete representations of graphs and we have to be careful to distinguish these from abstract graphs. Specifically, a *concrete graph* will be taken to be a graph of the form $(\{1,\ldots,n\}, E)$ with $n > 0$. The *trivial graph* $(\{1\}, \varnothing)$ will be denoted by **1**. Given G a concrete graph, we write $[G]$ for its isomorphism class, that is to say the corresponding abstract graph. Next, we define two operations on concrete graphs:

Definition 3. *Let $G_1 = (V_1, E_1)$ and $G_2 = (V_2, E_2)$ be concrete graphs, the* join *of G_1 and G_2 via $u \in V_1$, $v \in V_2$, written $G_1.u \oplus G_2.v$, is defined as:*
— $V = \{1,\ldots,|V_1|+|V_2|\}$, *and*
— $E = E_1 \cup \{\{u, v + |V_1|\}\} \cup \{\{a + |V_1|, b + |V_1|\} \mid \{a,b\} \in E_2\}$
and the self-join *of G_1 via u, $v \in V_1$, written $G_1.(u,v)$, is defined as $V = V_1$, and $E = E_1 \cup \{\{u,v\}\}$.*

Note that in the binary join operation, the nodes of G_2 are shifted by $|V_1|$, and as a consequence the result is again a concrete graph. These operations naturally extend to abstract graphs, and they define together a partial order, written $<$, on concrete as well as on abstract graphs.

Definition 4. *A* growth scenario *$\mathcal{G}$ is a set of abstract graphs such that for all non trivial $G \in \mathcal{G}$, either $G = G_1.u \oplus G_2.v$, for some G_1, $G_2 \in \mathcal{G}$, or $G = G_1.(u,v)$ for some $G_1 \in \mathcal{G}$.*

All graphs within a scenario are connected, and conversely any connected graph is obviously contained in some scenario.
   Here is an example of a growth scenario $\mathcal{G} = \{G_1,\ldots,G_9\}$:

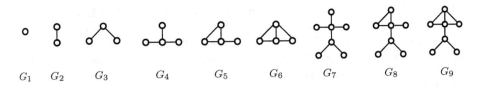

G_1 G_2 G_3 G_4 G_5 G_6 G_7 G_8 G_9

   The idea is that agents wants to self-assemble to reach the target graph G_9, and $\mathcal{G}$ specifies all intermediate graphs they are allowed to construct in so doing. The figure below, where nodes stand for graphs in $\mathcal{G}$, bi-edges correspond to joins, and mono-edges to self-joins gives a visual proof that $\mathcal{G}$ is indeed a growth scenario. Note that we do note require scenarios to be downward closed with respect to $<$, and indeed $\mathcal{G}$ is not.

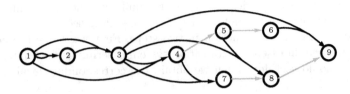

To each scenario $\mathcal{G}$ corresponds naturally a set of reactions, denoted by $\mathfrak{R}(\mathcal{G})$, obtained by translating as reactions the joins and self-joins under which $\mathcal{G}$ is closed. These reactions are quite specific, since they create only one name, and delete none. If we return to the example, the bi-edge linking G_1 and G_2 to G_3 corresponds to the following join reaction in $\mathfrak{R}(\mathcal{G})$:

$$\langle x \rangle, \langle x \rangle, \langle \rangle \to (\nu y)(\langle x \rangle, \langle x, y \rangle, \langle y \rangle)$$

3.2   The Local Transition System

With these definitions behind us, and supposing given a target graph G, and a scenario $\mathcal{G}$ with G as only maximal graph, the self-assembly problem can now be rephrased as the problem of finding a set of local reactions that will simulate $\mathfrak{R}(\mathcal{G})$.

Let us begin with a still informal description of our algorithm and its local reactions. Each agent has in its internal state an integer called the *role* meant to describe its coordinate on the map being circulated between the agents in a same component. By allowing at any time only one agent to modify a component, we prevent concurrent modifications of the structure. This agent is chosen to be the only one holding a map of the current component, since only him needs it. So to speak, the map is itself the activity token. This makes both the internal state of an agent and the update phase simpler. Update is then reduced to the propagation of the new role played by the agents in the new enlarged component.

Going back to definition 3, we see that joins affect only the roles in one of the two connected components. Therefore, role updates are only required in one component which we take in the implementation to be the smaller one. Second, we see also that the new role is easily determined from the old one, since it is just a shift of the old one by the number of nodes, say N of the other component. Agents involved in an update phase will then simply transmit this integer N.

The case of a self-join is easier, since roles are all unchanged. Role updates are not needed, it is just the map of the active agent which is modified.

To avoid conflicts arising during update in case the graph is cyclic, we use a tree spanning the component to transmit the shift. This spanning tree is itself a dynamic structure that grows along with the component. A new edge is added to it at each join. To reflect this structure in their states, enriched agents partition their name set in two classes, written respectively S (for span) and C (for cycles), depending on whether the corresponding edge belongs to the spanning tree or not. Take note that this tree is undirected. The direction of transmission is also dynamically determined and varies over time.

Following these informal explanations, we define the interface of an *enriched agent* as a tuple $[S, C, g, r, m]$ where:

— S is a set of names which represents neighbours in the spanning tree
— C is a set of names which represents the set of the remaining neighbours
— g is a name used as a group identifier for agents in a same component
— r is an integer referring to the role played by the agent in the component

— m is the agent *running mode*:

    ○ P when the agent is in *passive* mode

    ○ $Act(G)$ when the agent is in *active* mode, with G a concrete graph

    ○ $Up(L, N)$ when the agent is in *update* mode, with L a set of names referring to the neighbours still to be updated, and N the shift

3.3   Local Reactions

Given $\mathcal{G}$ a growth scenario, we may now define our family of local reactions $\mathfrak{R}^l(\mathcal{G})$ simulating $\mathfrak{R}(\mathcal{G})$. First come the *join reactions*, with $[G_1.r_1 \oplus G_2.r_2] \in \mathcal{G}$ and $g_1 \neq g_2$:

$$\begin{array}{l} [S_1, C_1, g_1, r_1, Act(G_1)], \\ [S_2, C_2, g_2, r_2, Act(G_2)] \end{array} \longrightarrow (\nu x) \begin{array}{l} [S_1 + \{x\}, C_1, g_1, r_1, Act(G.(r_1, r_2))], \\ [S_2 + \{x\}, C_2, g_1, r_2 + |G_1|, Up(S_2, |G_1|)] \end{array}$$

Next come the *self-join reactions*, with $[G.(r_1, r_2)] \in \mathcal{G}$:

$$\begin{array}{l} [S_1, C_1, g, r_1, Act(G)], \\ [S_2, C_2, g, r_2, P] \end{array} \longrightarrow (\nu x) \begin{array}{l} [S_1, C_1 + \{x\}, g, r_1, Act(G.(r_1, r_2))], \\ [S_2, C_2 + \{x\}, g, r_2, P] \end{array}$$

Then the *update reactions*:

$$\begin{array}{l} [S_1, C_1, g_1, r_1, Up(L + \{x\}, N)], \\ [S_2 + \{x\}, C_2, g_2, r_2, P] \end{array} \longrightarrow \begin{array}{l} [S_1, C_1, g_1, r_1, Up(L, N)], \\ [S_2 + \{x\}, C_2, g_1, r_2 + N, Up(S_2, N)] \end{array}$$

and the *end-of-update reactions*:

$$[S, C, g, r, Up(\varnothing, N)] \longrightarrow [S, C, g, r, P]$$

Finally we need the *switch reactions*:

$$\begin{array}{l} [S_1 + \{x\}, C_1, g, r_1, Act(G)], \\ [S_2 + \{x\}, C_2, g, r_2, P] \end{array} \longrightarrow \begin{array}{l} [S_1 + \{x\}, C_1, g, r_1, P], \\ [S_2 + \{x\}, C_2, g, r_2, Act(G)] \end{array}$$

This last switch reaction, which allows the activity to circulate around in a component, can only be fired if the agents share the same group identifier. We may also note that in the update reactions, the updated agent simultaneously changes its role, and group, while entering himself in update mode. In the end-of-update reaction, the agent goes passive because his contact list is empty.

We may now complete our definition:

Definition 5. *Given a growth scenario $\mathcal{G}$, one defines the local transition systems associated to $\mathcal{G}$ as $(\theta(\langle\rangle_n), \mathfrak{R}^l(\mathcal{G}))$, where $\mathfrak{R}^l(\mathcal{G})$ is defined above, and the initial state $\theta(\langle\rangle_n)$ is given by $g_1\langle\varnothing, \varnothing, g_1, 1, Act(1)\rangle, \ldots, (g_n)\langle\varnothing, \varnothing, g_n, 1, Act(1)\rangle$.*

4   Correctness

We have completed the formal definition of our algorithm, and it remains to explain in which sense our local transition systems, $\mathfrak{R}^l(\mathcal{G})$, behave as the global ones, $\mathfrak{R}(\mathcal{G})$. We use the notion of barbed bisimulation to do this.

Given a network G, we will write $G \downarrow_n C$, if the connected component C occurs n times in the network G.

Definition 6. *Two transition systems $(G_0, \mathfrak{R})$ and $(G'_0, \mathfrak{R}')$ are bisimilar if there exists a binary relation $\sim$ over $\mathcal{N}$ such that:*

- $G_0 \sim G'_0$,
- *if $G \sim G'$, then $G \downarrow_n C$ if and only if $G' \downarrow_n C$;*
- *if $G \sim G'$ and $G \to_{\mathfrak{R}} H$, there exists H' such that $G' \to^*_{\mathfrak{R}'} H'$ and $H \sim H'$;*
- *if $G \sim G'$ and $G' \to_{\mathfrak{R}'} H'$, there exists H such that $G \to^*_{\mathfrak{R}} H$ and $H \sim H'$.*

We can now state the correctness of our distributed algorithm. Note that the obtained property is slightly more general than the one we wanted, since it does not assume that there is a unique maximal graph in the scenario $\mathcal{G}$.

Proposition 1. *For all n and all growth scenarios $\mathcal{G}$, the local and global transition systems, $(\langle\rangle_n, \mathfrak{R}(\mathcal{G}))$ and $(\theta(\langle\rangle_n), \mathfrak{R}^l(\mathcal{G}))$, are bisimilar.*

The key to proving this proposition is to prove that active agents always have a *consistent* view of their components. That is to say, for any G which is reachable from the initial state $\theta(\langle\rangle_n)$, and any active agent in G, both the map and the role occurring in the interface of this agent are the actual ones in G. Once this is done, it is relatively easy to construct a bisimulation between the global and local systems.

5 Escaping Deadlocks

The local transition systems defined above are monotonic in the sense that edges can only be added. Different components representing partially grown target graphs could compete and be deprived of resources.

Klavins suggests a simple timeout method to grow, starting with a given population of agents, the maximum possible number of copies of the target graph. We can easily incorporate an abstract version of this deadlock escape mechanism in our algorithm.

First we extend our notion of growth scenario by allowing each connected component to dislocate, with the constraint that as soon as a it has begun to do so, it has no choice but keeping on breaking down to unconnected agents. To enforce this, we allow only active agents to fire an alarm. Second, we introduce a new *alarm mode*, written Al. And third, we add the accompanying reactions, starting with the *breaking-loose reactions*:

$$[S, C, g, r, Act(G)] \longrightarrow [S, C, g, r, Al]$$

and the *alarm propagation reactions*:

$$\begin{array}{l} [S_1 + \{x\}, C_1, g_1, r_1, Al], \\ [S_2 + \{x\}, C_2, g_2, r_2, \_] \end{array} \longrightarrow \begin{array}{l} [S_1, C_1, g_1, r_1, Al], \\ [S_2, C_1, g_2, r_2, Al] \end{array}$$

$$\begin{array}{l} [S_1, C_1 + \{x\}, g_1, r_1, Al], \\ [S_2, C_2 + \{x\}, g_2, r_2, \_] \end{array} \longrightarrow \begin{array}{l} [S_1, C_1, g_1, r_1, Al], \\ [S_2, C_1, g_2, r_2, Al] \end{array}$$

and finally the *alarm-end reactions*:

$$[\varnothing, \varnothing, g, c, Al] \longrightarrow (\nu g)[\varnothing, \varnothing, g, 1, Act(\mathbf{1})]$$

Note that one has two reactions to propagate the alarm depending on whether the alarm is shipped along the spanning tree or not. There are no longer any conditions on g_2 and r_2, since consistency is lost during dislocation. An agent goes active again only when it has broken all connections and spread the alarm to all its neighbours. Thus, active agents still view correctly their components, and we can easily extend proposition 1 to include the case of dislocation.

6   Conclusion

We have presented an abstract self assembly protocol by which a population of autonomous agents can grow an arbitrary network of connections in a distributed way. The target network is built incrementally. In any given connected component, a map of the current component circulates between the agents allowing them to make decisions concerning whether an edge should be added and where. One specificity of our approach is to cast the problem in the language of concurrent processes, actually in a simplified version of π-calculus, and be able to subsequently give a precise statement of the correctness of the proposed protocol. An implementation of the protocol extended with an alarm propagation meachanism avoiding deadlocks is available online.[2] A decidedly interesting extension would be to incorporate true space, if only because no ordinary process algebra models true space.

This protocol was inspired by self assembly questions in biological systems. Although we assumed computationally strong agents, we took care to keep their internals and interaction capacities, as embodied by the local reactions, to a minimum. It is agreed that they are still way stronger than anything that could be implemented today in the combinatorics of molecular biology. Nevertheless exploring such self-assembly procedures could help in building a working engineering intuition of biological self assembly.

References

1. Eric Klavins. Automatic synthesis of controllers for assembly and formation forming. In *Proceedings of the International Conference on Robotics and Automation*, 2002.
2. Eric Drexler and Richard Smalley. Controversy about molecular assemblers. Available at www.foresight.org/NanoRev/Letter.html, 2003.
3. Radhika Nagpal. Programmable self-assembly using biologically-inspired multiagent control. In *Autonomous Agents and Multiagent Systems Conference (AAMAS)*, July 2002.
4. Jeff Hasty, David McMillen, and James J. Collins. Engineered gene circuits. *Nature*, 420:224–230, November 2002.

---

[2] http://www.pps.jussieu.fr/~tarissan/self

5. Robin Milner. *Communicating and mobile systems: the π-calculus.* Cambridge University Press, Cambridge, 1999.
6. Robin Milner and Davide Sangiorgi. Barbed bisimulation. In W. Kuich, editor, *Nineteenth Colloquium on Automata, Languages and Programming (ICALP) (Wien, Austria)*, volume 623 of *LNCS*, pages 685–695. Springer, 1992.
7. Vincent Danos and Cosimo Laneve. Core formal molecular biology. In *Proceedings of the 12th European Symposium on Programming (ESOP'03, Warsaw, Poland)*, volume 2618 of *LNCS*, pages 302–318. Springer, April 2003.
8. Vincent Danos and Cosimo Laneve. Formal molecular biology. *Theoretical Computer Science*, 325(1):69–110, September 2004.

Knowledge Modeling for the Traffic Sign Recognition Task

M. Rincón[1], S. Lafuente-Arroyo[2], and S. Maldonado-Bascón[2]

[1]Dpto. de Inteligencia Artificial, ETSI Informatica, UNED,
Juan del Rosal 16, 28040 Madrid, Spain
mrincon@dia.uned.es
[2]Dpto. de Teoría de la Señal y Comunicaciones,
Escuela Politécnica Superior, Universidad de Alcalá,
Campus Universitario, Alcalá de Henares, 28871 Madrid, Spain
{sergio.lafuente, saturnino.maldonado}@uah.es

Abstract. In this paper we analyse the problem of traffic sign recognition at the knowledge level. Due to the complexity of the task, our approach decomposes it into simpler subtasks until the primitive level is reached. The task has been modeled at the knowledge level as a hierarchical classification task. This has allowed to discover a simple and robust Problem Solving Method (PSM) for the classification task which is reused along different classification stages of the process.

The resulting system is divided into three main subtasks: image segmentation according to color, classification of the geometry of the candidate blobs and identification of the specific type of traffic sign. Finally, the system has been evaluated and the results are presented.

Index Terms: Knowledge Modeling, Hierarchical Classification, PSM, Traffic Sign Detection and Recognition, Color Space, Bounding Box, Distance to BoundingBox Borders (DtBB), Support Vector Machine (SVM).

1 Introduction

During the last years, detection and recognition of traffic signs with automatic systems have become an important subject of study in engineering. This is due to the important information provided by traffic signs for safety and successful driving, which has derived in applications such as support to intelligent vehicles, road security, inventory of traffic signs for highway maintenance, etc.

Traffic signs present simple forms and are usually designed with some special colors. In fact, it is common to detect and isolate them from the background using color features [3],[8]. A first classification of road signs according to color distinguishes between red and blue ones. Nevertheless, the segmentation according the color has two difficulties: 1) Other elements in the scene can show similar colors to traffic signs and 2) illumination changes modify the color features. As an alternative to color criteria, other investigation lines have been followed using techniques based on textures [9],[11] or artificial neural networks [2],[7]. However,

J. Mira and J.R. Álvarez (Eds.): IWINAC 2005, LNCS 3561, pp. 508–517, 2005.

independently of the technique chosen for our purpose, additional problems are found like occlusions, rotations and deterioration by atmospheric aggression or vandalism action.

Once the traffic sign is located, the most commonly used techniques for traffic sign recognition are based on neural networks [1] and mask searching [3]. Techniques based on masks present low reliability when faced to partial occlusion and rotation. Even when the shape to classify is not completed only in one of its vertices, the mask searching algorithm isn't able to recognize it. In the last years, genetic algorithms have been the method proposed in some works [1]. However, this technique has some disadvantages like premature convergences or the risk associated to find local maximal/minimal points.

In this paper, we present a new method for detection and recognition of traffic signs which is strongly robust to the problems pointed out above. This work has been done following the well-established foundations of Knowledge Engineering that prescribe the maintenance of the conceptual structure from the modeling stage at the knowledge level down to the implementation. First, at the knowledge level, we have decomposed the task following the CommonKads Expertise Model [10] as a classification task and it has been solved using a hierarchical classification PSM.

In the first classification step, the image regions belonging to traffic signs are identified by its color feature. In the second step, considering the color, the traffic signs are classified by shape. And, in the last classification step, the specific traffic sign type is obtained. Decomposing the problem into simpler subtasks derives in simplified partial goals, so we can use simpler and robust mechanisms for each subtask. Knowledge level modeling of the problem has allowed to discover a simple and robust mechanism which is reused along different classification stages of the process.

This paper is organized in the following way. Section 2 describes the proposed PSM for a classification task. Section 3 describes how to address the task which, because of its complexity, has to be broken down into simpler sub-tasks. In order to evaluate the system, section 4 contains experimental results of its application to a traffic sign data set. Finally, the findings obtained in this work are set out in Section 5.

2   Classification-by-Descriptors-and-Winner-Takes-All-Decision PSM (CD-WTA PSM)

The structure of a generic classifier consists of three phases: feature extraction, generation of metrics and decision. In this paper, a simple and robust classification method is proposed, which is based on the following problem assumptions:

1. Descriptors are significative enough to provide by themselves, independently, a significant estimation of the output class.
2. Output classes are mutually exclusive.

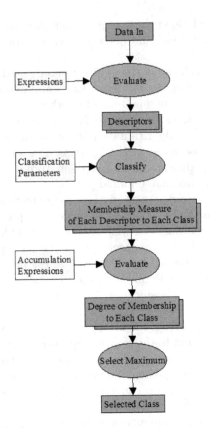

Fig. 1. Inference diagram for the *CD-WTA* PSM

In Figure 1, the inference diagram of the proposed PSM is shown. First, like in any classification, the descriptors from the input data are extracted (*Evaluate* inference). Next, based on assumption 1), an estimation of the degree of membership of each descriptor to each output class is realized in the metric generation phase (*Classify* inference). And finally, based on assumption 2), the output class is decided using a simple decision mechanism as winner-takes-all (WTA) (*Evaluate* and *Select-Maximum* inferences). To speed up the process, a first decision is made based on an aproximation to the degree of membership and, if indetermination exits because of various classes are simultaneously active, a second decision is made using more precise data.

To operationalize the inferences of this *CD-WTA* PSM we have used the following operators:

- The *Evaluate* inference in the feature extraction phase consists of the evaluation of a set of application specific expressions.
- The *Classify* inference has been operationalized using the SVM-macrooperator, which is based on the SVM (support vector machine) tech-

nique. This macrooperator can be interpreted as a supervised learning method based on statistics for the design of a feedforward neural network with a single hidden layer of linear or nonlinear units.

- Each SVM returns a real value. As positive values indecate membership to the class and negative values indicate non-membership, the sign of the value can be a first aproximation to estimate the membership of a descriptor to a class. In order to accumulate evidences we use the number of positive values (`count (i+)`) to operationalize the *evaluate* inference and the decision consists of the selection of the maximum. In case of tie between various classes, we will use the real value returned by each SVM (`sum(i)`) and the final decision will be again the selection of the maximum.

3   The Traffic Sign Recognition Task

In Spain, blue traffic signs have circular (duty) or rectangular shapes (information) and red traffic signs present triangular (danger), circular (forbidden action or special restriction) or octagonal (stop) shapes. Starting off with this domain knowledge, if we organize all the types of traffic signs in subgroups by color and shape, the recognition problem of a specific traffic sign is simplified, because the set of possible classes in the final stage is reduced and, in addition, the differences in the characteristics of the traffic signs pertaining to the same subgroup are more significant.

We have modeled the task at the knowledge level as a classification task and it has been solved using hierarchical classification. The structure of subtasks can be summarized as shown in Fig. 2. The process begins with the segmentation of the image. Pixels are classified according to their color feature (*Select&Classify-by-color* subtask) and the image regions belonging to traffic signs are identified, filtering those regions which doesn't present the required features (*Filter* subtask). Next, considering the color of the blob, a second classification step is performed and traffic signs are classified by shape (*Classify-by-shape* subtask). Finally, in the last classification step, the proper traffic sign type is identified from the subtype preselected (*Classify-Traffic-Sign* subtask). In the following sections each one of the process phases is explained in more detail.

3.1   Image Segmentation

Segmentation of images is the first step in traffic sign detection. It is known that the HSI (Hue, Saturation and Intensity) color model is the more invariant color space to illumination changes. And, as we perform the segmentation according to color features, our application only works with two components of this space: Hue and Saturation. The third component, Intensity, is not used because the chromatic information doesn't depend on intensity and it is known that intensity is very variant to illumination changes.

For each color, Equation 3.1 represents the segmentation function that is evaluated in the *Select&Classify-by-color* subtask, where parameters H_a and

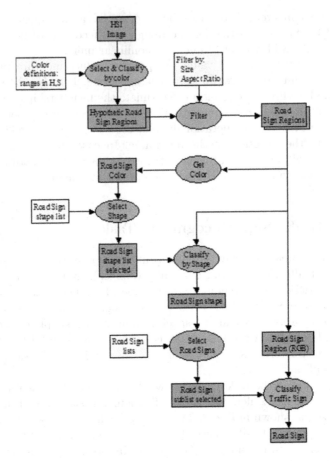

Fig. 2. Diagram of subtasks for the traffic sign recognition task

H_b are the minimal and maximal thresholds of the Hue component, S_a corresponds to the minimal threshold of the Saturation component and f_H and f_S are, respectively, the values for the Hue and Saturation components of every pixel of coordinates (x, y) in the digital image. After testing a bank of 300 images under different lightly conditions, we have choosen fixed thresholds for both components.

$$g(x,y) = \begin{cases} 255 & \text{if } \begin{cases} H_a \leq f_H(x,y) \leq H_b \text{ with } H_a \leq H_b \\ f_S(x,y) \geq S_a \end{cases} \\ 0 & \text{otherwise} \end{cases}$$

So, using Hue and Saturation components every image is segmented looking for red and blue zones within the allowed range. In order to eliminate noise blobs with the same color as the traffic signs, we pay attention to their size and aspect

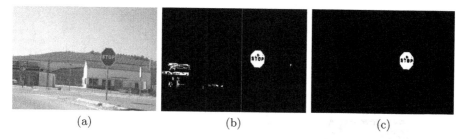

Fig. 3. (a) Original Image; (b) results after *Select&Classify-by-color* subtask: segmented regions belong to a red traffic sign and red noise blobs; (c) results after the *filter* subtask: red noise blobs are eliminated

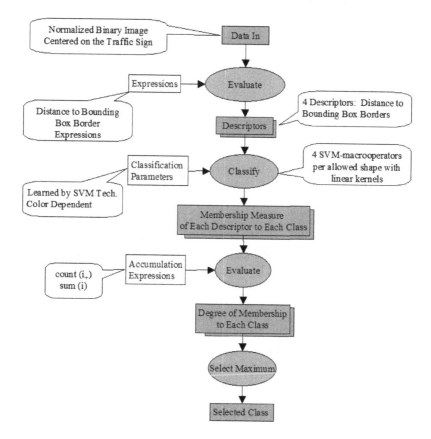

Fig. 4. Configuration of *CD-WTA PSM for traffic sign shape analysis*

ratio. Traffic signs may present different sizes according to the distance to the camera, however the variation of size is limited. For this reason, we can filter those blobs with an area out of range (parameters A_{max} and A_{min}). On the other

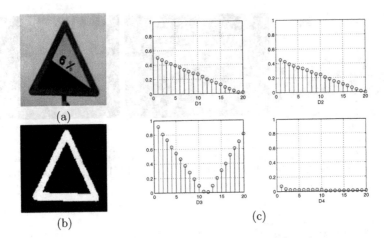

Fig. 5. (a) Red traffic sign original image; (b) candidate blob after the segmentation process; (c) shape descriptors: distances to bounding box borders

side, the traffic sign shapes of the study (triangular, circular, rectangular and octagonal) have aspect ratio near unity, so we can discard any other blob with different aspect ratio (parameters AR_{max} and AR_{min}). So, the *Filter* subtask implements a filter by size and aspect ratio. An example of the partial results obtained by each of these subtasks can be seen in Figure 3.

3.2   Shape Analysis

In order to simplify the process we apply a rotation to every blob, so that every shape is always orientated in the same way. Once standardized the region of the image corresponding to the traffic sign, we use the *CD-WTA* PSM configured of the following way (Figure 4): The description of shape used in this work is the feature named as "Distance to bounding box borders" that stores the minimal distance from the blob to the sides of the bounding box. We have 4 descriptors, one for each side of the bounding box. Each descriptor is a vector of 20 components normalized along each side of the bounding-box.

In order to classify each side descriptor a SVM is used, so we have 4 SVM for each possible shape (circular, triangular and octagonal shapes for red color traffic signs and circular and rectangular shapes for blue color traffic signs). A Linear Kernel SVM is used in order to classify each descriptor. The main reason to select the linear kernel is its low computational requirements with good results.

Figure 5 presents graphically the four blob descriptors. The values are normalized respect to the dimensions of the bounding box of the blob.

3.3   Traffic Sign Final Recognition

In this last stage, knowing beforehand its color and shape, the specific traffic sign type is identified. We utilize for it the *CD-WTA* PSM described in section 2 again.

RGB image format is used in this classification step. In order to eliminate spureous levels, a uniform quantification of the channels is performed, converting each pixel value from 256 to 16 levels. Besides, the region of interest is normalized to 30x30 pixels. The normalization decision is a compromise between minimum computational load and enough resolution for the recognition task. So, in this step, we have three chromatic descriptors (the three color channels) of a constant size of 900 pixels.

Because recognition of traffic signs on the basis of its chromatic channels isn't a linearly separable problem, SVMs with gaussian kernel are used for operationalizing the classify inference.

4   Results

In this section, we present the results obtained along the diferent phases of the proposed method for the recognition of traffic signs.

The training and test for the shape classification have been made in the following way:

- Training the SVMs with 30 road signs of each shape (circular, triangular, octagonal and rectangular).
- Testing over a bank of 250 images with multiple road signs.

Figure 6 shows the classification of results where we distinguish the red circular traffic signs from the blue circular ones according to the structure of classification. Octagonal shapes have the worst results due to the high similarity to circular ones in medium/high distances to the camera. The average of success is more than 94%.

	Triang.	Circ.-Red	Octog.	Rectang.	Circ.-Blue	Unclassif.	success %
Triang.	201	-	-	-	-	8	96.17
Circ.-Red	-	141	8	-	-	1	94.00
Octog.	-	5	69	-	-	2	90.78
Rectang.	-	-	-	79	2	4	92.94
Circ.-Blue	-	-	-	-	64	2	96.96

Fig. 6. Shape Classification Results

In order to evaluate the recognition module of the system, scenes that contain one of the eight circular traffic sign subtypes shown in Fig. 7 have been used. As shown in Fig. 8, the recognition rate is high and, however, the average number of training images is as low as 13. It is important to emphasize that some of the test images include problematic regions such as imperfect segmentations and occlussions.

Fig. 7. Set of Circular Traffic Signs used for the final recognition stage

	Type 1	Type 2	Type 3	Type 4	Type 5	Type 6	Type 7	Type 8	Unrecogn.	success %
Type 1	29	1	-	-	-	-	-	-	2	90.62
Type 2	-	30	-	-	-	-	-	-	2	93.75
Type 3	-	1	21	-	-	-	-	-	1	91.30
Type 4	-	-	-	27	-	-	-	-	2	93.10
Type 5	-	-	-	-	22	-	-	-	1	95.65
Type 6	-	-	-	-	1	14	-	-	-	93.33
Type 7	-	1	-	-	-	-	22	-	1	91.67
Type 8	-	-	-	-	-	-	-	43	3	93.47

Fig. 8. Circular Traffic Sign Recognition Results

Recognition errors are mainly due to three factors:

- Poor traffic sign visibility caused by long distance to the camera, fog, deterioration, etc.
- Noise in the segmentation process that diminishes the degree of correlation between the chromatic vectors and the reference patterns.
- Oclussions: it can be said that occlussions bigger than 25% of the traffic sign area prevent its recognition.

5 Conclusions

This paper presents a new procedure for identification and recognition of traffic signs as a result of a methodological study: prior analysis of the task, decomposition of the task into subtasks and the use of current image processing techniques and application-specific domain knowledge to operationalize each subtask. This has allowed the definition of a simple PSM for the classification task that has been reused along different subtasks of the task decomposition.

The system defined has been successfully applied over an extensive set of images that contain problems such as partial occlusions, rotations, shadows or changes of illumination. The obtained results show that the system is specially invariant to translation, change of scale and rotation. So, the system can be used over different images with acceptable results. In addition, due to the domain knowledge introduced in the decomposition of the problem in simple partial classifications, it is possible to use a reduced training set.

Future work includes the use of much more domain knowledge to improve the system so that it is more tolerant to bigger occlusions and illumination changes and in order to reduce its complexity, centering the analysis only in those areas of the frame where traffic signs are expected.

Acknowledgments

This work was supported by the project of the Ministerio de Educacin y Ciencia de Espaa number TEC2004/03511/TCM.

References

[1] Y. Aoyagi, T. Akasura. *A study on traffic sign recognition in scene image using genetic algorithms and neuronal networks*. Proceedings of the 22nd Int. Conf. On Industrial Electronics, Control and Instrumentation, pages. 1838-43, Taipeis, August 1996.

[2] N. Bartneck, W. Ritter. *Colour segmentation with polynomial classification*. Proc. of 11ht Int. Conf. on Pattern Recognition, vol. II, pp. 635-638, August 1992.

[3] A. de la Escalera, L. Moreno, M.A. Salichs, J.M. Armingol. *Road Traffic Sign Detection and Classification*. IEEE Transaction on Industrial Electronics 44 (6): 848-859, December 1997.

[4] D. Gavrila, V. Philomin. *Real-time object detection using distance transforms*. Proceedings of the Seventh IEEE International Conference on Computer Vision, pages 87-93, Kerkyra, Septembre, 1999.

[5] S.-H. Hsu and C.-L. Huang. *Road sign detection and recognition using matching pursuit method*. Image and Vision Computing, 19: 119-129, 2001.

[6] Hu, M. K. *Visual Pattern Recognition by Moment Invariants*. IRE Trans. Info., Theory, vol. IT-8, pp. 179-187, 1962.

[7] R. Janssen, W. Ritter, J. Stein, S. Ott. *Hybrid approach for traffic sign recognition*. Proc. Of 1993. IEEE Intelligent Vehicles Symposium, pp. 678-685, July 1993.

[8] D.S. Kang, N.C. Griswold, N. Kehtarnavaz. *An invariant traffic sign recognition system based on sequential color processing and geometrical transformation*. Proc. Of the IEEE Southwest Symposium on Image Analysis and Interpretation, pp.88-93, Dallas, TX, April 1994.

[9] R. C. Luo, H. Potlapally, D. Hislop, *Natural scene segmentation using fractal based autocorrelation*. IEEE Int. Conf. on Ind. Elect., Control, Instrum. and Automation, vol. 2, pp. 700-705, November 1992.

[10] Schreiber G., Akkermans H. , Anjewierden A. , de Hoog R., Shadbolt N., Van de Velde W. and Wielinga B. *Knowledge Engineering and Management: The CommonKADS Methodology*. The MIT Press. 1999.

[11] A. C. She, Huang. *Segmentation of road scenes using color and fractal based texture classification*. IEEE Int. Conf. on Image Processing, vol. 3, pp. 1026-1030, November 1994.

Interval-Valued Neural
Multi-adjoint Logic Programs*

J. Medina, E. Mérida-Casermeiro, and M. Ojeda-Aciego

Dept. Matemática Aplicada, Universidad de Málaga
{jmedina, merida, aciego}@ctima.uma.es

Abstract. The framework of multi-adjoint logic programming has shown
to cover a number of approaches to reason under uncertainty, imprecise
data or incomplete information. In previous works, we have presented a
neural implementation of its fix-point semantics for a signature in which
conjunctors are built as an ordinal sum of a finite family of basic conjunctors (Gödel and Łukasiewicz t-norms). Taking into account that a
number of approaches to reasoning under uncertainty consider the set of
subintervals of the unit interval as the underlying lattice of truth-values,
in this paper we pursue an extension of the previous approach in order
to accomodate calculation with truth-intervals.

1   Introduction

A number of different approaches have been proposed with the aim of better explaining observed facts, specifying statements, reasoning and/or executing
programs under some type of uncertainty whatever it might be. The frameworks
of *fuzzy logic programming* [10] and *residuated logic programming* [1] abstract
the details of several well-known approaches to generalized logic programming.

Multi-adjoint logic programs were introduced as a common umbrella to cover
a number of approaches to reason under uncertainty, imprecise data or incomplete information and, in particular, can be instantiated as both fuzzy logic
programming and residuated logic programming. The handling of uncertainty in
the multi-adjoint approach is based on the use of a generalised set of truth-values
as an extension of fuzzy logic programming. On the other hand, multi-adjoint
logic programming generalizes residuated logic programming in that several different implications are allowed in the same program, as a means to facilitate the
task of the specification.

The recent paradigm of soft computing promotes the use and integration of
different approaches for problem solving. The approach presented in [7, 9] introduced a hybrid framework to handling uncertainty, expressed in the language
of multi-adjoint logic but implemented by using ideas from the world of neural
networks.

---

* Partially supported by Spanish DGI project TIC2003-09001-C02-01.

J. Mira and J.R. Álvarez (Eds.): IWINAC 2005, LNCS 3561, pp. 518–527, 2005.

Several semantics have been proposed for multi-adjoint logic programs but, regarding the implementation, the fix-point semantics was the chosen one: given a multi-adjoint logic program $\mathbb{P}$ its meaning (the minimal model) is obtained by iterating the $T_\mathbb{P}$ operator. At least theoretically, by computing the sequence of iterations of $T_\mathbb{P}$ one could answer in parallel all the possible queries to $\mathbb{P}$; in order to take advantage of this potential parallelism, a recurrent neural network implementation of $T_\mathbb{P}$ was introduced in [7], where the truth values belonged to the unit interval, the connectives were the usual t-norms Gödel, product and Łukasiewicz, together with any weighted sum. Later, this neural net was improved in order to be able to use any finite ordinal sum of Gödel, product and Łukasiewicz t-norms in [8].

The machinery underlying multi-adjoint programs is that of adjoint pairs, which abstracts out the behaviour of classical conjunction and implication and provides a convenient version of the modus-ponens rule to be used on sets of truth-values more general than $\{0, 1\}$. A possible extension of the previous approach consists in implementing an interval-based semantics. This generalization is not unnatural, since when defining a fuzzy set sometimes it is not easy to associate a value in the unit interval to any element in the set, but we'd rather associate an interval instead; this generalization of fuzzy set is called an interval-valued fuzzy set.

The method of using intervals, either symbolic (inf, sup) or numerical $[a, b]$, to describe uncertain information has been adopted in several mechanisms, and they are useful in applications such as decision and risk analysis, engineering design, and scheduling. Intervals are also used in some frameworks for generalized logic programming such as the *hybrid probabilistic logic programs* [2] and the *probabilistic deductive databases* [6]. Just note that, in the latter framework, one could write rules like:

$$paper\_accepted \xleftarrow{\langle [0.7, 0.95], [0.03, 0.2] \rangle} good\_work, good\_referees$$

where we have a complex confidence value containing two probability intervals, one for the case where *paper_accepted* is true and other for the case where *paper_accepted* is false (there can exist some lack of information, or undefinedness, in these intervals).

The purpose of this paper is to present a refined version of the neural implementation of [7, 8] in order to cope with interval-valued data. The main difference with the previous version is that the type of conjunctors considered in the neuron are pairs of conjunctors on the unit interval, the first (second) one intended to make the calculations on the initial (final) point of the truth-intervals; in addition, the procedure to handle the aggregators, usually weighted sums, has been conveniently adapted to work with truth-intervals.

The structure of the paper is as follows: In Section 2, the syntax and semantics of multi-adjoint logic programs are introduced; in Section 3, the new proposed neural model for homogeneous multi-adjoint programs is presented in order to work with truth-intervals, a high level implementation is introduced and proven to be sound. The paper finishes with some conclusions and future work.

2   Preliminary Definitions

Multi-adjoint logic programming is a general theory of logic programming which allows the simultaneous use of different implications in the rules and rather general connectives in the bodies. To make this paper as self-contained as possible, the necessary definitions about multi-adjoint structures are included in this section. The basic definition is the generalization of residuated lattice given below:

Definition 1. *A* multi-adjoint lattice *$\mathcal{L}$ is a tuple $(L, \preceq, \leftarrow_1, \&_1, \ldots, \leftarrow_n, \&_n)$ satisfying the following items:*

1. *$\langle L, \preceq \rangle$ is a bounded lattice, i.e. it has bottom and top elements;*
2. *$\top \&_i \vartheta = \vartheta \&_i \top = \vartheta$ for all $\vartheta \in L$ for $i = 1, \ldots, n$;*
3. *$(\&_i, \leftarrow_i)$ is an adjoint pair in $\langle L, \preceq \rangle$ for $i = 1, \ldots, n$; i.e.*
   (a) *Operation $\&_i$ is increasing in both arguments,*
   (b) *Operation $\leftarrow_i$ is increasing in the first argument and decreasing in the second argument,*
   (c) *For any $x, y, z \in P$, we have that $x \preceq (y \leftarrow_i z)$ holds if and only if $(x \&_i z) \preceq y$ holds.*

2.1   Syntax and Semantics

Definition 2. *A* multi-adjoint program *is a set of weighted rules $\langle F, \vartheta \rangle$ satisfying the following conditions:*

1. *F is a formula of the form $A \leftarrow_i B$ where A is a propositional symbol called the* head *of the rule, and B is a well-formed formula, which is called the* body, *built from propositional symbols $B_1, \ldots, B_n$ ($n \geq 0$) by the use of monotone operators.*
2. *The weight ϑ is an element of the underlying truth-values lattice L.*

Facts *are rules with body*[1] *$\top$ and a* query *(or* goal*) is a propositional symbol intended as a question ?A prompting the system.*

Once presented the syntax of multi-adjoint programs, the semantics is given below.

Definition 3. *An* interpretation *is a mapping I from the set of propositional symbols Π to the lattice $\langle L, \preceq \rangle$.*

Note that each of these interpretations can be uniquely extended to the whole set of formulas, and this extension is denoted as $\hat{I}$. The set of all the interpretations is denoted $\mathcal{I}_{\mathcal{L}}$.

The ordering $\preceq$ of the truth-values L can be easily extended to $\mathcal{I}_{\mathcal{L}}$, which also inherits the structure of complete lattice and is denoted $\sqsubseteq$. The minimum element of the lattice $\mathcal{I}_{\mathcal{L}}$, which assigns $\bot$ to any propositional symbol, will be denoted $\triangle$.

---

[1] It is also customary not to write any body.

Definition 4.

1. *An interpretation $I \in \mathcal{I}_{\mathfrak{L}}$ satisfies $\langle A \leftarrow_i \mathcal{B}, \vartheta \rangle$ if and only if $\vartheta \preceq \hat{I}(A \leftarrow_i \mathcal{B})$.*
2. *An interpretation $I \in \mathcal{I}_{\mathfrak{L}}$ is a* model *of a multi-adjoint logic program $\mathbb{P}$ iff all weighted rules in $\mathbb{P}$ are satisfied by I.*

The operational approach to multi-adjoint logic programs used in this paper will be based on the fixpoint semantics provided by the immediate consequences operator, given in the classical case by van Emden and Kowalski, which can be generalised to the multi-adjoint framework by means of the adjoint property, as shown below:

Definition 5. *Let $\mathbb{P}$ be a multi-adjoint program; the* immediate consequences *operator, $T_{\mathbb{P}} \colon \mathcal{I}_{\mathfrak{L}} \to \mathcal{I}_{\mathfrak{L}}$, maps interpretations to interpretations, and for $I \in \mathcal{I}_{\mathfrak{L}}$ and $A \in \Pi$ is given by*

$$T_{\mathbb{P}}(I)(A) = \sup \left\{ \vartheta \mathbin{\&_i} \hat{I}(\mathcal{B}) \mid \langle A \leftarrow_i \mathcal{B}, \vartheta \rangle \in \mathbb{P} \right\}$$

As usual, it is possible to characterise the semantics of a multi-adjoint logic program by the post-fixpoints of $T_{\mathbb{P}}$; that is, an interpretation I is a model of a multi-adjoint logic program $\mathbb{P}$ iff $T_{\mathbb{P}}(I) \sqsubseteq I$. The $T_{\mathbb{P}}$ operator is proved to be monotonic and continuous under very general hypotheses.

Once one knows that $T_{\mathbb{P}}$ can be continuous under very general hypotheses, then the least model can be reached in at most countably many iterations beginning with the least interpretation, that is, the least model is $T_{\mathbb{P}} \uparrow \omega(\triangle)$.

3   An Interval-Valued Network for Multi-adjoint Logic Programming

Regarding the implementation as a neural network, the introduction of the so-called *homogeneous rules* given in [7], provided a simpler and standard representation for any multi-adjoint program.

Definition 6. *A weighted formula is said to be* homogeneous *if it has one of the following forms:*

- $\langle A \leftarrow_i \mathbin{\&_i}(B_1, \ldots, B_n), \vartheta \rangle$
- $\langle A \leftarrow_i @(B_1, \ldots, B_n), \top \rangle$
- $\langle A \leftarrow_i B_1, \vartheta \rangle$

where $A, B_1, \ldots, B_n$ are propositional symbols.

The homogeneous rules represent exactly the simplest type of (proper) rules one can have in a program. In some sense, homogeneous rules allow a straightforward generalization of the standard logic programming framework, in that no operators other than $\leftarrow_i$ and $\mathbin{\&_i}$ are used. The way in which a general multi-adjoint program is homogenized is irrelevant for the purposes of this paper;

anyway, it is worth mentioning that it is a model-preserving procedure with linear complexity.

In this section, we introduce a neural network that implements an extension of the previous approaches [7, 8] with the enhancements stated in the introduction: namely, the calculation with truth-intervals and the possibility of considering ordinal sums. This extension generates an overloaded use of intervals, on the one hand, as truth-values and, on the other hand, as the data needed in the definition of an ordinal sum which, aiming at self-contention, is recalled below:

Definition 7. Let $(\&_i)_{i \in A}$ be a family of t-norms and a family of non-empty pairwise disjoint subintervals $[x_i, y_i]$ of $[0, 1]$. The ordinal sum of the summands $(x_i, y_i, \&_i)$, $i \in A$ is the t-norm $\&$ defined as

$$\&(x, y) = \begin{cases} x_i + (y_i - x_i) \&_i(\frac{x - x_i}{y_i - x_i}, \frac{y - x_i}{y_i - x_i}) & \text{if } x, y \in [x_i, y_i] \\ \min(x, y) & \text{otherwise} \end{cases}$$

For the handling of truth-intervals we will consider the lattice $\langle \mathcal{I}([0, 1]), \leq \rangle$, where $\mathcal{I}([0, 1])$ is the set of all subintervals of $[0, 1]$, and given $[a, b], [c, d] \in \mathcal{I}([0, 1])$, we have

$$[a, b] \leq [c, d] \text{ if and only if } a \leq c \text{ and } b \leq d$$

$$\sup\{[a, b], [c, d]\} = [\sup\{a, c\}, \sup\{b, d\}]$$

$$\inf\{[a, b], [c, d]\} = [\inf\{a, c\}, \inf\{b, d\}]$$

3.1   Description of the Net

Each process unit is associated either to a propositional symbol of the initial program or to an homogeneous rule of the transformed program $\mathbb{P}$. The state of the i-th neuron in the instant t is expressed by its output vector, $S_i(t) = (S_i^1(t), S_i^2(t))$, which denotes an interval. Thus, the state of the network can be expressed by means of a state matrix $S(t)$, whose rows are the output of the neurons forming the network. In the initial state matrix, $S(0)$, all the rows are null, that is $(0, 0)$, but those corresponding to neurons associated to a propositional symbol A, in which case $S_A(0)$ is defined to be:

$$S_A(0) = (S_A^1(0), S_A^2(0)) = \begin{cases} (\vartheta_A^1, \vartheta_A^2) & \text{if } \langle A \leftarrow \top, [\vartheta_A^1, \vartheta_A^2] \rangle \in \mathbb{P}, \\ (0, 0) & \text{otherwise.} \end{cases}$$

The connection between neurons is denoted by a matrix of weights W, in which w_{ij} indicates the existence (value 1) or absence (value 0) of connection from unit j to unit i; if the neuron represents a weighted sum, then the weights are also represented in the entries associated to any of its inputs. The weights of the connections related to neuron i (that is, the i-th row of the matrix W) are represented by a vector w_i, and are allocated in an internal vector register of the neuron.

Four more internal registers v_i, x_i, y_i, m_i are defined in any neuron:

- The initial truth-interval $[v_i^1, v_i^2]$ of a propositional symbol or homogeneous rule is loaded in the internal register v_i.
- The registers x_i, y_i are used to restrict the domain of primitive conjunctors in order to enable the possibility of defining conjunctors as *ordinal sums* by using the technique presented in [8]. Note that if $x_i = (x_i^1, x_i^2)$, then x_i^1 denotes the initial point of the domain of the first conjunctor, whereas x_i^2 denotes the initial point of the domain of the second conjunctor.
- Vector $m_i = (m_i^1, m_i^2)$ indicates the functioning mode of the neuron, the possible pairs are $(1,1)$ to denote a propositional symbol, or $(5,5)$ to denote an aggregator rule, or (x,y) with $2 \le x, y \le 4$, to denote a rule where x (resp. y) denotes the t-norm acting in the beginning (resp. end) of the intervals.

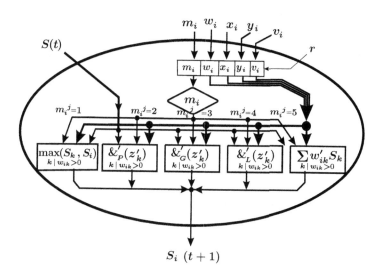

Fig. 1. The proposed generic neuron

3.2 On the Output of a Neuron

Taking into account the values of the registers and the state matrix S at instant t the output of a given neuron is computed as follows:

If $m_i = (1,1)$, then its next state is the maximum value among all the operators involved in its input and its previous state in each component. More formally, let us denote $K_i = \{k \mid w_{ik} > 0\}$ then we have:[2]

$$S_i(t+1) = \max\{S_i(t), \max\{S_k(t) \mid k \in K_i\}\}$$

---

[2] Note that the computation involves the join operation in the lattice $\mathcal{I}([0,1])$.

When $\boldsymbol{m_i} = (m_i^1, m_i^2)$, with $m_i^1, m_i^2 \in \{2, 3, 4\}$, the neuron i is associated to a homogenous rule where the connective is built from two basic conjunctors. The description of the output of the neuron will be given in terms of the vectors $\boldsymbol{z_k}(t) = (z_k^1(t), z_k^2(t))$, which accommodate the value of $S_k^j(t)$, originally in the subinterval $[x_i^j, y_i^j]$, to the unit interval and, for each $k \in K_i$ and $j \in \{1, 2\}$, are computed as follows:

$$
z_k^j(t) = \begin{cases} 1 & \text{if } S_k^j(t) \geq y_i^j \\ \dfrac{S_k^j(t) - x_i^j}{y_i^j - x_i^j} & \text{if } x_i^j \leq S_k^j(t) < y_i^j \\ 0 & \text{if } S_k^j(t) < x_i^j \end{cases}
$$

Once the vectors $\boldsymbol{z_k}(t)$ have been computed, then the corresponding conjunctor (product, Gödel or Łukasiewicz) is applied, and the result is again accommodate in the interval $[x_i^j, y_i^j]$. Thus, the output of the neuron is[3]

$$
S_i^j(t+1) = \begin{cases} x_i^j + (y_i^j - x_i^j) \cdot (v_i^j \&_P z_1^j \&_P \cdots \&_P z_{N_i}^j) & \text{if } m_i^j = 2 \\ x_i^j + (y_i^j - x_i^j) \cdot (v_i^j \&_G z_1^j \&_G \cdots \&_G z_{N_i}^j) & \text{if } m_i^j = 3 \\ x_i^j + (y_i^j - x_i^j) \cdot (v_i^j \&_L z_1^j \&_L \cdots \&_L z_{N_i}^j) & \text{if } m_i^j = 4 \end{cases} \quad (1)
$$

where N_i is the cardinal of $K_i = \{k \mid w_{ik} > 0\}$.

Note that, for instance, if $\boldsymbol{m_i} = (2, 3)$, the components $S_i^j(t)$ of the output mimic the behaviour of the product (for $j = 1$) and Gödel (for $j = 2$) implications, respectively, in terms of the adjoint property.

Finally, a neuron associated to an aggregator has $\boldsymbol{m_i} = (5, 5)$, and its output is

$$
S_i(t+1) = \sum_{k \in K_i} w'_{ik} S_k(t) \quad \text{where} \quad w'_{ik} = \frac{w_{ik}}{\displaystyle\sum_{r \in K_i} w_{ir}}
$$

3.3   Implementation

A number of simulations have been obtained through a MATLAB implementation in a conventional sequential computer. A high level description of the implementation is given below:

1. **Initialize** the network is with the appropriate values of the matrices V, X, Y, M, W and, in addition, a tolerance value *tol* to be used as a stop criterion. The output $\boldsymbol{S_i}(t)$ of the neurons associated to facts are initialized with its truth-value $\boldsymbol{v_i} = (v_i^1, v_i^2)$.
2. **Find** the neurons k (if any) which operate on the neuron i, that is, construct the set $K_i = \{k \mid w_{ik} > 0\}$ and calculate $N_i = \sum_{K_i} w_{ik}$. When $w_{ik} = 1$ for all $k \in K_i$ then N_i is the cardinal of K_i.

---

[3] The functions corresponding to each case are represented in Fig. 1 as $\&'_P$, $\&'_G$, $\&'_L$, resp.

3. **Repeat.** Update all the states $S_i(t) = (S_i^1(t), S_i^2(t))$ of the neurons of the network:

   (a) If $m_i = (1,1)$, then update the state of neuron i as follows, for $j \in \{1,2\}$

   $$S_i^j(t+1) = \begin{cases} \max\{S_i^j(t), \max_{K_i} S_k^j(t)\} & \text{if } K_i \neq \varnothing \\ v_i^j & \text{otherwise} \end{cases}$$

   (b) If $m_i^j = 2$, 3 or 4, then update the state of neuron i to $S_i^j(t+1)$ as defined in Eq (1).

   (c) If $m_i^j = 5$, then the neuron corresponds to an aggregator, and is updated by:

   $$S_i^j(t+1) = \frac{1}{N_i} \sum_{k \in K_i} w_{ik} \cdot S_k^j(t)$$

   Until the stop criterion $\|S(t+1) - S(t)\|_\infty < tol$ is fulfilled.

We introduce some toy examples below in order to show the behavior of the network.

Example 1. Consider the homogenous program with the fact $\langle r \leftarrow \top, [0.1, 0.2]\rangle$ and the two rules $\langle p \leftarrow @_{1,2}(r,q), [1,1]\rangle, \langle q \leftarrow_{PG} p, [0.6, 0.7]\rangle$.

The net for the program will consist of five neurons, three of which represent propositional symbols p, q, r, and the rest are needed to represent the rules (one for the aggregator and another for the product-Gödel rule).

Note that no ordinal sum occurs in the program, therefore the matrices X and Y are constantly 0 and 1, respectively.

The initial values for the rest of matrices are:

$$V = \begin{pmatrix} 0.0 & 0.0 \\ 0.0 & 0.0 \\ 0.1 & 0.2 \\ 1.0 & 1.0 \\ 0.6 & 0.7 \end{pmatrix}, \quad M = \begin{pmatrix} 1 & 1 \\ 1 & 1 \\ 1 & 1 \\ 5 & 5 \\ 2 & 3 \end{pmatrix}, \quad W = \begin{pmatrix} \cdot & \cdot & \cdot & 1 & \cdot \\ \cdot & \cdot & \cdot & \cdot & 1 \\ \cdot & \cdot & \cdot & \cdot & \cdot \\ \cdot & 1 & 2 & \cdot & \cdot \\ \cdot & 1 & \cdot & \cdot & \cdot \end{pmatrix}$$

After running the net, it gets stabilized after 328 iterations providing the following truth-intervals for p, q and r (only three decimal digits are given):

$$p = [0.055, 0.200], \qquad q = [0.033, 0.200], \qquad r = [0.100, 0.200] \qquad \square$$

Example 2. Consider the fact $\langle r, [0.3, 0.5]\rangle$ and the rules $\langle p \leftarrow @_{1,2}(r,q), [1,1]\rangle$ and $\langle q \leftarrow_{TG} p \&_{TG} r, [0.7, 0.8]\rangle$, where T is the ordinal sum given by the Łukasiewicz conjunction on $[0.2, 0.4]$ and product conjunction on $[0.6, 0.9]$.

The net for the program consists of eight neurons, three of which represent variables p, q, r, and the rest are needed to represent the rules (one for the aggregator and four for the TG one).

The initial values of the net are the matrices:

$$V = \begin{pmatrix} 0.0 & 0.0 \\ 0.0 & 0.0 \\ 0.3 & 0.5 \\ 1.0 & 1.0 \\ 0.7 & 0.8 \\ 0.7 & 0.8 \\ 0.7 & 0.8 \\ 1.0 & 1.0 \end{pmatrix}, M = \begin{pmatrix} 1 & 1 \\ 1 & 1 \\ 1 & 1 \\ 5 & 5 \\ 4 & 3 \\ 2 & 3 \\ 3 & 3 \\ 3 & 3 \end{pmatrix}, X = \begin{pmatrix} 0.0 & 0.0 \\ 0.0 & 0.0 \\ 0.0 & 0.0 \\ 0.0 & 0.0 \\ 0.2 & 0.0 \\ 0.6 & 0.0 \\ 0.0 & 0.0 \\ 0.0 & 0.0 \end{pmatrix}, Y = \begin{pmatrix} 1.0 & 1.0 \\ 1.0 & 1.0 \\ 1.0 & 1.0 \\ 1.0 & 1.0 \\ 0.4 & 1.0 \\ 0.9 & 1.0 \\ 1.0 & 1.0 \\ 1.0 & 1.0 \end{pmatrix}$$

together with the following matrix of weights

$$W = \begin{pmatrix} \cdot & \cdot & \cdot & 1 & \cdot & \cdot & \cdot & \cdot \\ \cdot & \cdot & \cdot & \cdot & \cdot & \cdot & \cdot & 1 \\ \cdot & \cdot & \cdot & \cdot & \cdot & \cdot & \cdot & \cdot \\ \cdot & 2 & 1 & \cdot & \cdot & \cdot & \cdot & \cdot \\ 1 & \cdot & 1 & \cdot & \cdot & \cdot & \cdot & \cdot \\ 1 & \cdot & 1 & \cdot & \cdot & \cdot & \cdot & \cdot \\ 1 & \cdot & 1 & \cdot & \cdot & \cdot & \cdot & \cdot \\ \cdot & \cdot & \cdot & \cdot & 1 & 1 & 1 & \cdot \end{pmatrix}$$

After running the net, it gets stabilized at 430 iterations giving the following output for p, q and r:

$$p = [0.233, 0.500], \qquad q = [0.355, 0.500], \qquad r = [0.400, 0.500] \qquad \square$$

Regarding the soundness of the implementation sketched above, the following theorem can be obtained, although space restrictions do not allow to include the proof.

Theorem 1. *Given a homogeneous program $\mathbb{P}$ and a propositional symbol A, then the sequence $S_A(n)$ approximates the value of the least model of $\mathbb{P}$ in A up to any prescribed level of precision.*

4   Conclusions and Future Work

A new neural-like model has been proposed which extends that recently given to multi-adjoint logic programming in such a way that it is possible both to do calculations with truth-intervals and obtaining the computed truth-values of all propositional symbols involved in the program in a parallel way. This extended approach considers, in addition to the three most important adjoint pairs in the unit interval (product, Gödel, and Łukasiewicz) and weighted sums, the combinations as finite ordinal sums of the previous conjunctors.

The original model of neuron could have been extended by considering simpler units, each one dedicated to represent a different type of homogeneous rule or propositional symbol; however, we have decided to extend the original generic model of neuron, capable of adapting to perform different functions according with its inputs. The advantage of this choice is related to the uniform (although more complex) description of the units, and the attainment of a clearer network than using simpler units. An analysis of the compromise between simplicity of the units and complexity of the network will be the subject of future work.

References

1. C.V. Damásio and L. Moniz Pereira. Monotonic and residuated logic programs. *Lect. Notes in Artificial Intelligence* 2143:748–759, 2001.
2. A. Dekhtyar and V.S. Subrahmanian. Hybrid Probabilistic Programs, *Journal of Logic Programming* 43(3):187–250, 2000
3. P. Eklund and F. Klawonn. Neural fuzzy logic programming. *IEEE Tr. on Neural Networks*, 3(5):815–818, 1992.
4. S. Hölldobler and Y. Kalinke. Towards a new massively parallel computational model for logic programming. In *ECAI'94 workshop on Combining Symbolic and Connectioninst Processing*, pages 68–77, 1994.
5. S. Hölldobler, Y. Kalinke, and H.-P. Störr. Approximating the semantics of logic programs by recurrent neural networks. *Applied Intelligence*, 11(1):45–58, 1999.
6. L. V. S. Lakshmanan and F. Sadri. On a theory of probabilistic deductive databases. *Theory and Practice of Logic Progr.*, 1(1):5–42, 2001.
7. J. Medina, E. Mérida-Casermeiro, and M. Ojeda-Aciego. A neural implementation of multi-adjoint logic programming. *Journal of Applied Logic*, 2(3):301-324, 2004.
8. J. Medina, E. Mérida-Casermeiro, and M. Ojeda-Aciego. Decomposing Ordinal Sums in Neural Multi-Adjoint Logic Programs. *Lect. Notes in Artificial Intelligence* 3315:717–726, 2004.
9. J. Medina, E. Mérida-Casermeiro, and M. Ojeda-Aciego. A neural approach to extended logic programs. *Lect. Notes in Computer Science* 2686:654–661, 2003.
10. P. Vojtáš. Fuzzy logic programming. *Fuzzy Sets and Systems*, 124(3):361–370, 2001.

Author Index

Lecture Notes in Computer Science

For information about Vols. 1–3452

please contact your bookseller or Springer